must be slightly higher than the adjacent parts of the field.

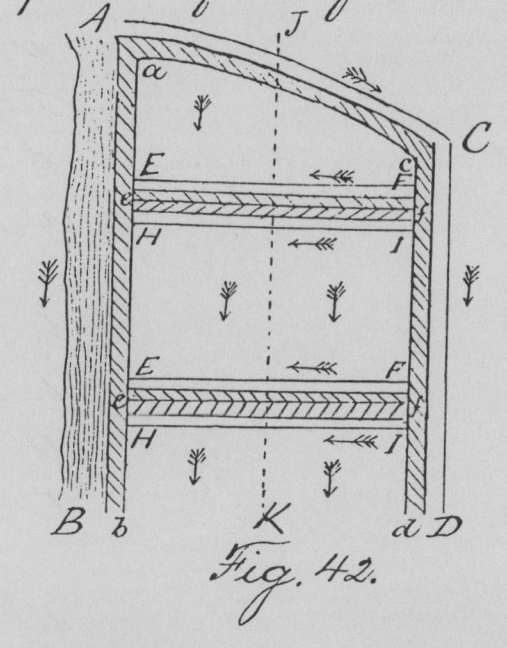

Fig. 42.

Fig. 42 represent a plan of a water meadow in which AB is the stream or canal; ab, an embankment on the edge of the stream; ACD, a canal which takes water from the stream AB; acd, an embankment on the edge of this canal; EF, a drain for catching the superfluous water from the portion of land above it; ef, an embankment between successive sections of land; HI, a distributing canal leading from the canal ACD.

Fig. 43.

This method of irrigation can be used only when the slope of the land is slight and uniform. The location of the

Lectures on Agriculture
by William P. Brooks, Professor of Agriculture

ブルックス札幌農学校講義

髙井 宗宏 編

北海道大学図書刊行会

本書は社団法人札幌農学振興会の出版助成を得て刊行された

W. P. ブルックス。2000年にブルックスの子孫より北大に寄贈された写真〈北海道大学附属図書館蔵〉

札幌農学校教授時代のブルックス(明治12年)とサイン〈同左蔵〉

卒業を前にした札幌農学校第2期生たち(明治14年) 前列左から町村金弥, 1人おいてピーボディ, ブルックス, カッター, 宮部金吾, 後列左から2人目が新渡戸稲造, その隣内村鑑三, 2人おいて南鷹次郎〈同上蔵〉

札幌農学校第3期生たち(明治15年) 前列中央ブルックス, その右カッター, 後列左から2人目がノート者の伊吹鎗造〈同上蔵〉

明治10年頃の模範家畜房(モデルバーン)〔左〕と穀物庫(コーンバーン)〔右〕。現在の北10〜11条西5丁目付近
〈北海道大学大学院農学研究科畜産学教室蔵〉

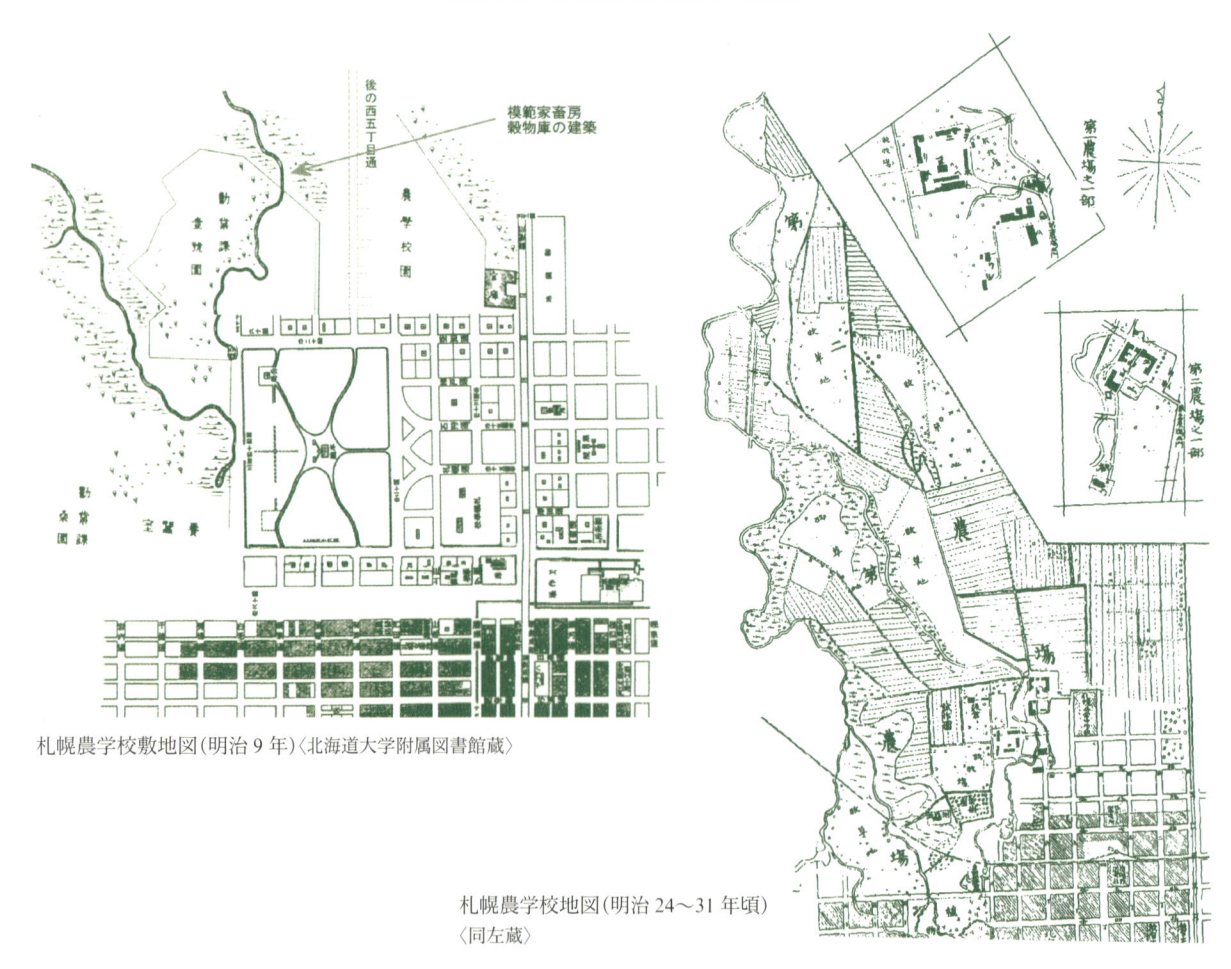

札幌農学校敷地地図(明治9年)〈北海道大学附属図書館蔵〉

札幌農学校地図(明治24〜31年頃)
〈同左蔵〉

現在の模範家畜房(後左)とブルックスの輸入した農機具

北海道大学農学部附属農場平面図(昭和44年)

明治11年米国からの輸送船上で産まれたエアシャー種牡「太平海号」
〈北海道大学大学院農学研究科畜産学教室蔵〉

ブルックス設計の穀物庫(現在)

ブルックスの創設以来，現在も使用されている第1農場とポプラ並木(2001年)

北側から見た農学部(2001年)

冬の農学部(2001年)

旧昆虫学教室と楡の大樹(2001年)

花木園内にある新渡戸稲造像(1996年設置)

まえがき

　本書は，1877-80（明治10-13）年に北海道大学農学部の前身となる札幌農学校で行なわれたブルックス教授による「農学」講義を2期生新渡戸稲造らの講義ノートから再現し，これに章ごとの解説とその学問分野の継承発展の経過を述べた解題を加えたものである。

　当時の日本は，300年間も外国との交流を禁止した鎖国政策を外圧によって終わらせられ，政治体制を良くも悪しくも近代的に変えて，多くのお雇い外国人を招請して西欧の文化・技術の導入を図り始めた時である。加えて北海道は，原始の森に覆われた未開拓地であるばかりか，日本領土としての帰属も不明確であったため，府県からの農業移住を推奨して国境紛争（北辺鎖鑰）に備えたが，稲作主体の日本人には寒冷地である北海道の開拓手法を海外の畑作酪農技術に学ばねばならなかった。

　1876（明治9）年に開校した札幌農学校は，北海道開拓の手法を策定したケプロン構想に基づいて開拓の高級技術者を養成する日本最初の官立高等教育機関として開設され，マサチューセッツ農科大学から招請したクラーク博士を代表とする外国人教授陣によるアメリカ式一貫教育にすべてを任せた。したがって，本書が扱う「農学」講義は，当時の在来農業を示す「農書」類とは根本から違う内容であり，日本人には異質のアメリカの畜力畑作農業である。

　農学講義を担当したブルックス教授は，学内での教育に止まらず，学外に出て丘珠タマネギの導入や農家の技術向上を図ったり，江別市の産業となった暗渠土管の製造敷設の技術を紹介したり，開拓使を助けて開拓農家の基本設計を行なったり，産業振興展示会を提案して継続開催につなぐなどによって，泰西農業と呼んだ当時の西欧農業技術を適切に伝播したため，明治末までにいわゆる「北海道農法」が完成し，ひいては日本農業の近代化に大きな影響を及ぼしている。したがって本書は，当時受容した泰西農業の全貌を明らかにし，それを日本がどのように普及させたかを示す貴重な記録として提示するばかりか，これまで忘れられがちであったブルックス教授の功績を明らかにすることにある。

　しかしながら，当時では進んだ農業技術であるが，元素数は1869年にメンデレーエフが周期律を発見した頃の70元素ほどであるし，1865年に発表されたメンデルの遺伝法則ですら1900年まで価値が認められなかった時代である。さらに18世紀後半にイギリスで始まった産業革命が19世紀前半にフランスやドイツに広まり，ようやくアメリカに伝わった時代であるから，講義録を今日の視点から見ると評価に堪えない内容も少なくない。加えて札幌農学校の生物学や化学講義は，農作物を題材にして生態・機能・成分など農学分野の基礎事項を扱っているため，本書の農学講義は農場での栽培管理を中心にせざるを得なかった。このことは，ブルックス教授が農学講義と農場実習のすべてをひとりで担当しただけでなく，

これを継承した2期生南鷹次郎(第2代総長)も農業経済学の佐藤昌介(初代総長)と生物学の宮部金吾(ともに1期生)の分野を除いて，作物学，畜産・獣医学，農芸化学，農芸物理学の全分野を担当し，その教え子から学科目や講座の教授を輩出させたという経過からも裏づけられる。さらに本書の解題は，関係分野の10余名が研究会を組織して執筆したが，大学院改革前の50余講座の2割にすぎない領域の担当者で分担できたこともその狭さを裏づける。

　本書は英文の講義録に和文の解題と注を配した構成となっている。創立当初の札幌農学校では，すべて通訳なしの英語で講義を行なったことから，残された講義録もすべて英文である。本書作製に際して北大附属図書館北方資料室蔵の講義ノートのなかから保存状態のよいものを選んで底本とした。次いで内容の吟味と整理にあたっては，ノート原文を和訳して当時の農学を再現するよりも，その内容をどのように継承・発展させたかを示す解題を付け，現代農学の起点となった泰西農学を理解した方が意義深いと考えた。さらに近年，領域ごとに研究手法が急速に高度化して遺伝子レベルの研究が主力になり，本書を最新の農学から解説するには無理があるため，研究分野が細分化される前の1955年頃までの視点で評価することとした。このような経過から本書では英語と日本語が混在することとなった。また，英文で収録した講義録は，口述筆記によって進められた講義ノートから採録しているため，口述の区切りごとにカンマがあるなど英文としての体裁に疑問があった。そこで英語を母語とする校閲者に依頼して修正を施したが，その際に19世紀末の英語か，現代英語の観点から修正するかの議論があった。本書は，当時の泰西農業の全貌を紹介するのが本旨であり，原文を尊重するものの，部分的には当時の英文で理解を難しくするより，現代文で分かりやすくすべきであるという意見を採用してノートの原文を改変した部分もある。これらの点について了解を得たい。

　北海道大学は，2001(平成13)年秋に創基125周年を祝う記念行事を行なった。これは札幌農学校が開校式を行なった1876年8月からの年数であるから，本書で扱う講義録も125周年を迎える。本書によって，日本農学の濫觴を振り返り，先人の労苦や業績を考える契機となれば幸いである。

　なお，執筆者のひとりであるユステス・藤居恒子さんは，ブルックス教授の長女レイチェル・ブルックスが北海道で生まれた初めての外国女性となった奇縁にちなみ，北大の創基80周年事業のなかでブルックス家から奨学金を受けて文学部哲学科を卒業した人である。詳しくは本人の文章を見てほしいが，その後恒子さんは，見知らぬ「あしながおじさん」を訪ねてアメリカに留学して永住した。それが執筆者群に加わった理由は，2000年の春にブルックスの玄孫夫妻が日本観光を企画した際，恒子さんが「せっかくの訪日だから北大を訪れて，ブルックスゆかりの建物や記念物を見たら」と勧めて北大との連絡を担当したことと，本書にかかわっていた髙井が案内者に選ばれたという偶然の一致があったためである。125年前にブルックス教授によって基礎杭が打ち込まれた北海道農業が，今日では日本の食糧基地として確固たる地位を築いて発展を続けていると同様に，農学校開校時に在籍したブルックス夫妻の愛着の念がブルックス家と北海道大学を今も強い絆で結んでいる。

　また，クラーク構想に従ってブルックス教授が開設し，北海道開拓のモデル農場として運

営された「札幌農学校第2農場」は，1877(明治10)年建築の模範家畜房(モデルバーン)とブルックス教授設計の穀物貯蔵庫(コーンバーン)を中心とした9棟の事務所畜舎等施設と，その中に開学以来の農機具を収容して今日まで残ったため，北海道農業発足の記念物としての価値と希有な洋式農業建築物との価値が認められて，国から「重要文化財」に，北海道から「北海道遺産」として指定された。そこで北大評議会は，2000年春に本書の筆者に加わる髙井，大久保らに一般公開への準備を命じたため，退職後も学内の関係者と力を合わせ，年次計画によって整備を進めている。ここでの展示を通して，ブルックス教授による活動の永遠なることを示すとともに，本書を公開資料に位置づけて活用を図っていく予定である。

終わりに，刊行の意義を理解して資金面で絶大のご支援を頂いた北海道大学農学振興会理事長諏訪正明農学研究科長，その道を開いてくださった同前理事長太田原高昭北大名誉教授に深甚なる謝意を表する。同振興会による最初の支援は，1980年頃に専門分野ごとの同好者を募って発足した「ブルックス農学研究会」による，手書きの講義ノートをワープロ入力して英文講義録を作る企画に対して行なわれた。解題執筆者の約半数は，この作業過程で講義録の意義を再発見して補充調査・研究に入り，「ブルックス農学研究会」を活性化させているため，最初の支援が本書の土台を築いている。そして第2回目の支援は，本書の400頁を超える膨大な図書の刊行費に充てられ，単なる同好会の記録として残るかも危ぶまれた資料を上梓させて頂いた。本書の成立は，すべて農学振興会の支援のたまものであることを述べ，あらためて感謝する。また，梅田安治・佐久間敏雄両北大名誉教授には，単に執筆者としてだけでなく，素案段階の約20年前から励ましを受け，ついに刊行への道筋を開いて頂いたことに厚くお礼申し上げる。さらに刊行決定後は，英文の校閲を担当されたマイケル(Michael Van Remortel)さん，予定が遅れて煩雑になった業務を担当された北海道大学図書刊行会の前田次郎さんらに厚くお礼申し上げる。

2004年6月

ブルックス農学研究会

代表　髙井宗宏

目　次

まえがき

第1編　札幌農学校とブルックス

第1章　札幌農学校と農学講義 ……………………………………… 3
　　1. Catalogue of Sapporo Agricultural College in 1876-77 ……… 3
　　2. Agricultural Improvements needed in Japan ……… 4
　　3. Progressive and Non-progressive Education in Japan ……… 6
　　4. Lectures on Agriculture ……… 8
　　5. Expansion of Subject and Experimental Work ……… 10
　　6. Questions of the Final Examination Exercises ……… 13
　　解　題 ………………………………………………………髙井宗宏… 16

第2章　「ブル」先生 ……………………………………常瑤居士(新渡戸稲造)… 23

第3章　ブルックス家のひとびと ……………………ユステス・藤居恒子… 25

第2編　ブルックス農学講義録

Chapter 1　Agriculture ……………………………………………… 39
　　1-1. The Importance of Agriculture ……… 39
　　1-2. The Influence of Agriculture on National Prosperity ……… 40
　　1-3. The Necessity for a Variety of Industries ……… 40
　　1-4. Advantages of Agriculture as an Occupation ……… 40
　　1-5. Development of Modern Agriculture ……… 41
　　1-6. Knowledge Necessary to the Farmer ……… 41
　　解　題 ………………………………………………………髙井宗宏… 42

Chapter 2　Soil …………………………………………………… 49
　　2-1. The Manner in which Rocks have been Converted into Soils ……… 49
　　2-2. Organic Matter ……… 50
　　2-3. Classes of Soils ……… 52
　　2-4. Physical Properties of Soils ……… 54

2-5. Adaptation of the Soil to Crops	*60*
解　　題 ……………………………………佐久間敏雄…	*62*

Chapter 3　Farm Drainage and Irrigation　*71*

3-1. Effects of Drainage	*71*
3-2. Open Drains and Ditches	*73*
3-3. Different Methods of Making Underdrains	*75*
3-4. Tile Drains	*81*
3-5. Direction, Distance and Depth of Drains	*83*
3-6. Necessity of System in Drainage	*84*
3-7. Draining Implements	*88*
3-8. General Directions for Tile Drainage	*89*
3-9. Irrigation in Several Countries	*91*
3-10. Kinds of Water which are best for Irrigation	*93*
3-11. The Amount of Water Needed for Irrigation	*94*
3-12. Methods of Irrigation	*98*
3-13. Water Meadows	*102*
3-14. Management of Irrigated Fields	*106*
3-15. Crops most Benefited by Irrigation	*108*
3-16. Means of Getting a Supply of Water	*109*
解　　題 ……………………………………梅田安治…	*114*

Chapter 4　Pulverization of Soils (Tillage)　*123*

4-1. Methods of Pulverizing the Soil	*124*
4-2. The History of the Plow	*126*
4-3. The Mechanical Principles Involved in the Plow	*127*
4-4. Varieties of Plows	*128*
4-5. Harrows and their Uses	*132*
4-6. Cultivators	*134*
4-7. Rollers	*135*
4-8. Simple Tools	*136*
解　　題 ……………………………………髙井宗宏…	*139*

Chapter 5　Manures (and Fertilizer)　*147*

5-1. Classification of Manures	*147*
5-2. Potash	*150*
5-3. Lime	*151*
5-4. Soda	*153*
5-5. Magnesia	*154*
5-6. Phosphoric Acid	*154*
5-7. Organic Manures	*156*

5-8. Application and Physical Properties of Manure ·· *163*
5-9. The Atmosphere as Related to Vegetation ·· *164*
解　　題 ···但野利秋··· *173*

Chapter 6　Farm Economy (Management) ················ *183*

6-1. System Necessary to Prevent Loss or Waste ·· *183*
6-2. Farm Management ·· *186*
6-3. Kind of Farming ··· *188*
6-4. Management of Mowing Land ·· *193*
6-5. Management of Permanent Mowing Land ·· *199*
6-6. Implement Necessary for Harvesting Hay ·· *201*
6-7. Pasture Land and its Management ··· *203*
6-8. Management of Tillage Lands ··· *205*
6-9. Forests on a General Farm ·· *206*
6-10. Orchard on a General Farm ·· *208*
6-11. Farm Road and Fence ·· *216*
6-12. Winter Work upon the General Farm ·· *223*
6-13. The Routine and Time of Doing Spring Work ······································ *229*
6-14. Summer Work on the General Farm ·· *230*
6-15. Fall Work upon the General Farm ··· *234*
解　　題 ··太田原高昭・中嶋　博・髙井宗宏・近藤誠司··· *235*

Chapter 7　Crop Cultivation ·· *251*

7-1. Wheat (*Triticum aestivum*) ·· *251*
7-2. Rye (*Secale cereale*) ·· *259*
7-3. Oats (*Avena sativa*) ·· *261*
7-4. Barley (*Hordeum*) ··· *264*
7-5. Indian Corn (*Zea mays*) ··· *265*
7-6. Buckwheat (*Fagopyrum esculentum* Moench) ·· *271*
7-7. Millet (*Setaria* and *Panicum*) ·· *272*
7-8. Hungarian Grass ·· *273*
7-9. Potato (*Solanum tuberosum*) ·· *274*
7-10. Beans (*Phaseolus*) ··· *278*
7-11. Peas (*Pisum sativum*) ·· *280*
7-12. Turnip (*Brassica campestris*) ··· *282*
7-13. Cabbage (*Brassica oleracea*) ··· *285*
7-14. Carrot (*Daucus carota*) ·· *291*
7-15. Parsnip (*Pastinaca sativa*) ·· *294*
7-16. Onion (*Allium cepa*, "Negi") ·· *294*
7-17. Squash (*Cucurbita*) ·· *298*
7-18. Pumpkin (*Cucurbita pepo*, "Tonasu") ··· *301*

7-19.	Melon	301
7-20.	Cucumber (*Cucumis sativus*)	302
7-21.	Celery (*Apium graveolens*)	303
7-22.	Tomato (*Lycopersicom esculentum*)	304
7-23.	Asparagus (*Asparagus officinalis*)	305
7-24.	Lettuce (*Lactuca sativa*)	306
7-25.	Eggplant (*Solanum melongena*, "Nasu")	307
7-26.	Endive (*Cichorium endivia*)	307
7-27.	The Culture of Vegetable Seeds	308
7-28.	Beet (*Beta vulgaris*)	314
7-29.	Sorghum and Imphee (*Sorghum vulgare* var. *saccharatum* and var. *imphee*)	318
7-30.	Broom Corn (*Sorghum vulgare*, "Morokoshi")	320
7-31.	Hemp (*Cannabis sativa*)	320
7-32.	Flax (*Linum usitatissimum*)	321
7-33.	Hops (*Humulus lupulus*)	322
7-34.	Rape or Colza (*Brassica napus*, *B. campestris*, *B. oleracea*)	323
	解題 中世古公男・由田宏一・原田 隆・中嶋 博・生越 明	324

Chapter 8 Stock-Farming 341

8-1.	Importance of Stock-Farming	341
8-2.	Climate and Soil Best for Stock Farming	341
8-3.	Locations Best for Stock-Farming	342
8-4.	Hokkaido as Adapted for Stock-Farming	343
8-5.	Neat Cattle	343
8-6.	Cattle Breeding	359
8-7.	Rearing Calves	367
8-8.	Care of Neat Cattle	369
8-9.	Sheep	372
8-10.	Cattle Feeding	382
8-11.	Digestion in Animals	388
8-12.	Decomposition of Nutrients in the Body	391
8-13.	Excretion in Animals	391
8-14.	Determinations of the Gain and Loss of Flesh	392
8-15.	Protein Consumption	394
	解題 大久保正彦	398

編者・執筆者・校閲者紹介 405

第1編　札幌農学校とブルックス

　札幌農学校は，未開地の北海道を西欧の農業技術によって開拓するための技術者を養成する高等専門教育機関として1876(明治9)年に開校した。

　第1章では，1877年より刊行が開始された6冊の *Annual Report of Sapporo Agricultural College*(『札幌農黌年報』：第6年報を除いて和文・英文の各冊があるが，和文は用語の不統一と古文調できわめて難解のため，本書では英文を抄録)から，外国人教授陣が書いた札幌農学校の紹介，日本農業の改革案，農学講義手法などについて述べた文章を引用し，農学校の教育姿勢や講義内容と日本の在来教育との差異などを明らかにして，本書第2編に収録した農学講義録の解説に代えた。なお，引用に際し，一部省略および編者による見出しを付け加えた箇所がある。

　第2章は，ブルックスの農学講義を受講し，本書の底本となったノートを筆記した新渡戸(当時は太田)稲造によるブルックス回顧録である。これは新渡戸が札幌農学校教授であった時，学生組織である札幌農学校学芸会の求めに応じ，機関誌『恵林』に教えを受けた先生ごとに想い出を綴ったシリーズ「忘れ奴草」中の一編である。なお，他の外国人教師について，ホイーラーは「ゴート先生」，ペンハローは「ペン先生」と題し，髭や頭の形が山羊や鋳鉄薪ストーブ(開拓使製造のストーブ名に転化して今日も残る)に似ているとする親しみを込めた渾名で執筆している。

　第3章は，まえがきに述べた藤居恒子さんの文であり，今日もブルックス家と深いつながりを持つ北大OBによるブルックス再発見の記録である。ここに掲載されるブルックスの私信は，初公開の文書であるばかりか，クラークに関して新事実を指摘するなど，随所に興味深い記述がある。

第1章　札幌農学校と農学講義

1. Catalogue of Sapporo Agricultural College in 1876-77

1) Officers
   Gondaishiokikuwan Dzushio Hirotake, Japanese Director.
   William S. Clark, Ph. D., LL. D., President and Director of College Farm.
   William Wheeler, B. S., Professor of Mathematics and Civil Engineering.
   David P. Penhallow, B. S., Professor of Botany and Chemistry.
   William P. Brooks, B. S., Professor of Agriculture.
   Hori Seitaro, Secretary and Interpreter.
   Yoshida Kiyonori, Farm Overseer.
2) Object and Plan of Organization

　The Sapporo Agricultural College was founded by the Colonial Department for the education and practical training of young men from all parts of the Empire who are expected to become its employs after graduation, and to remain in its service for the term of five years. The number of students is limited to fifty, and all their expenses while in College are defrayed by the Government. The course of instruction will occupy four years, and those who complete it in a satisfactory manner will receive the degree of Bachelor of Science.

　The following branches of knowledge will be regarded as important parts of the curriculum: namely, the Japanese and English Languages; Elocution, Debate, Composition and Drawing; Book-keeping and the Forms of Business; Algebra, Geometry, Trigonometry, Surveying, Civil Engineering, so far as required in the construction of ordinary roads and railroads, and of works for drainage and irrigation; Physics, with particular attention to Mechanics; Astronomy; Chemistry, with especial regard to Agriculture and Metallurgy; Structural, Physiological and Systematic Botany; Zoology, Human and Comparative Anatomy and Physiology; Geology; Political Economy; Mental and Moral Science; Physical Culture; Military Science and Tactics; and the most thorough instruction in the theory and practice of Agriculture and Horticulture, the various topics being discussed with constant reference to the circumstances and necessities of the farmers of Hokkaido.

3) Calendar

　Each collegiate year will begin on the fourth Thursday of August, and close on the first Wednesday of July. It will be divided into two terms, the first continuing from the fourth Thursday of August to the fourth Wednesday of December inclusive, and the second, from the fourth Thursday of January to the first Wednesday of July.

4) Terms of Admission

Candidates for admission to the Freshman Class will be examined orally and in writing upon the following subjects, namely: The Japanese and English Languages, which they should be able to read, write and speak with correctness and facility; Arithmetic, Geography and Universal History, the knowledge required being equivalent to that contained in the common higher text-books for public schools. Candidates for examination must be at least sixteen years of age, of sound constitution and good character. They must also sign the prescribed form of contract, with the Government, and furnish a satisfactory surety, residing either in Tokyo or Hokkaido.

5) Rank and Dismission

An accurate record will be permanently kept of the deportment, attendance and scholarship of every student. At the end of each term an examination in writing will be held and the papers marked upon the scale of one hundred for perfection. The average mark for the daily recitations of the term will be added to the examination mark and the sum divided by two. The quotient thus obtained will be the mark for the term in the subject under consideration. The sum of the marks in the several departments of instruction, divided by the number of such departments in which the student has been taught during the term, will give his general average mark, and by this his rank in the class will be determined. When, however, the total number of exercises in the departments for a term differs materially, the general average mark shall be obtained by the following rule: namely, multiply the number of exercises in each department by the average mark of that department, and divide the sum of the products thus resulting by the total number of exercises in all departments for the term. The quotient will be the general average mark. Only such students will be allowed to go on with their respective classes as have made satisfactory progress in the studies of the preceding term, and have been regular in attendance and exemplary in conduct.

Any student who may be dismissed from College on account of deficient scholarship, or misconduct, or who may leave voluntarily before the expiration of the time stipulated in his contract, will be required to pay the cost of his education as specified in said contract up to the date of his leaving. But no payment will be expected in the case of any one who may die, or who may be discharged on account of ill health, or by the Government without fault on his part.

Any student, desiring temporary leave of absence on account of the sickness or death of a relative, is required to present an application from hls parent or guardian accompanied by a certificate from a physician and the chief magistrate of the town where the said relative resides.

〔*First Annual Report of Sapporo Agricultural College, 1877,* published by the Kaitakushi, pp. 39-43 より〕

2. Agricultural Improvements needed in Japan

It has been well said, "A country is nothing without men, men are nothing without mind, and mind is little without culture. It follows that the cultivated mind is the most important product of a nation. The products of the farm, the shop, the mill, the mine, are of incomparably less value than the products of the schools. If the schools of a people are well taught, all else will prosper. Wherever schools are neglected it is a sure sign of national degradation and decay. The central point of every wisely administered government is its system of education. The education of youth well cared for by a nation, out of it will grow science, art, wealth, strength, and all else that is esteemed great in the judgment of men." A profound respect for learning has long characterized the more cultivated nations of the East, but never has greater enthusiasm in the pursuit of useful knowledge and the establishment of educational institutions been manifested by any people than by the Japanese under the intelligent government of His Imperial Majesty Mutsuhito. Actuated by this spirit, Your Excellency has founded the Sapporo Agricultural College at the capital of your vast province of Hokkaido in order that the young men who are educated for officers may become familiar with its climate, soil and resources, and be qualified to aid efficiently in the development of its various productive industries. Though it contains exhaustless supplies of valuable timber and excellent coal, marble and other minerals, and though the salmon, herring, cod and oilier fisheries are of immense value, if properly conducted, yet the greatest wealth of the province is to be derived from its fertile soil. Agriculture is the surest foundation of national prosperity. It feeds the people, converts the elements into property, and furnishes most of the material for manufactures, transportation and trade. The business of a country can be most profitably done by resident citizens who are intelligently and earnestly devoted to its welfare, and they alone can be relied on for its defense in time of foreign invasion. As soon as practicable, therefore, the migratory fishermen of Hokkaido should be converted into permanent settlers.

The agriculture of Japan greatly needs improvement, and the value of its agricultural products should soon be largely increased. The most important step forward consists in the general introduction and use of domestic animals of suitable breeds. These are necessary for a great variety of purposes. Horses and oxen are required for farm labor and for draft, good milk cows for dairy products, beef, veal and hides, sheep for wool and mutton, and swine for lard and pork. The food thus produced should be chiefly consumed by the people for the increase of their comfort and ability to labor; but if they do not desire it, it may be exported to foreign countries and exchanged for gold or other desirable articles. The wool is needed for the manufacture of cloth and the hides for leather, of which at present there is no adequate supply. An active, energetic people cannot long be contented with clumsy wooden clogs. Moreover, a scientific agriculture should both increase the products and improve the fertility of the soil, and this can only be accomplished in a large way by the abundant use of fertilizers, the most useful of which for all purposes is the excrement of farm animals.

The general use of horses and oxen would necessitate the manufacture of harnesses, vehicles, and agricultural implements, and so give employment to a great number of artisans. It would also enable farmers possessing teams to cultivate much more land and to do it more economically than under the old system.

A proper mode of feeding would require the extensive cultivation of Indian corn and foreign grasses, which are the most abundant and profitable crops of the northern portion of the United States. When now it is considered that all these sources of agricultural wealth are practically ignored in Japan, it will be readily admitted that there is room for improvement.

But Your Excellency has appreciated the importance of introducing domestic animals, and already a successful beginning has been made in breeding horses, neat cattle, sheep and swine at the agricultural establishments of the Department at Sapporo and Nanaye. The difficulties, however, in the way of rapid progress are by no means inconsiderable. The natural vegetation of Hokkaido is rank and coarse, and well suited neither for pasturage nor hay. Experience in America shows that while the natural grasses of the country are in some instances of excellent quality, yet in almost all cases, it is necessary for success in stock raising to depend chiefly upon cultivated plants for winter feeding. While, therefore, wild Hokkaido hay may be often profitably employed, especially if reduced to chaff and steamed, it is nevertheless very essential for progress that forage plants be introduced as rapidly as possible. In like manner it may be wise to use for fodder beans, barley and rice bran, but the main dependence for grain for stock in Hokkaido should be upon Indian corn, for which the soil and climate are very favorable, at least about Sapporo.

Another obstacle in the way of successful stock-raising is found in the poverty of the farmers and their wretched modes of building. Well-bred animals, which are the most profitable, could hardly survive the winter storms with such shelter as the country people deem sufficient for themselves. Good barns, therefore, are required for the protection of animals, crops and manure from the injurious effects of the weather. Your Excellency has so clearly apprehended this fact as to order the erection upon the College Farm at Sapporo of a model barn, the plan adopted being a slight modification of the one used on the estate of the Massachusetts Agricultural College.

〔*First Annual Report of Sapporo Agricultural College, 1877,* written by W. S. Clark, President of Sapporo Agricultural College, published by the Kaitakushi, pp. 1-5 より〕

3. Progressive and Non-progressive Education in Japan

It has been remarked by many, both natives and foreigners, whose opportunities for observation have covered a much longer period than do my own, that Japanese students have shown greater aptitude for learning than is manifested by those of American and European institutions; but that subsequently, in the active walks of life, they have almost invariably fallen behind the heirs of that practical, progressive, self-asserting spirit which has been the impulse and the fruit of western civilization.

This is said as well, perhaps chiefly, of those sent to America, who afterwards returned to serve their country; and hence cannot be attributed to the relative inefficiency of institutions founded at home for the promotion of foreign learning. Single-handed against a hundred-fold their number; they have frequently carried off honors of the highest rank at foreign institutions.

We need not look far to discern some of the causes which have produced this anomaly. Evidently the principal are these: first, inherited qualities of mind accruing from peculiar systems of learning and conditions of society which have existed for ages; and secondly, influence of environment upon powers of application and habits of thought.

A profound regard for learning during many centuries had left unlearned the most potent laws and principles of nature, of society, of human capacity. Utter dependence, even to veneration, upon the ancient classics of equally inert China as the fount of all knowledge, gave no impulse to rise above its source. The time necessary to master the formidable array of characters, which were the sole weapons of intellectual warfare, left few men enough of strength or years to make higher conquests. While a few aesthetic arts were carried to a wonderful degree of perfection, the practical resources of design and invention were strangely neglected. The shifting wealth and power of barons and commanders confirmed the poverty and helplessness of the masses. Multiple centers of dominion necessarily co-existed with the usurpation of governmental authority, rendering the state impotent to promote the general welfare. The maintenance of nearly one-half the people—the ablest of the nation—in unproductive plenty, subjected the other half to hardship and poverty.

The ancient learning of Japan, vast but inert, was such as to cultivate the memory to the highest degree and to neglect the powers of thought. Her art and industries, the same for ages, were productive of a remarkable degree of imitative power and hand skill, at the expense of the faculty for invention and original design. The helpless masses in yielding their substance to the powerful, learned to be content with a mere life-sustaining pittance. The powerful, whose strength was in the subjection of the weak, planted deep the root of helplessness in themselves.

Now, fortunately, the entire regime is changed; but the fruits of the past cannot be eradicated at once. Traits which have been ages in the making must need generations in the breaking.

Memory is imitative. It is the agent of all primary instruction, enabling the learner to follow readily where others have led. Original thought is progressive, and makes the possessor a leader where there be few can even follow. The thinker is a poor copyist, though possessed of creative and life-giving power. The imitator is incapable of surpassing his model just in proportion to the singleness and perfection of his power to equal it. The one makes leaders, the other, only followers.

An excellent memory will enable a student to outshine one who must stamp every fact and principle with the seal of perfect comprehension ere it can claim a place in the archives of the mind, so long as the work involved consists in taking in the elements of truth as commonly taught. But when the exigencies of life call for their use, not in the few elemental forms in which they were stored away, but by combination and permutation to be applied to an infinity of cases, not one of which, perhaps, can the best memorizer or imitator recall or identify, then are their powers displayed in a widely different light.

So the habit and faculty of adapting and applying in life even the simple facts which come to one who never saw the interior of college walls, often enable him to outstrip others who have received the best instruction the institutions of the world affords; and he is accredited with being "self-made", when in truth he is indebted, beyond the average of his fellows, to the Creator of all men.

Now in view of the peculiar traits which have descended to the present generation, it seems clear that if those systems of scientific and practical instruction which have been so widely adopted in the higher institutions of the western world rest upon true principles, still more are they meet for the needs of Japan. The deductions of science are both logical and practical; and the value of its study as a mental discipline, far from being impaired when allied with processes for deducing its applications, should be increased. The logical connection between facts and principles -the subordinate and paramount in learning- ought always to be clearly maintained; and their application to the general concerns of life serves not only to render them of double and ten-fold the service to man, but fixes them more securely in the mind. To know how is to be a peer in a single art. To know wherefore is to be a master in all its own and dependent possibilities. Knowledge without thought is impotent to go beyond itself; but thought and knowledge have unlimited resources.

An old writer observes that "we fatten a sheep with grass, not in order to obtain a crop of hay from his back, but in the hope that he will feed us with mutton and clothe us with wool". Paraphrasing the comment of another, we may apply this to the feeding of the mind: we teach a young man mathematics, logic, and natural science, not that he may simply take his equations and diagrams into a government office, or carry his borrowed propositions and demonstrations into the national council chamber, or that he shall hang his formulas and classifications high above the door of nature's laboratory; but that he may bring into all these places a mind so well stored with the sound principles of truth and reason as to be able to distinguish between fact and appearance, argument and sophism; to mould the facts of science to the intelligent purposes of his will in the ever changing problems and circumstances of life; to make the end to be accomplished, the key to its accomplishment, even as the simple yet wonderful effects in nature shadow forth their potent cause.

〔*Second Annual Report of Sapporo Agricultural College, 1878,* written by Wm. Wheeler, President of Sapporo Agricultural College, published by the Kaitakushi, pp. 11-15 より〕

4. Lectures on Agriculture

It seems befitting that, in this the first annual report published since my active connection with the college as professor of agriculture, I should present a brief account of the methods of instruction pursued and proposed, in order that all interested may be able to form some idea as to the scope and value of the course of study and training offered. The education in this department is both theoretical and practical, and is intended to fit those who faithfully complete the course to successfully manage the business of farming in all its branches according to the most advanced scientific and economic methods. It is also the intention to teach them how to conduct accurate agricultural experiments for the solution of questions requiring the application of scientific knowledge.

The theoretical instruction is given entirely by lectures, upon which the students are required to take notes that they afterwards copy upon loose sheets of paper. These papers are critically examined, and such corrections made as are needed both as to matter and style, after which they are neatly copied into convenient record books furnished for the purpose, in which shape the lectures are preserved in a good form for future reference and use. The students are carefully questioned at each exercise upon the subjects discussed in the last; and the greatest pains is taken to make certain that all thoroughly understand every topic treated of. Questions of practical nature are frequently introduced, which require for solution an application of the principles under consideration. By means of this careful questioning and frequent reviews, the subject matter passed over becomes very firmly fixed in the minds of all. In all my instruction, I am particular to give the reasons for every statement made, and I try to lead the students in their recitations to state these reasons plainly and forcibly, believing it to be of the utmost importance to train them to think logically and clearly. Wherever it is possible for them to deduce the reasons for anything stated from principles already taught them, I endeavor to lead them to do so, aiming thus, in every possible way, to teach them to think consecutively and with mathematical precision and method. The faculty of putting observed facts and known principles together, and deducing there from the legitimate conclusions which a trained and logical mind is capable of forming is, in my opinion, worth much more to a man than an encyclopedic knowledge of dry facts without this power, which is, indeed, the animating essence which alone is capable of making a man's knowledge of use either to himself or his fellows. The course of lectures given is systematically arranged according to the order which seems to me most natural, and embraces a very wide range of topics. To mention them all would occupy too much space, neither would it serve any good purpose, and therefore I will enumerate only a few of the most important.〔以下には本書第2編の目次相当分を列記，引用省略〕

The aim of all instruction in this department, while it is specially applied to the circumstances of this province, is to impart a knowledge of correct principles, which, well understood, will enable a man to succeed under whatever conditions he may be placed. Knowing these principles a man can vary his practice as the occasion may require, and a little experience will make success certain in all cases. Special training will be given in experimental farming during portions of the Sophomore and Junior years. To this end questions will be proposed to each member of the class at the beginning of the second term of the Sophomore year, which require in their solution the exercise of painstaking exactness and more or less scientific knowledge. Each student will be required to write out a careful account of an experiment that he thinks will answer the question. This he will submit to me for criticism and revision; in which criticism no pains will be spared to point out his errors, if there should be any, and to show him clearly wherein and why his proposed experiment would fail to answer the question under consideration. Having thus seen that each is prepared to carry out an experiment which promises a correct and reliable answer to the question asked, we shall be in readiness to begin out-of-door work as soon as the season will allow. Each will then be allotted sufficient land for his purpose, and furnished with all the necessary material as it is needed. He will be required to keep an accurate account of all material furnished him from the farm and of all labor of teams, his own

labor and expenses of all sorts. 〔実験手法の詳細を中略〕 To gain the necessary instruction in the methods of doing farm work, the students during the earlier portion of the course labor in the field six hours each week under my direct superintendence, whenever the weather will allow. Their work will be made to cover as wide a ground as the time will permit, and will include the planting and harvesting of crops, the making of permanent improvements, the use of farm machinery and implements, the care of animals and the driving of teams both of horses and oxen. I have adopted the practice of paying for such labor at the rate of five cents per hour, more for the sake of cultivating habits of industry, and economy in the use of money than as wages. The students are thus made to see that, to get money, it requires exertion on their part, and hence they more fully appreciate its value. That they may become useful citizens and officers, it is essential that habits of industry and of making a proper use of money be acquired in their youth; I have therefore employed such as were not otherwise engaged to work upon the farm during vacation, whenever they have chosen to do so.
〔*Second Annual Report of Sapporo Agricultural College, 1878,* written by Wm. P. Brooks, Professor of Agriculture, published by the Kaitakushi, pp. 39-48 より〕

5. Expansion of Subject and Experimental Work

1) Expansion of Subject

In this department no particular change has been made in the methods of instruction pursued but the course has been gradually extended until it embraces a considerably wider range of subjects than formerly, and the rules and methods according to which the student's experimental work is carried on have been systematized and perfected.

These rules will be found in subsequent paragraphs of this report. The course in agriculture now embraces lectures upon the following subjects: —soils; their formation, physical and chemical characteristics, and methods of improvement, including drainage, paring and burning, liming, irrigation &c., plants; elementary composition, food requirements and sources of food, manures; their use, methods of application &c., crop rotation: relations of the atmosphere to vegetable life; utilization of sewage, machines and implements and the chief operations in agriculture; selection, division and general management of the farm, farm buildings; mowing and pasture lands, their selection, preparation and management: tillage land and its general management: the chief crops of temperate climates, origin and history, botanical characteristics, soil, climatic and manurial adaptation, selection of seed, time and methods of planting: cultivation and harvesting; plant diseases; history, causes, diagnosis, remedies—general preventive and special: animals of the farm; breeds, their history, description and special adoptations: stock breeding; cattle feeding, dairy farming: bee-farming and forestry. Besides the above, the subjects agricultural economy and rural law have been somewhat briefly treated and in so far as my observation permitted, instruction in the habits of injurious insects and the methods of preventing their ravages and destroying them has been given. 〔以下，引用略〕

2) Results of Student's Experiments

Before summarizing the results of the experiments tried, I must call attention to the well known fact that before conclusions of much importance can be drawn from agricultural field experiments they must be continued through a number of years. This is due to the impossibility of controlling the conditions under which they mast be carried out. In the first place it is impossible to obtain perfect uniformity of soil; and in the second place amount of rain fall, the temperature, sun-shine, insect ravages & c. vary greatly from year to year. Differences in yield not due to the various treatments under trial are therefore the inevitable consequence: and the results of one year may be contradicted by the results of the next. Moreover it must be remembered that, while an effort is made to make the experiments of practical value, they are mainly designed as a means of educating students, who neither from previous training nor experience are qualified to carry out different experiments, even did the time at their disposal allow which it most certainly does not. The educational discipline which the students get, however, is most valuable; and their results are not infrequently of practical importance to the agriculture of the Hokkaido.

I. Plant-food Requirements of the College Farm Soil.

Experiments to determine in which of the elements of plant-food the soil of the College Farm is most deficient have been tried by several students and with various crops. In these experiments chemical fertilizers of known composition imported from America have been employed, viz:—Potassium nitrate, sodium nitrate, potassium sulphate, potassium chloride, ammonium sulphate, magnesium sulphate and superphosphate. The importance of this question, in its bearing upon an economical use of manures, is self-evident; but such is the natural fertility of the land upon which our experiments have been tried, that I cannot report very decisive results. The general indications, however, are that potash, is most deficient, phosphoric acid next and nitrogen least. The addition of all these however produces the largest, though not the most profitable crop. It is hoped that in a few years our experiments will enable us to give definite practical advice concerning the use of manures.

II. Methods of preventing Plant Disease.

Experiments having this object in view have been carried out in two principal ways, viz:—it has been attempted to prevent such diseases by treatment of the seed of wheat, oats and barley with solutions of various kinds; and second by applying to the soil certain fertilizers. Treatment of the seed with solutions depends upon the idea that thus it is possible to destroy the vitality of the spores of the parasites always accompanying the diseases of the cereal grains above mentioned; and the solutions selected for our experiments were copper sulphate and lye. The results indicated that both solutions were about equally effective in preventing smut and bunt; but had no effect upon rust. No doubt the smut and bunt may be mostly prevented by the judicious treatment of the seed with either of the solutions mentioned. The mineral manures experimented with supplied potash, lime and phosphoric acid abundantly. They were variously compounded and applied in different amounts to different plots; but, though most of the experimenters reported a lessened amount of smut and bunt in consequence of their use, I regard the reported differences as too small to warrant this conclusion. Such differences may be merely incidental. Rust was not diminished in amount by any method of manuring adopted.

　III. The Experiments for the Comparison of Foreign Drill with the Japanese Drill system.

Results showed no great differences in yield though the plots sown according to the latter were cultivated while the former were not. In several instances the yield under the Japanese system was slightly superior; but the experimenters generally reported that considering the labor they found the foreign machine-drill system the better.

　IV. Comparative Trials of Manures.

In these experiments the manures readily obtainable here were used, being variously combined —with a view generally to supplying a complete manure and used in varying quantities. The manures experimented with were barnyard manure, herring guano, rape cake, nuka (rice-cleanings), wood ashes and lime. Barnyard manure has usually produced, the largest crop, though its superiority over herring guano and. ashes used together is not great. It is of course the cheapest, but on the other hand the labor of applying is considerable. Ashes while very cheap here, always prove highly beneficial and every effort should be made to promote their use in connection with herring guano.

　V. Suckering Indian Corn.

This was reported by the student trying the experiment to increase the yield; but the difference reported was small, and I do not regard the question as settled.

　VI. The Proper Distances between Drills and Plants of Indian Corn.

This experiment was rendered nearly worthless by the ravages of the Farm cows which broke into the field; but so far as could be judged the experiment indicated, that corn is often planted here too thickly.

　VII. Best Method of Cutting Seed Potatoes.

Tubers for planting were prepared in four different ways and planted in alternate rows over the entire area under cultivation. The methods of preparation were as follows: —1st. whole tubers from which all except three or four eyes were cut out; —2nd. tubers cut longitudinally to pieces containing about the same number of eyes (3 or 4); —3d. tubers cut transversely large ends with 3 or 4 eyes; —and 4th similarly cut, small ends with 3 or 4 eyes. The potatoes prepared in the third manner yielded the best crop, though not much superior to those prepared in the second method. Considering that the latter is least laborious and involves no waste of seed, it must be regarded as the best method for practical adoption. The weight of the crop from tubers prepared in the first method was a little greater than from any of the other methods; but size and quality were much inferior. It appears to me likely that some of the eyes supposed to be removed were not completedly cut out, and hence sprouted. This would account for the greater number of tubers and their small size.

　VIII. Yield of Sugar in the Beet as affected by Manuring.

No very decisive results were obtained; but the use of barnyard manure gave the most satisfactory result, while herring-guano and ashes with lime ranked next.

　IX. Comparison of Varieties.

The beets from the "Vilmorin's Improved Sugar" seed proved slightly superior to those from the German seed obtained from Monbetsu.

X. Pruning Squash Vines.

Results not very decisive; but showed a slight advantage in favor of pruning; not enough, however, to repay the cost of the labor last year.

XI. Artificial Fertilization of the Squash Flowers.

This experiment gave no very decisive results through the fact that the attempted cross fertilizations were mostly failures. This was due to the inexperience and lack of skill on the part of the operator.

XII. Protection of Swedish Turnips from Insects.

This experiment was a failure; the crop being entirely destroyed. Some useful observations were however made which may constitute the first step towards ultimate success.

XIII. Ridge and Level Culture of Corn.

This experiment showed a slight superiority, but not enough to repay the extra labor in favor of ridge culture.

XIV. Ridge and Level Culture of Beans.

Result the same as in the preceding. This experimenter made a number of observations upon insects injuring the crop which may prove useful.

[*Sixth Report of the Sapporo Agricultural College, covering the Years 1881-1886 incrusive,* written by Wm. P. Brooks, President ad interim of Sapporo Agricultural College, published by the Hokkaido Cho, pp. 12-25 より]

6. Questions of the Final Examination Exercises

Chapter 1
  1. Define Agriculture—Its Importance as an Occupation. (2-118)
  2. The Importance to a Farmer of a Knowledge of Chemistry, —Botany —Entomology —Veterinary Medicine —Enumerate the other Branches of Knowledge which it is Important for Them to Know. (2-118)
  3. The Influence of Agriculture on National Prosperity. (3-106)
  4. The Importance of Knowledge of Entomology and Veterinary Medicine to the Farmer. (3-106)

Chapter 2
  1. Define Soil, Surface Soil or Tilth, Subsoil —Hardpan. (2-118)
  2. Enumerate the Agents which have been Chiefly Instrumental in the Pulverization of Books, and Describe the Mode of Action of Glaciers. (2-118)
  3. Influence of Organic on Inorganic Matter in the Soil. (2-118)
  4. Enumerate the Different Classes of Soils as to Their Place of Origin; and State the Characteristics of Each Class, and how Each may have been Formed. (2-118)
  5. Absorption of Vapor of Water by Soils. (2-118)
  6. Power of Soils to Remove Dissolved Solids from Their Solutions. (2-118)
  7. Sources from which the Surface Soil Derives its Heat, Influence of Color on the Power of Soils to Absorb Heat, and the Influence of Texture on Their Power to Retain Heat. (2-118)
  8. Crop Adaptation of Sandy Soil—of Alluvial Soil. (2-118)
  9. What is Soil, and What Classes of Matter, Does it contain. (3-106)
 10. The Agencies which have been, and Are Instrumental in Breaking up Rocks. Describe the Mode of Action of any one of them. (3-106)
 11. Influence of Organic Matter upon the Mineral Constituents of Soils. (3-106)
 12. The Products Resulting from the Decay of Vegetation with Full access of Air. (3-106)
 13. The Form in Which Plants Take Their Mineral Food. (3-106)

14. Importance of a Variety of Elements in an Agricultural Soil. (3-106)
15. The Manner of Formation of Alluvial Soils, and Their Chemical and Physical Characteristics. (3-106)
16. What is Sub-soil, and What Are the Differences Commonly Found between It and Surface Soil ? (3-106)
17. In what condition as to solubility, should manures applied to sandy soils be and why ? (4-93)
18. Influence of color and physical condition upon the power of soils to absorb and retain heat. (4-93)
19. Crop adaptation of sandy loam, clayey loam and alluvial soils. (4-93)

Chapter 3-1
1. The Beneficial Effects of Drainage on Soils. (3-93)
2. The Relative Merits of Open, Box, and Tile Drains. (3-93)
3. The Manner of Laying Tile Drains. (3-93)
4. Indications by which one can tell whether land needs artificial drainage. (4-93)
5. Beneficial effects of drainage. (4-93)
6. The merits of stone drains and the manner of making them. (4-93)
7. Manner of Laying tile drains. (4-93)
8. Benefits to be derived from irrigation. (4-93)
9. A method of irrigating a garden. (4-93)
10. Method of making a water meadow on the banks of a stream in a nearly level valley. (4-93)

Chapter 3-2
1. The Benefits of Irrigation, the Kinds of Water Used, and Their Relative Values for this purpose. (3-93)
2. The Method of Making a Water Meadow, where the Surface of the Land is nearly Level, the Water to be Taken from a Stream Flowing through It. (3-93)
3. The Construction and Use of a Wing Dam. (3-93)

Chapter 4
1. The Best Method of Pulverizing Soil, and the Good Points in Swivel and Gang Plows. (3-93)
2. The Different Kinds of Harrows, and the Merits of Each. (3-93)
3. The physical condition in which land should be when ploughed. (4-107)
4. Give a description of the ordinary plough, and tell the uses of its parts. (4-107)
5. The construction and use of the roller. (4-107)

Chapter 5
1. Carbonic Acid Gas—Its Relations to Vegetation. (2-116)
2. Ozone—Its Relations to Vegetable Nutrition. (2-116)
3. Diseases of Plants Caused by Atmospheric Influences. (2-116)
4. Principles upon Which Crop Rotation Depends. (2-116)
5. Sulphate of Lime—Its Value and Manner of Action as Manure. (2-116)
6. Phosphate of Lime—Its Value as Manure. Method of Manufacture of Superphosphate. (2-116)
7. Super phosphate of Lime. —Its Importance as a Fertilizer, and the Method of Making It. (3-93)
8. The Relative Importance of Mineral and Organic Manures. (3-93)
9. Muck, —Its Chemical and Physical Characteristics, and Soil Adaptation. (3-102)
10. Formation of muck, and its constituents. (4-107)
11. Soil adaptation of ashes, and their mode of action. (4-107)
12. Mention the conditions which affect the value of animal excrements. (4-107)
13. Fish guano; its manufacture, composition and uses. (4-107)
14. Possible sources of waste in manures, and the means of preventing such wastes. (4-107)
15. Relations of the carbonic acid of the atmosphere to plant growth. (4-107)
16. Give a concise statement of the atmospheric supplies of plantfood. (4-107)

Chapter 6
1. Value of Animal Excrements as Affected by Age, Condition, Work and Food of Animals. (2-116)
2. Farm Economy—Importance of System—Economy with Regard to Labor. (2-116)

3. Advantages of Special Farming. (2-116)
4. Principles which Should Guide in the Selection of a Farm for General Farming. (2-116)
5. The Best Season for Sowing Grass-seed, the Selection of varieties to be Sown for Permanent Mowing, and the Manner of Doing the Work. (3-90)
6. Principles which should Guide in the Selection of Pasture Land, and the Manner of Converting Forest Land into Pasture. (3-90)
7. The Best Place upon the General Farm for a Forest, the Manner of Planting It, and the Kind of Trees to be Planted. (3-90)
8. The Location and Quantity of Fences on the General Farm, and the Statement of a Just Law in Regard to Fences between Neighboring Land-owners, and along the Public Roads. (3-90)
9. Farm Roads.—Their Importance, and the Manner of Constructing Them. (3-90)
10. The Location and Soil most suitable for an Orchard, the Manner of Raising Apple Trees, Transplanting them, and Their Subsequent Management. (3-90)
11. Winter Work on the General Farm. (3-90)
12. Manner of Determining What Elements of Plant Food will be in the Excrements from Animals Fed on a Certain Amount of Food, and the Practical Use of Such Knowledge. (3-90)
13. The Method of Ascertaining Experimentally, what Elements of Plant Food are Most Wanting in the Soil of a Farm, and the Practical Value of Such Knowledge. (3-90)
14. The Routine of Spring Work on a General Farm, and the Proper Time for Planting the Different Crops. (3-90)
15. The Value of Animal Excrements as Affected by the Food, Age, and Condition of the Animal. (3-102)
16. The Nitrogen of the Atmosphere, and Its Relations to Vegetable Nutrition. (3-102)
17. The Economy of the Use of Brains in Farm Management. (3-102)
18. General Farming,—Its Advantages and Disadvantages as an Occupation. (3-102)
19. Location and Selection of a General Farm. (3-102)
20. Division of the General Farm into Mowing, Tillage, etc. (3-102)
21. Points Which Should Be Secured in the Construction of a Barn for General Purposes. (3-102)
22. Timothy and Red Glover,—the characteristics and Adaptation of Each. (3-102)
23. The Manner of Sowing Grass Seed with Grain. Time allowed, two hours. (3-102)
24. Considerations which should guide the farmer in the selection of a location for a pasture on a general farm. (4-88)
25. Conversion of wooded land into pasture. (4-88)
26. Selection and management of tillage land. (4-88)
27. Location and construction of farm roads. (4-88)
28. A just law with regard to fences between adjoining land owners. (4-88)
29. Method of determining approximately, without Analysis, the elements of plant food in the manure from certain animals: use of such knowledge. (4-88)
30. Winter work on the general farm. (4-88)
31. Improvements best made in summer. (4-88)
32. Culture as affecting the amount of available moisture in the soil. (4-88)
33. The implements of culture and their relative importance. (4-88)

Chapter 7
No list

Chapter 8
1. The requisites for keeping up to its original standard of excellence, and of improving any variety of vegetable belonging to the biennial class. (4-97)
2. The advantages to be derived in this country from stock farming. (4-97)
3. General characteristics and adaptation of the Short-horn cattle. (4-97)
4. The points of excellence in a good working animal, and the breed possessing most of them. (4-97)
5. The breed of cattle best for milk, and the reasons for its superiority. (4-97)
6. Points desirable in cattle of any breed, and the selection of breeding animals. (4-97)

7. Care of the cow at and immediately after parturition. (4-97)
8. Food and care of young cattle. (4-97)
9. Characteristics and adaptation of merino sheep. (4-97)
10. Care of sheep during the lambing season. (4-97)

Appendix; Fruitculture.
1. Name the Different Parts of Trees, and State the Functions of Leaves. (3-99)
2. Production of New Varieties by Crossing. (3-99)
3. Propagation of Varieties. (3-99)
4. Soil and Situation Best for an Orchard. (3-99)
5. The Method of Pruning young Trees to Insure Their Assuming a Desired Form. (3-99)
6. The Curculio. —Its Habits and the Best Means of Preventing its Ravages. (3-99)
7. Root-Grafting Apple Trees. (3-99)
8. The Causes which Predispose the Peach Tree to Injury from Gold. (3-99)
9. Propagation of Grapes. Describe the Method most Practiced. (3-99)
10. The Pruning and Training of the Grape Vine. Time allowed, two hours. (3-99)
11. The process of growth in exogenous fruit trees. (4-102)
12. The methods of producing new varieties of fruit. (4-102)
13. The means of propagating varieties of fruit and statement of the essentials for successful grafting. (4-102)
14. The points which must be observed for success in transplanting trees. (4-102)
15. Preparation of apples for a distant market. (4-102)
16. The apple worm (*Carpocapsa pomanella*): its habits and the best means for its destruction. (4-102)
17. The characteristics which recommend dwarf pear trees for cultivation, and the method of making them. (4-102)
18. Pruning the peach tree—the cherry tree. (4-102)
19. Soil and location best for a vine yard, and the training of the vines. (4-102)
20. The propagation, culture and pruning of raspberries. (4-102)

解　　題

1．農学校の概要

　第1節 1) は農学校の教職員名簿である。*First Annual Report of Sapporo Agricultural College, 1877* を和訳した『札幌農黌教頭米人クラーク氏原撰 札幌農黌第一年報』は，開拓権大書記官調所廣丈を黌長（校長），招請したクラークを教頭兼黌園長（農場長）と明記しているが，引用した英文ではクラークを President としており，混乱しているように見える。これは，クラーク雇用契約書ですら President と書いていることから，実質的な校長はクラークとするが，開拓使の学校のため権大書記官を名目上の校長とし，英文では Japanese Director と記載して区別したのである。なお，調所廣丈は，鹿児島出身で参議兼開拓長官黒田清隆と同年齢であり，戊辰戦争の政府軍参謀である黒田の下で榎本武揚らの旧幕府軍と戦って名を上げ，1873(明治6)年から高等教育進学者を養成する開拓使仮学校の校長であった。札幌農学校が同仮学校の専門課程に位置づけられるため，調所が校長に就任するのは当然である。

クラーク(1826-86年)はアメリカ，マサチューセッツ州に生まれ，アマースト大学を卒業してドイツに留学，化学と鉱物学を学んで学位を取得し，帰国後に母校の教授になった。南北戦争時には義勇兵を募って北軍に参加し，幾多の戦績をあげる活躍をして大佐，准将に推挙された。その後，モリル法(1862年)ができるとマサチューセッツ農科大学の創設に奔走して学長に就任し，農科大学が全国紙に紹介されるほどに評価を高めた。札幌農学校への誘いを受けると，他人が2年かかる仕事を1年で成し遂げるとして，現職のまま1年間の休職をとって来日した。一行が札幌に到着した1876(明治9)年7月31日は，偶然にもクラーク50歳の誕生日であった。クラークは，アマースト大学在学中に神の存在を実感する回心体験をして敬虔なピューリタンとなったことから，農学校教育の基本にキリスト教をおき，「細々した規則は要らない。ただひとつ"Be Gentleman"のみが校則である」と言って人格教育を進めた。このキリスト教の採用にあたっては，東京から小樽に向かう玄武丸上で黒田長官と激しい論争を行なっているが，1872(明治5)年まで禁制としてきたキリスト教を官立学校教育の基本におくことを認めた黒田の度量にも驚きを感ずる。さらに，それを受容した学生が1882(明治15)年に札幌基督教会を設立し，日本のキリスト教発祥地のひとつとなったことから，教育の強さを知らされる。また，9カ月後の翌年4月にクラークが帰国する際，島松沢(現北広島市島松)まで見送った生徒に'Boys, be ambitious!'と叫んだ話は，目的に向けての努力，科学技術に対する献身などの意義を加えて広められ，青少年の教育や育成面での象徴的な用語となった。

　クラークが伴ってきたホイーラー，ペンハローとブルックスの3教授は，来日時にそれぞれ25，22，25歳と若いが，ともにマサチューセッツ農科大学の1・3・5期生としてクラークの教え子である。最年長のホイーラー(1851-1932年)は，卒業後に鉄道技師を経て土木工事の事務所を経営していたが，クラークの誘いに先生の行動力を見出して来日し，物理学，土木工学などを担当するとともに，演武場(現札幌時計台)と農場にあるモデルバーンの設計をはじめ，開拓使の依頼を受けて豊平川の治水と札幌—茨戸間の運河開削，小樽—札幌間の道路と鉄道敷設などの測量を手がけた。クラークの帰国後は，教頭となって役人との煩わしい交渉が増えたため，マサチューセッツ工科大学出身のピーボディーとハーヴァード大学医学部出身のカッターを雇用して校務を減らし，全道各地の道路建設測量を行なったほか，わが国で2番目に古い観測開始記録となる気象観測を手がけるなどの活躍をし，3年後の1879(明治12)年12月に帰国した。なお，ピーボディー(1855-1934年)は，物理学，土木工学などを担当してホイーラーの後を継ぎ，1881(明治14)年7月に帰国している。また，生理学を担当したカッター(1851-1910年)は，マサチューセッツ農科大学からハーヴァード大学医学部を卒業した医学博士でもあったから，開拓使顧問として医療の発展に貢献したばかりか，1880(明治13)年に開講した動物医学は獣医学，魚類学と養魚学は水産学の本邦における開祖である。また，札幌をこよなく愛し，衛生面から札幌大通に現存する水飲み場を設けたというエピソードもあり，これら功績によって1887(明治20)年の帰国時に勲四等旭日小綬章を受けている。

　博物学と化学を担当したペンハロー(1854-1910年)は，最初に道内の動植物・鉱物の博物

標本を収集して実験室を完備させ，人跡未踏の空知芦別まで学生を引率して自然を教科書とする自立的な態度を養い，第3節に掲載するように棒暗記を重視する日本の学習法に改革を迫った。ホイーラーの帰国後に教頭心得となると，試験研究機関を持たない開拓使の依頼によって各地の鉱物や土質分析，石油試験，ビール分析，麦芽病の原因分析など，化学分野で北海道開拓に貢献したが，夫人の病気が伝えられて1880(明治13)年8月に帰国した。

　本書の中心となるブルックス(1851-1938年)は，後述の農学講義の記事，第3章のユステス・藤居恒子氏の記事で述べるため，クラーク一行から半年遅れの1877(明治10)年1月に来日し，1888年10月に帰国するまで外国人教師で最長の12年間も滞在し，日本に泰西農学と北海道農業の基礎を定着させた功績が大きいこと，それらによって1919(大正8)年には日本農学博士会から，1932(昭和7)年には北大から名誉博士号を授与されていることを特記するに止める。なお，ブルックスの後継者にはブリガムが雇用され，州会議員や農会長を歴任した経験豊かな農業専門家として北海道農業のソフト面の確立に尽くしていたが，1893(明治26)年に札幌農学校の文部省への移管に備えて解雇され，最後の外国人教師となった。

　書記の堀誠太郎は，クラークが校長であるマサチューセッツ農科大学に留学し，4人の外国人教師とかかわりがあったから，開拓使などとの交渉の際に通訳や翻訳官として大変有能であったし，農学校への進学生を教育する予科では，井川洌とともに教師を務めた。

　第1節2)の学校の目的では，「札幌農学校は，卒業後に開拓使に勤める高度技術者を教育し習練させるために開拓使が設けた学校であり，4年制で卒業生に学士号を与える。生徒数50名，学費は官費とする」と明記しているが，翌2年次には2名の私費生を加え，開拓使の廃止が明らかになった第4期生から開拓使へ勤務の条件が廃止されている。

　学年暦は，1学期が8月の第4木曜日から12月の第4水曜日まで，2学期が1月の第4木曜日から7月の第1水曜日までとし，抄録部分以外であるが，毎日の講義は8時30分に始まって正午に終わり，午後は13時30分から1時間の武芸を行ない，次いで2時間の農場実習や画学などを行なうよう定めている。講義は英語で行なうため，講義中に筆記をさせた事項は，放課後に浄書をして教師の校閲を受けた。学業の評価は，平常の評価点を100点，期末試験の成績を100点とし，それを加算して科目点とし，その受講科目の平均点でクラス中の席次を決めた。なお，「学業の優秀と進歩を公認し，鼓舞させんがため」として，科目ごとに最優秀の生徒には賞金8円，第2位には賞金4円が与えられたし，農場実習では勤労の精神を養うために1時間5銭の賃金が支払われた。なお，この頃の物価は1ドル＝0.997円であり，チモシー乾草10円/t，バレイショ8.60円/kl，ダイズ28.50円/kl，トウモロコシ17.10円/kl，予科授業料6円/年であるから，賞金額は相当に高価であった。なお，黒田長官の給料は年俸6000円，それに対しクラーク博士は旅費や宿舎経費を除いて7200円であった。このような講義法，評価法や賞罰制度などは，それまで日本の私塾や大学予備門などで学んでいた生徒にとって驚きであったに違いない。

2．畜産の導入論

　第2節は，クラーク博士の日本農業改革論である。冒頭に「先人が「国に人民なければ国

にあらず，人に志なくば人にあらず，而して人の心田も耕さざれば，心あれども無きに等し」と言うように，開拓使が機械製造所，製粉所や鉱山を開いた成果は大きいものの，学校教育ほど大切なものはない。学制を整えて教育が良好に進めば，学術，技芸，富国強兵，人知を要する事項のすべてが隆興して国が栄える」と書き，開拓に先立って札幌農学校を作った先見の明を褒めている。次いで「北海道の水産・山林資源を適切に生産すればさらに利益が高まるが，北海道の最大の富は，肥沃な土地を使った農業の振興である」として，日本で行なっていない畜産導入の重要性を説いている。

近年の北海道における家畜飼養頭数は，乳牛が 88 万頭(全国シェア 47%，以下同じ)内外，肉牛が 41 万頭(15%)内外，豚が 54 万頭(5%)内外，綿羊が 8000 頭(52%)内外と言われて畜産王国を自負しているが，その発端は未開の北海道を調査して開拓の根本方針を提言したケプロン(1804-85 年，米国農務長官の職から 65 歳で来日)による『ケプロン報文』，ここに引用したクラーク論文を契機にしており，その歴史は 120 余年にすぎない。一方，府県での歴史を見ても，幕末に神奈川の生麦地域で乳牛を飼い，横浜に駐在する外国人に牛乳を供給したのが日本酪農の始まりとされるが，日本人のために牛乳を生産し始めるのは，明治維新後に海外の農業技術を受容してからである。したがって，ケプロンの指示によって 1873(明治 6)年に短角牛 20 頭(別船便で牛 20 頭，羊 91 頭)とともに来日したエドウィン・ダン(1848-1931 年，そのうち 1875 年から 82 年の間畜産技師として北海道に在住，その後も駐日米国公使等に就任して日本で生涯を送る)の指導を受けた七重や札幌の官園，その流れのなかでクラークを代表として普及・実業化に尽くした札幌農学校の農黌園，府県の研修生を集めて畜産の普及を図った千葉県の御料牧場の 3 カ所が日本酪農の発祥地となる。その後北海道では，ダンの指導を最初に受けて北海道酪農の基礎を築いた町村金弥(札幌農学校 2 期生)と宇都宮仙太郎，子供の頃から宇都宮農場に出入りして札幌農学校を卒業し，北海道酪農の父と尊称される金弥の子の町村敬貴，みずから酪農を営みながら製酪販売組合を率いた黒澤酉蔵ら多くの先覚者の活躍によって牛乳を生産する酪農と加工販売する乳業業界が定着する。それが全国に広まったのは，明治末から大正期頃の町の「ミルクホール」に契機を求めることもできるが，本当に大衆化したのは第 2 次大戦後にアメリカの援助で学校給食に粉乳が出されてからであろう。

それ以前のことを大日本農会編『明治維新前農業史』で調べると「明治維新までの日本農業の特色は，すべて人力労働に依存して畜力を用いないことであり，世界に例がない国である」と記している。しかし，馬は戦国時代に大きな働きをしているし，牛は牛車を引いて貴族の足になっているばかりか，歴史を細かく見ると牛乳をまったく知らなかったとか，牛馬で引く畜力農具が伝来しなかったと言うわけでない。例えば 700(文武天皇 4)年 10 月に，牛乳を低温で 10 分の 1 ほどに煮詰めて黄白色の「蘇」を作らせ，内蔵寮に保管して大臣大饗や二宮大饗など重要な行事で使ったとか，1792(寛政 4)年に将軍吉宗が白牛 3 頭から蘇をさらに精製した「白牛酪」を作らせたとの記録がある。これらは，乳牛がいたものの，殿様や上流階級の人々が牛乳を精製して薬感覚で飲んだのみであり，広く牛乳を飲む風習まで広がらなかったと理解される。また，奈良朝以前に畜力で田畑を耕す農具の犂(和犂)が伝来して

正倉院の宝物となり，さらに荘園等で実際に使われながら，その後土地生産性を高めるために人力の鍬，備中鍬で深起こしを行なって畜力耕起を排除し，一部に馬鍬による代掻きがあるものの，明治維新まで牛馬の役畜がほとんど普及しなかったという経緯もある。そのため前述の書は，零細な農家規模，土地生産性の重視，仏教による牛の崇拝思想などが農業に畜力が使われなかった理由であると指摘している。

東南アジアのタイでは牛乳を飲む風習がなかったが，約25年前から国民の栄養改善と農業振興を合わせて乳牛の飼育と牛乳の飲用に国を挙げて取り組み，今日では1人当たりの消費量が日本の約半分まで拡大している。この機会に日本畜産の発達が，国民の栄養や生活にどのように貢献してきたかを考えるのも有意義であろう。

3. 教育改革論

第3節は，日本の教育と米欧の教育を比較し，日本の教育理念の改革を提言している。クラークの帰国後に第2代教頭に就任したホイーラーは，年報の冒頭に本文を掲載して就任の挨拶に代えたと考えられる。その文は「日本人は世界一記憶力を鍛え，勤勉で素晴らしいが，永年の風習または思想が災いして中国古典の暗記を重視し過ぎ，物事の道理や真理をきわめる学問が欠けている」とし，封建制度の不合理を列挙するばかりか，これまでの評価基準である記憶や模倣よりも，学習結果からの創造や臨機応変の活動力の養成が大切だと言っている。したがって日本の学校は，西欧の学校が学術と実業教育を組み合わせて理論と実験によって真理を教え，幅の広い人格形成をしているのを理解し，できるだけ早く見習うべきだと提言する。

1871（明治4）年から1年半かけて世界を一周した岩倉使節団は，明治期の政治社会体制を作る上に大きな影響を与えているが，その際の公式報告書である『米欧回覧実記』(1878年刊，岩波文庫5巻本で収録)は，為政者側の考えを知るに都合のよい資料である。その実記は，イギリスの巨大な製粉，羊毛・ラシャ，造船工場などを視察して「欧洲の経済説に於て，民業を3種に分つ。化形なり，変形なり，変位なり」。「化形とは，物の形質を，造化力によりて変化させるを謂ふ，土地糞培の力に因て，種子を艸木の花実に化し，秣草を牛羊にあたへて，乳酪肉毛に化」す農業である，変形は農産物を製造によって人工的に「種種の形に変する」工業を言い，変位は物品の「位地を変する」商業であるとし，農業が社会的分業の基本環であると述べるが，「日本の風俗は，只穀〔水稲〕のみを重んし，国の貧富も穀を収量する多寡にて較するに至る，是工商未た興らす，生意の未開なるに因る」(5巻179-181頁)と彼我の差異を強調している。当時の日本では，生糸を作る製糸業のみが盛んであるが，それすら買付けから貿易まで外国人に支配されているため，商工業が未発達だと断定したのであろう。続いて「独逸は，勧農のことに就て，最も欧洲中に超越す。州郡に於て勧農社の流行盛ん〔中略〕，農学校の建立も増加したり」，さらに「理論実験両備の農学校を興し〔中略〕，理術並完からしむ」(5巻194-197頁)と書いていて，使節団が本書のホイーラーの主張を視察過程で習得したことを述べている。

そのため，使節団副使を務めた大久保利通は，帰国早々に農商務省を作り，産業振興から

農学校作りを推進している。佐賀の乱から西南戦争，そして大久保の暗殺などによって実現が遅れるが，しかし，「脱亜入欧」という言葉通りに日本の近代国家建設が進んだ。これらとの関係を考えながら第3節を読むと意義深いものがある。

4. 農学講義の概要

　第4節以降はブルックスによる農学講義の報告書である。最初の第4節は，講義の初年のため，農学講義に対する考え方を述べているが，本文があまり長いため，全体の約半分を引用した。さらに農学講義の成果は，『札幌農黌年報』の初年次を除く全巻に登場するが，ここでも第5節に第6年報を引用するに止めた。

　まず「農学講義は，農業の諸事項を学術と経済の法に則って解決して農業経営ができるように理論と実際とを併置して教えるし，さらに学術や現実の農業の発展に寄与できるように精密な試験を行なう技術も習得させる」とした上で，講義で述べる理論は正確に筆記し，教師の校閲を受けて浄書し，将来の座右の書となるようにさせた。すなわち本書講義録の英文中に〈 〉で記載した欄外の書き込みがあるが，これは講義録内の相互対照記号であるから，ノートを事項ごとに前後参照しながら利用した経過がうかがわれる。また，理論に対する現実の課題を提示して回答させて理解を深めさせ，多様な課題に応用できるように仕向けてもいるが，この詳細は分からない。次いで上級に進むと農場実験課題を与え十分に紙上検討を行なった後に，実際に栽培をさせて結論を出させる訓練（次節の実験）も行なう予定であった。

　講義とは別に週6時間の農場実習を課し，作物栽培，牛馬の扱い，農具を使った作業など一連の農作業を行なわせて実技をも習得させたが，若い年代から勤労の習慣と金銭価値を理解し，官吏や農場管理者となる際の知識とするため，1時間に5銭の賃金を支払った。

　なお，その後の引用省略部分には，苫小牧から室蘭，長万部，岩内，余市と小樽を巡回して各地の農業や原野の様子を描いている。また，引用しなかった第3年報には，講義で述べる参考書は，本によって詳細または粗雑な記述があって利用目的に合わず，むしろ講義の口述ノートのみを吟味した方が勉学によいと指摘した上で，学生に与えた農場実験の個々の成果について述べている。さらに第4年報では，4年級に農業事項の討論会を追加したことを述べたほかには，前年とほぼ同様のことが書かれている。なお，この年から学生定員が50名に増加したことと，クラークとホイーラーの帰国によってペンハローが教頭心得となり，新たにカッターとピーボディーが雇用されたほか，予科教授を兼ねた宮崎道正，市郷弘義を採用して教授陣を強化した。第5年報は，農学校を作った開拓使の廃止が予定されて混乱し，学校運営の全体を総括した文はあっても農学講義の内容は詳述していない。

5. 農業実験

　第5節は，前節の第5年報までが1年ごとの報告書であるのに，1881-86年の5年間を総括した第6年報から引用している。このようになった理由は，国内的には西南戦争後の財政難と民権運動などの社会不安があった上に，1882(明治15)年2月に北海道開拓使が廃止されて札幌農学校が存立基盤を失ったことにある。開拓使の廃止に際して札幌農学校は，文部

省からの移管要請に反して農商務省(1年間)や北海道事業管理局(3年間)の管轄に置かれ、その官制によって400 haに広がった農場をほとんど取り上げられたばかりか、ついに農学校廃止論まで浮上した。札幌農学校に戻った佐藤昌介や南鷹次郎らは、廃校論に反対する活動を始め、1886(明治19)年に新設された北海道庁の管理として、廃校の危惧を解消した。なお、文部省管轄に移るのは1895(明治28)年4月のことである。このような経過から、5年間の記録をまとめた第6年報が1888(明治21)年にブルックスによって作成された。ブルックスは、1880年にペンハローが帰国した後を受けて教頭心得に就任したものであり、その後1888年に母校マサチューセッツ農科大学の教授として帰国するまで、丸8年間も教授陣の中心として札幌農学校を率いながら北海道開発に貢献した。なお、1888年秋に『札幌農学校一覧』が刊行され、その後今日の『北海道大学一覧』に至るまで同一形式が継承されたため、年報は第6次で終了した。

　第6年報は、まず5年間の所管官庁の変化、教授陣の交代、施設の充実や送り出した47名の卒業生などについて述べ、続いて科目と部門別の報告となっている。本節は、その農学報告を引用しているが、まず農学講義の内容の変化について述べ、次いで引用を省いた博物場の充実、展示品の内容(この展示品の多くは、現在の植物園内の博物館と重要文化財「札幌農学校第2農場」に保管されている)などについて述べている。最後に農学講義で実施した農場を使った学生実験について膨大な紙面を割いているが、ここでは手法と経過の引用を省いて、もっぱら成果のみを引用した。そのなかには、1884(明治17)年にブルックスの居室が火災で焼失し、学生レポートが灰燼に帰したために残念ながら詳細が書けなかったものもある。引用した実験事例は、施肥、播種、防除法等の14件が書かれている。そこではバレイショ種子の切断法、コーンの穂から位置別に採取した種子の差異、平畦と高畦の効果、和洋の播種機の性能の違い、土性・作物と所要施肥量、病害虫観察など北海道農業のなかで実験しなければ普及につなげない項目が選ばれており、引用省略部分に年次ごとの気候や降水の影響を受けて恒常的な結論が得られないという欠点が指摘されているものの、学生教育ばかりか、実際面でも有益な結果が得られている。

6. 農学の期末試験問題集(1876-80年)

　年報には、科目ごとの期末試験問題と学生の成績、表彰者等が細かく記載されている。ここでは、全年報から農学科目に出された問題について拾い上げ、重複をいとわず本書の章別に整理した。なお、問題文末の(　)内は『札幌農黌年報』の号数-頁を示す。ほとんどが論文形式であることから、長文のノートをどのように整理して解答するか、すなわち要点をしっかり理解していないと満点にならないが、最優秀の成績は毎年98点ほどとなり、学生がいかに猛勉強していたかを示している。

［髙井宗宏］

第2章 「ブル」先生

常瑤居士(新渡戸稲造)

　ブルックス氏は先の二氏に後れて来札せられ，年歯こそ勝りたれ，家柄は他の二氏に劣りてけり。マサチュセッツ農学校を卒業せしばかりにて，学理には深く通ぜざりしかども，氏の宅はもと農家にして，子供の頃より農事には経験もあり，年齢も長け居りしこととて，他の卒業ホヤホヤの輩とは些か異なるところなきにしもあらざりき。性質まことに実着にして，物事を気に掛くる程精密ならんことを願ひ，吾々血気の書生輩より見しときは，如何にも小胆なる様に思はれたり。教場にて生徒等のあまりに学科出来ざる時は，――あはれ，いたはしき事なりしよ！――鼻をすすりて涙ぐみ玉ひ，且つ平素不勉強の学生輩は，後に残していとねもごろに説き聞かせられけり。或は病気のため欠席せしものある時は，自ら寄宿舎まで見舞に来りませり。これとても敢て学生の歓心を買はん為めにはあらで，真底より親切心のありしやうに見受けられたり。この周到なる老婆心は何れの方へも及ぼして，教場などにて先生の尋ぬる問題の，いと細やかなるには，吾々殆んど閉口しあへり。先生曾て丁寧にも果樹を運搬するに用ふる縄の直径を問ひ玉ひしことあり。その他農具を使用するに指の捻り様，節の曲げ様などに至るまで，いひ残す隙もなく説明せられしに，一同これはとばかり驚きあきれ，百姓になるには，かくばかり困難に及ぶものかよとて，互に云ひはやしけり。かく精密なるにも拘らず，弁舌に至ては至極下手にて，一生懸命に説明せらるれど，滑らかに言を発すること能はざりき。学理に至ては，最近の発見などは，到底先生の口より聴聞すること能はざりしが，卒業後実地農業に従事せし人の話によれば，当時「ブル」先生の講義は，まことに有益なりしとなり。

　先生は兼ねて農園の管轄もなし居たりしが，いづれの局に当りても，非常に注意行き届き，一より十に至る迄，能く落着をつけたりけり。人と交はるにも，かくの如くにて，特に金銭上の事に立ち至りては，細かき事，篩の目の如く，吾々書生輩より見る時分には，吝嗇とより外は思はれざりき。

　右の次第にして，うまれつきての実業家なりしより，文才少なく，吾々書生すらも，なほ講義の文章を見て，その拙きこと，シツこい事に驚けり。故に氏の報文の成りし時は，京浜に在る外人等は，これを見て冷評せりとなん。されど人を評するに，その最弱点のみを以てするは，不当の甚しきものと云はざるを得ざれば，「ブル」先生を批評せんとするものは，その文を以てせずして，その実業に方りたる事蹟を以てするこそ適当なめれ。

　氏は来札後，細君を米国より呼び寄せ，十余年間当地に留まれり。小児は当地にて二人までも生み落し，夫妻共に，札幌の気候と，札幌の社交とを愛し，この地を去ることをいみじ

う惜みたりしが，マサチュセッツ農学校の聘に応じ終に帰国したりけり。現時も該校の教師を務め居れりと云ふ。

　この人は，外人に言はしむれば美男子なりとの事なれど，吾々日本人より見る時は，毛髪非常に赤く，且つ縮れ，眼玉引ッ込み，且つ垂れ，鼻高くして顔色赤く，所謂般若然として居たりしが，子供を愛せしことは尋常にはあらざりけり。市中を散歩するときも，小学校の終る時分には，すかさず校前に至りて小供を見る。吾々の眼より汚穢なりと思ふ子供も，男子なりせば先生のお気に入りけり。その極終に，坂本某といふ十一許の小児を，親に乞ふて己れの家に養ひたり。先生の名は有名なるものにて，此頃，市中の小児等は「ブル」先生あるを知らざる者無かりけり。先生は帰国後当地を慕ふこと甚しく，或人曾て同氏を訪ひ，談たまたま札幌の事に及ひしときは，悄然として自失し，細君の如きは殆ど涙を流さんばかりなりしとなん。

　　　　　（札幌農学校学芸会機関誌『恵林』第14号［1895(明治28)年2月刊］所収，雑録「忘れ奴草」より）

第3章　ブルックス家のひとびと

ユステス・藤居恒子

　北海道大学の前身となった札幌農学校が設立された時，アメリカのマサチューセッツ農科大学から教頭として招聘されたクラーク博士のことは，「青年よ，大志を抱け」の言葉とともに，日本人なら誰でも知っている。しかし，クラーク博士の後を引き継いで，同じ大学から卒業したばかりで赴任されたウィリアム・ブルックス博士のことは，あまりにも知られていない。12年もの間北海道に住み，北海道を愛し，札幌農学校の発展に尽くされたブルックス先生が，なぜクラーク博士の名前のかげにいつも隠れてしまうのか。心苦しくてその疑問をもらすと，ブルックス博士の曾孫娘にあたるセーラは，「それは仕方がないわよ。クラーク博士はもう50代だったし，すでに大学の学長だったわ。私の曾祖父さんは，大学出たてのほやほやだったんですもの」と事もなげに言ってくれた。

　私は1960年の春，当時の杉野目晴貞北大学長（任期1954-66年）とブルックス博士の3人の孫との間に取り交わされた会話から，北海道で生まれた白人女児第1号のレイチェル・ブルックス・ドルー奨学資金を頂くことになった。北海道生まれの女子学生10人の応募者から選ばれて頂いた当時の500ドル（18万円）は，北大4年間の授業料と生活費となったのみならず，アメリカ留学へのはるかな夢へ導いてくれた。その頃のアメリカは信じられないほど遠い異国であったのだ。

　ブルックス先生の3人の孫息子とその家族と文通することを義務づけられた奨学資金を頂いたことは，私の人生をまったく変えてしまった。そのひとりのベン・ドルー氏を「あしながおじさん」と呼びながら，留学への道をひたすら歩んだ私の青春に悔いはない。北大文学部卒業と同時に幸いにミシガン大学大学院からリサーチ・アシスタントシップを得たことから，ドルー家から渡航費を頂いて，船便で太平洋を，バスで大陸を横断し，一夏をマサチューセッツのウェストフォードの，ベン・ドルー家で過ごさせて頂くというアメリカ生活の第一歩を印した。そこの長女セーラはスミスカレッジを卒業後，1年間のドイツ留学から帰って，婚約者でハーヴァード大学神学部を終えたデイヴィッド・リーヴスと結婚を控えていた。その夏のドルー家は結婚式の支度で忙しかった。私にとっては，いざ出陣とも言える修学前の一夏を，そんな華やぎのなかで過ごしたことは，かけがえのない経験であった。結婚式には，ブルックス家のゆかりのひとびとが多く集まって，北海道で生まれたレイチェルを記念する私を暖かく迎えてくださった。

　ブルックス博士の3人の孫息子のうち，レイチェルの長男のウィリアム・ドルー博士は，ハーヴァード大学出身でミシガン州立大学の農学部の学部長，植物病理学の権威で，北大に

も夫人同伴で度々訪れておられた。私も北大時代に何度かご一緒したし，アメリカ大陸の旅行中に泊めて頂いたりもした。その後私の結婚式にも出席してくださった。次男ベン・ドルー氏は私の「あしながおじさん」で，ボストン郊外に広い果樹園を経営し，荒れた手でリンゴの樹を剪定しながら「楽しき農夫」と自称していたが，ダートマスカレッジを卒業し，ドイツ留学もし，町長を長年務めた知的な苦労人であった。三男のジョージ氏は，ベルモントに住み，教会の仕事を手伝い，鳥類学に造詣が深かった。病弱であったが，遠慮がちで優しい人柄を偲んで，ベルモントの教会に記念のプレートが残っている。

その孫世代の懐かしい方たちはもうほとんど他界されたが，その次のセーラたち曾孫の7人も，その次の玄孫世代もそれぞれに活躍しておられ，北海道大学の基礎を固められたブルックス博士の後裔は，セーラの孫たちの6代目を迎えてますます盛んである。

さてブルックス博士の実像は125年の歴史の彼方に遠のいて，曾孫と言えども捉えがたい。それでも私なりにこの42年間で得たさまざまな思い出や，切れ切れの知識をもとに，問わず語りを続けたい。

1. セーラとデイヴィッド

セーラとデイヴィッド夫妻の家は，ダートマスカレッジから川を隔てたヴァーモントの丘の上にある。見渡す限りの新緑のなかにはるかに川霧がたちのぼり，眺望のなかに他の家は一軒も見えない。セーラと会うのは久方ぶりのことである。その前の面談は，彼女の父親であり，私の「あしながおじさん」であったベン・ドルー氏が亡くなった直後で11年も前であった。彼女は私と同い年で，しばらくぶりであっても血を分けた姉に会ったような気にさせる。私の夫ジェフと私を，2人はとても暖かく迎えてくれた。

2000年の春，セーラ夫妻は桜の盛りに日本を訪問し，東京と京都，奈良を歩いた後で，祖母の生まれた北海道，札幌，そして曾祖父の足跡を回顧するべく北大を訪問している。この時に応接したのが本書の編者であったため，この原稿を書く契機となった。セーラ夫妻は，私が本書の著者に加わる仲介者と言えるが，編者も私もまったく想像しないことになって不思議な因縁を感じる。2人は，その時に受けた歓待をとても喜んでいた。旅行のこと，家族のこと，ヴァーモントの暮らしのこと，異文化圏での生活経験のさまざま……。話は尽きずに夜は更ける。

私がいちばん興味を持つのは，125年前に飛び込んだウィリアム・ブルックス博士の，日本という異文化圏での体験である。ほぼ2カ月近くかけて札幌に赴く道々，何を考えたか。1963年の私の旅は，同じ行程を逆にたどって，10日間の船旅を入れて約20日かかっている。札幌—横浜間に鉄道のない時代である。どんな旅だったのか。日本語を知らずにどうやって学生の教育をしたのか。何を面白くあるいはつらく感じたのか。航空便のない時代の文通はどうであったのか。

セーラは，父母から譲り受けたもののなかから見つけた一つの箱を私に見せてくれた。その中に潜んでいたのは，20～30通にも余る，ウィリアム・ブルックスから，その姉に宛てた125年前に書かれた私信であった。その箱は，ベン・ドルー氏が40年前に見せてくれて

見覚えがあったが，手紙を読むのは初めてである。それらは，家族の健康を気遣いながら，美しい筆跡で愛情深く息づいていた。一般に明治初頭の文章は，われわれには読みづらいものであるのに，これらの手紙が，現代でもまったく違和感のないことに驚いた。2週間に1度しか太平洋に船便のない時代に，家族からの手紙を焦がれて待つ彼が，いちばん親しい姉に書いた手紙である。

　宝庫を見つけたような気がした。しかし，文章に違和感がなくとも，活字に慣れた私たちには，英語の肉筆を読むのは難しい。翌朝2時間費やしたけれど，5％も読めなかった。複写機を持参して再訪問することにした。全部写させてくれとは言いにくい。少しずつ丁寧に読ませてほしいと思った。ようやく1年前にセーラの家を再訪してコピーしたが，セーラの曾祖父の姉宛の手紙のみに止めた。

2．クラーク博士のこと
　いつものことながらブルックス博士の紹介には，必ず前置きとしてクラーク博士のことに触れなければならない。なぜならば日本歴史のなかで〈偉人〉としての位置を占めるクラーク博士の存在を無視して，ブルックス博士の位置づけは，現代ではあまりにも困難だからである。
　すでにマサチューセッツ農科大学の学長であり，人生経験豊かで熱情家のクラーク博士は，1876年夏から翌春のわずか1年で札幌農学校の第1期生を完全に魅了した。1期生と起居をともにして，農学のみならず，キリスト教を基調とした精神教育を施したことは広く知られている。アメリカの大学にはTenure（終身在職権）の制度があって，それを持つ教授なら，7〜10年に1度，有給で1年間自分の行きたいところで過ごすことができる。たいていは海外の大学や研究所に招聘されて，普段できない経験を積むことを目的とする。125年前にその制度があったかどうか知らないが，単身で赴任されたクラーク博士が，母国での地位を踏まえたまま，この1年間の異国での教育経験に，ゆとりをもって当たられたことは想像に難くない。
　人間クラーク博士は，初めて接する外国人を師と仰ぐ1期生たちの期待のまなざしを，どう受け止めただろうか。袴姿の1期生の写真を見たことがある。彼らは維新後の日本で，新天地を目指して各地からやってきた士族の子弟たちであった。北大附属図書館で見たことがあるが，彼らは英語での講義をノートにとり，その筆跡はみごとであった。それだけでも，すでに当時としては異常なほどの語学力を身につけた教養人であったらしいことが分かる。ただ，今とは比較にならぬ飲酒文化から，問題を起こす学生もいたようである。それを憂いた博士はたった1本残っていたワインの壜を割って，学生たちと禁酒を誓ったのは，有名な話である。帰国後およそ10年後に，酔って事故で最期を迎えられたという記事を昔読んだことがある。衝撃的とも人生の皮肉とも思われて，真偽のほどを確めたこともなかったが，今の私には，本当は好きだった飲酒を絶って，学生の心をつかんだ博士のあり方に共鳴できるし，謹厳居士ではなかったところに，人間的な魅力を感じてもいるのである。
　1877年4月16日，帰国するクラーク博士を慕って学生も教官も島松まで馬で見送りに

いった。さらについてこようとする学生たちに向かって冒頭の有名な言葉を残して，クラーク博士は去っていかれた。見送る学生たちに混じって，ペンハロー，ホイーラー，そして新着のブルックス先生も手を振っていた。

3．ブルックス先生の活動

　自分の出身校の学長の後を引き継ぐことは，若い学究肌でかつ実務家のブルックス先生には，至難の業であったろう。札幌農学校におけるクラーク博士の直弟子たちばかりか，直後に入学してクラーク精神を学んだ者たちも，卒業後それぞれに大志を抱いて，北海道を離れ，農学を離れていった者がいた。日本キリスト教史のなかで大きな位置を占める内村鑑三，国際連盟の事務次長まで務めた後，日米開戦前のアメリカで，死に至るまで日米間の相互理解のために全米を説いて回り，初心の「太平洋に架ける橋」を実行した新渡戸稲造など，この弟子たちのなかから輩出した数々の人物には驚くものがある。

　その傑出した学生たちとあまり年も違わない若いブルックス先生は，「ブル先生」の愛称で呼ばれながら，週に6講というかなりきつい授業を受け持ち，北大農場の経営と発展に尽くしながら，その後12年間も踏み留まって，北海道の広い大地に，着実にアメリカの農学と技術を導入していった。日本にタマネギやジャガイモを紹介したのも，北大の第2農場を設計して建てたのも彼であった。学問上のことは他の章節で述べて頂くことにして，ここではむしろブルックス先生の人間的側面に絞りたい。

　彼の名に思いがけないところで出遭うことがある。日本にスケートを紹介したのも彼であった。本国から持参したスケートを滑走して披露したのが始まりであった。札幌のタマネギ生産農家のなかでブルックス先生のもたらしたタマネギの話が今も伝承されている。日本で初めての運動会はクラーク博士の提唱で，札幌農学校が1878年5月に，今の大通公園で開いている。これはその後長い間伝統となって札幌市民を楽しませた。ブルックス先生も当然関与しておられたであろう。また日本の西洋料理史を研究しておられる方からブルックス夫人について示唆を受けたことがある。「パン食い競走」なるものの発祥地は札幌であって，この運動会で誰かの発想で18年後に始められたものであるが，どうもブルックス夫人の影響によるものでないだろうかと。なぜ餅食い競走や饅頭食い競走ではなくて，パン食い競走だったのであろうと思うと，札幌に根を下ろした西欧文化の深さが知られる。

　ブルックス先生は，着任1年後にアメリカに一時帰国して結婚し，新妻を伴って札幌に帰任している。その後札幌に滞在した11年間に，北海道初の白人女児となった長女のレイチェルと，長男のサムエルが誕生している。1888年にアメリカに帰国しているが，それは苦しい決断であったようだ。1887年の手紙では，心情的にも現実的にも札幌を離れたくない理由が書いてある。安楽な住処を得て，経済的にも心配ない生活をしている自分が，文化果つるところに居住するわけでは決してないことを強調している。

　帰国時には，明治天皇により，その功績を称えて叙勲されている。その勲章は北海道大学のどこかに収められているはずである。北海道に心を残しながら，母校マサチューセッツ農科大学に帰った彼は，その後学位を取られて教授となり，さらに農学では全米でも屈指の農

学部の学部長になられた。帰国後幾十年経っても，北海道からの訪問者を迎える時，ブルックス教授は北海道を懐かしんで落涙なさったと聞く。

4. ブルックス夫人について

　さてサウジアラビアで8年間の育児経験を持つ私は，ブルックス先生もさることながら，125年もの昔に，異国で初産を迎えた若いブルックス夫人の勇気に頭が下がる。当時の札幌にアメリカ人の医師がいたとは聞いていない。どんな女性が，助産の役目を果たしたのだろうか。当時の普通の女性と同じように，経験を豊かに積んだ「産婆さん」に頼って子供の産声を聴いたであろうか。出産は男性のまったく関与しない領域であったはずである。そうするからには日本語で意思疎通ができたか，通訳のできるよほどしっかりした女性がついていたであろう。何年に撮られたのであろうか，「お雇い外国人教師」の写真のなかには2人の女性が混じっている。ブルックス家に北海道で初めての白人の女の児が産まれる時，お互いが助け合ったことは想像に難くない。

　子供が病気の時などはどう対処したのだろう。漢方薬を飲ませたのだろうか。多少はアメリカから薬を持参していたと思うが，風邪をひいたり熱を出したり，はしかに罹った時など不安だったに違いない。幼い子供たちを抱える家族持ちのブルックス先生について，学生たちは辛辣である。新渡戸稲造の「子供たちがピイピイ泣いて」という描写を読んだことがある。それだけで心が縮まった自分の若さを思い出して，今の私はおかしくなる。当たり前ではないか。電話もない時代に，前触れもなく訪ねてくる学生たちは，子供にとっては，親の歓心を競う相手以外の何者でもない。赤子なりの自己主張である泣き声に当惑する学生が，欲求不満をもらした言葉が後世に残ってしまった。そんな時，取り繕う暇もなかったブルックス夫人に，心から同情する私でもある。

　ブルックス先生が精魂を傾けている農場で生産される乳牛のミルクは，どう配分されたか知らないが，ブルックス夫人はミルクを使ってバターを作り，パンやクッキーの作り方を教えたとも聞く。夫人は，当時の日本人になじみのないバターの味を札幌に少しずつ浸透させ，北海道と酪農のつながりに役立ったと考えるに無理がない。夫人がアメリカに帰国して6年後の運動会で「パン食い競走」の発祥地となった札幌は，夫人のおかげで手作りバターで焼くパンのおいしさを知っている街に育ったのである。

　私の想像は尽きない。パンを焼く時，薪や炭火のオーブンなどを発明したのは誰だったのだろう。それは手に入ったのだろうか。瀬戸枕や蕎麦枕に驚き，大きな枕を作って布団に横たわったのだろうか。子供たちの靴は買えたのだろうか。学校教育は？

　10歳前後で帰国した，北海道生まれの長女レイチェルは，その後名門ウェルズレー女子大学を卒業している。ウェルズレーの隣まちに住んでいる私は，ヒラリー・クリントンも通った全米でいちばん美しい広いキャンパスを散策しながら，全米から集まる学生の顔の知的な輝きを眺める。当時いくら優秀でも，女性はハーヴァードやMITに籍を置くことはできなかった。このことだけでも，ブルックス夫妻の帰国子女の教育ぶりに，あらためて脱帽させられる。

当時の外国人教師や技師たちが，いかに優遇されたとしても，現代ならモンゴルの奥地に家族とともに12年間過ごして大学を建てるに匹敵する。私には，このブルックス夫妻の札幌での生活ぶりに，限りない感銘を覚えるのである。

クラーク博士をかくも神格化させ，他のひとびとの業績をかくも忘れさせたのは，明治の教育・思想界の重鎮となった彼の直弟子たちの著作であろう。けれど北海道開発に尽くした外国人の業績を，すべてクラーク博士のかげに隠してしまっていいのだろうか。私の「あしながおじさん」は，11年前に他界した。祖父の働きが日本人の心から消えてしまったことを悲しみながら。

5．ブルックス青年の見た日本

125年前にブルックス博士が，姉に向けて書いた手紙を何通か紹介したい。以下は青年ブルックスの見た日本であり北海道である。タイムトンネルをくぐったような気がする。北海道開拓使の意気込みが感じられるし，当時の陸の交通が，まさしく馬であったことの実感も伝わる。札幌建設わずか10年に満たないのに，劇場があったことも面白い。学生たち6人とともに，馬で北海道を踏査して回る教師像も，今考えてみるとすごいと思う。全訳できないのが惜しい。なお，文中に金銭価格が登場するが，当時の1ドルは約1円であった。

〈1877年2月26日付〉

昨日の午前11時ついに目的地に着きました。旅の長さに心底あきあきしていたので，本当にうれしかったです。日本に到着してから今までのことをかいつまんでここに書きますが，サンフランシスコから横浜までの経過については，マーサ宛の手紙に書いてあります。

東京と横浜は，とても面白かったです。目新しく耳新しく，興味をそそられるものばかりでした。日本の一般家屋は小さくて貧弱です。東京と横浜では，ほとんどの家は面白い形をしたタイル〔瓦〕で屋根を葺いていて，それがなんとも言えない絵画的な情趣を生み出しています。これらの家は，たった1室か2室で，小さい部屋があるだけです。部屋の仕切りは紙でできた障子や襖で仕切られています。ドアも窓も紙でできています。家は，簡単に組み立てられていて，どうかすると柱を縄で縛りつけて組み立てられているのもあります。家の前面は，壁を少なくして雨戸や障子を取り付けているため，日中はまったく開け放されていて，通りがかりの人に一目で中の様子が全部見えるようになっています。家具もあまり置いていませんし，食卓と座布団があるだけです。火鉢があって湯を沸かしたり，煮炊きをしているようです。男も女も，子供も，ひっきりなしに煙草をすっているようです。床は畳になっていて，とてもきれいにしてあります。この上に土足で踏み込むことはしませんが，一種のスリッパ〔足袋？〕を履いている人もいます。外国風の家もありますが，ごくごくわずかです。われわれだったら惨めに感じるに違いない環境でも，ひとびとはとても幸せそうに満足して見えます。

凧上げが国民的関心事であるらしく，老いも若きも熱中しています。通り過ぎる村の空にいくつも上がっている凧の数を数えられます。赤子を背負った6歳くらいの男の子が，夢中

になって凧上げをしていて，それを赤子も一心に見上げている光景もありました。着ているものは色もスタイルもさまざまです。ほとんどの人が，この辺ですら〔手紙執筆時の厳冬期2月の札幌，以下数行も〕，帽子をかぶりませんし，しばしば裸足です。快適という観念などないのかと思うほどです。この地方の家々に煙突というものがないということは，あの小さな火鉢の中の炭で暖をとっていることになります。でも，ここで日本について語る時間がありません。この辺にしておきましょう。

マーサ宛の手紙を書いたのは，僕のために開拓使庁〔東京麻布〕で開かれたパーティーに出かける時でした。予想をはるかに上回るほど，楽しい時間を過ごしました。日本側の来賓は12名で，そのうちなんとか英語を話せる人が2人いました。ご馳走は大変に凝ったもので，テーブルは日本の生花で品よく飾られていました。献立表を同封しますが，フランス語と日本語の両方で書かれていました。ワインもたくさん出ていて，僕以外はみんな飲んでいました。

翌日の15日(木)は，見物して回り，日常品などの買い物をした後，函館丸に乗り，その翌朝出帆しました。乗組員も他の乗客もみな日本人でした。少しだけ英語が分かる人がいたので，何か頼む時，分かってもらうのにとても愉快なことになりました。僕は，なるべく日本語を使おうとしました。例えば「クラッカーが欲しい」とボーイに言うと，ボーイはありとあらゆるものを持ってくるのですが，クラッカーは出てきません。そこで私はやおら辞書を取り出して，みんながあんぐりと口をあけて打ち眺める前で，「カタイパンヲ，クダサイ」と言います。函館までの3日半をそのように楽しく過ごしましたが，最後の晩だけは時化ました。ほとんどいつも陸が見えていて風光明媚でした。函館には19日月曜日の昼に着きました。

着くとすぐ上陸して，開拓使庁の役人に会った後，ロシアホテルに案内されて宿泊しました。そこに水曜日の朝までいて，そこここを見て歩きました。そこからオタモイ〔小樽港〕を目指して出航したのですが，海峡を出たとたんに海がひどく荒れていて，仕方なしにまた函館に戻り，丸1日荒れた海上で過ごしました。それからまたホテルに戻りました。翌22日，湯地〔定基，札幌農学校の構想を策定したひとり，後に七重勧業試験場長〕という人と，そこから10マイルほどのところにある七重に行きました。彼はアマーストにいたことがあり，僕をそこに連れていくために函館まで来てくれたのです。そこには開拓使の立派な農場〔七重官園〕があり，彼はその支配人でした。そこで楽しい1日を過ごした後，夜の10時に函館に戻り，函館丸に乗船しました。翌朝3時，海が凪いでいる時にまた出航しましたが，しばらくするとまた風が強まりました。そして夕方には，完全に強風状態でした。山のように高い波が渦巻いて，蒸気船はもまれもまれて大波が絶えず甲板を洗い，船側にまるで雷が落ちるようにぶち当たりました。僕は一睡もできませんでした。水夫がひとり海に落ちて行方不明になり，船も損傷しました。でもなんとか切り抜けて，翌24日(土)に小樽内に着きました。ここは札幌から25マイルほどのところにある港です。札幌からの公式の迎えの人が来ていて，会うとすぐにテキパキと荷物やら手続きなど，煩わしいことをすべてやってくれました。素敵なボートに乗せて上陸させてくれて，すぐに日本式の旅館に行きました。そこで

僕は，ナイフとフォークを持参していましたが，それを使わずに初めて箸で食べてみました。案内の佐藤氏はさっそく札幌に電報を打って僕の到着を知らせ，そこから10マイルのところにある当夜宿泊予定のホテルにコックを手配しました。小憩をとってから，馬に乗って銭函まで行って泊まりました。日本式の布団はとても心地よいものでしたが，枕がないのが困りました。

　翌朝8時に出発し，札幌から5マイルのところまで行くと，クラーク学長，ペンハロー教授，ホイーラー教授と斉藤氏が出迎えてくださいました。そこから札幌の外国人教師館まで，愉快な仲間たちと一緒の楽しい道程になりました。家具が調って暖房のきいた大きな2部屋が，僕のために用意されていました。居心地満点で，すぐに，そしてついに自分の家に落ち着いた気分になりました。先任の3人の紳士たちと家族のように住んで，実に気に入っています。農学校の学生みんなに紹介されました。彼らの様子も，とても気に入りました。

　夜も遅くなりましたので，これくらいにしておきます。家族のみんなによろしく伝えてください。そしてみんなに長い手紙をしばしば書いてくれるように伝えてください。僕の友達が訪ねてくれたら，みんなによろしく。

　　　　　　　　　　　　　愛をこめて　　ウィリアム P. ブルックス

〈1877年4月22日付〉

　1週間前ここから25マイルほどのところにある小樽まで，馬で往復しました。通訳もガイドもなしにひとりで行ってきました。どうしても必要なものを詰めた箱〔貨物〕を取りに行ったのです。目的の箱を探しあて，開いて，馬の背袋に詰め直して戻ってくるまでひとりでできるほど，僕の日本語が上達したことになるでしょう。

　道程の半分くらいは海と断崖絶壁の間を通っていて，絶景でした。海沿いのひとびとは，男女，子供もみんな，最盛期の鰊(にしん)漁におおわらわでした。彼らは，年間に何百万ブッシェルもの鰊を獲るのです。長さ20～30 m，幅15 m，高さ2 mもの鰊の山をいくつも見ました。海は鰊でいっぱいで，ひとびとはただ打ち寄せる波のなかに立っていて，波が引くと何匹か摑んで水の届かない岩の上に投げるだけでいいのです。鰊の切り屑が30 cmも岩の上に積もっていたりしました。どこもかしこも鰊，鰊，鰊でした。男も女も，子供もみんな鰊と一緒に転げ回り，頭から足先まで鰊と一緒になってはね回っていました。鰊は彼らの肉であり，高くてあまり手に入らないのか塩をそれほど使わずに保存処理をして，食料にします。油や肥料として加工することもしています。

　先週の月曜日〔4月16日〕にクラーク学長が帰国されました。全校生と教授全員が，12マイルほど馬で見送りに行きました。彼は僕の家族を訪問しましょうと言ってくれました。たぶん8月になるでしょうけれど，彼が到着したら，そちらから招待の手紙を書いて，ご都合のいい時にSS駅までお迎えする旨を伝えるとよいと思います。ボストンを朝の汽車で発って日帰りするのがよいと言われると思います。来年の秋に，彼は家族を連れてここに戻られると思いますが，これは内緒です。

　僕は週に講義を4回受け持っていますので，講義の準備，学生の答案の添削，農場の監理

で多忙です。またすぐ便りをください。ではまた。

あなたを愛する弟より　　ウィリアム P. ブルックス

〈1877年5月28日付〉〔私が今これを訳しているこの日から，きっかり125年前に青年ブルックスが書いていた文である。筆者〕

　親愛なる姉上様，あなたの4月1日付の手紙が，水曜日に届きました。手紙をもらってとてもうれしかったのですが，ミルトンの容態が悪いというのは悲しいことです。彼を入院させる前に医者のアドバイスを聞いてください。必ずしもわれわれが望む結果にならないかもしれないけれど。

　そうです。横浜からの手紙は急いでしたためたものでした。海上での僕の体験を書いたつもりでした。もしそうでなかったとしても，今はもう夢だったように思えますが，4, 5日も僕の体調は変だったと言えば十分でしょう。食べたものは，じっくりと胃に止まって消化するのではなくて，先進しようとする傾向にあったようです。

　すでに1, 2度書いたと思いますが，サンフランシスコを出る前に，あのズボンは無事に受け取りました。大事な行事があるたびに，僕はあれをはいて，さっそうと出かけます。

　この前の月曜日に，大枚27ドルを払って馬1頭に投資しました。彼は栗毛色で，こちらの馬としては平均以上の大きさですが，マサチューセッツでは小さい方です。彼は気が優しくて丈夫な8歳馬です。駿足で，ホイーラー教授やペンハロー教授の馬をはるかに引き離すことができます。僕の愛馬になることでしょう。土曜日の午後にイサリ〔恵庭市漁〕までの17マイルを往復しました。仕事上，馬に乗って行き来する機会が多いのですが，そうでなくとも1日に6, 7マイルは馬に乗ることにしています。運動のためでもありますが，楽しみのためでもあります。

　水曜日には，ホイーラー教授，ペンハロー教授と一緒に，通訳同伴で劇場に行き，日本の演劇を観ました。非常に優れた演技もあるのですが，多くはとても気取ったスタイルのものだと思いました。でも日本人の観客は素晴らしいと思っているようでした。ほとんどの観客は，1階と2階の桟敷の床に畳を敷いて座って観ていました。舞台の仕掛けや風景はとても簡素でしたが，俳優の着る衣装は素晴らしく凝っていました。日本の劇で女性が役割を演じることはきわめて稀で，女性の役を男が演じます。日本の演劇にはひとつとても珍しい特徴があります。舞台から上のところに席があって，男がひとり座っていて，悲しげな声を張り上げて，舞台の解説をし続け，それをひとびとは「うた」と呼んでいることです。演劇の内容はたいてい史実に基づいており，その歴史を物語るのが，この男の仕事なのです。そして男優たちは台詞を交わしているだけです。

　天候はとてもよく，暖かですが暑くはありません。マサチューセッツのように，植物が一斉に勢いづくこともありません。僕はまだトウモロコシを植えていません。これを書いている机から，6マイルのところにある山〔手稲山〕の頂上に雪が見えます。

　学期は7月4日に終了します。終了式があって，日本語と英語での式辞や演説などがあります。夏休みには山登りと探検をするつもりで楽しみにしています。長い手紙をたくさん書

いて送ってください。どんなことでももらさず知りたいのです。

　そろそろ講義の時間ですので，このあたりでおしまいにしましょう。あなたが午後10時に寝るとすれば，それはこちらでは翌日の昼の12時です。われわれは，そちらよりも14時間先を行きます。

　　　　　　　　　　　　　愛をこめて　　ウィリアム P. ブルックス

〈1877年7月29日付〉

　あなたからの5月18日付の手紙をしばらく前に受け取りました。この前の船便では，家の誰からも手紙が来なかったのでがっかりでした。でも次の便が2, 3日後ですから期待しています。現在は蒸気船はかなり不定期ですが，月に3, 4回というところです。けれど，これもあと2, 3カ月のことで，つまりTea Season〔日本茶の輸出最盛期には船便が多かった〕の間だけです。毎月1日と15日に蒸気船が出る普段の時期になれば，そちらから手紙を出すのによい日は，前に書いた通りです。

　家から手紙が来るたびに，必ずあのズボンは届いたかと聞かれます。そして僕が返事を書くたびに，無事受け取ったと書くのですが。けれど，もう一度だけ言いましょう。5月18日の手紙では，まだ着いたかどうか分からなかったようですが，あの衣服は確かに受け取りました〔この頃のズボンはすべて誂えか，家族の誰かが縫ったものであろう。ミシンはまだ普及していない〕。

　学期終了の7月4日には終了式があり，日本語と英語とでスピーチなどがありました。すべてうまく行きました。晩にはホイーラー教授とペンハロー教授と僕の3人が，役人と市民のひとびとを招待して独立記念日を祝って花火を打ち上げました。けっこう素敵な見世物になったのですが，途中で夕立が降ったのは残念でした。花火は現地の職人たちが作って上げました。われわれの教師館は，おびただしい提灯で飾られて，そのなかのいくつかはオリジナルの作品でした。その1つは2 mの幅で60 cmの高さでした。それにはアメリカン・イーグルと日本の竜が描かれていました。もう1つは，マサチューセッツの紋章で，他の1つは自由の女神，日本の神聖な鳥である鶴の絵もありました。

　7月4日の次の月曜には，ホイーラー教授とペンハロー教授の2人が札幌を発ちました。前者は測量踏査に，後者は石狩川の踏査のためでした。ペン教授は学生6人，ホ教授は4人を連れて出発し，残る6人は僕と一緒に残りました。彼らは8月半ばまで出かけています。僕はそれ以後，小樽と銭函まで所用で学生たちと行ってきました。その他の時期は札幌の農場に従事していました。次の水曜日には，僕も学生と一緒に最終旅行に2週間ほど出かけます。

　7月29日。昨日の朝の便でアメリカからの便りがありました。家族，友達，それからマーサからまたいつもの質問が来ました。サンフランシスコでズボンを受け取ったと伝えてください。そのなかのニュースで，サウスウィックが，この3年間注目していた女性と結婚することになったというのがありました。従兄弟のデボラからの手紙では，従姉妹のクララが9月に結婚することになったと伝えてきました。その紳士には，僕も去年の夏フィラデル

フィアで会いました。

　今頃はクラーク学長がアマーストに着いておられることでしょう。新聞で見る限り，州議会はこの冬，彼にひどい扱いをしたと思います。初めに彼の給料〔日本に渡航した休職中の〕を支払えなくするために，大学の歳費をすべて承認しないと言い，その後では彼に給料の支払いをしない条件で大学の費用を承認するというのですから。最初，彼が日本に行くことを承諾した時，彼は大学の評議会に辞職願を出したのです。でも評議会はそれを承諾しませんでした。そこで彼は，自分が受け持っていた授業を他の人にさせ，大学の管理は自分がするから，年俸のなかから1000ドル引いてくれるように交渉しました。評議会はそれに同意し，それで双方にとって公正だということになったのです。州議会も評議会も，その約束を破る権利はないはずですのに。

　われわれと一緒に住んでいる通訳の小島さんが，最近結婚しました。そのやり方はとてもユニークでした。日本人は，結婚したいということになると，誰か友達とか，あるいはまったく知らない人でも，仲人の仕事をする人に頼んで，まったく知らない人でも誰か適当な相手を探してもらいます。日本人の考えでは，誰かと親しくなってしまったというだけで，結婚には向かない相手ということになってしまいます。断る機会があるのは「床入り」の時だけ〔これはお見合いと混同しているのではないか？〕です。けれどもその段階で断るのは非常に稀なことです。2人は婚約から結婚式までの間に会うことを禁じられています。「式」と言いましたが，それらしいことはひとつもなくて，2人は友達や知人や親戚と一緒に会い，宴会をするのです。小島さんの奥さんは，日本女性として人並み以上に見かけもよく，慎ましくて，人あたりのいい人です。

　日本の女性が上半身裸でいるところを見ることは珍しいことではありません。道端で行水を使っていて「いちじくの葉」すら身に帯びていないのです。男性の裸などいつでも見られます。せいぜい腰布を着けているくらいで，それすら着けないで行水をしています。お役所ではそのような露出を止めさせようとしているのですが，庶民はまったく気にかけていないようです。

　　　友達が訪ねてきたらみんなによろしく。家族のみんなに愛をこめて　　　Brother Will

〈1877年8月19日付〉
　前に家に手紙を書いた時から，いつもより長く書かなかったので，いつ来るやらと思っていることでしょう。1週間前の金曜日に旅行から帰ってきたら，Old Colony Memorialの第1号と2号，それとあなたからの手紙が待っていました。ボストン・ヘラルド新聞も受け取りました。ありがとうございました。しばらく前に送ってくださった雑誌が届いたことを知らせたかどうか分かりませんが，どのみち知らせるべきですので，とてもうれしく読ませてもらっています。

　旅行で10日ほど留守にしましたが，よい天気に恵まれました。われわれは275マイル踏破しました。12〜15マイル行くたびに馬を乗り換えましたので，22頭の違った馬にまたがるという悦び（？）を経験しました。われわれの行程は，室蘭，有珠，長万部，歌棄，岩内，

余市，小樽，そして札幌でした。海沿いの道もあれば，岬を突き切って別の海辺の町まで行くこともありました。山，海，平野，川，滝，山中の急流など，ありとあらゆる風景に出会いました。素晴らしくきめの細かな砂浜もあれば，海水から330mほど屹立した断崖絶壁の岩浜もありました。まったく雄大でした。

　時には山越えの道は，馬で行くのは不可能に見えることもありましたが，ここの小馬〔ドサンコ馬〕はそれに慣れているのか，びっくりするほど何度もジグザグを繰り返して1マイルのところを2，3マイル歩きながら登ります。頂上までやっと着いても，そこでは馬の背にいるよりほかはありません。というのは，またすぐに同じくらい急勾配を下降しなければならないのです。もし腹帯がゆるんでいたり，ちょっとした不注意から，ある時学生のひとりがしたように，馬の首にまたがることになりかねません。けれど，こんな時でも，ここの馬は動じません。いちばん難儀だったのは有珠から礼文華にかけてでした。

　僕たちはその間いつも日本の宿屋に泊まりました。安くて，朝食付の1泊が20セント〔20銭，1ドル≒1円〕，夕食10セントであがりました。ナイフとフォークと少し食料を持参しましたが，ほとんど日本人と一緒に米飯を食べ，床に眠り，床に座るという現地スタイルでやりました。たいてい日本の宿屋はとてもこざっぱりしています。廊下の床はハンサムな黒人の肌のように光っていますし，部屋のマット〔畳？〕はきちんとしていて清潔です。けれど，ああ！　僕の経験ではどうかすると布団と蚤とは見知らぬ仲ではないのです。帰宅して椅子やら寝台を見た時には，ほっとしました。

　目的通り，各地の海岸で珍しい植物を見つけました。全員落馬はしましたが，重大事故には遭いませんでした。僕は僕の馬が，それ以上立っているのが嫌になって，高さ4尺の土手から落ちてみようとした時に落馬しただけです。そのケダモノは，みにくくて，僕を振り落とそうとしたりするので，土手を越えるのは落馬することだという危険を予測していました。どちらも全然怪我はしませんでしたが，僕はむしろ，そんなことをした馬の首の骨が折れてしまえばいいと思いました。

　次の木曜日に学期が始まって17週間続きます。2，3日のうちに新入生が17人〔2期生〕到着するでしょう。彼らは東京からやってきます。みんなにくれぐれもよろしく。わが家族のなかでも，若い連中に特によろしく。

　　　　　　　　　　　　　　　　　　　　　　愛をこめて　　　Brother Will

第2編　ブルックス農学講義録

　ブルックスが行なった農学の講義ノートは，北海道開拓記念館が所蔵する札幌農学校1期生内田瀞(きよし)(1880年7月卒業)，北海道大学附属図書館が所蔵する同2期生の新渡戸(当時は太田)稲造[*1]，町村金弥，南鷹次郎，宮部金吾，同3期生伊吹鎗造(そうぞう)[*2]による6冊がある。なお，同附属図書館は，1898年農業生物学科卒業湯地定彦，1908年農業経済学科卒業逢坂信(しん)忢(ご)のノートも所蔵しており，資料が少ない時代の農学の流れを理解するうえで貴重な資料となっている。

1. 底本：本編第1〜7章は2期生の新渡戸稲造，第7章の一部と第8章は3期生の伊吹鎗造の原ノートをまとめ，4分冊のコピー版として作成した「高井宗宏編・解題, *Lectures on Agriculture by Prof. Willium P. Brooks of Sapporo Agricultural College*」(1987)を底本とした。なお，原ノートはB5変型判であるが，コピー版作成に際し若干縮小した。

2. 校閲：本書底本の文章校閲にあたり，明らかに誤記または脱落と思われる箇所の訂正，さらには現代の使用法からみて誤解を受けやすい学術用語について，適切と思われる用語の補足等を行なった。

3. 章・節・項：原ノートは，章や節の区切りに相当する部分に大文字やひげ文字を用いているが，章や節の番号は付されていないため，底本作成段階で番号を付した。本書編集に際し，底本の第3章と第4章は同一分野のため一つにまとめて第3章とした。また第4, 5章と第7章では，講義の構成を現代の分類法に合わせるため，節の順序を入れ換えた箇所がある。その部分は，次項で付す頁番号が連続していないことから判別できる。

4. 頁番号：講義ノートは，学期毎に開講された全科目を1冊に綴じ，上記のノート者が通し頁番号を付している。したがって農学講義の原ノートは，5学期の5分冊となり，頁番号が連続していない。そのため編者が新たに底本の冒頭から通しの頁番号を付し，その頁番号を本書の対応する箇所に[p. ××]として示した。底本ではその位置から改頁されて次頁に移ることを示している。

5. 欄外記載事項：文中の〈　〉は，講義ノートの欄外に書かれた備考やインデックスを示す。その多くは，他章に記載した関連事項の参照頁であり，例えば〈See II p. 234〉などと原ノートの分冊番号と頁番号を記載している。そのため本書では，前項で付した底本の頁番号に置き換えて記載し，ノート者が作成した参照関係を明らかにした。

6. 表と図：文中に掲載する表は，明らかに表形式で記載されたもののほか，ノートに数値を列記してあるものも表形式に組んだ。したがって表番号・タイトルのあるものとないものとが混在しており，図でも Fig. 1 が重ねて出てくるなどの混乱が見られるが，講義ノートの性格ゆえとして原文のままとした。また底本に図版用の空白がありながら収録されていない場合も見られるため，文意に従って編者が図を主に下記の資料から転載した。さらには底本になくても図を追加して読者の理解を図ったものもある。図を転載した資料は，① 1898 年農業生物学科卒業湯地定彦による農学講義ノート，② W. P. Brooks, *Agriculture*, Vol. 1-3, 1901, ③札幌農学校図書館蔵書の主にマサチューセッツ農科大学教授陣が分担執筆した *Manual of Agriculture for the School, the Farm and Fireside*, 1885, および④札幌農学校当時のカタログ類である。

7. 農学講義ノート底本の閲覧：北海道大学附属図書館北方資料室が所蔵する新渡戸稲造らのノートは，大学史の貴重資料のため一般公開されていない。しかし，編者の髙井は，前述のコピー版を作成した際に同北方資料室に寄贈したので，コピー版は随時閲覧できる。また，解題作成にあたって同時期の農学講義録を探し求めた際，東京大学農学部附属図書館で「帝国大学農科大学農学科を 1895 年に卒業した廣瀬次郎氏ノート」のコピーを入手した経緯から，同図書館にも寄贈してあるため閲覧可能である。

[*Lecture for the first semester*]
Chapter 1 Agriculture

Webster[1] defines Agriculture as the art or science of cultivating the ground, especially in fields or in large quantities, including the preparation of the soil, the planting of the seed, the raising and harvesting of crops, and the breeding and management of cattle. Walker[2] in his *Science of Wealth*, includes fisheries and mining, saying that every occupation, the labor of which produces valuable articles from the earth—either from the ground or the sea—is really agriculture. However, we shall not consider the subject in its broadest sense, but agriculture in the usual sense of the term as defined by Webster. In commencing to consider the subject, it is well to devote a short time to the consideration of its importance as an occupation. [p. 1]

1-1. The Importance of Agriculture

Agriculture is important; first, because it is the basis of every other occupation. Without agriculture, there can be no great manufactures, for it produces raw materials, with which the manufacture works. Without agriculture, there would be no considerable commerce, for the merchant deals in articles either directly or originally produced by farmers. Without agriculture, there could be no large mines or fisheries; for those engaged in these avocations must be fed by the produce of the soil, to a greater or less extent. I have said, extensive manufactures, considerable commerce, and large mines and fisheries; because all of these occupations may be carried on, to a limited extent, by such savage nations as have no agriculture.

The government of Japan seems to have apprehended that agriculture is the basis of all other occupations, and if the people can [p. 2] only be made to improve the methods of carrying on their farming operations, all other improvements will necessarily follow. Machines, carriages, harnesses, tools, and convenient clothes with which to work will be needed by the farmer, and as these can only be made here, manufactures will spring up to supply the demand.

Japan is capable of producing much more of many articles than her own people can consume, and when her soil is made to produce these by the labor of the husbandman. Merchants will be needed to buy and export them, taking in return either those things which this country does not produce, or those which can be bought cheaper in some foreign countries.

There may, of course, be prosperity for a time without agriculture; but this can only last till the natural products of the country are exhausted. [p. 3] So long as the immense supply of timber, and the great quantities of wild game such as the bears, deer, *etc.,* on the island of Yesso[3] last, so long can its people carry on a flourishing commerce and profitable manufactures; but the supplies of these will soon be exhausted, and then, unless the farmer is ready to produce raw materials, its commerce and manufactures must die. (1) The state of the agriculture of a country, as the mercury in a thermometer measures the temperature, is an unfailing index of its advancement in civilization. (2) Agriculture is important to a country because it furnishes the most healthful occupation, both physically and morally, to its inhabitants.

Manufactures and commerce, from their nature, render it necessary for those engaged in them to be much indoors and to be crowded into villages or towns, which are not only destructive to the [p. 4] health of the body, but make the spread of vice much more easy, and thus lower the moral tone of the people. But there is a stream of healthy, moral, and physical life flowing from the farm to the centers of manufactures and trade, while those worn out and sickened by the unhealthy life which they have lived in towns generally retire to the countryside, and there they strengthen themselves. Country life produces men of sterling character, who are after all the surest basis of national prosperity. Many of the greatest names in the history of America are those of men born and reared on farms; Washington[4], Webster[5], and Lincoln[6] were all farmers. [p. 5], 〈「不見公与相起身自犂鋤」韓退之, 符読書城南〉[7]

1-2. The Influence of Agriculture on National Prosperity
The prosperity of a nation depends almost entirely upon agriculture. Unless the crops of a country are abundant, the price of all articles of food will become high, so that the poor people, if they do not actually suffer for the want of their money for food, and so on, will not have anything with which to buy other necessities of life.

In such conditions, manufacturers will be obliged to pay higher wages to those people whom they hire, in order that they (the hired people) may have money enough to purchase those things which are essential for their existence. It follows from this that the cost of manufactured articles will necessarily become greater in such seasons other than when the crops are abundant. [p. 6]

1-3. The Necessity for a Variety of Industries
It is possible for a man, in almost any country, to supply all his absolute wants by the labor of his own hands, but he cannot enjoy the comforts which he could had there been a variety of industries carried on in his country. The reason why he cannot enjoy so many comforts is due to this: that he cannot become skillful in all the different kinds of work which are necessary to be carried on.

1-4. Advantages of Agriculture as an Occupation
(1) It is the most healthful occupation of mankind. This statistics abundantly prove. The reason is because its active out-of-doors duties, and its various kinds of work, develop all [p. 7] parts of the body. Then, too, it is free from the excitements and many of the anxieties of commercial or manufacturing business, and these are the greatest sources of diseases of both mind and body. (2) Agriculture is the safest business, if it is rightly carried on, and the one by which a man is the surest of obtaining a subsistence or, perhaps, even the comforts of life. If one crop fails wholly or in part, the farmer is almost sure to have another crop which will yield him a good return. (3) Agriculture is the most independent

occupation, for the farmer raises his own food largely, and hence is not obliged to depend upon others.

1-5. Development of Modern Agriculture
The primary conditions of all [p. 8] people, of whom we have any record, were essentially those in which the American Indians were found at the time of the discovery of the Western Continent. They subsisted almost entirely upon the natural products of their country, hunting and fishing for their animal food, and gathering the wild products of the forests. Their houses were very rude; their clothing, so far as they wore any, was the skins of wild beasts. So long as the number of inhabitants to a given area is small, they can get a living in this way, but when the number increases, they must either cultivate the soil or raise flocks and herds of sheep and cattle. The cultivation of the soil seems to have been the next step in Japan, but in many [p. 9] countries the next method of subsisting was by means of domestic animals. Immense droves of cattle, sheep, swine, *etc.,* are owned in common, and the owners pasture them upon natural grasses, moving about from place to place, whenever they find good pasturage. Such a people are called nomadic, and their mode of life is very rude. They live in tents, which are easily moved from place to place. This method of living, however, requires a large amount of land for the support of a given number of people; and as population still further increases, it is found necessary to settle permanently in one place, and to pursue a mixed husbandry, raising both plants and animals. [p. 10] Then, for a time, the virgin soil is so rich that little is required in order to raise crops, except to sow the seed and to keep down the weeds. This is simply drudgery and requires but little intellect. Probably from this fact, it has been found that the farmer, in most countries, has stood, and in many does still stand low in the social scale. But as population becomes still more dense, and the soil loses its original fertility, it requires the exercise of the greatest knowledge and skill to produce enough food for the consumption of the people.

This is the condition of the greater part of Europe today, and of the earlier settled portions of the United States; and in those countries, the agriculturist must be a man of much and varied knowledge, if he would attain to the greatest success. [p. 11]

Now, although the soil of Hokkaido is still fertile and may produce large crops for many years without the exercise of much knowledge, yet it should be the aim of the inhabitants to so conduct their agricultural operations as to keep it in this condition and not to exhaust its fertility, as is almost invariably done in newly-settled countries. Though the eastern parts of the United States were settled early in the seventeenth century, no agricultural schools were established for more than two hundred years. Japan has begun more wisely than this in Hokkaido, and has thus, early in its settlement, founded an agricultural college from which will go forth men who can do much to prevent her from following the ruinous policy of other nations in their history. [p. 12]

1-6. Knowledge Necessary to the Farmer
The farmer should know:
1. Geology, in order that he may understand how soils were formed, and the

physical, and to a certain extent the chemical, characteristics of the soil of any given locality;
2. Chemistry, that he may be able to ascertain the constituents of soils, plants, and manure, and more especially that he may be able to judge of the effects of certain manure upon certain soils;
3. Meteorology, that he may be able to judge whether certain crops will, in a country of which he knows the climates, and also that he may have the power to tell in advance what the weather is likely to be;
4. Mechanics, that he may understand the principles involved in machinery;
5. Physics, that he may understand the action of the various agencies [p. 13] of nature on rocks, soils, building materials, *etc.,* so that he may avail himself of these when they will help him, and avoid their action when it would be deleterious to his interests;
6. Botany, that he may know the structure and habits of the growth of plants, and thus be able to cultivate them intelligently, and thus to gain the greatest profit;
7. Zoology, Anatomy, and Physiology, that he may be able to know the structure, habits, and needs of the animals which he has under his care;
8. Entomology, that he may know the habits of the various insects and whether they are beneficial or injurious;
9. Veterinary medicine, that he may be qualified to care for his animals when they are sick;
10. Social and Political Economy, and the Laws of Trade, that he may be able to judge intelligently what crops it is [p. 14] most necessary to raise, and how to sell them advantageously;
11. Accounts or Bookkeeping, that he may be able to keep correct accounts of expenditures and of various matters and circumstances in regard to his farm;
12. Law, especially that which relates to rural affairs, in order that he may conduct his business transactions correctly;
13. Lastly, and most indispensable of all, he must know how to conduct all his practical farming operations in the best manner and with the smallest expenditure of time and labor; on this, his success will mainly depend; without this knowledge, he cannot succeed. [p. 15]

解　　題

1. 日本農学の発達

　第1章は農業の定義から意義，農業史と農業者に必要な知識などを述べている。冒頭のウェブスター辞典による農業の定義は，日本で「作物栽培または家畜飼養によって有用な食糧と一部の工業原料を生産する産業」とする農業の定義（狭義）と変わりない。一方，ウオーカーの定義は，食品が農産物，畜産物，水産物，その他（山菜類，嗜好飲料，菓子類，醸造食品など）に分けられること，日本の大学農学部に林学，水産学，獣医学などの学科がある

から，日本でも広義の定義と言えるが，鉱業まで含む事例はない．この農業にかかわる科学を総称して農学があり，近代農学の祖と言われるテーア(1752-1828年)は，その著書『合理的農業の原理』に「農業を対象とする学問分野は，生産技術と経営の2分野があるが，終局的には多収をあげて最大利潤を得ることを目的とする」と記して農学の目標を示した．

　この農学について日本での流れを概観すると次のようである．なお，発展契機の区分法は，川田信一郎の「日本の農学」(『世界大百科事典』第2版，日立デジタル平凡社)を参考・引用しつつ，著者の考えを加えたものである．

(1) 農学の起点：商品作物に注目し始めた元禄時代(1700年頃)から，作物栽培法や農家の作業暦などを書いた各種の「農書」が作られ，カンショ(サツマイモ)の普及に大きな貢献をした青木昆陽(1698-1769年)のような本草家などが活躍しているが，これらは社会的に大きな動きとなっていることから，日本の農学の起点と考えてよいであろう．しかし，多くの農書は，幕藩体制の下で藩の石高を超える生産を目指した藩内に限定した指導書とか，高い年貢に対処する篤農技術を一族に限って継承する遺言書などが多く，市井の人たちによる地域内での経験と観察による直観的な成果を整理し，広く民衆に伝えようとする農書は数少ない．しかしながら，当時の農法は，今日の過剰に生産性を追求する資源浪費型農法と異なり，自然や環境との調和，資源の循環，人力労働の適正分担などを基本とした農法であるため，あらためて見直して今日の農法に生かすべきであるという意見が高まっている．

(2) 農学の近代化：明治維新になると政府は，実証的，実験的な欧米の農業を「泰西(泰東＝東洋に対する用語，西の果ての意)農学」と呼び，殖産興業の基盤として積極的に導入するため，泰西農学の本を翻訳発行し，農学校を設置してお雇い外国人による農学教育を進めた．その意気込みは，1876(明治9)年8月14日の札幌農学校開校式で黒田清隆開拓長官が「我邦の農業たるや，多くは之を老農老圃の実験に得，法の取る可き無きに非ずと雖も，徒に慣習に因仍（いんじょう）し，格致の学を講じて以って闡明推広する能はざるを以って，遂に其の進歩の効を見ず．故に今農学校を設け，教師を海外に招し，英を育し，蒙を啓き，以って農学の面目を一新せんと欲す」と式辞しているし，また1878年1月24日の駒場農学校開校式で岩倉具視内務卿が「本邦の農事に於ける未だ専ら其学を講ずるを聞かず．陛下聡明叡哲農学の急務なるを知しめし給い，此校を創建し博く万国の実験を徴し，精く庶物の性質を究め，大に富民殖産の道を興隆せしめ給うは実に生民の大幸にして国家の洪福と謂うべきなり」と式辞したことから明らかになる．しかし，直輸入された泰西農法，作物，農学は，稲作技術がなくて老農による日本稲作の方が優れていたなど，日本農業の近代化に適さないとの批判も少なくなかった．幸いに札幌農学校は，クラークに率いられた師弟が同一精神で教育にあたって成果をあげたこと，日本人に未知の洋式畑作酪農を主体にして稲作を禁じていたこと，農学全般を担当したブルックスが農業実技にも優れており，学内外で大いに活躍したことなどで，幸先のよい出発をしている．

　これら不安定な受容期を経て日本人による近代農業，すなわち近代日本稲作は，1893(明治26)年に設置された東京の国立農事試験場と北海道の上白石・亀田・真駒内稲作試験地に始まると言ってよいだろう．そこに1901(明治34)年の八幡製鉄の開業を契機とする日本の

産業革命が進んだのと，日清・日露戦争および第1次大戦の影響によって，日本農業は明治末期から大正時代に大きく発展し，昭和の初期にひとつの農業形態として完成している。北海道稲作では，折衷苗代やいもち病防除技術が確立し，新たな発展を図ったが，戦時体制に入って停滞を余儀なくされた。

(3) 農学の発展：1945(昭和20)年の敗戦後の食糧不足が契機となり，駐留したアメリカの助言を受けて水稲作以外のムギ類作やカンショ，バレイショ，リンゴ，ミカンなどの栽培研究，土壌管理や雑草研究などと，技術の農村浸透のための普及体制が充実した。続いて「もはや戦後でない」と言われる頃になると一般工業が大きく発展して神武・いざなぎ景気などと呼んだ好景気を招くが，農業関係では，1962(昭和37)年以降の農業基本法農政，農業構造改善事業によって機械化，装置化，施設化や化学肥料中心の作物栽培，病害虫と雑草の化学防除などが盛んになり，農学というより農業科学とも言うべき様相を呈するに至った。

次いで農学が大きく展開するのは，異常気象による穀物の世界的な大減収，オイルショック(1973年)などが起きて世界的に食糧保護策がとられた時であり，食糧自給論，地力への関心を高めた省エネルギー農業論，無機質肥料や農薬を排除していこうとする有機農法が台頭するとともに，微生物化学，生理活性物質研究，生命科学，バイオテクノロジーなどの研究が急速に発展した。かつての農学の各分野は，細かな領域ごとに分割して発達するとともに，学際分野の研究も興って今日に至った。

しかし，今日では，農産物の国際障壁の撤廃，農産物増産よりも品質の重視，環境保護や自然との調和が主課題となり，分化しすぎた農学の再構築，新たな日本農業の展望に立った試験研究機関の再編成など第4の展開期を迎えていると考えられる。

2. アメリカと北海道の開拓の違い

本章1-1と1-2は，すべての産業の中心に農業があり，農業の繁栄はすべての産業の繁栄につながり，国家も富むと述べ，今日の農業観とは大きな隔たりがあるが，アメリカ開拓の経験から北海道開拓のあり方を敷衍していると理解される。

講義が行なわれた頃のアメリカは，ミシシッピ川を渡って西部開拓が本格化していた時である。メイフラワー号で移住して約100年後，東部13州でアメリカが建国された頃，アメリカの発展方向について大きく2つの考え方があった。1つは，大西洋を利用して海洋国家として発展し，商業，海運業，やがては工業を経済の中心とするという考え方であり，もう1つは，西部の広大な土地を利用して大陸国家として発展し，農業を経済の中心とするという考え方であった。1801年のジェファーソン大統領就任から南北戦争まで続いた民主党政権は，後者を政治の中心とし，既存の広大な土地と1903年のルイジアナ買収によって西進政策が進められ，プレーリから西部の農業開発が推奨された。しかし，1861年より4年間続いた南北戦争から1890年代までは，アメリカの産業革命期と言われ，共和党政権の下で急速に工業社会に転じ，農業のために外延的に拡大された広大な中西部も，工業のための市場として内包化されていった。ここで注目すべきことは，アメリカの工業化は脱農業化を意味せず，農業自体もその生産額，耕地面積を増大させ，工業はミシシッピ川流域に大きな製

粉業や製綿業，農機具工場を建てて，農業と工業とが相互補完的に発展したことである。また，そこで両者の相互補完を促進したのが大陸横断鉄道などの鉄道網と運河の設置であり，電話などの通信網の拡充であった。農学講義の内容は，これらの事実経過を裏づけたものと考えられる。

　一方，北海道の開拓は，アメリカ農務長官ケプロン[*3]らが北海道の実地調査から導いた「ケプロン報文」を基本方針にしている。そのケプロン一行は，1871(明治4)年8月(日本はまだ旧暦を用いていたが，本書ではすべて西暦に換算)に横浜港に到着し，ただちに「異邦の家畜，草木，果実などの新種を伝播する中継所」として東京に3つの官園(後の試験場)を開設するとともに，数年かけて北海道を踏査して開拓の基本方針を建議し，その後の実務をダン[*4]やクラークに引き継いで帰るが，「寒冷地の北海道は，稲作を廃し，畜力畑作酪農の混同農業を以って開拓する」という方針は継承され，その成果として明治末期までに「北海道農法」が定着した。1875(明治8)年3月には，ケプロンの帰国に際して明治天皇が接見され，「開拓使の重要な事業は，貴方の助言を受けて成功裏に実施され，進展を見ております。北海道の将来の発展は，貴方の労苦の結晶であり，大いに帝国全体の繁栄をもたらすことは疑う余地がありません」とのお言葉を受けているし，1884(明治17)年には勲二等旭日章を授与されている。

　このケプロンが来日して4カ月後の1871(明治4)年12月には，岩倉具視右大臣を全権大使とし，副使に木戸孝允，大久保利通，伊藤博文らを揃えた46名の「岩倉使節団」が横浜を出立し，1873(明治6)年9月まで米欧12カ国を回覧した。この使節団は，不平等条約の改正交渉こそ失敗に終わるが，見聞の成果を新たな日本の政治経済や産業振興の制度に色濃く反映させた。そこで公式報告書の『米欧回覧実記』から北海道開拓に関連する見聞内容を抽出する。岩倉使節団は太平洋を渡ってサンフランシスコに到着し，伊藤博文が有名な「日の丸演説」を行なった後にワシントンに向かった。この大陸横断の際にプレーリ開拓の実情を見聞して，「開拓には先ず交通を整備し，移民への車馬の援助を行なう」という植民政策を学んでいる。この使節団は，北辺鎖鑰のために北海道開拓を急ぐという大きな課題を開拓使の設置という方法で処理し，さらにケプロンらを歓待してから出発しているため，「シカゴやオマハの市街が原野から生じたことから，3日間走って来た無人の荒野もいずれの日にか移民の車が転走する時がくると確信」するなど，開拓方針の解説に殊更に興味を持って見聞している。したがって，帰国後に産業振興を担当した大久保大臣をはじめとする閣僚は，北海道開拓に交通網の整備が不可欠なことは十分に理解していたと考えてよい。しかしながら，ケプロンが北海道の道路と港湾の整備を提案したのに対し，1874(明治7)年10月の開拓使文書によると，ケプロンに対して「開拓と入植に関して多くの提案をいただき感謝しておりますが，政府の不可避の事情と限られた財源のため，すべての提案を実現できませんでした。これは従来の経験や教えに固執したり，提案を十分に理解しなかったわけでありません。貴方もご存知の通り，昨年の春早々に佐賀の乱があり，続いて台湾の事変が起きたため，新しい計画のほとんどを躊躇し，進展を許さない状況」であったと詫びている。この文意のなかに道路網と輸送手段の整備がどれほど含まれているか分からないが，実情を見ると，交

通輸送網の整備が不十分なため，農家の往来や生産資材の搬入はもとより，せっかくの生産物が港に運べない，府県に移出できないという二重の障壁となって現れて開拓を遅らせることとなった。

　1-5 では，最後の文節に興味深い文がある。ブルックスは「北海道の土地は未開地の開拓でまだまだ肥沃であるから，農学の知識がなくとも生産性は高いと思うのに札幌農学校を開校した。一方，アメリカでは移住後 200 年，建国 100 年目にようやく農学校を開校し，収奪農業で疲弊した農地を改善しようとしている。北海道の考え方は，土地が肥沃な時から農学を広めるという素晴らしい方法であり，持続的な営農を続ければ生産性の高い農地が永続できる」と絶賛している。その後の北海道農法は，ブルックスの予言通りにならなかったのはなぜだろうか。今日の農学者に与えられた課題であるように考えられる。

3．農学に必要な学科目

　1-6 では，農業および農学に必要な学問分野として，記載順に地質学，化学，気象学，力学，物理学，植物学，動物学，昆虫学，獣医学，社会経済学，簿記など多岐にわたることを教えている。事実，ペンハローの生物と化学講義では，繊維作物の組織構造と特性および化学特性，テンサイの糖分と日本酒のアルコール分析，博物調査で得た動植物と鉱物の標本リストなど農学に関連するテーマを扱っている。

　なお，日本語の学科目名が確定していないため，今日と異なるものがある。まず，Physics(物理)は「究理」と訳している。福沢諭吉著『学問のすゝめ』は，「究理とは天地万物の性質を見て其働を知る学問なり」と書いていて好例である。また，窮理と書いて物事の道理・真理をきわめる学問として哲学を言ったこともある。次いで，Chemistry(化学)は，オランダ語 chemie を音訳して舎蜜(セイミ)と訳している。また，農業と園芸，または「藝」の本来の意味を採用して，農業を行なうための技術を称する「農芸(藝)」という用語も生まれ，今日でも農芸化学，農芸物理で使われている。

　農業・農学が多分野にかかわるとの考え方は，今も変わらないと言えるが，今ひとつ考えなければならないのは，農業が自然環境下で行なわれるため，土性，水分，気象条件などが毎年変わり，教科書的な平準状況が現れず，常に応用動作を要求して長年の経験が重要になることであり，ここに親子の経営継承の効果が現れる。しかし，昨今の農業の企業化やコントラクタ等の受委託作業の発展とともに，これら応用動作を十分に配慮する態勢が失われないかと危惧するものである。

注
　*1　新渡戸(太田)稲造(1862-1933 年)。札幌農学校 2 期生。校則により開拓使御用掛を経てアメリカに留学し，ジョンズ・ホプキンズ大学で農業経済学を学び，1887(明治 20)年に母校の助教授，すぐにドイツに留学してハレ大学で学位を得帰り，1891(明治 24)年に農業経済学の教授となる。教育研究はもとより，遠友夜学校の創設，北鳴学校の経営などを手がけて多忙な生活を送り，1897(明治 30)年に病気となって休暇を得る。休養する間もなく 1900(明治 33)年に台湾総督府に招請されて台湾経済の安定に努力，さらに兼務で京都帝大教授，後年まで名を残す第一高等学校校長，東京帝大教授に就任する。1920(大正 9)年に国際連盟事務次長となり，終生の願いであった「われ，太平洋の橋とならん」

を実現した国際人であった。これらの活動により，帝国学士院会員，1926(大正15)年に貴族院議員となった。

＊2　伊吹鎗造。札幌農学校3期生。校則により北海道開拓使，間もなく改組によって北海道庁に勤務し，農商課長，農務課長，商工課長等を歴任後，農事試験場長，種畜場長となり，特に米作，繭作などの勧農行政を一手に担って北海道農業の基礎作りを行政側から支援した。

＊3　Horace Capron(1804-85年)。アメリカの農政家。南北戦争後に合衆国政府の農務長官を務めていたが，開拓使の依頼を受けて1871(明治4)年7月来日し，御雇教師頭取兼開拓顧問となった。彼はアンチセルやダンらお雇い外国人を同伴し，北海道を巡検して開拓諸事業の提言を行なった。その報告書をまとめたのが「ケプロン報文」である。1875年5月に帰国するが，その後も開拓使の依頼を受けて勧告や調査の協力をした。

＊4　Edwin Dun(1848-1931年)。オハイオ州の酪農家に生まれ，23歳で牧畜業を始めた矢先にケプロンに乞われ，1873(明治6)年に家畜を伴って来日した。彼は日本に初めての畜産の導入を担当し，真駒内・新冠牧場等を開設して牛馬の増殖を図り，併せて技術指導を行なって北海道酪農の基礎を築いた。それを顕彰して札幌市の真駒内牧場跡に「ダン記念館」がある。10年後に開拓使を退職するが，翌年に駐日アメリカ公使館の書記官となり，後に公使に昇格する。さらに退任後も日本の産業界で活躍して終生日本で暮らした。彼の著書『日本における半世紀の回想』は，当時の日本を的確に評価している。

1)　Webster辞典。編集者Noah Webster(1758-1843年)が，アメリカ合衆国の独立を契機に，文化的にも独立すべきだと唱えてアメリカ英語を定型化した辞典を指す。1828年刊の*An American Dictionary of the English Language*に始まる。なお，札幌農学校図書室は1840年刊の第2版を所有するが，Agricultureの定義文は講義とは異なる。

2)　Amasa Walker(1799-1875年)。アメリカの経済学者で，マサチューセッツ州会議委員，オーバリン大学(1942-48年)やアマースト大学(1860-69年)教授を務めた。自由貿易論と奴隷反対論を展開し，その主著が*Science of Wealth*である。

3)　蝦夷。それまで蝦夷地と呼んでいたが，1869(明治2)年に「北海道」と定めた。

4)　George Washington(1732-99年)。アメリカ建国の父と言われる合衆国の初代大統領(1789-97年)。祖父がイギリスから移住して成功し，彼はヴァージニア州ポトマック河畔の大地主の子として生を受けたが，12歳で父を失って苦難の道をたどっている。

5)　Daniel Webster(1782-18年)。19世紀前半のアメリカのニューイングランドを代表する指導的な上院議員(1827-41年)で，熱烈な保護関税論者。1830年代のホイッグ党指導者のひとり。

6)　Abraham Lincoln(1809-65年)。奴隷解放宣言を発してアメリカの民主化を進めた合衆国第16代大統領(1861-65年)。ケンタッキー州の開拓農民の子として生まれ，「丸太小屋からホワイトハウスへ」を実現した立身出世の人。

7)　ノートの欄外に落書きしてある唯一の漢詩。このほかには[140頁]にHydeの詩が書いてある。中国唐代の文学者，思想家，政治家である韓愈(768-824年，字は退之)が作った「符読書城南」と題する漢詩であり，別荘「城南」で息子「符」の教育のために書いた。詩の意味は「誰の力も借りずに自分の力で生きよ」である。

参考文献
1. 日本農書全集，第1期(第1～35巻)，第2期(第36～72巻)，農山漁村文化協会
2. 斎藤之男(1968)，日本農学史——近代農学形成期の研究，大成出版社
3. 日本科学史学会編(1970-71)，日本科学技術史大系(第22巻農学1，第23巻農学2)，第一法規
4. 北海道農業試験場編(1967)，北海道農業技術研究史，同試験場
5. 北海道農業試験場編(1982)，北海道農業技術研究史：1966～1980，同試験場
6. 記念誌出版委員会編(2002)，北海道農業技術研究史：1981～2000，北海道農業研究センター
7. メリット・スター著，西島照男訳(1986)，ホーレス・ケプロン将軍——北海道開拓の父の人間像，北海道出版企画センター
8. 久米邦武編・田中　彰校注(1977-82)，米欧回覧実記1～5，岩波文庫
9. 北海道大学編(1980-82)，北大百年史〈部局史〉〈札幌農学校史料(一)(二)〉〈通説〉
10. 東京大学百年史編集委員会編(1984-86)，東京大学百年史

［髙井宗宏］

Chapter 2   Soil

We understand by the word soil that portion of the earth's crust which has been broken and decomposed till it consists of a large portion of fine particles[1]; an agricultural soil may be defined as one capable of supporting vegetation. ⟨See p. 254⟩

There are two classes of matter in all agricultural soil: organic and inorganic. Organic matter is that which either is now, or has been in some past time, part of some plant or animal, and which still possesses some trace of organic structure[2]. Inorganic matter is that which neither is nor has been a part of any plant or animal, and which has no organic structure. ⟨See p. 244⟩

2-1. The Manner in which Rocks have been Converted into Soils

The first step in the process is the pulverization[3] of the rocks. This change has been effected by four principal causes, which are as follows: [p. 16]

1) Changes of temperature

The globe on which we now live was undoubtedly an immense molten ball, the supposition of which is quite reasonable. And the cooling of the earth, contraction, and hence cracking. As rocks are made up of different minerals, the rate of expansion and contraction of its different particles is unequal. Cracks are formed on the rocks when they are subjected to changes of temperature. Crystals expand and contract unequally in their different diameters and, as many rocks are often made up of crystals, this fact also causes cracks to be made in the rocks when they are subjected to changes of temperature. [p. 17] By the penetration of water into the cracks, which have been formed on the rocks in one of the ways of which we have been speaking, the rocks are generally burst asunder, owing to the immense power of the water's expansion when frozen[4].

2) Moving of Water or Ice

Water, in moving over any thing with which it comes in contact, wears it away more or less rapidly according to the degree of hardness of the substance. Thus water, even when flowing over the solid rock, wears it away quite rapidly. The amount which is worn away in one year, or even in a hundred years, might seem quite small; yet in the countless ages which have elapsed since the creation of the world, running water has accomplished very great results. Rivers, at the present time, are exerting a great influence in this direction. The waters of the Rhone are so muddy [p. 18] that they may be traced a distance of five or six miles into the ocean[5]. The delta of the Mississippi River annually carries approximately 28,000,000,000 cubic feet of sediment to its mouth. The waters of the Amazon River can be traced 300 miles from its mouth. Moving ice, in the form of glaciers, has in the past exercised a tremendous influence in the grinding up of rocks in all the temperate and frigid zones of the earth. These glaciers, in the past, have been in motion from the north in a southerly direction over all the frigid and northern temperate zones, as far south as the latitude of Philadelphia[6]. And as they proceeded on and on, they ground up more or less of the rocks with which they came in contact. It seems to be true that glaciers in olden times exercised, and still exercise, a great influence upon the surface of the earth. This can be proved by examining the summits of [p. 19] mountains or even plains, where marks consisting of many parallel lines, generally tending from north to south, may be seen. They are supposed to have been made by the passage of glaciers.

3) Chemical Action of Water and Air

Water has the property of uniting chemically with the various elements, forming compounds which are called hydrates. This chemical action is called hydration, and it has doubtless been a very important agent in the formation of soils from rocks. The hydrates of the various elements with which water unites are, most of them, softer than the original elements; hence, after an element has been hydrated, it can be more easily pulverized than before. Water acts also in another way: namely, by solution. [p. 20] Water dissolves the rocks in the same manner as it does sugar. Water also usually contains carbonic acid, and this greatly increases its power to dissolve other substances. The quantity of carbonic acid in rainwater varies from 8 to 90 parts in 10,000. The water of rivers and its affluents contains somewhat more, and the capillary water of soils which are rich in organic matter contains a still greater quantity. The capillary water of soils rich in organic matter contains even more carbonic acid. The reason why it contains more carbonic acid is this: organic matter always contains carbon, and this carbon, when the organic matter decomposes, unites with oxygen, forming carbonic acid. Hence the capillary water of soils rich in organic matter has a very great solvent power. [p. 21]

4) Action of Oxygen

Oxygen has the power[7] of uniting with many elements, changing them into oxides, and these oxides occupy more space[8], and are usually more soluble, than the original elements. The conjoined influence of water, carbonic acid, and oxygen is expressed by the word "weathering"[9]. ⟨p. 79⟩ When rocks weather, they are decomposed or dissolved, and new compounds or new forms of original matter are the result. Different rocks weather with different degrees of rapidity. Quartz rock weathers very slowly. Feldspar rocks[10] vary much as to the rapidity with which they decompose. Lime feldspar weathers very rapidly, while potash feldspar weathers more slowly. Most volcanic lavas [p. 22] weather very rapidly, and form very fertile soils.

5) Heterogeneous Kind of Soils

On account of the manner in which most soils have been formed, they are composed of a great variety of elements. A drift soil is the best example of the heterogeneous kind. Homogeneous soils[11] are formed by the deposit of sediment by still water, such as a clayey soil.

2-2. Organic Matter

1) Organic matter its origin and manner of formation and mixture with inorganic matter in the Soil

The first order of plants which grew upon the earth was undoubtedly lichens. Lichens can grow upon the face of the hardest rocks, so that even before the rocks were ground [p. 23] into soils, lichens were probably growing upon them. By the decay of these lichens, the face of the rocks would be covered with dead organic matter, and so gradually became fit for the growth of the higher orders of plants. The next plants which grew were perhaps ferns, which also have the power of growing where there is but a very small quantity of soil. By the decay of these ferns, the face of the rocks would be rendered a still better place for the growth of still higher species of plants. Then the next plants which grew would perhaps be herbs and shrubs, and probably grasses too, and these, by their decay, would furnish still more organic matter, and render the face of the rocks a suitable place for the growth of the highest orders of plants, such as the trees and various flowers which [p. 24] we can see around us at the present time. So, by the decay of successive generations of plants, the soil gradually acquires a certain content of dead organic matter. Falling leaves, seeds, and stems do not generally waste as rapidly as they are renewed. This accumulation of organic matter is called "humus". Leaf-mold, swamp-muck, and peat are good examples of humus. The roots of plants penetrate the soil, and by their decay, dead organic matter becomes mixed with the soil. This is the only way in which mixture will take place in nature[12].

2) Influence of Organic or Inorganic Matter and its Function in Forming Agricultural Soils [p. 25]

(1) The disintegration of rocks is aided by the presence of organic matter in a decaying state insofar as they make the soil more moist. This is a fact in regard to organic substances that they increase the power of the soil to absorb and retain water. Now we have seen, in speaking of water, that it aids in the disintegration of rocks by dissolving certain elements contained in them. Hence, since organic matter makes the soil more moist, and since water is a powerful agent in disintegrating rocks, the presence of organic matter will cause the rocks to disintegrate more rapidly. (2) Organic matter supplies a great deal of carbonic acid. A large part of all plants is composed of carbon, on the average 44%. Oxidation of this carbon furnishes a great deal of carbonic acid to the soil. The carbonic acid in 10,000 parts [p. 26] of ordinary atmosphere is, by weight, 6 parts; in the air contained in sandy subsoil of a forest, 38 parts; in the air contained in surface soil of a forest, 130 parts; in the air contained in surface soil rich in humus, 543 parts; and in the air contained in surface soil newly manured during wet weather, 1413 parts. (3) Organic matter, by its decomposition, furnishes various organic acids. Among these are humic and ulmic acids[13], and these act directly upon the mineral elements in the soil, making them more readily soluble. (4) Organic matter, by its decomposition, gives rise to various salts of ammonium and then to nitrate[14], as well as the organic acids: oxalic, tartaric, *etc*. All these substances, upon being dissolved in water, increase its solvent power. [p. 27]

3) Changes in Organic Matter Necessary to make Fit It for Plant Food

Before organic matter can be made use of by plants as food[15], it must be thoroughly oxidized or decomposed[16]. ⟨See p. 243⟩ This can take place only when the air has free access to the organic matter. Under these conditions, oxygen oxidizes or burns it up. Without free access of the air, such compounds as humic and ulmic acids are formed; these cannot serve as the food of plants. ⟨See p. 318⟩

4) Action of Living Plants on Mineral of the Soil

Living plants furnish to the soil both moisture and carbonic acid; these, as we have seen, are powerful agents in the disintegration of rocks. So, by furnishing these substances, living plants aid in the decomposition of rocks. [p. 28] It is probable also that the roots of plants contain organic acids, which enable them to extract nutriments in a slight degree even from the solid phase.

5) Changes in Inorganic Matter by which it is made Fit for Plant Food

The elements in inorganic matter can be used by plants as food only when in solution in water[17]. Inorganic matter must be decomposed before it serves as food. Inorganic matter in the soil may be said to have three important offices as regards the growth of plants: (1) It furnishes to the plants the elements which constitute their ashes, (2) It serves as the home for the roots of plants, the roots penetrating through the pores of the soil both downward and laterally, and thus holding the plants in their natural position. [p. 29] (3) The inorganic matter in the soil stores up and tempers the heat of the sun, supplying to the plants warmth as may be necessary. It is important that the inorganic matter in the soil be in a fine mechanical condition. This is important for the following reason:— as I have already stated, plants cannot appropriate inorganic matter unless it is in solution in water, and things which are finely pulverized are much more easily or readily dissolved than those which are not. The principal mineral elements which plants feed upon, and which are derived from the soil, are phosphorus, sulfur, silicon, chlorine, potassium, sodium, calcium, magnesium, and iron. Plants cannot grow without the aid of the elements already mentioned. We see that it is important to know the number of elements which are necessary to the growth of plants. And we must also know the variety of elements a soil should contain. Lime is most important for the growth of plants[18]. [p. 30]

2-3. Classes of Soils

1) Drift soil, Alluvial Soil, Colluvial Soil

As regards the place of their origin, soils are divided into two classes, Sedentary and Transported[19]. A Sedentary Soil is one which has been formed from the rocks in the place where it exists at the present time. Sedentary soils are usually underlaid by rocks of the same kind as those from which they have been formed, and are not usually very deep. Thus we can tell, from the character of these underlying rocks, what the character of the soil is whether it is fertile or not. If a soil is underlaid by a rock which contains mineral elements necessary for plant food, we can at once conclude that the soil itself contains those elements and hence will be fertile[20]. If, on the contrary, the soil is underlaid by a rock which does not contain those elements, it is said to be probably barren. A Transported Soil is one which has been transported a greater or [p. 31] less distance from the place where it was originally.

There are three kinds of transported soil: Drift, Alluvial, and Colluvial. A Drift Soil is that which is formed by the action of moving ice; owing to this reason, it usually contains stones which vary in size from small pebbles to immense masses weighing many tons. One peculiarity of these stones is that they are always more or less rounded, their corners having been worn off. This is caused by their own motion.

The surface of drift soil is usually hilly, and the soil itself is a heterogeneous mixture of stones of various sizes and fine particles of soil. It is not usually stratified. Drift soils were formed during the glacial period. It is possible to determine, from the nature of the stones contained in [p. 32] the drift soil, something in regard to the fertility of the land. If these stones are composed of the elements which are necessary for plant food, the finer part of the soil will be likely to contain these same elements. If, on the contrary, the stones do not contain the elements necessary for plant food, the soil itself will not be likely to contain them, and hence will not be fertile.

Alluvial Soil consists of more or less rounded particles, which are usually very small. Alluvial soil is one which has been deposited by water and is usually more or less stratified. This kind of soil has been formed at all times, and is still forming even at the present time. The reason why it is stratified, is that it has been formed in water. [p. 33] Colluvial Soils consist in part of drift or alluvial soil, but have mostly originated from rocks, very near the place where they are found. That they contain angular stones and rocks shows that they have not been carried very far.

2) Variety of Soils; Alluvial Soil, Clay, Sand, Loam, Gravelly Soil, Clayey Loam, Sandy Loam, Gravelly Soil, and Calcareous Soil[21]

Alluvial Soil is composed of particles of all the kinds of soils found in the country, drained by the stream which formed it. It is therefore usually very fertile, because it contains a great variety of elements and their particles are very small. [p. 34]

A Clayey Soil has been deposited by water at rest, and consists of Kaolinite, a hydrated silicate of alumina, with a variety of other substances such as free silica, oxides, and silicates of iron and manganese, carbonate of lime, and fine powder of various other minerals. The particles of a clay are the most finely pulverized of any other particle, and also the smallest. A clay is usually very hard and compact, and almost impervious to water. Notwithstanding the fact that a clay possesses one of the requisites of a fertile soil, it is so impervious to water and to air, and also to the roots of plants, as to be almost valueless for agricultural purposes[22].

A Sandy Soil is composed of visible granular particles of rocks, which are in most cases quartz. [p. 35] A sandy soil is the heaviest of any other kind of soil. A pure sandy soil is almost valueless for agricultural purposes.

A Loamy Soil consists of a mixture of sand, clay, and organic matter in varying proportions. Although there are many kinds of loam, it is usually valuable for agricultural purposes[23].

A Sandy Loam is one in which the sand is the predominating substance. There are two classes of this kind of soil: light and heavy sandy loam.

A Clayey Loam is one in which clay is the predominating substance. Such a soil as this is so valuable as to rank next to alluvial soil. [p. 36]

All the different kinds of soils which are mentioned above vary in their characteristics, according to the sand, clay, or loam which predominates in them (Table shown at right).

Gravelly Soil: A Gravelly Soil consists of a mixture of

Kind of Soil	Cray or Imparable Matter	Sand
Heavy Clay	75-90%	10-25%
Clayey Loam	60-75%	25-40%
Loam	40-60%	40-60%
Sandy Loam	25-40%	60-75%
Light Sandy Loam	10-25%	75-90%
Sandy Soil	0-10%	90-100%

stones of various sizes with fine particles, which may be either clay, loam or sand; or it might contain all these. The value of a gravelly soil depends upon the constitution of its finer portion, and this in turn, varies much in different gravels. The value of gravelly soil can possibly be assayed by inspection of the stones which are contained in them. Thus, if the [p. 37] stones contain plant food, the finer portion of the soil will be likely to contain the same kind of plant food, since it has been formed from rocks similar in composition to the stones. If, for example, the stones happen to contain feldspar, we may conclude that the soil is rich in potash. If, on the contrary, the stones contain a great deal of quartz, we are sure to find that the finer portion will also contain quartz, and as quartz is a substance which is not necessary for plant food, such a soil will not be fertile. Such a soil as this usually contains a great deal of quartz, and hence is not very valuable.

A Calcareous Soil is one which contains a large quantity of carbonate of lime, which is usually mixed with either clay or sand. These soils, however, are not very common, and not usually of great value. [p. 38] We can determine whether a given soil is calcareous or not by taking a small quantity of it, and adding some acid or vinegar to it; when the soil is calcareous, it will effervesce more or less according to the amount of carbonate of lime in it[24].

3) Surface, Subsoil, Hard-pan

A Surface soil or tilth[25] is that part of a soil lying near the surface which has been acted upon, more or less, by the atmosphere, which usually contains more or less organic matter, and which usually differs in color from the soil lying beneath it, being usually darker.

A Subsoil is that part of the soil lying beneath the surface soil, from which it usually differs in color, since it contains little or no organic matter. ⟨See p. 254⟩ The plant food contained in the subsoil is not usually in an available form, [p. 39] because the subsoil has not been acted upon by atmospheric influences.

Hard-pan is the appropriate name of a hard, impenetrable layer of soil often formed at no great distance below the surface. It is really soil which is being changed back into rock by being cemented together by some agencies of nature, but especially by means of ochery clay[26]. The thickness of this layer of hard-pan is variable, -sometimes not more than a few inches, and sometimes many feet. It is often very injurious to the land, especially when it lies near the surface, because it is quite or almost impervious to water, and hence, the water falling upon such land, cannot soak below the place where the roots of plants are deposited.

2-4. Physical Properties of Soils

The physical characteristics of the soil[27] [p. 40] are those which concern the form and arrangement of its visible or palpable particles, and likewise include the relations of these particles to each other, and to the air and water, as well as to the forces of heat and gravitation. ⟨See p. 308⟩ Of these physical characteristics, we shall notice the following: The weight of soils, state of division, absorbent power for water vapor (hygroscopic capacity), property of condensing gases, power of fixing solid matters from their solution, permeability to liquid water, capillary power, changes in bulk by drying, adhesiveness, and relation to heat.

1) The Weight and Specific Gravity of Soils

The weight of soils varies with their porosity and is greater, the more sand or gravel they contain. If a soil is very porous, it will weigh much less than that which is not porous. [p. 41] According to Shübler's[28] calculation, the following table gives the weight per cubic foot of several kinds of soils shown in above table: The specific gravity of all soils is nearly alike, but it varies from 2.53 to 2.71.

KIND OF SOIL	WEIGHT lb/ft^3
Dry Silicon or Calcareous Sand	100 lbs
Half Sand and Half Clay Soil	96 lbs
Common Arable Land	80–90 lbs
Heavy Clay	75 lbs
Garden Soil rich in organic matter	70 lbs
Peat Soil	30 lbs

⟨The specific gravity of any substance is the weight of a given volume as compared with the weight of the same volume of some other substance taken as a standard.⟩

2) State of Division of Soil[29] and its Influence on Fertility

On the surface of a block of granite, only a few mosses and lichens can grow. [p. 42] Crush this granite into coarse pieces, and then a few more plants can grow on its surfaces. Reduce this granite into fine powder and water it plentifully, then even cereal grains can be made to grow upon it. As a general rule, all fertile soils contain a large amount of fine matter. The valley of the Scioto River in Ohio is remarkably fertile; it has produced large crops for more than sixty years, and is still very rich. That soil is remarkable for the fineness of its particles. ⟨See p. 246⟩ But if the particles of soil are too-fine, the soil becomes compact, and hence the growth of roots, and the germination of seeds, are arrested.

3) Absorption of Water Vapor by Soils

Soil has the power to draw water vapor from the air and to condense that water in its pores; in other words, it is hygroscopic[30]. ⟨See p. 319⟩ This property of soil is considered to be of the utmost importance in an [p. 43] agricultural soil for the following reasons: (1) Because it supplies the vegetation with necessary moisture[31], (2) Because a soil which is hygroscopic has the power to absorb other gases or vapors; the atmosphere always contains more or less water vapor, and when but a little rain falls, a soil which has the power to absorb this water vapor will become far less dry than that which does not have this power.

The right table shows the number of parts of hygroscopic moisture absorbed by 1,000 parts [‰] each of several previously dried soils, upon being exposed for 24 hours to an atmosphere nearly saturated with moisture. [p. 44]

Kind of Soil[32]	Quantity of Water Absorbed (‰)
Coarse Quartz Sand	0
Gypsum	1
Lime Sand	3
Plough Land	23
Clayey Soil (60% Clay)	28
Slaty Marl	33
Loam	35
Fine Carbonate of Lime	35
Heavy Clay (80% Clay)	41
Garden Mold (7% Humus)	52
Pure Clay (Fine Powder)	49
Carbonate of Magnesia	82
Pure Humus	120

The lime sand, which is coarse carbonate of lime, absorbs 3 parts of moisture, while fine carbonate of lime absorbs 35 parts. Hence, we see that pulverization increases the absorptive power. A clayey soil which contains

60% clay absorbs 28 parts of moisture. The 80% pure Clay absorbs 41 parts of moisture, while pure clay absorbs 49 parts. Hence, we see that clay increases the absorptive power of soil. Ordinary plow land which contains but little humus absorbs 23 parts of moisture. Garden mould (7% humus) absorbs 52 parts of moisture, while pure humus absorbs 120 parts. Hence, we see that [p. 45] humus increases the absorptive power of soil. There is present in the atmosphere at nearly all times a greater or less quantity of ammonia, which as we know is an important element of plant food on account of the nitrogen it contains. ⟨See p. 317⟩ Now, a soil which has the capacity to absorb a great amount of moisture from the atmosphere will also have the power to absorb this gas.

We will now draw a practical conclusion in regard to mixing different kinds of soil. We see from the previous table[33] that sand has little hygroscopic power, while clay and humus have a great deal; therefore, it will be a good practice to mix clay or humus with a sandy soil for the purpose of increasing its capacity to absorb moisture and *vice versa:*—clay and humus absorb a great deal of moisture and are very often too wet; therefore it will be [p. 46] a good practice to mix sand with such soils for the purpose of decreasing their hygroscopic capacity, and hence making them drier.

Every body has a definite power of condensing moisture upon its surface, or in its pores; hence, if its surface area is increased by pulverization, its capacity for

SANDY SOIL	Amount of Hygroscopic Water Absorbed (‰)	Temperature of Hygroscopic Water (F)
1,000	13	55
1,000	11.9	66
1,000	10.2	77
1,000	8.7	88

absorbing moisture will also be increased. The rapidity of absorption depends upon the amount of vapor which is accessible, but the amount of vapor absorbed by a soil depends upon the atmosphere. The amount of vapor in the atmosphere at different times varies very much, usually being greater the warmer the atmosphere. When there is a great deal of moisture in the air, it will be absorbed by the soil very rapidly; when there is less moisture, less rapidly. The amount of moisture absorbed by the soil depends upon its temperature, [p. 47] and is less the greater the temperature. A table showing the amount of moisture absorbed by a sandy soil at different temperatures is shown above.

4) Property of Condensing Gases by the Soil

In the fact that soils and porous bodies generally have a physical power of absorption for water vapor, we have an illustration of a principle of very wide application that the surface of liquid and solid matter attracts the particles of other kinds of matter. This force is called Adhesion[34]. When it acts upon gaseous bodies, it overcomes to a greater or less degree their expansive tendency and brings them into a smaller space; that is, it condenses [p. 48] them. Charcoal has a very great power to attract and condense gases; in general, any body which contains a great deal of carbonaceous matter has this power to a greater or less extent, according to the amount of carbonaceous matter it contains.

Those figures which I have just given you are the quantities of gas found in the substances named, after having been exposed to an ordinary atmosphere. This gas is found to differ in composition from ordinary air. Nitrogen was found to be present in a greater proportion than in the air. Oxygen is often found to be nearly or quite absent. Carbonic acid gas is usually found to be [p. 49] abundant. When moist, a substance is capable of containing less gas than when dry. Chemical

AMOUNT OF CHARCOAL	AMOUNT OF GAS ABSORBED FROM THE ATMOSPHERE[35]
100 grams	164 cubic centimeters
100 grams	14 cubic centimeters
100 grams	38 cubic centimeters

action is often brought about by this adhesion of gases to substances by which they are absorbed. Soil, as well as charcoal, absorbs putrid effluvia, and undoubtedly often oxidizes them as charcoal does. These gases themselves are sometimes changed into plant food by the chemical actions which they cause. Hence, the importance of this power of soil, for it often furnishes plant food.

5) Power of Fixing Solid Matters from their Solutions[36]

All soils have more or less power to remove dissolved solids from their solutions. They possess this power in a great degree, the more carbonaceous material or clay they contain; in general, the finer the soil, the greater will be this power. [p. 50] Coarse sandy soils have but little of this power, while soils which contain clay or carbonaceous material have a great deal of this power. It is important that the soil should have this power in order that when rain falls upon it, carrying down with it the soluble elements of plant food, the soil shall be able to fix and retain these elements[37]. If we apply manure to a sandy soil, the first heavy rainfall will be likely to carry down, out of the reach of the roots of plants, a large part of the soluble elements of plant food which the manure contained. Such a soil is said to be leachy. If, on the other hand, manure is applied to a clayey soil, or to one which contains a great deal of organic matter, this loss will not occur, because such soils have the power to fix and retain the soluble plant food. Hence, it is often advantageous to mix clay or organic matter with sandy soil for the purpose of increasing its power to remove the dissolved substances from their solutions[38]. [p. 51]

6) Permeability of Soils to Liquid Water and Capillary Power

Any thing is said to be permeable when it will allow a liquid to pass freely through it. The permeability of any body is dependent on the number and size of its pores. If a body has large pores, liquid will pass through it very rapidly; when it has many and small pores, less rapidly. A soil which has small pores has great capillary power[39]; (1) Capillary power is important to soils, since in seasons of drought it helps to supply plants with the moisture necessary for their growth. (2) It is important for the fact that it helps to supply plants with food, the water rising by capillarity, bringing with it those elements of plant food which are soluble[40]. A soil which has great capillary power has the capacity to retain a great amount of moisture; sometimes, this capillarity is too great, causing the soils to be too wet. [See next page table] [p. 52] The lack of capillary retentive power, is undoubtedly one of the greatest causes of the barrenness[41] of sandy soils.

7) Changes in Bulk by Drying and Frost

Many soils, upon being dried, shrink a great deal. In general, those soils which absorb a great deal of moisture shrink most upon [p. 53] being dried. Thus clayey soils shrink a great deal. Often in a dry time, such soils are seen to be full of cracks, running in all directions. Those cracks were made by the shrinkage of the soils. Cracking often does a great deal of damage to the roots of plants growing in such soil sometimes breaking them very much. All soils upon being frozen, on account of the water which they contain, expand more or less. In general, those soils which have the power to absorb and retain the most moisture expand most when frozen frost heave. This expansion often does great damage, by throwing the roots of plants growing in the soil out of the ground

KIND OF SOIL	% of Water Absorbed	% of Water Evaporated in 4hr.
Quartz Sand	25	88.4
Gypsum	27	71.7
Lime Sand	29	75.9
Slaty Marl	34	68
Clayey Soil (60% Clay)	40	52
Loam	51	45.7
Heavy Clay (80% Clay)	61	34.9
Pure Clay	70	31.9
Fine Carbonate of Lime	85	28
Garden Mould	89	24.3
Humus	181	25.5
Fine Carbonate of Magnesia	256	10.8

during the winter or spring. When thus thrown out, the plants are often completely killed.

8) The Adhesiveness of Soils

Adhesiveness[42] of soils is that property [p. 54] which causes their particles to adhere or stick to each other. Those soils which have this property to a great degree are called heavy soils, though they may be lighter than those which are called light. A clayey soil is an example of a heavy soil, and sand of a light soil. In general, those soils which have great adhesive power are cultivated with greater difficulty than those which have less adhesiveness.

9) Relation of the Soil to Heat[43]

The temperature of a soil varies to a certain depth in relation to that of the air, yet its changes occur more slowly, are confined to a considerably narrower range, and diminish downward in rapidity and range, until at a certain depth, a point is reached where the temperature is invariable. In summer, the temperature of the soil in the daytime is sometimes higher than that of the air, but at night the [p. 55] temperature of the surface rapidly falls, especially if the sky is clear. In temperate climates, the temperature of the soil remains unchanged from day to night at the depth of 3 feet below the surface; the temperature remains unchanged from day to night at the depth of 20 ft. the annual temperature varies from 1 to 2; the temperature remains unchanged from day to night at the depth of 75 ft. the temperature is always the same. In the tropics, the point of nearly unvarying temperature is reached at the depth of one foot. The mean annual temperature of the soil is the same as that of the air, or in higher latitudes, it may be 1 or 2°F more. The nature and condition of the soil must considerably influence its temperature.

10) Sources of the Heat of the Soil

There are three sources from which soils derive their heat: from the original heat in the interior of the earth, [p. 56] from chemical reactions, and from the sun. (1) The Original Heat of the Earth: the surface of the earth is still partially dependent upon heat from the interior. In temperate zones, that heat increases as we go downward into the earth, 1°F for each 45 feet. It is not, however, sufficient to support vegetation. (2) Chemical Processes as a Source of Heat: most chemical processes, especially oxidation, are ongoing in the soil, where they cause an increase in temperature, but these are not sufficient to support vegetation. (3) The Sun as a Source of Heat: [p. 57] the sun is the great source of the heat received upon the earth. Without it, all animal life[44] must soon die. [p. 58]

11) Color and Texture of the Soil[45], as Influencing the Absorption and Retention of Heat

The following substances (Table shown below) were subjected to the same conditions as regards the supply of heat, but in one instance were whitened and in the other blackened by sprinkling upon it powders of these different colors. [p. 59]

KIND OF SOIL[46]	SURFACE WHITENED	SURFACE BLACKENED	DIFFERENCE °F	SURFACE WET	SURFACE DRY	DIFFERENCE °F
Magnesia (pure white)	108.7	121.3	12.6	95.2	108.7	13.5
Fine Carbonate of Lime (white)	109.2	122.9	13.7	96.1	109.4	13.3
Gypsum (light clay)	110.3	124.3	14	97.3	110.5	13.2
Ploughland (gray)	107.6	122	14.4	97.7	111.7	14
Sandy Clay (yellowish)	108.3	121.6	13.3	98.2	111.4	13.2
Quartz Sand	109.9	123.6	13.7	99.1	112.6	13.5
Loam (yellowish)	107.8	121.1	13.3	99.1	126.1	13
Lime Sand (whitish gray)	109.9	124	14.1	99.3	112.1	12.8
Heavy Clay (yellowish gray)	107.4	120.4	13	99.3	112.3	13
Pure Clay (bluish gray)	106.3	120	13.7	99.5	113	13.5
Garden Mold (blackish gray)	108.3	122.5	14.2	99.5	113.5	14
Slaty Marl (brownish gray)	108.3	123.4	15.1	101.9	115.3	13.5
Humus (brownish black)	108.5	120.9	12.4	103.6	117.3	13.7

12) Rapidity of Temperature Change

Some soils have the power to retain heat much longer than others, and in general, those soils which contain the most sand or gravel retain heat the longest.

The following Table (Table shown below) shows the time required by several different soils to cool from 145°F to 70°F, the temperature of the air being 61 degrees. [p. 60] The finer the soil, the less is its power to retain heat. Advantage is often taken of this property[47] of sand in this way. It is frequently spread upon and mixed with soils consisting largely of humus or clay for the purpose of increasing their capacity to retain heat.

13) Influence of Moisture upon the Temperature of Soils

We see by examination of the above table that those soils which were wet absorbed less heat than those which were dry, the difference varying from about 12 to 14°F[48]. Water, during evaporation, uses [p. 61] up or renders latent a large amount of heat; this is the reason why a soil which is wet is always cooler than one which is dry, provided they are subject otherwise to the same conditions.

14) The Angle at which the Sun's Rays Strike the Soil, as Influencing the Temperature

The more closely to a right angle perpendicular the rays of the sun strike the earth, the warmer will it be. For this reason, a soil, the surface of which slopes toward the south is always warmer than one which is level or slopes in any other direction. In this latitude (Hokkaido, 43 degrees N), an inclination of about 30 degrees from the plane of the horizon would insure the absorption of the greatest amount of heat[49]. It is very important, to ensure success in the growth of many farm-crops, to select a soil which is as warm as possible. [p. 62]

15) Influence of a Wall on the Temperature of the Soil

KIND OF SOIL[46]	Time Required to Cool from 145 to 70°F	Relative Capacity to Retain Heat
Lime Sand	3 hours 30 minutes	100
Quartz Sand	3h. 27min.	95.6
Potter's Clay	2h. 41min.	76.9
Gypsum	2h. 34min.	73.8
Clayey Loam	2h. 30min.	71.8
Clayey Ploughland	2h. 27min.	70.1
Heavy Clay	2h. 24min.	68.4
Pure Gray Clay	2h. 19min.	66.7
Garden Earth	2h. 16min.	64.8
Fine Carbonate of Lime	2h. 10min.	61.3
Humus	1h. 43min.	49
Magnesium	1h. 20min.	38

The temperature of soil on the south-side of a wall is considerably higher when the sun is shining than that of another not so situated. This is because the rays of the sun are reflected from the wall upon the soil. The average temperature on the south-side of such a wall during daylight hours will be 8 degrees Fahrenheit higher than that of the soil otherwise similarly situated.

2-5. Adaptation of the Soil to Crops

It is true of nearly all plants that they will thrive best upon some particular kind of soil. This we see in nature. Certain plants are always found in certain localities. There is a great difference among plants, however, as regards their ability to grow in different kinds of soils. At the present time, we are acquainted with about 200,000 different species of plants. Of these, 276 species are found only in deep, damp, dark swamps; 170 only along seashores or where washed by salt water or spray; 128 only in cultivated regions; 121 only in meadows; 78 only in sandy soil; 126 only in forests or in leaf-molds; 70 in strong limestone soil; 64 in heaths; 30 on stone or brick walls; 29 only on rocks; and 19 only in salt marshes; the rest are found in different [p. 63] kinds of soils.

Another fact that sustains the theory that plants are adapted to certain kinds of soil is the homogeneous growth of plants often found over a great extent of country. The pine regions in North and South Carolina, in the United States, are a good illustration. In some regions, nothing but sedges or rushes abound. Aquatic plants will not grow on the land, nor terrestrial plants in water, neither the air plants[50]. [p. 64]

1) Crop Adaptation of Sand

Pure sand is not well adapted for the growth of many kinds of plants, and it may be said in regard to those plants which do grow upon it that they are not those which are well adapted for the food of either man or animal. Some of the plants which are particularly adapted to sandy soils are beachgrass or *Calamogrostis Arenaria*, locust-trees belonging to the genus Robinia, pine-trees belonging to the genus Pinus, peanuts (*Arachis Hypogaea*), rye (*Secale Cereale*)[51], and birch belonging to the genus Betula.

2) Crop Adaptation of Sandy Loam

Sandy Loam is warm, and crops grow upon it very quickly, and it is therefore well adapted for market gardening. Such a soil is not very well adapted for crops which make their best growth late in the season.

3) Crop Adaptation of Clay

Owing to its compactness, clay is [p. 65] impervious to the natural influences which develop plant-food[52]. It produces naturally only the coarser grasses, sedges, and rushes. But with persistent and thorough cultivation, and when thoroughly drained, it may be made to produce grasses and spring grain.

4) Crop Adaptation of Clayey Loam

Clayey Loam is strong and retentive, and well adapted to most crops. All the grains except rye flourish on it, and it is the best soil for grasses, onions, beets, and root crops.

5) Crop Adaptation of Loam or Peat

A soil which consists entirely of peat is not well fitted for the production of crops[53]. When, however, the organic matter has been decomposed thoroughly and broken up by cultivation, it produces good crops of grasses, potatoes, and cabbages. [p. 66]

6) Crop Adaptation of Alluvial Soil

Alluvial soils contain a very great variety of elements, and such a soil is usually very finely pulverized; hence, it is well adapted for the production of nearly all kinds of crops.

7) Crop Adaptation of Gravelly Soil

No very definite rule can be given in regard to the crop adaptation of gravels, since their productive capacity is dependent largely upon the composition of their finer portion, and this finer portion varies very much. If the finer portion is sandy in its nature, the gravel will be adapted to such crops as will grow upon sand; if clayey, to such crops as will grow upon clay; if full of organic matter, to such crops as will grow upon loam.

8) Crop Adaptation of Calcareous Soil

Calcareous soils are usually very strong and retentive, and are said to be particularly [p. 67] well adapted to the production of the cabbage and other similar crops.

9) Physical Faults of Soils and their Remedies

A soil which contains all the necessary elements of plant food is sometimes in such a poor condition physically as to make it almost impossible to raise crops successfully. Indeed, the physical condition exerts as much influence upon plant growth as the chemical constituents of a soil. It is, therefore, very important to understand how to put a soil into proper physical condition. The character of the growing vegetation on any tract of land may sometimes entirely be changed simply by changing the physical condition of the soil. An excess of water in a soil is a very serious fault and can be remedied only by drainage.

10) Reasons why a Soil Saturated at All Times with Moisture is Useless for Agricultural Purposes [p. 68]

(1) Because it is so wet that valuable plants will not grow upon it. (2) Because it cannot be permeated by the air. (3) Because the roots of plants cannot penetrate sufficiently deep to obtain nourishment. (4) Because the plants growing upon such

a wet soil are liable to be killed by frost, both later in the spring and earlier in the autumn than the plants growing upon a drier one.

11) Indications by which You may Know whether the Land Needs Artificial Drainage

Any land, upon the surface of which water stands for any great length of time at any season of the year, needs drainage. No land, however, upon which water seldom or never stands, need drainage. So, perhaps the surest sign of this need is the character of the vegetation naturally growing upon them. Wherever you see sedges or rushes growing, you may be certain that the land needs drainage. Coarse water grasses are equally sure evidences of this need. [p. 69]

解 題

1. ブルックスと農学講義の時代

　19世紀は，農学にとっても，土の科学にとっても，大変革の時代であった。前世紀末のマルサスによる不吉な予言(『人口論』1798年)は，農学者たちに，17〜18世紀とは違った視点から農業生産システムと土地生産性の維持を再検討することを迫っていた。そして19世紀の半ばからは，コペルニクス的転回とも言うべき認識の大転換が立て続けに起こった。1つは植物栄養の有機栄養説から無機栄養説への転換，もう1つは地質学・農芸化学的な土壌観から生態的な土壌観への転換である。アマーストのマサチューセッツ農科大学を卒業して1年たらずの若者ブルックスが札幌農学校に着任した1877(明治10)年は，まさに大転換の真っ直中の時期であった。

　前者の発端は，ドイツの化学者・生理学者ユスタス・リービッヒが1840年に発表した論文であったが，これにいち早く反応したのは，イギリスの農学者たちだった。1843年には，ロザムステッド農事試験場において，組織的な圃場実験をスタートさせ，1852年頃までには，無機栄養説の正しさを大筋で確認した。マメ科植物の窒素栄養問題などに疑問点が残ったが，コミュニケーションの手段が限られていた当時としては，電光石火の対応と言うべきであった。地力問題の解決は，19世紀の農学にとって焦眉の急務だったのだ。

　無機栄養説の確立によって，土の科学は，養分元素の絶対的含量に影響する地質に注目する分野と農業上の特性およびその変化に注目する農芸物理・化学の分野にまたがって進められることになり，19世紀後半のドイツを中心に一時代(地質学・農芸化学的土壌観の時代)を画することになった。この種の研究は，19世紀末までには莫大な地質学的・化学的データを蓄積したが，それらの情報を土壌被の生成・発達段階に即して正しく理解し，農学上の問題に適切に結びつけていくための方法論はいまだ確立したとは言えなかった。ロシアのV. V. ドクチャエフが1883年に発表した論文「ロシアのチェルノジョーム」が，その停滞を打ち破るきっかけになった。初期ドクチャエフ学派の土壌観(注1参照)は，言葉の壁のために，欧米に知られるのに相当な年月を要したが，それでも彼の論文から四半世紀後の

1909 年には第 1 回国際土壌学会の開催にこぎつけた。アメリカの土壌学者マーバットは後年(1936 年)、Joffe が著した *Pedology* の序文において、初期ドクチャエフ学派の土壌観について、「ロシアの研究者たちは、土壌学を地質学・農芸化学的土壌観によるカオスと混乱から解放し、自前の基準、見方、方法を持った独自の科学として確立した」と書いた。

19 世紀半ばのアメリカでは、ヒルガードら(注 27 参照)が先駆的な研究を行なっていたが、学会の主流からほど遠く、新大陸の土の研究者の多くが、ドイツ留学を目指した。アマーストの W. S. クラークもそのひとりで、1850-52 年にゲッチンゲンのゲオルギア・アウグスタ大学に留学し、「隕石中金属の化学構造に関する一研究」によって博士の学位を取得した(マキ、1978)。土木専攻のホイーラーを別にすれば、札幌農学校の創設に尽力したアマーストの農学者たちの科学者としてのルーツは地質学・農芸化学的土壌観にあった。そのしんがりを担ったブルックスは、実務に詳しい農学者として、開拓使の廃止(1882(明治 15)年)から北海道庁の発足(1886(明治 19)年)に至る激動の時代に、12 年間にわたって札幌農学校を支えた。クラーク博士はよき後継者を得たと言うべきだろう。

2. 農学講義における土

ここでは農学講義の第 2 章で注目した事項について簡単に解説する。なお、個々の表現上の問題や現在と異なる用語法の解説等は注にゆずる。

(1) 土の位置づけ：農学講義は土から始めて、土地改良と耕耘に説き進め、肥料、施肥論、農家経済など、作物栽培の環境条件を概説した後、作物栽培の各論に至るという実学的な構成になっている。ブルックスは、帰国後の大著 *Agriculture*(全 3 巻、1901 年)でも同じ構成を採用し、土の農業上の適性を判定し、耕耘、改良、施肥方針を見極め、その上に立って栽培作物や農法を選択するのが農学の第一歩だとの姿勢をつらぬいた。

ブルックスの土に対する考え方が、地質学・農芸化学的土壌観に基づくことは、冒頭の土の定義(注 1 参照)から明らかである。また、「地殻の構成物を土に変える地質学的な変化とそれに影響を与える物理・化学的原理」の説明に相当な紙数を費やしているのは、土の評価・分類や農業立地としての適性、改良および耕耘法などを考える基礎として、地質学的および物理化学的原理を理解することが不可欠だと考えていたからだろう。

(2) 風化・有機物集積と植物養分：ここでは、植物の養分は無機物で、土から供給される、養分は水に溶けた状態になって初めて有効になる、有機物は直接利用できないなどのことを明示しており、要点を押さえて無機栄養説の基本的な考え方を要領よく説明していると言える。ただし、植物による有機物の還元を別にすれば、土の形成における生物の役割についてはほとんど触れておらず、土壌侵食の防止についても注意を喚起していない。

(3) 土の分類：土の分類に堆積様式と母材を用いているのは、地質を重視する当時の考えを反映したものだが、堆積様式に風積成の範疇が欠如していた(注 19、21 参照)。初期の殖民地撰定調査では、これと類似の凡例が用いられ、火山灰に覆われた土地をどう評価・分類するかについて現場に意見の対立が生じた。この問題の本格的検討は、後の北海道農事試験場による特殊土壌調査に持ち越された。火山灰土壌は欧米の土の科学に共通する盲点だった。

(4) 土の物理性とその改良：土の物理性に関する研究のルーツは，イギリスのタル（注27参照）による耕耘の研究にまでさかのぼることができ，19世紀のシュブラー，シューマッハー，ウォルニーらの研究（注27, 28参照）に発展した。しかし，この時期には，農芸化学の華々しい進展に覆い隠されて，傍流の観をまぬがれなかった。ところが，本講義では，物理性とその改良についての論考が第2章の半分以上を占め，広い領域をカバーしているだけでなく，当時としては最先端の実験事実を豊富に引用しており，作物に対する土の適性や改良の要否判定に生かされている。これは本講義の著しい特徴のひとつで，農場長として開墾と土地改良を指導したブルックスならではの構成であり，物理性とその改良に注目して論旨を整理することによって，土壌観の地質・化学偏重による「カオスと混乱」を意識的に回避しようとしているようにも見える。

(5) 土の作物培地としての相対的適性：土の農業適性をどう評価するかはこの章の主な目的のひとつであり，初期の殖民地撰定調査に大きな影響を及ぼしたと思われるが，土性の如何にかかわらず，河成沖積土に高い評価を与える点で一貫している。これに対して，粘土や砂土は，いずれも農業上は almost valueless だとしている。この評価はわれわれには極端に思えるが，一等地から順に選べばよかった拓殖初期の特殊な評価法だったと言うべきだろう。

3. 土性調査とその後の土壌学

(1) 殖民地撰定調査と土性調査：土壌学史は1909年の第1回国際土壌学会開催を境にして2つの時代に大別できる。わが国では，フェスカ(1883-94(明治16-27)年滞日)による土性調査の推進が前期のハイライトだった。フェスカの土性調査はヨーロッパの伝統的手法をわが国の学者に伝えたが，新しい学理を伝えたとは必ずしも言えなかった。フェスカの土性調査は1907(明治40)年の青森県をもって終了したが，北海道については予察程度にしか実施されなかった(北海道農業試験場, 1967)。

その理由は明らかにされていないが，当時の北海道には，すでにある程度の土地情報の蓄積があり，フェスカ式の土性調査を一からやりなおす必要はなかったのではないかとも思われる。地質についてはライマンの日本蝦夷地質要略の図・同説明書が1876-77(明治9-10)年に完成していたし，1882(明治15)年の札幌農学校生活を描写した志賀重昂の日記(亀井・松木, 1998)にも見えるように，ブルックスの直弟子たち(札幌農学校1期生の内田瀞，田内捨六，柳本通義，同3期生の福原鉄之輔など)による殖民地撰定調査も1880年代初頭には本格化していた。1891(明治24)年に出版されたその最初の報文は全道の大原野をほぼカバーしていたし，第2, 3報文では，フェスカ法の凡例を導入して調査法を大改訂するとともに，宮部金吾，大島金太郎の協力を得て，精度も大幅に向上した(北海道庁, 1891, 1897)。

その後，1910(明治43)年にスタートした第1期拓殖事業計画では，殖民地撰定調査は北海道農事試験場の担当になり，1912-28(大正1-昭和3)年の特殊土壌(火山灰および泥炭)調査を経て同場の土性調査になった。すなわち，北海道の土性調査は殖民地撰定調査の考え方を継承・発展させたものであった。この調査は，その後ロシアの土壌断面形態調査法を採用するとともに，下層土の土性や乾湿を土性図に盛り込むなど実用性を重視したもので，筆者

が知るようになってからでも北海道モンロー主義と評されるほどの独自性を保っていた。この背景には，北海道農事試験場のモノグラフ『火山灰地と其農業』(1913(大正 2)年)，『泥炭地と其農業』(1920(大正 9)年)あるいは時任一彦(東北帝国大学農科大学農学科農芸物理学講座の初代教授)による『泥炭地改良及利用論』(1914(大正 3)年)，時任らによる『樺太島ポドゾル土壌の特性とその農業』(1929(昭和 4)年)などに示されるように，土の特性とその利用・改良に関する実用研究の蓄積とそれを農業に生かす地道な努力が重ねられていたことがあった。

(2) 土壌学の自立と展開：後期になると，国際土壌学会が発足し，わが国でも土壌学と題した著書の出版が相次いだ。鈴木重礼(1910(明治 43)年に東北帝国大学農科大学農芸化学科第 1 講座(土壌肥料学専攻)の教授に就任したが，在任 2 年たらずで急逝した)は 1917(大正 6)年に出版された『土壌生成論』(鈴木の死後遺稿を整理して出版された)において「ドクチャエフ学派の思想が受け入れられれば，土壌学は地質・鉱物学から離れて独立すべきもの」であるとして，ドクチャエフ学派の思想を手がかりにして土壌学が自立するであろうとの見通しを示した。その後継者三宅康次は，酸性土壌の研究を進めるとともに，サハリン島南部の土壌予察調査(1926-29(大正 15-昭和 4)年)で，ロシアの土壌調査・分類に接触する機会を持ち，北海道の酸性土壌はポドゾルに近い生成的特徴を示し，北海道北部は土壌地理学的にポドゾル地帯への移行帯にあたるとの認識を強めた。この問題意識がドクチャエフ学派の気候的土壌帯説を積極的に導入する素地になった。

注

1) 地質学・農芸化学的な土の定義。現在の土壌の定義はドクチャエフによる「土壌とは腐植によって多少とも色のついた，地表の鉱質および有機質層の集まりであり，これらの層は，常にそれに作用した生物因子，すなわち生きたもの，死んだものを含む生物(植物および動物)，母材，気候および地形の複合した作用の結果を具現する」に基づくもので，生物の役割が明確にされ，土壌自身が生成因子と動的平衡を保ちながら生成，発展，衰退する博物学的自然物であると理解されるようになった。この講義が行なわれた時点では，このような土壌観は成立していなかった。

2) organic structure という術語は誤解を招くおそれがあるかもしれない。土壌有機物の化学構造は現在でも完全には解明されていない。ここでの organic structure は生物遺体の物理的な組織や構造を意味するものであろう。

3) pulverization：細粒化。近年では物理的な崩壊 disintegration と化学的変質 decomposition を厳密に分けて用いる。この pulverization は細粒化とそれに伴う変質を一部含んでいるようだ。

4) 岩石はさまざまな鉱物結晶とガラス質(非晶質)の石基からなり，温度変化による膨張・収縮に違いがある。このため，日射などによる温度変化に伴って内部にひずみが生じ，剝げたり，ひびが入ったり，砕けたりする。温熱風化である。生成した隙間に水がしみこむと，その凍結・融解によって岩石内部に大きな応力が発生し，その崩壊が加速される。これは凍結風化と呼ばれる。

5) ローヌ川は南西アルプスの奥地に源を発し，レマン湖を経てリヨン付近でソーヌ川を合わせ，マルセイユの西で地中海に注ぐ。それゆえ，ここでの ocean は Mediterranean Sea である。ローヌ川は欧州有数の急流河川で，懸濁物質濃度が高い。

6) フィラデルフィアの緯度は北緯 40°である。ローレンタイド氷床はオハイオ川，ミシシッピ川の谷にかけて最も南に広がり，イリノイ氷期には北緯 36°40′まで達したとされる。

7) この講義では power という用語がしばしば現れるが，その物理・化学的内容は必ずしも統一されていない。この文の power としては，親和力 affinity が適切。しかし，今日のような化学親和力の概念はファント・ホッフによって初めて示された。1883 年のことである。このほか，現在の用語としては，ability, capacity, potential などが適切と思われるケースがある(注 30, 36, 39 参照)。

66   Chapter 2   Soil

8) 鉱物の結晶は珪酸イオンとアルミニウムその他の金属イオンからなるイオン結晶である。例えば，珪素イオンは酸素との親和性が強く，ほとんどすべてが4個の酸素イオンと結合した珪酸イオン(SiO_4^{-4})になっている。珪素のイオン半径が 0.26 Å であるのに対し，酸素イオンのそれは約 1.4 Å で，酸素イオンと結合することによって結晶中で占める容積は著しく大きくなる。

9) 風化 weathering は現在では「高温高圧で水と酸素に欠乏した条件下で生成した岩石が，低温・低圧で水と酸素が存在する地表の環境に適応した状態に変化する過程」として広く捉えられ，物理的，化学的，生物的作用が複合して地表で進行するものとされる。

10) quartz rock, feldspar rocks は鉱物としての石英，長石の意味と考えられるから rock は不必要。また，灰長石 lime feldspar＝anorthite は斜長石の Ca 端成分，同じく Na 端成分は曹長石 albite である。斜長石類では灰長石が最も不安定で風化を受けやすく，曹長石に近づくほど風化抵抗性が高くなる。カリウム長石 potassium feldspar には 4 種の多形があるが，最も普通なのは正長石 orthoclase である。正長石は長石類のなかで最も風化抵抗性が大きい。本文の potash feldspar がこれにあたる。

11) 土の物性概念が変化して，土壌は物理的に「不均一な多成分分散系」として捉えられるようになり，現在では，heterogenous, homogeneous の区別はあまり強調されなくなった。

12) 腐植と堆積腐植 leaf-mold, peat, swamp-muck は現在の用語体系では厳密に区別される。leaf-mold は，陸成(モル，モダー，ムルに細分)，peat は半水成，swamp-muck は水成堆積腐植に相当するかと思われる。また，腐植 humus は堆積腐植が土壌生物により分解され，下層の鉱質物と混合，変質してそれぞれの土壌の生成条件を反映する暗色の有機物に変化したもので，酸・アルカリへの溶解性によりフルボ酸，腐植酸およびヒューミンに 3 大別される。これらの概念が確立したのは 20 世紀前半のことである。なお，本講義では，混合者としての土壌生物の役割についての言及がない。19 世紀末までは土壌生物学は未発達で，それについての情報は少なかった。

13) フミン酸 ulmic acid，フミン ulmin ともに主として木質成分が嫌気分解して生じた黒色物質の主成分。元来，石炭の主成分として定義されたが，泥炭でも全重の 60% 以上を占めることがある。

14) アンモニア化成，硝酸化成はいずれも微生物が関与する反応で，そのメカニズムはこの講義の時点では，正確には理解されていなかった。原文が thoroughly oxidized or decomposed と無機的な表現になっているのはそのためだろう。土壌有機物の分解・変質には土壌生物，特に微生物が重要な働きをしているが，これについての知見が整理されるには，19 世紀末〜20 世紀初頭の土壌微生物学の体系化を待たなければならなかった。

15) plant food という用語がしばしば使われている。現在の術語としては nutrient が普通。この講義でも nutrient が使われているところがあるが，どう使い分けているのかは分からない。

16) 腐植酸やウルミン酸がそのままでは植物養分として有効でないことを指摘している。

17) 無機物の養分としての有効性について，「水に溶けた状態になって初めて有効になる」という無機栄養説の基本に触れている。

18) 肥料の 4 要素として，窒素，リン酸，カリおよび石灰が古くからあげられている。カルシウム(石灰＝Lime)が植物の生長にとって重要な要素であることは明らかだが，窒素，リン酸，カリをあげないで，石灰だけを強調する理由はよく分からない。文脈のつながりもよくない。なにか脱落があるのかもしれない。

19) この項のテーマは土の分類だが，現在の用語からすれば堆積様式の分類である。より一般的には，残積成，運積成(氷積成，水積成(河成，河海成，海成，湖成)，崩積成，風積成など)に分類される。ただし，この講義では風積成には触れていない。その後の殖民地撰定調査でも類似の凡例が用いられ，風積成火山灰からなる十勝・根釧地方などのいわゆる高丘地が広く殖民適地とされた。これについては拓殖関係者の間にも意見の相違が表面化し，これをどう評価・分類するかが大きな問題になった。これが，北海道農事試験場の特殊土壌調査で火山灰土が特殊土壌とされた遠因になっているかもしれない。なお，Alluvial はここでは水積成の意味である。

20) 必ずしも本文のようには言えない。土壌生成過程が極度に進んだ場合には，母岩とは性質が非常に異なる肥沃度の低い土になることがある。例えば，ポドゾル化が極度に進行した森林土壌は，表層部が塩基や粘土分を失って，強酸性の石英質砂になるとともに Bhs 層が腐植や三二酸化物で固結され，自然肥沃度がきわめて低い。熱帯，亜熱帯にも，土壌生成過程が極度に進んだ結果，母材とは非常に異なる肥沃度の低い土壌(赤色土，赤黄色土など)になった例が広く見られる。本文のような母岩重視の考え方は，当時は，世界の主流であったが，母岩の特性から土の肥沃度を推察できるという記述は誤りの可能性を含んでいる。

21) 沖積土は堆積様式による区分，clay〜sandy loam は手触り(粒径組成)による区分，calcareous と peat は母材の質による区分で，分類の概念に不統一が見られる(注 19 参照)。ただし，この項の表題

は Variety of Soils になっているので，厳密な分類を示そうとしたものではないかもしれない．なお，地質を重視した 19 世紀中頃のドイツの土壌分類では，特殊な母岩と土性による区分を組み合わせた分類法がしばしば見られた．

22) 粘土に対する評価が極端に低い．当時の技術レベルでの利用可能性の低さをはっきりさせるために，意識的に誇張した表現をしているかと思われる．同様に，砂土 sandy soil が「農業目的にほとんど無価値」だという断定も，現在の知識からすれば，やや極端（注 41 参照）である．

23) loam は rich soil を意味し，肥沃な土の代名詞であった（注 53 参照）．この講義の時点では，粒径およびそれによる土性区分についての合意は形成されていなかった．アッターベルグによる粒径区分の提案は 1902 年のことであるが，それが広く受け入れられるまでには四半世紀以上かかった．

24) 石灰質土壌 calcareous soil はカルシウムに富む母材から生成した乾燥地，半乾燥地の土壌，下層に炭酸カルシウムの集積層を持ち，酸を加えると二酸化炭素ガスが発生して盛んに発泡する．これは野外で石灰質土壌を判定する簡便法として広く用いられる．

25) tilth はここでは，作土あるいは耕土 top soil の意味で使われているが，現在は表土の耕しやすさ，播種床の細かさ，軟らかさ，砕けやすさなど，一般的にはコンシステンシーを示す用語として定着している．

26) ochery clay は酸化鉄の沈澱によって非常に硬くなった粘土層．

27) 土の物理性に関する研究のルーツは J. タルの『馬耙農業』(1733 年) までさかのぼることができるが，実験的研究は 1810 年代初頭にスタートし，デービー，シュブラーを経て，1850 年代から 19 世紀の末にかけてのシューマッハー，ウォルニーらの研究に発展した．アメリカでも 1870 年代に公表されたヒルガードの研究をはじめ，いくつかの研究が発表されたが，多くはブルックス来日以後に公表されたものである．なお，ここでは，土壌構造 soil structure という用語は使われていないが，物理性が土粒子の形態と配列にかかわる性質であるとして，現在の土壌構造概念が明確に述べられており，それが粒子の相互関係，空気および水との関係，熱との関係にもかかわる性質であることを指摘している（注 28 参照）．

28) シュブラーは世界で最初に土壌物理の実験的研究を行ない，1833 年に出版した *Grundzätze der Agricultur Chemie II* において土の生産性にかかわる物理的性質について幅広い研究成果を公表した．彼の研究は，①土の比重，乾湿土の容積重，②重量および容積ベースでの土の水分率，③乾土と湿土の粘性と剛性，④空気中での乾燥に対する容量の違い，⑤乾燥による容積減少，⑥大気からの水分の吸収，⑦大気からの酸素の吸収，⑧土の熱容量，⑨太陽光からの熱吸収容量，⑩潤熱，⑪電気の極性と導電性などに及んだ．

シュブラーは種々の砂・粘土，炭酸塩および石膏の粉末，腐植，生産性の高い土など 13 種の物質を供試してこれらの物理的性質を測定した結果，物理性と粒径組成の間に密接な関係があることを明らかにし，砂：粘土のいろいろな割合の混合物を作って土壌構造，水分，コンシステンシー，温度などの物理性をさらに比較研究した．彼の研究では，気相容量は直接測定されなかったが，水保持容量を体積ベースで表し，容積重を測定することによって孔隙量を数値化することに成功した．また，水蒸気吸収，潤熱，乾燥速度，乾燥に伴う容積変化などの研究は土粒子表面の特性を明らかにすることの重要性を示した．さらに，土の機械的な処理に際してのコンシステンシーの重要性は粘性，剛性，付着性の研究によって強調された．耕耘用具に使われる木，鉄に対する付着性の測定はこの意味で特筆すべき仕事であった．一方，直射日光の下で土の温度を比較した研究では，湿った淡色の土はよく排水された暗色の土に比べて温まりにくいことを見出し，土壌面の色と水分率が土の温度にかかわる重要な因子であることを明らかにした．

29) 分割 division に伴って比表面積が増加し，吸着および毛管力による保持水量が増え，溶存物質濃度も高まる．ただし，比表面積 specific surface area という用語はこの講義では使用されていない．

30) この項の「大気から水蒸気を吸収 absorption する power」の absorption は，単に，ある物質が他の物質の系内に取り込まれる現象で，保持される物質が 2 つの相の界面で両相の内部とは異なった濃度で平衡に達する吸着 adsorption とは別ものだが，この講義では，全体を通じて adsorption という用語は用いられていない．吸着式はギブスが 1876 年に液々界面について熱力学的に誘導したのが始まりで，表面状態が複雑な固体表面への吸着が科学的研究の対象になったのは 20 世紀に入ってからであった．

微細な土粒子表面への吸着による吸湿性 hygroscopicity，それによって保持された吸湿水 hygroscopic water とその最大量＝吸湿係数 hygroscopic coefficient などの物理的概念は，1901-02 年のミッチェルリッヒ／ロードワルドの研究によって確立された．特に，ロードワルドの研究はクラウジウスが 1865 年に発表した熱力学式を応用して，土の水吸着と潤熱を統一的に説明したもので，これによってミッチェルリッヒが「もはや潤熱を発生しない水分状態」として吸湿係数を定義したことの意味が物

理的に裏づけられた。この時期，土粒子の水蒸気吸収の研究は，熱力学の最先端の研究に直結していた。

31) 吸湿水は固相表面に強く吸着されているために植物はこれを吸収できない。植物の水吸収に関連して土壌水を化合水，吸湿水，毛管水，重力水に分類し，その間に植物の水吸収に関連した特異点があること，すなわち水分恒数の概念が確立されたのは，ブリッグスによる「土壌水の力学」(注 39 参照)とそれに続く 20 世紀初頭の研究による。

32) 表中の用語は，現在では Lime sand＝calcareous sand, carbonate of lime＝calcium carbonate, carbonte of magnesia＝magnesium carbonate である。

33) 粘土質の土は吸湿性が強いだけでなく，凝集力が大きいために，畑土壌としては問題がある。北海道の洪積台地には重粘土が広く分布するが，初期の殖民地撰定調査では，この種の土壌は農業適地とは見なされなかった。これに砂を混合することは吸湿性だけでなく，凝集性を減らして tilth を改善できるため，その改良法のひとつとして砂客土が国の事業として行なわれた。

34) 駆動力とそれによる現象が同じ意味に使われている。force を phenomenon にするか，adhesion を adhesion tention にするべきである。

35) 表の Amount of gas absorbed from the atmosphere 欄の数値の違いの由来が分からない。atmosphere の状態についてもう少し説明がいる？

36) この表題は，solid matter を ions, fixing を ion exchange or fixation, power を capacity に読みかえれば，現代の読者にも理解できよう。ファラデーがギリシャ語の「行くもの」にちなんでイオンを命名したのは 1833 年のことであり，トンプソンとウエイが土の養分保持機能を最初に報告し，その農業上の重要性を指摘したのは 1850 年のことであるが，その時点では両者とも absorptive power of soils あるいは power of soils to absorb manure の表題を掲げており，表立っては「イオン」も「交換」も使われなかった。19 世紀後半には，この現象は吸収と考えられ，この原因になる力 power が土壌粒子表面の静電力であることも，正しくは理解されていなかった。

37) 溶脱を防止する機構として「固定」と「保持」の重要性を指摘しているが，内容の違いには触れていない。現在では，イオン交換現象(注 36 参照)による一時的な吸着を保持，オキシアニオンの土粒子表面への配位結合またはそれらと鉄，アルミニウムなどの結合による難溶性化合物の形成などを固定と呼んでいる。

38) 有機物の多量投与や土性改良対策としての粘土客土・ベントナイト投与などは，その後も砂土の改良にしばしば用いられた。

39) 毛管力 capillary power は現在も用いられるが，ここは毛管ポテンシャル capillary potential の方がよい。液体中に毛細管を立てた時に管内の液面が管外よりも上がりまたは下がる現象が毛管現象 capillarity である。液体が管を濡らす時に管内の液面が管外のそれより高くなり，毛細管の半径を r とすれば，管内液面の上昇高さ h は，h＝$2\gamma\cos\theta/r\rho g$($\gamma$＝液体の表面張力，$\theta$＝液体と毛細管の接触角，$\rho$＝液体の密度，g＝重力の加速度)で表されるので，毛管力は水の表面張力と水と毛細管の付着張力に由来する。この機構と土壌水の移動および平衡における重要性は 1870 年代にはある程度認識されていたが，土壌を毛細管の束と見なす毛管モデルを明示的に導入したのはブリッグスが 1897 年に発表した「土壌水の力学」が最初とされる。毛管ポテンシャルという表現は今も慣用されるが，一般には matrix potential の方がよい。粘土質土壌では，微細な粘土粒子間の静電的斥力により粒子同士がある距離を保とうとする傾向が強いため，これに由来する保水能を含めた概念である。

40) 土中の水の移動と毛管ポテンシャルの関係，水の移動と溶存物質の移動・集積の関係に関する基本概念に言及している。

41) 土壌によって保持される水の量は土の孔隙分布，毛管上昇速度などによって異なる。砂土は表層の毛管連絡が断たれやすいために，乾燥地では蒸発損失が少なくなり，土層が深ければある程度の水量を保持でき，かつ地表に塩類集積を起こしにくいなどの利点もある。砂土の「不毛性」という断定は現在の技術レベルではやや極端(注 22 参照)。

42) 付着性 adhesiveness は，現在では，農具などへの付着しやすさを意味する術語になっているが，ここでは，土粒子相互の付着しやすさを示すのに用いている。現在の用語では凝集性 cohesiveness に相当する。なお，...are called heavy soil, though they may be lighter than those which are called light は，初めの lighter が容積重が小さい意味の軽さを示し，後の light が凝集力が小さい意味の軽さを示すと考えれば，重さによる軽重と凝集性による軽重の使い分けがあることを示したものと理解できよう。

43) 土壌温度の記述が多いので soil temperature を表題に加えて考えておく方がよい。

44) animals のラテン語の原義は広く Living being を指したようだが，一般には，植物，微生物以外の生き物とされる。もっと広く「生き物」全体と受け取っておくのがよさそうだ。太陽光がなければ最

解　題　69

45) Texture は，今の土壌学で多用される土性の意味ではなく，単に「組織」の意味。
46) 表の実験はいずれも，土の含水率，色以外の土壌面の状態，大気の乾湿など，必要な条件が示されていない。
47) 前表のデータと合わせて砂マルチあるいは砂客土の有用性を指摘したものと言える。
48) ここでは，厳密には，比熱，熱容量，蒸発量などについて触れる必要がある。
49) 札幌は北緯43°03′に位置し，春〜秋分間の日南中時の平均太陽高度から計算すると南に30.7°傾いた面が太陽光に直交し，単位面積当たり日射量が最大になる。ただし，傾斜面が受ける日射量は緯度だけでなく，季節，斜面の向き，斜面の傾斜度，時刻によって著しく異なるので，現在では，日変化，季節変化を考慮して施設や斜面圃場が設計されている。
50) 水生植物は aquatic plant，陸生植物は terrestrial plant がよい。最後の部分は「着生植物(air plant または epiphyte)は水中でも陸上でも生長できない」の意味と解される。
51) beachgrass or *Calamogrostis Arenaria*：ノガリヤス属の海浜性草本(和名不詳)，locust-trees belonging to the genus Robinia：ハリエンジュ属のニセアカシア，pine-trees blonging to the genus Pinus：マツ属のマツ，peanuts(*Arachis Hypogaea*)：ナンキンマメ，rye(*Secale Cereale*)：ライムギ。
52) 「植物養分を発現させる自然の影響」は漠然としすぎる表現である。その内容として，①鉱物風化の促進，②有機物の分解と土壌生物活性，③窒素の無機化・有機化，④その他の有機態養分の無機化，⑤酸素の供給，水の供給(通気・透水)，⑥土壌温度と根の生活環境などが考えられる。この講義が行なわれた時点では，土壌生物，特に微生物に関する理解は十分ではなく，②③④の機構については情報がなかったと思われる。漠然とした表現をとらざるを得なかったのであろう。
53) 項のタイトルに Loam と Peat が同格で入っており，本文には，loam についての特別なコメントがないことに注意。loam は古くは，a rich soil composed of clay, sand and some organic matter として，また一般用語としては rich, dark soil を示す言葉として用いられてきた(注23参照)。なお，泥炭は構成植物や鉱質土の混じり具合にもよるが，細かい木質泥炭などは野菜栽培に適する。

参考文献
1．北海道庁第二部拓殖課，北海道殖民地撰定報文(1891)，同第二報文(1897)，同第三報文(1897)；復刻版(1986)，北海道出版企画センター
2．北海道農業試験場編(1967)，北海道農業技術研究史，同試験場
3．亀井秀雄・松木　博編著(1998)，朝天虹ヲ吐ク──志賀重昂『在札幌農學校第貳年期中日記』，北海道大学図書刊行会
4．J. M. マキ著・高久真一訳(1978)，W. S. クラーク──その栄光と挫折，北海道大学図書刊行会
5．佐藤昌介・稲田昌植(1935)，世界農業史論；復刻版明治大正農政経済名著集20(1976)，農山漁村文化協会

［佐久間敏雄］

Chapter 3 Farm Drainage and Irrigation

3-1. Effects of Drainage

All land, that it may be cultivated, should be in such a condition that it can at pleasure be made so dry that it may be worked upon conveniently; for most farm crops, it should be in this condition throughout the year. Rice and cranberries are, however, exceptions to this rule. The reason for the necessity of having the soil dry is that our valuable agricultural plants will thrive only when growing upon such lands; hence, all land must be drained either naturally or, in case there is no adequate natural drainage, artificially. The reason for this is that the annual rainfall is greater than can be dissipated by vegetable growth and evaporation. In Massachusetts, the annual rainfall is about 40 inches and the evaporation from the surface of the land is only about 20 inches; hence, if there is no natural drainage, there will remain about 20 inches of water per year which must be carried off in some manner in [p. 70] order that the soil may be fitted for the production of plants. The amount of water which must be carried off by artificial drainage will vary according to the climate of the country, the nature of the underlying soil, and the slope of the land.

The climate of this portion of Sapporo is so moist that the necessity for drainage of some sort, either natural or artificial, is very great; but most of the soil in this locality seems to be underlaid by a diluvial formation of very variable materials. In many cases, however, the subsoil is of loose sand or gravel, and hence the necessity for drainage is not so great as it would otherwise be, but in many places, the slope of the land is such that drainage must be resorted to in order that the water may be carried off with sufficient rapidity.

(1) Drainage deepens the soil. It renders it, to a greater depth, a fit medium for the home of the roots of plants. Since the water line is lowered by drainage [p. 71] as the line AB is to A'B' in Fig. 1, it enables plants to gather their food from a greater amount of soil as M does more than N in Fig. 1, and thus their growth will be more luxurious and the soil itself will not be so quickly exhausted. If the soil is undrained either naturally or artificially at no great depth as is K in Fig. 1 below the surface, and in extreme cases, on the surface, the water will be found stagnant, and aquatic plants can be grown upon the soil.

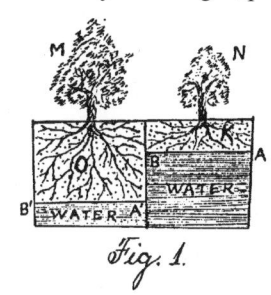

(2) Drainage lengthens the season both for work and lant growth, and in this latitude (in Hokkaido) this is very important, for our seasons are rather short for the perfection of many crops which are desirable to raise here. Drainage makes a field dry enough to be worked upon much earlier in the spring than an un-drained one. It also remains in this condition later into the autumn. It is highly important that the land should become dry early in spring, because even then we are very

much [p. 72] hurried in planting our seed. If the land remains wet for a long time, we are so much hurried in our planting operations as to be obliged to do the work imperfectly in order to finish it in season, so that the plants may have time to ripen before the coming of autumnal frost. Drainage lengthens the season for plant growth by making the soil warm enough and drying it up to promote cultivation earlier in the spring, and by keeping in this condition later into the autumn.

(3) Drainage promotes aeration of the land. This is important, because some of the elements in the atmosphere, by their chemical action on both the organic and inorganic constituents of the soil, render them available for plant food. They do this by changing them into forms which are soluble in water.

(4) Drainage renders the pulverization of the soil much easier. In fact, a soil which is wet cannot be ulverized, however much labor may be bestowed upon it. ⟨See p. 258⟩ [p. 73] Plowing land, when wet, only serves to throw the soil up into compact masses which, when they become somewhat dried, are hard and rocklike in their nature.

(5) Drainage promotes the germination of seeds planted in the soil. Seeds must have, in order that they may germinate, oxygen, warmth, and moisture. Now, we have seen that drainage makes the soil warmer, and also that it places them into such a condition that the air can freely penetrate it. It is evident that, so far as the warmth and the presence of oxygen are concerned, a drained soil will be more favorable for the germination of seeds than an undrained one. You may think that perhaps drainage will make the soil so dry as to prevent germination, but this is not true. It is impossible, by drainage, to remove so much water as so to render the soil too dry, provided it be in a proper physical condition. The reason for this is that [p. 74] the water which the soil absorbs by virtue of its hygroscopic power cannot be extracted by any ordinary means. Let H, I, K, *etc.* in Fig. 2 represent particles of earth; the small dots inside them, their pores; and the blank spaces, the interstices. Suppose the water was at first at the line AB, and then, by drainage, that it is lowered to A'B'; this soil will not be too dry because, even by drainage, the hydroscopic water or that which is contained in the pores cannot be drawn out.

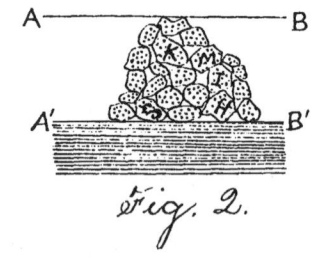

(6) Drainage prevents "freezing out". By freezing out, we mean the throwing out of the roots of plants from the soil, their freezing and the consequent death of the plants. Plants can thus be frozen out only upon lands which contain a great deal of water, since it is by the freezing and expansion of water that their roots are drawn out of the ground. It is, therefore, very evident that drainage, since it [p. 75] removes suerfluous water, will prevent the freezing out of plants.

(7) Drainage prevents "surface washing". The surface washing mean the washing away of the fine particles of the soil and the soluble elements of plant food by flowing water. If a soil is saturated with moisture, when rain falls upon the surface, it will run off over the surface provided there is any considerable slope, since it cannot sink into the ground because the ground is already full of water. If such a soil is well drained, water falling upon it will soak down into the ground and thus pass out of the way without doing injury, but in case if it flows over the

surface, it takes with it the finer portions and carries off all the soluble elements of plant food. If the soil is in proper physical condition, though the water sinking down through it will dissolve the soluble elements of plant food, yet these elements will be fixed and retained by the soil. [p. 76]

(8) Weeds can be much more easily destroyed upon drained land than upon undrained land. If the land is undrained, the surface will usually be quite moist and you may pull up the weeds growing upon it, there roots will still be able to get sufficient moisture to keep the plant alive. In many cases, they will even penetrate the soil and grow with as much vigor as before being pulled up.

(9) Drainage promotes the absorption of fertilizing substances from the air. ⟨See p. 286⟩ It does this partly because it places the soil into better physical condition for the absorption and retention of such substances, and partly because it places the soil into such a condition that the air can more freely penetrate it. ⟨See p. 282⟩ As you have seen in the condition of the soil, the air often contains substances which can be used as plant food, such as ammonia, nitric acid, etcetera; therefore, it is very important that the soil can absorb into such a condition that it can absorb and retain these elements freely. [p. 77]

(10) Drainage supplies air to the plants. It has been proved by experiments that plants will not thrive unless their roots have access to oxygen; therefore, since oxygen is essential, it is important that the soil be placed in such a condition that air can reach the roots. ⟨See p. 264⟩

(11) Drainage warms the soil, often by as much as 15°F, and in this latitude it is important, for most crops, to make the soil as warm as possible. Therefore, if by drainage we can raise the temperature a number of degrees, it will be wise to do so.

(12) The quality of crops is much improved by drainage. Plants growing upon undrained soil are usually coarse and rank, and contain a great deal of water. The farmer calls such crops "sour". Hay grown upon wet land is not nearly as nutritious as that grown upon dry land, and the same is true of most farm crops.

(13) We may draw heavy loads [p. 78] without injury on drained lands. This consideration is especially important with respect to permanent grass land. It is also, of course, very convenient to have all the lands upon which you work sufficiently hard that you may drive over them with loaded teams easily.

(14) Drainage prevents drought. It does this because it enables us to bring the land into such a physical condition that it will have the capacity to absorb a large amount of moisture from the atmosphere. ⟨See p. 279⟩ It also increases it capillary power, thus making it possible for moisture to rise from below whenever the surface becomes somewhat dry.

3-2. Open Drains and Ditches

The most obvious method of getting rid of superfluous water is by digging an open channel through which the water may flow away, and this was the method first adopted in practice. In most cases, however, it is the poorest style of drainage which can be used; such drains, under some circumstances, [p. 79] such as for catch waters and for outlets for systems of under-drainage, are useful. A catch water, as its name indicates, is something to catch water. At the foot of slopes or wherever water is liable to flow over the surface to places where we do not wish

74 *Chapter 3 Farm Drainage and Irrigation*

to have it, open ditches are very useful in collecting and retaining such water. Let A in Fig. 3 represent a hill, from the top of which water flows downward to the farm F. In this case, this farm will be good for nothing; hence, to prevent the coming of water, an open ditch D should be constructed. Open ditches are also frequently used as outlets to systems of under-drainage. The principal objections to the use of open ditches as a means of drainage are as follows:

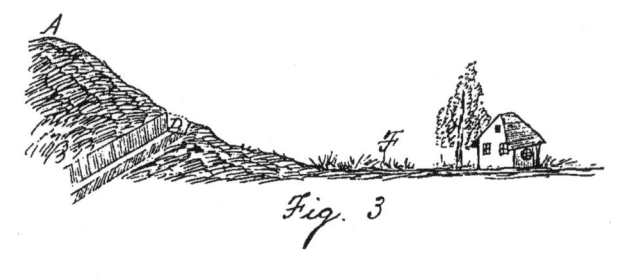

(1) They are expensive in construction, [p. 80] and it requires a large amount of money or labor to keep them in repair after they have been constructed. It costs more to make open ditches for the drainage of land than to construct a system of under-drainage, because they must be made much larger, and the principal expense in any system of drainage is usually the cost of excavating the ditch. In order that the size of an open ditch may be made as permanent as possible, it is necessary to give them a considerable slope, the amount being varied according to the nature of the land. If the soil is stiff heavy clay, the ditch will stay in place with much less slope than one which is sandy in its nature, but in general practice a slope of from 30° to 40° is necessary. This great degree of slope is required because, by means of running water and the tramping of horses and cattle near the banks of the ditch, and from many other causes which may be mentioned, there is a constant tendency for the earth along the banks of the ditch to fall into it. To prevent [p. 81] this, we must excavate much more earth in order to have a ditch of a certain width at the bottom, if the ditch is to remain open, than would be necessary if it were to be filled again, as it would be in any system of under-drainage. There is still another item of drainage, which is this that if the ditch is to remain open, the earth which has been thrown out of it must either be carried away or spread over the surrounding land; but this requires much more labor than would be necessary in under-drainage, for in the latter case, the earth which is taken out is simply thrown again into the hole.

(2) Although you make your open ditches as permanent as possible by giving their banks a great slope, yet they are constantly vulnerable to obstructions of various sorts. Water flowing into ditches over the surface of the land will carry with it considerable amount of earth and other material which may be found upon the land that can be moved by flowing water. Water grasses, rushes, sedges, [p. 82] and various aquatic plants will grow in open ditches, and by their decay through successive generations, considerable dead organic matter will accumulate in the ditches, thus obstructing the flow of water.

(3) Open ditches obstruct good husbandry. The intelligent farmer uses implements and machines drawn by horses or oxen, such as plows, harrows, cultivators, mowing machines, horse rakes, *etc.* Now, if the field in which he wishes to use these implements is divided by open ditches, he cannot use these implements to nearly as good advantage as he could if it were not so divided.

(4) Open ditches occupy too much land. In any system of under-drainage, no land at all is rendered unfit for cultivation, but with open drains a considerable land

is occupied by ditches, to the exclusion of everything else.

(5) On land drained by open ditches, manure washes off and is lost, but [p. 83] no such loss can occur in any system of under-drainage if the land is in proper physical condition, because in this case the water, before it can escape, must pass through the earth above the drains. On the other hand, if open ditches have been constructed, water may enter them by flowing directly over the land, and it will of course carry with it anything which it has brought into solution.

(6) Covered drains of the same depth will drain a greater breadth of wet land than open drains. In open ditches, the sides and bottom become obstructed by growing plants, and are also liable in seasons of drought to become sun-baked, hard, and more or less impervious to water.

3-3. Different Methods of Making Underdrains

1) Brush Drains

The "brush" mean the smaller [p. 84] limbs of trees with all the small twigs attached. It is possible to make a very good drain by the use of this material, although it is the poorest and least permanent of all systems of under-drainage. These drains will last the longest in those soils which are quite impervious to air and which will remain moist throughout the entire year. In a heavy, clayey soil such drains have been known to last about 15 years. The manner of construction is as follows:

(1) Excavate the ditch to the desired depth and width, taking care to have the slope of the bottom uniform throughout the entire length of the ditch.

(2) Having excavated the ditch, place the brush which you propose to use in the bottom, laying them in shingle fashion with the larger ends upstream. Enough brush should be placed in the bottom of the ditch to make a layer one foot in thickness and they should be placed as compactly as possible, [p. 85] because when the earth is thrown back into the ditch, the brush will be pressed down by its weight.

(3) After having placed the brush in position, it should be covered with two or three inches of coarse hay or straw, wood shavings, or spent tan bark to prevent fine earth from being carried into the brush.

(4) Fill the remainder of the ditch with earth; in this operation, the earth should be thrown in carefully and trod down as firm as possible. While this system of drainage is not as perfect as many others, it is very cheap due to the material used in its construction; under the circumstances, it may be advisable to drain land in this manner. I would not, however, advise you to use these drains in land which is very sandy in its nature, since the sand would very soon fill all the spaces, and the drain becomes useless. [p. 86] In heavy clay, these drains may be quite permanent and serve a good purpose.

2) Wedge and Shoulder Drains

These drains consist simply of an underground channel through the earth. The sides of this channel are entirely unsupported. Figure 4 represents a wedge

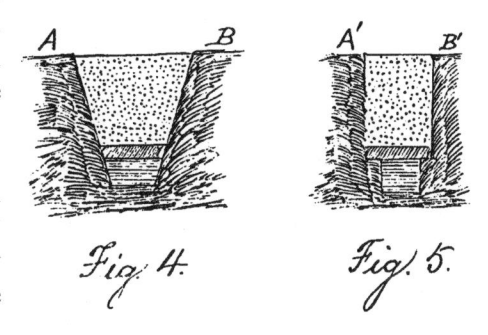

Fig. 4.   *Fig. 5.*

drain and Figure 5 represents a shoulder drain. These systems of drainage can be used only in heavy. clayey land, and only when the land is permanently in grass. These cannot be used in sandy land, because the earth which is sandy in its nature, will fall into the channel. These cannot be used on cultivated land because the surface of such land is usually soft, and animals traveling over it will press the earth down into the underground passage. [p. 87] Thus, both in sandy land and in land devoted to the culture of hoed crops, these kinds of drains would very likely be filled up, and hence they should not be used in such cases. The shoulder drain differs from the wedge drain only in the shape given to the bottom of the ditch. The wedge drain is so called from the shape of the ditch excavated resembling that of a wedge, while the name shoulder drain is applied to a ditch which has two "shoulders". The method of constructing a wedge drain is as follows:

(1) The sod is carefully cut and thrown to one side, and the earth is then thrown out on the other side of the ditch. The side walls of the ditch should slope quite rapidly down from top to bottom.

(2) The sod is then cut into pieces of such size that, when placed in the ditch, they will extend nearly to the bottom. [p. 88] You may make them larger or smaller to give a channel of any desired dimensions. The grassy side should always face downward in placing the sods in the ditch.

(3) We should be careful to make it the sods fit the sides of the ditch closely, and to have the ends of the sods touch each other at all points.

(4) Having thus placed the sods in position, the ditch is then filled with the earth which was thrown out of it. It is important that this earth be packed down firmly in order that it does not settle much after the drain is constructed; but in this case as in all the other the ditch is then filled to a little above its original level, in order that when earth sinks down, then may be no hollow.

3) Plug Drains

A plug drain consists, like the wedge drain and shoulder drain, of an underground channel, the sides of which are entirely unsupported. Like the wedge and shoulder drains, the plug drain can be used only in heavy, clayey lands. [p. 89] Figure 6 shows a rough drawing of a plug drain. The method of construction is as follows:

(1) Excavate the ditch to the required depth, making the bottom of such a shape and size that the plug which you propose to use will exactly fit and fill it. Four or six pieces of wood, each about one foot in length and of any size and shape desired, are joined together by means of little pieces of iron in such a manner as to make a slightly flexible chain (Fig. 7). An oval-shaped channel (Fig. 6) is the best.

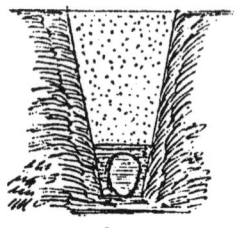

(2) Having prepared this implement, place it in the bottom of the ditch and cover all, except a short piece at one end, with earth which you have [p. 90] previously thrown out of the ditch. This earth should be tamped down very solidly.

(3) After having placed about one foot of earth above your plugs, they

may be drawn forward in the ditch by the use of strong links with the lever attached at the end. In drawing the plugs forward, they should not be carried far enough to leave them entirely uncovered, but the hind end should be left covered in order to insure continuity throughout the entire length of the channel. In order to make the plugs slide more readily, that is to lessen friction, they may be wetted with water before being covered. This will make the amount of friction quite small, and they can be drawn forward very easily.

(4) After the plugs have been drawn forward, the remainder of the ditch should be filled, enough earth being used to bring it somewhat above the surrounding level, as directed in the case for other drains. [p. 91] Since in this kind of drainage as in the wedge drain and shoulder drain systems, there is nothing to support the walls of the channel, these drains, even in clayey soil, are extremely liable to become obstructed; therefore. I can recommend their use except under conditions where it is impossible to drain land in any more permanent manner. The cost of excavation, which in all systems of drainage is usually the largest item of expense, is as great in the last three systems described as it would be if some material were to be used for the formation of the underground channel; therefore it will be wise, after having incurred the expense of digging ditches, to use some material which will make a permanent channel.

4) Mole Drains

Mole drains consist, like those last described, of a simple under-ground channel [p. 92] which is made by the use of an implement made for the purpose and is called a "mole plow" (Fig. 8). The beam AB is a rectangular piece of wood 4 to 5 feet in length, and 4×5 inches square. The standard FG is a flat plate of iron which should be about 3½ feet in length, 4 to 5 inches wide, and about 5/8 of an inch in thickness. The front edge of the plate of iron standard, which is called a coulter, should be sharp; it is past to the beam in such a manner that it can be raised or lowered at pleasure. To the bottom of the coulter is fastened a piece of iron which may be called a "mole" DL; this may be of any size or shape which it is desired that the underground channel shall have. It should be about 18 inches in length, and the folded end which should be brought to a point in order that it may penetrate the ground readily. C represents the handles, which should be like those of an ordinary plow. [p. 93] The underside of the beam AB should be covered with sheet iron, in order that it does not wear out too rapidly. To give additional strength, a thin plate of iron, the edge of which should be sharp, is sometimes attached to the point of the mole. At the forward end of the beam, a clevis should be fastened, by which the implement may be drawn. The use of this implement is as follows:

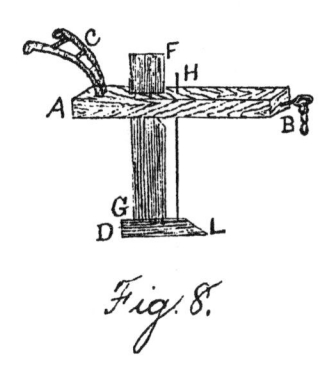

(1) The standard is moved up or down in order to bring the mole to such a distance from the bottom of the beam as the depth which you wish your drain so have. The implement is then set in the ground in such a manner that the bottom of the beam rests upon the surface.

(2) The implement is then drawn through the beam in such places and directions

78   Chapter 3   Farm Drainage and Irrigation

as you wish to have your drain. A man holds and guides it in the same manner as a [p. 94] plow is held and guided. It may be drawn directly by a strong team of horses or oxen, or by means of a windlass. As the implement moves through the soil, the mole leaves behind it a channel of the same size and shape as itself. The passage cut by the coulter, being exceedingly narrow, soon closes, but the one cut by the mole, being larger, remains. This system of drainage can be used only in heavy, clayey lands when they are permanently in grass, the reasons being the same as those given in the cases of the wedge and shoulder drains. This method of draining land is one of the cheapest ever devised, since the labor is performed almost entirely by horse or ox power; thus, although it is not a very permanent system of drainage method, it is often a good policy to resorted to under the circumstances specified above.

5) Pole Drains

The pole drain [p. 95] consists of an underground channel formed by the use of long, straight poles. The poles used should be 4 to 6 inches in diameter, and may be of any length convenient. The straighter and more regular and uniform in size the poles used, the better drain you can make. Figure 9 represents a pole drain. The method of construction is as follows:

(1) Excavate the ditch to the required depth and size, observing the same precautions as in all other systems of drainage; the width of the bottom of the ditch should be $2\frac{1}{2}$ times the diameter of the poles which you propose to use. In throwing out the earth (if the land is in grass), it will be wise to throw the sod to one side and the remainder of the earth to the other side, since the sod must be placed in the ditch first.
(2) Place two parallel rows of poles in the bottom of the ditch, placing [p. 96] each row against the outer edge.
(3) Place upon them a third pole, which will be above and between them. If the bottom of the ditch is very soft, it will be necessary to place boards upon it, since, if this were not done, the poles, after being covered with earth, will be likely to be pressed downward into the mud, thus obliterating the channel.

(4) After placing the poles in position, cover them carefully with inverted sods, straw, wood shavings, or tan-bark. Sods are perhaps the cheapest and most permanent of any of the materials above described. The object of covering with these materials is to prevent fine earth or sand from being carried into the channel by currents of water.

(5) Fill the ditch with earth in the same manner as in the case of any other systems of drainage. Such drains will last the longest in land which is impervious [p. 97] to air, or in places where the poles are always wet. This system of drainage would be the most permanent in clayey soils containing a great amount of organic matter, which will be moist throughout the year. In sandy soils, there is another reason that the poles will decay quickly, and why these drains will not be permanent—because the sand will be extremely likely to be carried by currents of water into the channel, thus filling it up and rendering the drain useless. In new countries where wood is cheap and tiles are expensive, especially if such countries are destitute of stones, it will often be wise to drain land in this manner.

6) Box Drains

The box drain consists of a continuous underground channel formed of successive boxes, which may be of any size or shape desired. The triangular shape (Fig. 10) is [p. 98] better than the square or rectangular one (Fig. 11). This is because in the triangular-shaped channel, a given amount of water will have greater depth and force than in the square- or rectangular-shaped one; hence, the channel will be less likely to be obstructed by fine earth or sand which may enter. These boxes, whatever their shape may be, should be made of wood which is not likely to decompose quickly—oak, ash, or elm is better than softer wood such as pine, *etc.*

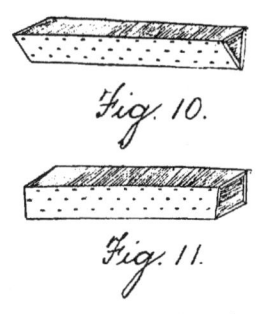

The boards used in the construction of these boxes should be from 1 to 1½ inches in thickness, according to the cost considerations. The thicker they are, the longer they last, all other things being equal. The width of the boards should be varied according to the size of the channel designed. If your boxes are to be triangular in shape, a width which will give an interior [p. 99] channel of 3 to 6 inches on each side will be sufficient. If the quantity of water to be carried away is not very large, a width of a few inches will be sufficient; in most cases, a channel the size of which are 4 inches wide will be amply large enough. In each side of these boxes, holes about 3/8 of an inch in diameter should be made at intervals of 3 inches. The number of holes necessary will depend upon the size of the boxes, a larger one requiring more than a smaller one. If you make more than one row of holes, they should be so placed that each hole of one row is opposite a space without a hole on the other side. If a square box is used, the same rules with regard to the thickness of boards should be remembered as in the case of triangular boxes. The boards should be of such a width as to give provide a channel of from 3 to 6 inches square, and holes should be made in the bottom and sides. The object here, as in the case of triangular boxes, should be to place the holes so that [p. 100] the water may have the best opportunity to enter the boxes.

After excavation of the ditch, the boxes are placed in the bottom with the apex of the right angle downward. If the soil in which this drain is to be constructed is clayey in its nature, it may be thrown directly upon boxes and there will be no danger of its being washed into them and filling up the channel. If, however, it is loose or sandy in its nature, there will be a danger of the drains becoming thus filled. In such soils, I would place around the boxes a small quantity of gravel, which would tend to prevent the sand from being carried into them. Such material as wood shavings, tan-bark, straw or hay, and even sods may be used, as recommended in the case of other drains. After having thus protected your boxes with one of these materials, the remainder of the ditch should be filled as usual. Some labor may sometimes be saved, if you propose to use [p. 101] triangular boxes, by excavating the bottom of the ditch in such a shape that the box will just fit it. In this system of drainage, care should be taken that the to place ends of channel formed by successive boxes shall be exactly continuous.

7) Stone Drains

In localities where stones of suitable shape and size can be obtained, it is possible to make very good drains by their use. There are many different kinds of

80 Chapter 3 Farm Drainage and Irrigation

stone drains, a few of which I will describe. The first is simply made of loose stones of variable sizes and shapes (Fig. 12) which, after the ditch is excavated, are thrown into it without any regard to the formation of a regular channel. These stones are then covered with sods, [p. 102] wood shavings, straw, or tan-bark, whichever may be most convenient; the remainder of the ditch being filled with earth, as in other systems. Since there is no regular continuous channel, water entering such drains cannot flow with any considerable force; therefore, any sand or sediment which is carried into these drains will remain there, with the consequence that in most soils they will soon become obstructed and useless. However, they will remain efficient for a long time in clayey soil, since the clay is not likely to be carried into them. In sandy soil they will last at a very little time. For these reasons, I would not advise their construction except in cases where the stones to be used in making them are present on the land, and must be carried off in some way in order to get rid of them, if they are not used for this purpose. If the stones which you propose to use are large enough, it will be much wiser [p. 103] to construct a regular channel by laying two parallel rows of stones in the bottom of the ditch, and placing over them another row of stones large enough to bridge the space between the first two rows (Fig. 13).

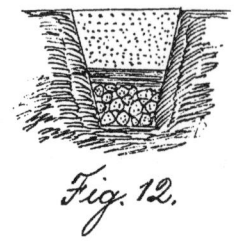

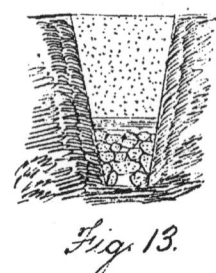

If you have a large quantity of stones to dispose of after you have constructed your channel, you may throw them in above the channel as is represented in Figure 13. Above these stones, you should place one of the materials already recommended above, to prevent sand from entering, and then fill the remainder of the ditch with earth as in the other systems. The more regular in shape the stones are, you can construct a channel into which sand will not to be likely to penetrate.

If you have upon your land, or can obtain at slight expense, [p. 104] stones of a regular shape, you can construct excellent drains with them. The channel formed by their use, those stones may take a great variety of forms, a few of which I will illustrate. You may construct a square or rectangular channel (Fig. 14), a triangular one with the apex upward (Fig. 15), a triangular one with the apex downward (Fig. 16), a triangular one with the apex against one side of the ditch (Fig. 17), etcetera. Of these various forms, the triangle downward would be the most efficient, for the reason already given in the case of box drains. It would, however, require [p. 105] stones of a more regular shape than is necessary for the construction of such forms as are shown in Figures 14 and 17.

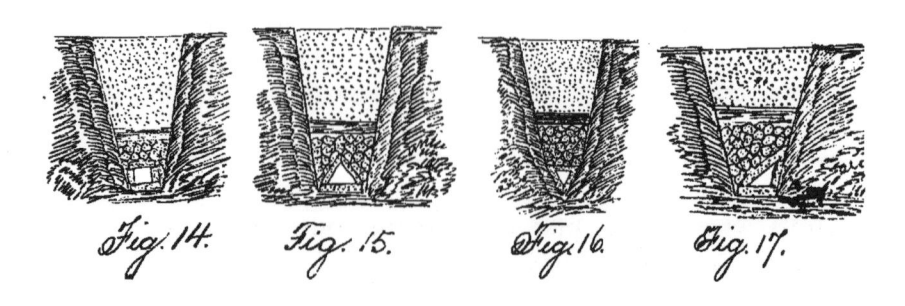

The materials out of which stone drains are made are, of course, almost indestructible; and the only method which sand or mud which such drains are likely to become useless, is by the filling up of the channel with sand or mud which may be carried into it by running water. In order to avoid this result, as much as possible, you should exercise the greatest care in the construction of the channel, and in protecting it by means of sod, tan-bark, or any other material of similar nature Stone drains are regarded as being the second best type, ranking next to tile drains; wherever stones of suitable nature can be obtained at slight expense, it will be wise to use them in the construction of this kind of drain; therefore, in a country where tiles are very expensive and stones are abundant and cheap, I would use stones [p. 106] rather than tiles or wood, provided good ones could be obtained with little labor or expense, since those will be much more durable.

3-4. Tile Drains

Tile drainage is, all things considered, the best system of drainage that can be used. It is the best for the following reasons: (1) Tile drains are more permanent than any other kind, because the material with which they are constructed is almost indestructible. (2) Tiles can be made of such regular shape that a more continuous channel can be made than with most other systems. (3) Tile drains are almost cheaper to build in most countries than any other kind of drains. The initial cost may, in some cases, be greater than with other systems, but—on account of their permanence—tile drains, would be cheaper in the end.

1) Shapes of Drainage Tiles

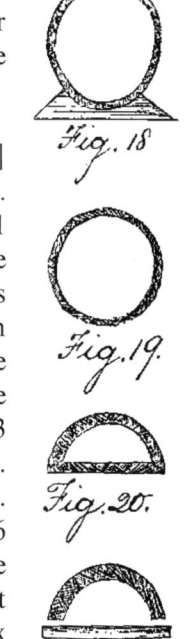

Drainage tiles have been made in several different forms. [p. 107] The following figures represent the cross-sections of some of them. Figure 20 shows a flat bottom with a semi-circular top. Figure 21 differs from Fig. 20, it only in that the top and the bottom are made in separate pieces. Figure 19 is circular in form with which collars are generally used as in Fig. 22. Figure 18 shows the cross-section of a tile, the orifice of which is usually oval in shape, and the bottom of which is flattened and expanded to give it a good base to rest upon. These tiles are usually made in sections about 13 inches in length. The thickness of the walls varies according to [p. 108] the size of the tiles, but is usually about 1/2 to 5/8 of an inch. The length of the collars used with the round tiles is from 4 to 6 inches. Of these several shapes the round tiles or the one with the oval orifice are the best. The round tile is preferable to others not only because of the reason given in the case of triangular box drains, but also because a channel of a certain capacity can be obtained by the use of less material than in any other form of tile. The tile represented in Figure 18 is called a sole tile, that in Figure 19 a round tile, and that in Figure 20 a horseshoe tile. Among these, the last named has the merit of remaining in place well; it also has the form of channel which is the best that has ever been devised. For these reasons, sole tiles are used more extensively in America than any other kind of drainage tile, although round tiles are coming [p. 109] into greater favor. In England, where much of the land is somewhat clayey, round tiles without collars are most extensively used. Horseshoe tiles, both those made in one piece and those with the tops and bottoms in separate pieces, were among the first forms used, but they have now gone out of general use, although some still continue to like them better than any other form.

2) Size of Tiles

The most obvious thing to be considered in deciding how large tiles it will be necessary to use is the amount of water to be carried; the greater the amount of water, of course, the larger in diameter must be the tiles. Those should be large enough to carry away, within a very short time, all the surplus water there may be present on the land. The velocity of the water flow must also be considered, because if water flows [p. 110] rapidly, a tile of any given capacity will carry away more water than if it flows slowly. Since velocity depends upon the slope of the drain, the greater the slope the smaller tiles may be used. Friction must also be considered. The amount of friction depends upon the smoothness of the channel, and also upon its size. If the channel is smooth, the friction between the water and the sides of the tiles will almost certainly be less than it would if the sides were rough. Friction also depends upon the size of the pipes. A certain amount of water spread over a large surface meets with more resistance than it would, if it were deeper. The amount of friction is less in proportion to the quantity of water carried in large tiles than in small ones.

The proper size of drainage tiles depends upon several considerations, some of which are: the amount of water to be carried away, the climate of the country, and the distances between the different lines of drains. The reasons for

Fall in 100 feet	Gallons of Water in 24 hours
3 inches	(2 inch tile) 10,575.4
3 inches	(3 inch tile) 24,687.2
3 inches	(4 inch tile) 43,697.6
3 inches	(5 inch tile) 99,584.2

the first consideration are so evident that it scarcely needs any explanation, since it must be clear to all that the greater the amount of water to be carried away, the larger must be the tiles. It is also very evident that in a country where the amount of rainfall is great, larger tiles will be necessary than in a country where little rain falls. It is also clear that if a large number of tiles are used, smaller ones will carry away all the superfluous water than in case where but few are used. It is impossible, therefore, to give [p. 112] any definite rule as to the size of tiles, but in general it may be stated that for a size of about 2 inches in diameter would be sufficient. The size used for minors varies from 1 to 3 inches. The size of tiles for main drains depends somewhat upon the number of minors which empty into them; the greater the number of minors, the larger the mains must be. A size of from of 3 to 4 inches is usually sufficient, although tiles as large as 10 inches are sometimes used.

3) How Water Enters Drainage Tiles

It is possible for water to pass quite freely through the porous tiles themselves, but it is stated on good authority that 500 times as much water enters through the crevices as through the pores.

4) Durability of Tiles

In order that tiles may be as durable as it is possible to make them, they should always be hard-baked. Soft-baked tiles will soon disintegrate, but hard ones will last a man's lifetime at least. [p. 113]

3-5. Direction, Distance and Depth of Drains
1) Direction

A question about which there is a considerable diversity of opinion is the direction with reference to the slope that drains should take. Some farmers argue that drains should run across the line of greatest slope, while others say that drains should run directly down the slope, and still others state that drains should run diagonal to the line of greatest slope. Here the various opinions are discussed.
(1) Across the line of greatest slope, water can enter drains only from places which are above them. If the drain runs across the line of slope, most of the water which enters must come from one side; therefore, since the distance from which it comes is greater than it would be if it entered from both sides. It will take a longer time to carry off [p. 114] a certain amount of water than it would if the drain would run directly down the slope. There is another objection to this method which is this, that water tends to flow out of this kind of drain on the downhill side.
(2) Diagonally across the line of greatest slope: drains running diagonally across the line of slope are often open to the same objections as those against drains running across the line of greatest slope. They would, however, be somewhat better than the first kind.
(3) Directly down the line of greatest slope: drains running directly down the line of greatest slope must be regarded as the best, as far as the direction is concerned, since they will carry the water away much faster than drains running in any other direction.

2) Distance between Drains

The distance between successive drains depends upon the nature of the soil, [p. 115] the depth of the drains, the climate of the country, and the comparative costs of labor and tiles. Here those factors are discussed.
(1) The nature of soils: if the soil is open and porous, water can flow through it freely and rapidly; therefore, in such soils, drains may be placed at greater distances than in those which are more or less impervious to water.
(2) The depth of drains: the deeper the drain, the greater will be the distance from which it will drain water; therefore, if the drains are deep, they may be placed at greater distances apart that it will in shallow drains.
(3) The climate of the country: if the climate of the country is moist, and if the country is one where a great deal of rain falls, it will be necessary to place drains at more shorter intervals than would be necessary in a drier climate.
(4) Comparative prices of labor and tiles: [p. 116] as already stated, the deeper the drains the fewer will be necessary, but to make a few and deep drains involves the expenditure of more labor than will be necessary to make more of shallow ones; however, in the former case, fewer tiles will be necessary than in the latter case. Therefore if you are situated in a country where labor is cheap and tiles are expensive, you should make a few and deep ones. On the other hand if tiles are

cheap and labor is expensive, it will be a good policy to make more shallow drains. Between 30 to 60 feet is the usual distance; it is, however, sometimes necessary to place them as frequently as once in 20 feet, and once in 70 or 80 feet may sometimes be sufficient; 40 feet may be regarded as the most common distance.

3) Depth of Drains

Underdrains must be placed deep enough to be below the reach of the ordinary plow, and in most cases [p. 117] they should be below the point where the subsoil plow reaches. The use of the subsoil plow would be likely to interfere with drains placed less than 3 feet below the surface.

Drains must also be placed so deep as to be below the reach of frost, since if filled with water (or even if the pores of the tiles are filled with water), freezing —by expanding the water—would be very likely to break the tiles. Even if freezing did not break them, being subjected to the influence of alternate freezing and thawing would tend to disintegrate the tiles and, therefore, they would if they were not subject. Frost might unequally expand the earth lying next to the tiles, thus throwing them out of position and destroying the continuity of the channel. The depth necessary to get below the reach of frost will, of course, vary according to the climate of the country [p. 118] and the amount of snowfall. If the country is a very cold one, greater depth will be necessary than in a warmer climate. If the country is covered each winter with a thick blanket of snow, frost cannot penetrate very deeply even if the climate is quite cold; therefore, in such countries, drains need not be placed at as great a depth as in countries where the snowfall is less.

Drains must also be placed deep enough to be below the reach of the roots of ordinary cultivated plants, since if they are not so placed, the roots of the plants will sometimes penetrate them, thus filling them and obstructing the flow of water. In general, a depth of about $2\frac{1}{2}$ or 3 feet will be sufficient to ensure safety from the effects of frost, and a depth of 3 to 4 feet will ensure comparative safety from obstructions by the roots of plants. [p. 119]

3-6. *Necessity of System in Drainage*

It is very important, before beginning to drain a piece of land, that you survey it carefully and, after due consideration, decide where, how deep, in what direction, and how many drains you will use. After having decided in regard to your drains, you should make a drawing upon paper showing the position, direction, and depth of the several lines. This plan will be very useful for recording in what parts of your field drains are located. It is wise, before undertaking any business, to carefully consider the cost and decide whether you can afford to undertake it or not.

1) The Amount of Fall Necessary

Within certain limits, the greater the fall the better; in most arable land, it may be safely asserted that you are not likely to have too great a fall. [p. 120] The question to be considered, then, is not how great a fall can be safely used, but how little it is prudent to rely upon. If drains are very carefully constructed, a fall of 2 or 3 inches in 100 feet is sufficient to ensure good drainage. Instances are on record where drains have been constructed with little more fall than this (2 or 3 in.) in 1,000 feet, but I would never advise locating a drain in a place where the amount of fall would be less than 3 inches in 1,000 feet. A greater fall than this would be much better, since if the amount of fall is considerable, sand or sediment will be carried away by the force of the current of water. If you are ever obliged to construct a drain with little fall, you must be especially careful to have the slope exactly uniform throughout the entire length, as also to construct and to lay the tiles, and to so cover them as to prevent, as much as possible, sand or sediment [p. 121] from being carried into them.

2) Outlets

In under-drainage, a good outlet is highly important. An outlet is the place where the under-drain empties or discharges its flow, and is usually an open ditch or brook. The water level in the outlets should be below the place where the underdrain empties into it. These outlets are exceedingly vulnerable to obstructions of various sorts. It is therefore very important to secure them as carefully as possible. Near the outlet, if tiles are used, they are likely to be thrown out of place in many different ways. Cattle walking in the ditch or brook for the purpose of getting water may knock the tiles out of position. Frost acting upon the soil may bring about the same result. In order to prevent this as much as possible, it is good to build a wall [p. 122] of brick or stone against the side of the ditch at the place where the tile empties. Wood may also be used in the construction of this wall. Its top and sides should project out a little, in order to further protect the end of the tile. Such animals as moles, field mice, and frogs are often in the habit of enjoying the coolness of the underdrain, and if there is not much water in them, in such cases, animals are very likely to advance so far that they can never get back; or, they may be so large that they cannot turn round and thus they die in the drain and their dead bodies obstruct the flow of water. In order to prevent such animals from entering drains, it is good to cover the end of the tile with wire netting screen which shall at the same time prevents their entrance but allows the free passage of water. [p. 123] The water in the open ditch or brook into which the underdrain may empty will not infrequently, in times of freshet, rise above the mouth of the tiles, thus preventing the water in them from flowing out; at times, the water of the brook itself may be forced into them. This brook water, especially at such times as I have mentioned, is often very muddy; if it flows into the drain, it will deposit a mass of sediment which will partially obstruct the channel. In order to prevent water from entering drains at their outlets, it is good to have a hinged cover fastened to the end of the outlet tile. This cover should open upstream (Fig. 23), and should fit the end [p. 124] of the tile so closely (Fig. 24) as to prevent the entrance of water. Some writers on the subject of drainage also recommend that a hale should be made in the ditch or brook at the place where the underdrain empties, and that the bottom and sides of this hole shall be covered with brick or stone. Thus, any sediment brought by the water emptying out of the drain will be

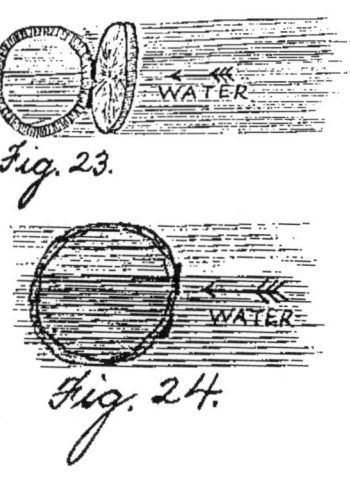

Fig. 23.

Fig. 24.

caught in this well and may be removed at pleasure.
3) Wells and Peep Holes

There is almost always more or less sediment carried by water flowing in underdrains; it is therefore advised by some to make, at important points, wells which shall sever to catch and retain this sediment. The wells may be made of brick, stone, or wood, and may be either round or circular. Brick or stone will, of course, be the most durable, and where either can be obtained at reasonable expense, [p. 125] those should be used in the construction of the wells. The depth of the well will be such that its bottom will be from 1 to 1½ feet below the mouth of the underdrain. The well's diameter should be large enough to allow it to be readily cleaned out. The drain from the upper side should enter the well about 16 inches above the bottom. The drain tile leading out of it should be about 3 inches lower. Referring to Figure 25, E should be about 16 inches from D, and O about 3 inches lower than E. These walls should be covered with some strong cover M, which should have sufficient strength to support the weight of heavy animals such as horses or cattle. The construction of wells which are large enough for a man to enter, is expensive as well as desirable; therefore, [p. 126] I will recommend it only on very important lines of drains. It is very convenient, however, to be able to see the flow of water in the drains at various points. In order to make this possible, and also for the purpose of collecting sediment, smaller wells are often made which are called "peep holes". These holes may be made of very large tiles or iron pipes or even wood. The depth should be as much as is recommended for larger wells. The diameter should be between 6 inches to a foot, and tiles should enter and leave them at points similarly situated as those recommended in the case of wells. The well cover should also be strong enough to support the weight of large animals. In case of both wells and peep holes, the cover should be so arranged that it can be easily removed. [p. 127] Peep holes should be placed, according to some writers, at intervals sufficiently frequent to enable a man, by inspection of them, to see readily whether the water is flowing freely in all the different lines of drains. I do not regard them as so important, and would not advise their being made in such large numbers. I would make only a few, if any, upon the principal lines of drains.

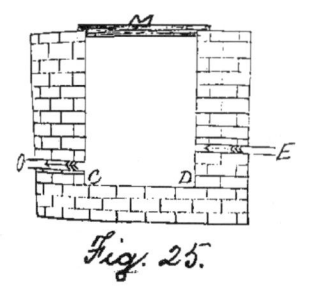

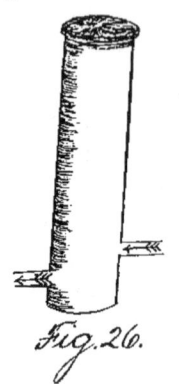

4) Mains, Submains and Minors

A main drain, as its name indicates, is a principal one whose office is to carry away the mater brought into it by submain or minors or by both at the same time. The mains are always located in the lowest parts of the field, and should follow the natural slope; however, the line may be straight or crooked. Mains are usually placed [p. 128] at the foot of slopes, if the land slopes in one direction. Submains are those drains which carry water received from minor into main drains. They are also placed in low parts of the field, but are not always necessary. While the principal function of mains and submains is to carry away water received from minors, they also receive water directly from the soil through which they pass. Minors are those drains whose function is to collect water directly from the soil and carry it into submains and mains.

In Figure 27, AB represents a main, CD a submain, and 1 & 2 the minors. [p. 129] O is an open ditch or a brook into which the main empties its water. Minors should generally situated at nearly right angles to the main drain. In land, the slope which is uniformly sloped and in soil which is homogeneous throughout, minors should be parallel to each other, and at equal distances apart. On land, the slope of which is not uniformly sloped, and where there may be subordinate hollows running from the principal hollow in which the main drain is located or in soil, some portions of which are wetter than others, the minors should sometimes follow the courses of these hollows or be placed in the wettest places. Although the general direction of the minors should usually be nearly at right angles to the main drain, they should be slightly curved just before joining it, in order that the current of flowing water, when entering [p. 130] with that in main drain, may have the same direction as that flowing in the latter. Water entering the main drain at right angles to it would check the current therein, and this might lead to the deposition of sediment at this point, which perhaps might eventually check the flow of water entirely.

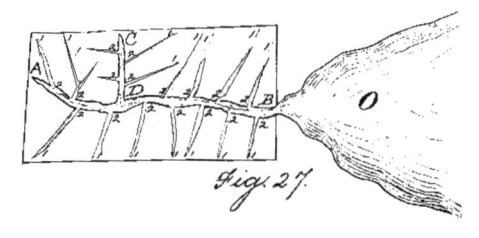

5) Junctions of Drains

The best manner in which to form a proper junction between two lines of drains is by the use of branch tiles. If you cannot procure branch tiles, a very good union may be effected by breaking out a piece of the tile used in the construction of the main and breaking the end of the tile used in the minor directly opposite the hole. In order to prevent the tile in the minor from being displaced at the place where it unites with the main, it is good to use bricks broken in the proper shape in the angles formed [p. 131] by the junction of the two lines of drains (Fig. 28). Even if this work at the junctions is done as carefully as possible, there will usually be larger crevices at these points than at others; special care should therefore be taken to cover these points with some material, to prevent the entrance of sand.

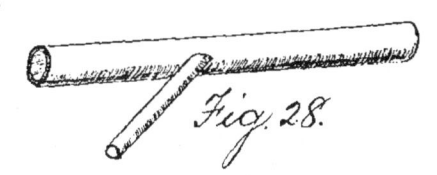

6) Draining into Wells or Swallow Holes

Soils are sometimes underlaid by impervious strata of the nature of a hard-pan which almost entirely prevents water from soaking down. Such soils, unless the land slopes considerably, will always be very wet. The hard-pan is sometimes of no great thickness, and is often underlaid by a porous soil which freely permits the passage of water. It is sometimes possible, by digging a well [p. 132] through this hard-pan and constructing drains in different parts of the field which will empty into it, to drain the land quite perfectly. The water, as it falls into the well, sinks into the pervious soil rapidly settle down, thus passing out of the way. Such a system of drainage, however, can only be resorted to, as you see, in soils of peculiar formation; I do not think it would be possible to get rid of a large amount of superfluous water in this way. The bottom of such a well is extremely vulnerable to be rendered more or less impervious by the deposition of sediment

88 Chapter 3 Farm Drainage and Irrigation

brought into it by the drains. It might, therefore, sometimes be necessary to remove the accumulations in the bottom of the well in order that it might remain serviceable. It is sometimes possible to drain land underlaid by a thin stratum of hard-pan simply by digging holes through it at frequent intervals; the water in the soil above the hard-pan will then find its way through natural channels to the holes. [p. 133] Such holes would, of course, be of most service if dug in the lowest part of the field. While the two systems of drainage last described may sometimes be effective, I would not advise resorting to them on a large scale without having previously determined, by experiment, that they will effect the designed objective. These holes are called "swallow holes".

3-7. Draining Implements

The implements most used in draining are spades, the blades of which should be of different widths, lengths, and shapes. These spades should also have handles of different lengths and shapes. There are also used various kinds of bottoming instruments, of peculiar structure, which are made expressly for the purpose, but good drains can be made without their use. Pick-axes are sometimes necessary for loosening soil before it can be thrown out with the spade or shovel. Shovels may be used for throwing out [p. 134] loose earth. The plow may be used for loosening the earth in the upper portions of the ditch. The subsoil plow is sometimes used with advantage.

It has long been a desire of practical farmers and inventors to invent a machine which could be used to excavate ditches, but thus far, although many have been invented, very few of them have been such as could be used with advantage; therefore, they have not been extensively used. One of the best is called "Paul's Ditching Machine", the essential part of which is a wheel, on the rim of which are cutting blades. This is made to revolve and, each blade cutting out a small portion, the earth is carried to the surface, where by a peculiar mechanism it is made to drop it on the side of the ditch. This machine is drawn across a field by means of chain and capstan, and it is made to do very good work. [p. 135] For determining the slope of land, levels are necessary in most places. The ordinary surveyor's level is the best implement for this purpose. For determining whether your ditches have a correct and uniform slope, levels of a simpler construction are very useful. One of the best is called a "span" or an "A Level", (Fig. 29). The method of construction of an A level is as follows;

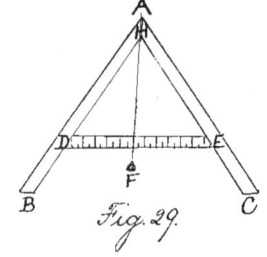

Two pieces of scantling (AB & AC in Fig. 29) of convenient size are fastened together at one end at an "angle A", a little less than a right angle. The two pieces should be of such a length that the space between their feet (between B & C) is some convenient factor of 100 feet. Across these, at a point about 2/3 of the distance from the apex to the bottom, should be nailed another piece of scantling DE. [p. 136] You should, then suspend a small weight F from the apex H by a strong cord HF. If the implement is placed upon a surface which is exactly level, the weight will hang down vertically, and the cord will pass by the crosspiece at its midpoint. Place under one foot of the implement a block 1/2 of an inch in thickness and mark the point where the string passes on the crosspiece. Next, place under it a block one inch in thickness, marking the point where the string passes as before, and so continue, adding 1/2 of an inch every time and marking the point across which the cord passes, until you have a graduated scale, the difference in each case being 1/2 of an inch with gradations. The other leg of the implement is then successively raised in the same manner, thus graduating the other half of the crosspiece. Having determined at what point the string should hang [p. 137] in the ditch which you propose to examine, place the implement lengthwise in the ditch, allowing each foot to rest upon the bottom; then move it along through the entire length of the ditch, whether, at all times, the string hangs at the proper place.

There is another kind of level called "Challoner's Level" (Fig. 30). This consists of two strips of wood (SA & EV), the shorter of which EV is fastened to the middle of the longer SA at right angles. The bottom piece may be of any convenient length, but it is better to make it of some number of feet which is an exact factor of 100 feet. The shorter piece should be 3 or 4 feet in length. From a point near its top should be suspended a weight by a cord EV. In order to use this implement, you must make another piece [p. 138] of wood of the same length IO. One end of this piece should be so thin that it will not affect the thickness of the implement, while the other should have the same thickness as the amount of fall in the number of feet represented on the bottom piece. In every system of drainage, if you propose to use this implement, you must make a wedge-shaped piece appropriate to the system; in order to render easy, this should be fastened to the main implement by means of screws. [p. 139]

⟨Sandy loam: (1) can be worked early in the spring; (2) can be worked immediately after a shower; (3) can be makes a much finer bed for the germination of seeds; (4) can be allows the roots to penetrate as they please; (5) can be allows air and heat to permeate to the roots; (6) does not expand, contract, and crack under the influence of heat and moisture. —A. Hyde⟩[p. 140]

3-8. *General Directions for Tile Drainage*

The first work is, as I have already said, to carefully survey the field and determine in regard to the position and fall of the various drains. The next work is to excavate the ditches, which should be done by the use of the implements which I have already described. In doing this work, it should be the aim to throw out as little earth as possible. Some writers recommend that the ditches be made only of such a width as will give a sufficient space at the bottom for the tiles to be laid. In digging ditches of this width, all the lower portions must be thrown out by the use of a long-handled implement, a workman standing on the bank of the ditch, since there is no place for him in it. If the ditches are excavated to this width, the tiles may be placed by the use of an implement called the pipe layer. [p. 141] This is simply a long handled implement, near the bottom of which an iron rod is fastened at right angles. To excavate ditches while standing upon the bank requires much more labor than to do so inside them; for this reason, what is saved in the amount of earth is partly, if not entirely, counterbalanced by the loss arising from the disadvantages at which the laborers work. I would, therefore, advise that ditches should generally be excavated of sufficient width that a man may stand in them and work conveniently, —believing that with most soils, it will be found to cost less to excavate the ditches of this width than to excavate very narrow ones. In any system of drainage, the main drain should be excavated first, beginning at its lower end; and next the submains, if there be any, beginning at the point where they empty into the main. The minors should be excavated next, always beginning with [p. 142] those which empty into the main at the points nearest the outlets, and beginning to excavate from the place where they unite with either the mains or submains. This order should be followed in order that water which enters the ditches during the process of construction may have a chance to flow away freely. After having dug the ditches have been dug, tiles should be brought and placed upon their banks in such a position that they can be easily reached by a man walking in inside the ditch. Material of some sort for protecting joints between successive tiles should also be prepared and brought to the field. The best material to use with sole tiles is tarred paper. This should be cut in strips about 3 inches in width and of such a length that each strip will just reach around the curved surface of the tile. Some of the paper should be left in large pieces for the purpose of covering the points of junctions. If you cannot obtain tarred paper, sod with the grass side downward is very useful for this purpose. With round tiles, collars may [p. 143] be used which may have a length vary from 2 or 3 to 5 or 6 inches in length. 2 or 3 inches will probably be sufficient. In soil that is soft, and hence where the tiles are likely to be pushed out of position, I would always recommend the use of collars with round tiles. In hard and compact soils, however, tarred paper or sods might be used. A brick, flat stone, or even a piece of wood should be placed against the end of the tiles at the beginning of every line of drains, for the purpose of preventing mud from falling into them. Having everything in readiness, two men should be employed in laying tiles, and they should begin at the upper end of the submain or main. Referring to Figure 31, [p. 144] begin from C to D, then E to D, then D to G, then F to G, etcetera. This order should always be followed so that water, after flowing through the ditches carrying with it sediment as it often does, may not be obliged to pass through the tiles, since it might deposit sediment in them, thus obstructing the flow of water. In laying the tiles, one man should stand inside the ditch with a bricklayer's trowel, while the other should stand upon the edge of the ditch with a shovel. The man inside the ditch should take the tiles from the bank and carefully place them in position, cutting them with the trowel, or smoothing the earth in the ditch if required, in order to make a level bed upon which the tiles may rest. Tiles are frequently of varying thickness, and if the worker finds one with a thicker or thinner bottom than usual, he should so change alter the level of the earth in the ditch with his trowel, so that the inner surface of the bottom of the tile is on the same level with the same line in succession of drainage. [p. 145] Having placed the tiles carefully in position, this

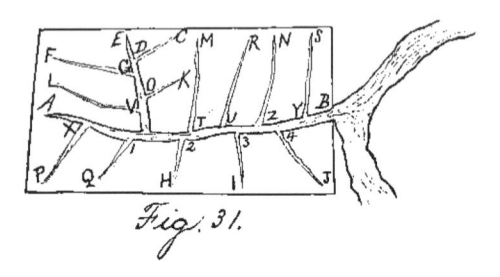

worker should take some tarred paper or sod (whichever he is using) and place it accurately over the joint, holding it in position until the other worker throws a shovelful of earth upon it. Besides always being in readiness to throw earth upon the tarred paper or sod which the man in the ditch places in position, the other man should throw in as much additional earth as he finds for. If you are using the collars with round tiles, the same rules should be obeyed. After the tiles have been laid and partially covered, as already directed, the remainder of the ditch should be filled as soon as possible, the earth being raised a little above the surrounding level as recommended in the case of other drains. This work should be done very promptly, because if a heavy rain should come, it would be very likely to carry sediment into the tiles, thus obstructing the channel. If it is intended to construct peepholes or wells, [p. 146] those should be made at the time you lay the tiles, since it can more easily be done at this time than on any other occasion. After having finished all other works, the outlets should be carefully protected as already described.

Underdrains are liable to be obstructed in the several different ways previously mentioned incidentally. Some of the principal causes of obstructions are: The washing in of sand, the penetration of the roots of plants, and the formation of iron oxide in the interior of the tiles. In soils where iron is abundant, soluble forms of it are sometimes carried with water into the interior of the tiles; there, being acted upon by oxygen of the air, these are changed into the higher oxides, thus being deposited in the interior of the tiles. It may be necessary, in soils [p. 147] which contain a great deal of iron, to use tiles of a somewhat larger size than in other soils, though the deposition of iron oxide will seldom entirely check the flow of water. In order to remove smaller obstructions, it is very convenient to be able to flush the drains. This is done by simply checking the flow of water from the tiles at the outlets, or at wells or peepholes, until the tile has become filled, and then opening them again to allow the water to pass rapidly and quickly out. In spite of all these precautions, however, drains are sometimes entirely obstructed; therefore, it is highly important to know how to detect the precise position at which the principal obstruction is located. If any drain becomes obstructed, you will notice that the land about it becomes wet, but you cannot tell—since this wetness is distributed over an extensive area—the exact point, [p. 148] where the obstruction is by superficial examination. You can detect that point, however, by taking a bar of iron and passing along the line of the drain, making holes at frequent intervals and observing how high the water is in each. The water will be found to stand the highest at the point directly above the obstruction. [p. 149]

3-9. Irrigation in Several Countries

Irrigation is the activity of watering or moistening the land, especially for the purpose of furnishing nutriment to plants. It has been practiced from very remote times. In many of the eastern Asiatic countries where the human family first dwelt, the climate was such as to render irrigation absolutely necessary in order to produce plants. Accordingly, we find China, Damascus and the surrounding countries such as Syria, Italy, and the Barbary States; have practiced irrigation for many thousand years. In the Bible, we find irrigation mentioned by some of the earliest writers. Four thousand years ago, one of the Egyptian Pharaohs contrived to irrigate 444,000 acres by damming up the waters of the Nile River. He made a dike 30 feet high and 200 feet thick across a valley, thus forming a lake which held 3,694,000,000 cu.-yd of water. By this means, the Egyptians were able to irrigate a vast extent of the country, thus producing [p. 150] food for many people which could not otherwise have been produced in such a comparatively rainless region. (1) Italy: in no other country in the world has irrigation received such careful scientific study and investigation as in Italy. In none has there been such an enormous expenditure exclusively upon this agricultural art as in Italy. In none has there been such splendid results produced from it as in Italy. There the government controls all the streams for the purpose of irrigation. There are a great many highly educated hydraulic engineers who make it their business to care for the immense systems of irrigation. The yield in hay produced by irrigation in that country is very large. In Piedmont, one farmer cut from an acre of land as follows: in February of a certain year, $4\frac{1}{4}$ tons; in April, $6\frac{1}{3}$ tons; in May, $6\frac{1}{2}$ tons; in July, $3\frac{3}{4}$ tons; and in September, 3 tons, making nearly 24 tons in the year. Where irrigation with [p. 151] sewerage water is practiced, much larger yields are expected. (2) Germany: irrigation has been tolerably tried, and trial has demonstrated its great value. The hay crop has been doubled by means of it. (3) South America: Peru which is a rainless country has been made fertile by means of guano fertilizer and irrigation. (4) United States: irrigation is quite largely practiced in some regions, the principal of which are Utah, California, and Colorado. In California, 90,000 acres are already under irrigation, and works are contemplated before this time sufficient to irrigate 600,000 additional acres. In some parts of the state, irrigation is especially important, since rainfall is very little and uncertain. The Great Valley of California contains 27,200 square miles of land and the smaller one contains 18,750 square miles, all of which is susceptible of irrigation and doubtless soon [p. 152] will be irrigated. In Utah, the Mormons have irrigated about 96 acres of land, most of it lying near the Great Salt Lake, and what was formerly a barren country, producing little but worthless shrubs, has now become very fertile and productive. There are many thousands of acres lying between the Rocky Mountains and the valley of the Mississippi River which are almost destitute of vegetation, but which will undoubtedly produce abundant crops if irrigated. In the state of Colorado, large quantities of land are now irrigated, and wherever irrigation has been introduced, its results have been highly beneficial. (5) France and Belgium: irrigation is extensively practiced and experience has shown that in these countries, also, it is of great advantage. You may perhaps think that it is only in countries where the amount of rainfall is small that irrigation is [p. 153] found useful, but this is not the case. Italy, Germany, France, and Belgium, and also England and Scotland, are countries where rain falls quite abundantly, and yet, in these countries, it has been found profitable to irrigate.

Growing plants contain from 70% to 95% water. The quantity of water necessary to produce a large crop is said to be equal to a layer of water 12 inches in deep over the entire surface of the land. This amount must mostly be furnished in a few summer months, when plants are growing; hence, although the amount of rain fall in Hokkaido is doubtless greater than 12 inches per year, it may sometimes happen that during the summer months the amount will not be sufficient to produce the maximum yield of crops. In the year 1876, little or no rain fall in Sapporo from April to July, and crops suffered severely [p. 154] for want of water. It is estimated that for every pound of dry matter in a crop of wheat, 200 pounds of water must have been used. For every pound of mineral matter in wheat, 2,000 pounds of water must have been used. Leguminous plants use even more water than wheat. With all crops, the amount produced bears a certain relation to the amount of water which has been used by the plant. It is not, however, for the sake of furnishing water alone that irrigation is resorted to. Most waters also furnish a large amount of fertilizing material. The amount of plant food contained in water, differs very much in different waters, but almost all water contains quite a large quantity. The following table gives an analysis of the water of the Delaware River: Total amount of solid matter in one gallon is 3.97 grains, consisting [p. 155] of the following substances in their respective proportions:

The water of this river is used for drinking purposes, and is regarded as being very pure, and yet you can see that it contains quite an appreciable quantity of plant food. A table showing the amount of solid matter in four American rivers is given above:

$CaOCO_2$	1.30 grains	$CaSO_4$	0.19 grains
$MgO\ CO_2$	0.89	$CaKPO_4$	0.14
$KO\ CO_2$	0.17	Si	0.50
NaCl	0.11	Fr_2O_3	0.03
KCl	1.10	Organic matter containing H_3N	0.63

Rivers	Passaic	Schuylkill	Croton	Hudson
Solid contents in 100,000 parts	12.75	9.41	18.71	18.41
Inorganic Matter	7.85	7.29	11.32	14.52
Organic matter	4.90	2.12	7.39	3.96

The waters of these rivers are considered [p. 156] to be unusually pure, and are used for drinking purposes. It is stated by a Frenchman, Herve Mangon by name, that each 200,000 m^3 of the water of the Seine River contain substances equal to one average ox. If this is true, the waters of that river carry into the ocean the equivalent of one fat ox every two minutes. This amounts to the equivalent of 720 oxen per day, or 262,000 oxen per year. This seems like a very large amount, yet it is small compared with the amount which must be contained in some of the largest rivers, such as the Mississippi and the Amazon. From the consideration of these facts, we see how greatly the amount of food produced in the world may be increased when the waters of its rivers are compelled, by irrigation, to yield up their fertilizing materials before emptying into the ocean. [p. 157]

3-10. Kinds of Water which are best for Irrigation

Sewerage water is by far the best, since it contains many elements of plant food. The waters of streams and rivers rank next in value, since they contain as is shown in the figures which I have given you quite a large amount of plant food. The waters of streams and rivers usually contain much more fertilizing material when in a swollen condition than at ordinary times.

(1) The water of springs may be made use of for purposes of irrigation, but such water is usually more pure than that of rivers or streams, and therefore does not contain much plant food. In exceptional cases, however, spring waters are strongly impregnated with salts of different kinds; in such cases, they may be either very

beneficial or very injurious if applied to crops. If the salts contained in solution are present only in very [p. 158] small quantities, as is usually the case, and if they are such as are useful for plant food, such water may be used with great advantage. If, on the other hand, salts are present in very large quantities or if present only in small proportions, if of such a nature as to be injurious to plants, their use should be avoided.

(2) The water from ditches along the sides of roads usually contains a large amount of plant food, since the soluble portion of the excrement of the animals passing over the road, and the fine particles of earth which have been ground up by heavily loaded teams are washed into the ditches. Wherever such water can be obtained, it is wise to use it for purposes of irrigation.

(3) Water from Wells: [p. 159] The purity of water in wells differs much in different localities, but even those wells which are regarded as quite pure usually contain a large amount of substances which are useful for plant food.

3-11. The Amount of Water Needed for Irrigation

The amount of water needed for irrigation varies according to the soil, the climate of the country, the nature of the subsoil, and the crop cultivated.

(1) According to the Soil; Those soils which are retentive of moisture, such as those which contain clay or organic matter, need the application of less water than those which are not retentive. Soils which are sandy or gravelly, and which contain little organic matter, must be irrigated more abundantly than those which have more retentive power. [p. 160]

(2) According to the Climate; It is very evident that the less the amount of rainfall in a country, the larger the amount of artificially applied water must be. In countries where the climate is very hot and dry, and the amount of sunshine during the growing season, are hot and dry, much more water will be evaporated than in countries where the climate is cooler and where the sky is clouded a large proportion of the time, or where the winds are few and moist. Hence, in deciding how much water must be artificially supplied to crops, you must carefully consider the climate of the country, both as to the amount of rainfall and points which may influence the amount of water evaporated.

(3) According to the Subsoil; If the subsoil is of such a nature that water will readily pass through it, more water may be supplied than in cases [p. 161] where the subsoil does not permit the water to pass through it readily.

(4) According to the crop cultivated; Some crops require much more water than others. Some crops are injured in quality if too large a supply of water is furnished, while others will not grow well if too much water is present in the soil. Hence, the amount of water which should be used in irrigation must depend largely upon the kind of crop cultivated. Notwithstanding the fact that all these various circumstances, of which I have just been speaking, should influence the amount of water used in irrigation, I will give you figures which show the amounts used in certain cases.

In one acre of land, there are 6,272,640 sq.-inches. One inch of water flowing at the rate of 4 miles/h will furnish 6,082,500 cu.-inches of water in 24 hours. Hence, in this length of time, [p. 162] a stream of one sq.-inch will cover an acre of land nearly one inch deep, if it flows at this rate. It is not usually necessary to water land more frequently than once a week. If this amount be supplied, therefore, a stream of water of the size of which we have been speaking will be sufficient to irrigate an area of about one acre; one inch of water flowing at 4 miles per hour will furnish 18 gallons per minute.

In Provence in France, where the climate is hot and dry, it has been found necessary to cover the whole surface of the arable land once in every 10 or 12 days with a layer of water from $3\frac{1}{2}$ to 4 inches in depth; this requires 24 in3 per second, continually flowing, for each acre of land. This amount of water is the government unit in that region. The extremes of water quantity necessary are stated by Herve Mangon to be from 1 pint to 2 quarts per second, continually flowing, which is sufficient to irrigate 8 acres of land. The Italian furnish 26 cu.-inch of water [p. 163] per second for each acre irrigated. In India, 1728 cu.-inch of water is made to serve for 200 acres of grain crops. This equals about 9 cu.-inch/sec, constantly flowing, for each acre. In Spain, the same amount of water is used to irrigate 10 to 260 acres of land. It is stated that rice cultivation requires 1 cu.-ft/sec for each 30 to 80 acres of land.

1) Garden Irrigation

In the cultivation of many garden crops, a large supply of water is essential to attainment of the highest success. Under this topic, I shall consider especially the irrigation of market gardens. Such crops are usually of large value per acre, the crops for a yield per acre of ground not infrequently being worth $1,000 in some of the market gardens in large towns of America. Since the crops are of such great price value, then, more costly systems of irrigation may be used with profit [p. 164] profitably in market gardens than can be used in ordinary field culture. There are three principal methods of irrigation used in market gardens; irrigation from canals, irrigation by means of pipes and tiles, and subterranean irrigation.

2) Irrigation from Canals

The manner of irrigating by means of canals must vary according to whether the surface of the land is level, slopes in one direction, or slopes in several different directions; in each of these cases, many different methods of arrangement, almost equally good, may be devised. I will not, therefore, try to describe all of them, as I have not time enough to do so, but will simply give a brief description of one good method of arrangement for each case. Whatever the system adopted may be, the object to be aimed at, is to distribute the water as equally and quickly [p. 165] as possible over the surface with the least amount of labor and the least waste of water.

Suppose that you have a level field (Fig. 32). It will be found profitable in most market gardens, if a field is large, to divide it into strips about 210 feet in width which may extend across the length of the field. Having divided the field in this manner, begin at a point from 15 to 30 feet from B; plow a furrow from F to E, then turn another furrow on that, and so continue to plow back and forth toward the middle each time. Repeat this operation twice. In plowing the second time, the furrows should grow shallower towards the outer edges [p. 166] of the ridge. The whole field

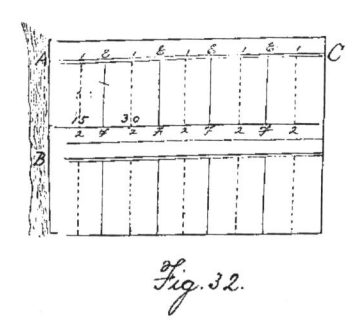

Fig. 32.

should be thrown into ridges in this manner. Then we should begin at A and plow towards C, and then back from C to A, turning the furrow in the same direction. This should be continued until we have plowed a strip about 15 or 20 feet in width. This strip should be plowed a second time, in the same manner, turning the furrow toward the ridges. The main supply canal should be made at the highest point of the longer ridge along the full length of the field. Small canals should then be made in the middle of each ridge, these canals leading from the main supply canal. In Figure 32, the main supply canal is represented by AC, and smaller canals by the lines EF. These will be the only permanent canals that will be required.

Whenever you wish to irrigate, [p. 167] you can make smaller furrows, with a plow or hoe, leading from the canal on the crown of the ridge to any part of the field. In the cross-section diagram (Fig. 33), the crown canal is represented by A. In any system of irrigation, the land must be well drained, since otherwise the water artificially supplied will become stagnant and will render the land cold, thus doing more harm than good. Tile drains are the best system of drainage that can be used, although you may use any other kind of which I have spoken. These drains should be located in the hollows between the ridges shown in the cross-section (Fig. 33) at the points marked B, and in Figure 32 by the dotted lines 1 & 2.

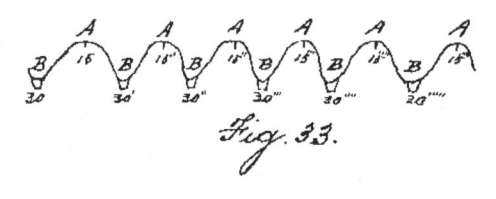

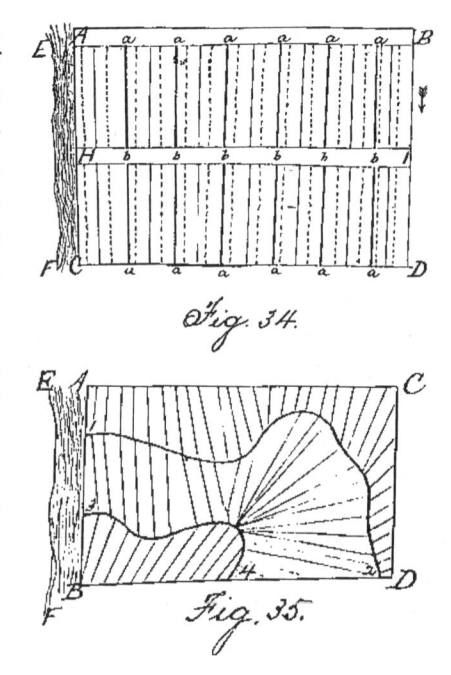

When the surface of the land slopes uniformly in one direction, a much simpler system of irrigation may be adopted. Suppose you have a field (ABCD, Fig. 34) which slopes uniformly from AB towards [p. 168] CD in the direction shown by the arrow. Suppose the double lines EF to be a main supply canal. The double lines AB represent a canal which is to supply water to irrigate the portion of the land enclosed within ABIH. This field may be of any length, but the width should be about 210 feet, as recommended in the system last described. HI represents another supply canal leading from the main canal EF which is designed to irrigate the land enclosed by the sides HIDC. The crops cultivated on land arranged in this manner are to be planted in drills, represented by heavier crossed lines. These drills should be raised a little above the surrounding level, and in irrigating, the water is [p. 169] allowed to flow between the several rows in the portions represented by dotted lines in Figure 34. If the slope of the land is irregular, the system of irrigation must be much more complex than those already described. In every case, however, you must lead the water through the canals which are located in the highest portions of the field: from these, you must make distributing furrows in direction according to the direction of the slope.

In either of the three systems of distributing water which I have described, the water is sometimes allowed, if the plants under cultivation admit of level culture, to overflow the banks of the supply canals in a thin sheet. [p. 170]

3) Irrigation by Means of Pipes and Hoses

Water is sometimes brought through pipes laid under the ground to the field which is to be irrigated. The water should be brought from a reservoir which is located higher than the field, in order that there may be sufficient head to distribute the water. The pipes should be laid at a sufficient depth below the reach of plows. There should be a pipe 1½ inches in diameter for each acre to be irrigated. In Figure 36 "ABCD" represents a field, in which the dotted lines FG represent the positions of underground pipes. At the points marked E, shown by the crosses which are so placed as to be 210 feet apart, are hydrants of such a shape that the end of a piece of hose can be readily attached to them. When the field is to be irrigated, [p. 171] a man attaches a hose to a hydrant and scatters the water over a circular area of land, as far as he can reach. The end of the hose should be of such a shape as to break up the stream of water, thus scattering it in a shower of spray like rain. This system of irrigation will be more expensive than those previously described, and I will not advise its adoption except on a limited scale, where it is impossible to irrigate in a different manner. The underground pipes may be made of lead, iron, or wood; but the last would last only a very short time. The hose may be made of either rubber or leather, but the former can be made lighter than the latter, and is therefore convenient.

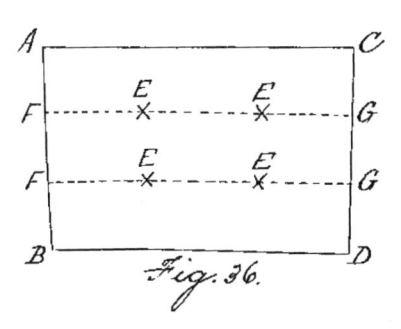

4) Irrigation by Means of Perforated Pipes

In Figure 37, ABCD represents a field, [p. 172] and the lines 1 & 2 show the positions of pipes laid upon the surface of the land on each side. Water, brought into these pipes from a reservoir situated above the level of the field, will find its way through these pipes in small jets which will be thrown a greater or less distance according to the amount of head. If the reservoir is laid considerably above the surface of the field, these jets will be thrown around a large area. On the other hand, If the difference in level is but slight, the water will be thrown but a short distance, the land should be a little higher midway between adjacent lines of pipe. It follows from what I have said concerning the distance to which the water will be thrown, that the distance between different lines of pipes must vary according to [p. 173] the amount of head. A distance of from 12 to 20 feet, however, will usually be as great as can be used. While this system of irrigation is somewhat expensive to begin with, water may be applied with a very slight expense after it has been constructed.

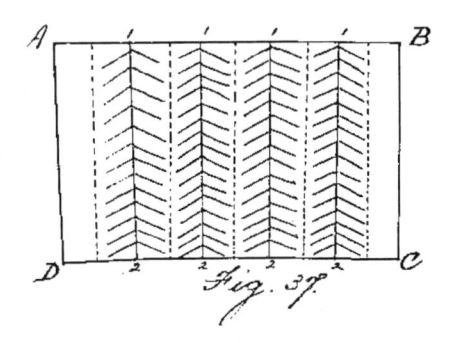

98   *Chapter 3   Farm Drainage and Irrigation*

Still, I do not regard this kind of irrigation as being often found to be advantageous to use.

5) Subterranean Irrigation with Tiles

Common drainage tiles may be laid a short distance below the surface of the land at frequent intervals, and the water which is allowed to flow through them will find its way out at the points between the tiles and, by capillary action, will be carried through and to different parts of the soil. The main pipe, which is used to bring water to the field, should be quite large, [p. 174] its size varying according to the amount of land which is to be irrigated. Secondary pipes of medium size should be extend from the main to different parts of the field; from the secondary pipes, small pipes, about 1 inch in diameter, must be laid at frequent intervals. The distance apart between these small pipes will depend somewhat upon the nature of the soil—the greater its capillarity, the greater the distance; in general, the pipes should not be more than 8 to 10 feet apart. The number of tiles of secondary size that will be necessary will depend very much upon the contour of the surface of the land. On land of uniform slope, fewer tiles will be necessary than on that which slopes irregularly. In Figure 38, let ABCD represent a field in which the double line [p. 175] EF is the principal supply pipe. The lines HI represent secondary tiles, and the dotted lines JK show the smaller tiles which are used to distribute the water across the surface. The tiles should be laid at a sufficient depth below the surface, beyond the reach of plows and other implements used in cultivation. A depth of 10 inches or one foot will ordinarily be sufficient. There are two objections to this manner of irrigation, first, that it is very expensive to lay the tiles; and second, that it is very wasteful of water.

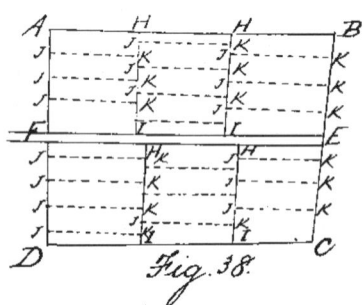

3-12. Methods of Irrigation

1) Cultivation or Disturbance of the Soil

After the soil of the garden has been irrigated, it should not be cultivated until it has become simply moist. Cultivated at this time, it may be pulverized, while if it were cultivated when wet, it would simply tend to make the soil into compact lumps [p. 176] which, when it becomes more dry, would be very hard. Cultivation after irrigation should, therefore, be deferred until you see that stirring the soil tends to pulverize it rather than to make it into lumps.

2) Time and Frequency of Applying Water

Gardens should always be irrigated either in the afternoon, in the evening, or on cloudy days. If land is irrigated when the sun is shining brightly, much of the water is evaporated and lost. A more important reason for not irrigating during bright sunshine is that the evaporation of so much water makes the soil very cool, and this checks the growth of plants. The frequency of irrigation should vary according to the nature of the soil, the climate, and the kind of crop cultivated, but I can give this general rule: it is better to irrigate so abundantly as to thoroughly moisten the soil [p. 177] to a considerable depth; therefore, the intervals between successive irrigations may be several days. Irrigation from every 8 to 14 days is the most suitable interval.

3) Garden Crops Benefited by Irrigation

Those crops which benefit most from irrigation are, Asparagus, Beans, Corn, Cabbages, Beets, Carrots, Onions, Potatoes, Peas, small crops, and garden fruits. These plants need but moderate irrigation. Irrigated too abundantly, they will not grow as well as with the less amount of moisture. Beans will thrive well with considerable moisture. The intervals confined between irrigations should be 7 days. Corn also grows well if irrigated quite abundantly, but so much water must not be supplied as to make the soil cold, because corn requires a warm soil. Cabbages [p. 178] may be supplied with a very large quantity of water, provided the soil in which they are growing is one which is well suited for them. Beets grow well if supplied with a large amount of water, but if you wish to raise sugar beets of good quality, you must irrigate less plentifully. Too abundant irrigation would produce a large beet poor in sugar. Carrots thrive well in light soils, if abundantly irrigated, but in soils which are clayey in their nature, care must be taken not to water too abundantly. Onions thrive well if very abundantly irrigated.

Potatoes: great care must be used in irrigating potatoes, since their quality is very likely to be injured by too abundant moisture. Until the time of blossoming, [p. 179] they may be irrigated at intervals of 9 or 10 days, if growing in light soil. After blossoming, they should be irrigated very sparingly, if at all. Peas: a considerable amount of water may be furnished to peas with advantage, but after the period of blossoming, less water should be given. Small crops such as Lettuce, Radishes, etcetera: for such crops as these, abundant irrigation is excellent, since it not only increases the quantity, but also improves the quality of the crops cultivated or produced. Garden fruits such as Strawberries, Currants, Gooseberries, Raspberries, Blackberries, etcetera: caution must be used in irrigating these small fruits, for while their quality is improved by a reasonable amount of irrigation, a supply of too much water would not only [p. 180] injure the quality, but would also lessen the quantity of the fruits produced.

4) Irrigation of Orchards and Vineyards

The supply of too much water to orchards and vineyards would not only lessen the quantity of fruits produced, but would also be highly detrimental to the growth of the trees, shortening their lifespan, lessening their vitality, and rendering them more likely to be injured by the cold of winter. Late in the season, very at this time, they will be exceedingly liable to be killed by the frosts of winter. When the fruit is ripened, water should be supplied only in small quantities. [p. 181] If this caution is not observed, the fruit will be very much injured in quality.

5) Irrigation with Liquid Manure

All manure, before it can be used by plants, must be brought into solution in water; hence, irrigation with liquid manure always produces beneficial results, since the plant food is in a condition to be used immediately. It is frequently possible to conduct liquid manure through open canals or pipes to the field to be irrigated, and there to spread it over the land with one of the methods already described as applicable in the case of ordinary water. There is another very good method of distributing much water which may be used to advantage on a small scale. That is by carrying it in a specially [p. 182] constructed carriage so arranged that, by opening a valve, water is thrown in a broad sheet upon the land passed over. This carriage usually has only two wheels and the receptacle in which the water is contained. It may either be a cask or a rectangular box. These carriages are made in sizes suitable to be drawn by a single horse, by a man, or even by two horses. These sometimes made of four wheels, and are so arranged that if they are drawn by horses, the driver rides on the carriage and controls the flow of water by opening or closing the valves.

6) Irrigation of Grass Lands

Grass is one of the crops which are very largely benefited by irrigation. In this latitude, if water can be used only for irrigation of a portion of the crops raised, it will be wisest [p. 183] to use it for the irrigation of grass lands. In all temperate climates, grass is one of the most valuable crops that are raised.

7) Value of the Grass Crop in the United States

As already stated, the yield in grass can be very much increased by irrigation. It is stated on good

The annual value of the hay crop	$ 273,000,000
The annual value of dairy products produced by grass	$ 400,000,000
The annual value of lambs and wool	$ 100,000,000
The increased value of other livestock such as horses, cattle	$ 200,000,000
Total	$ 973,000,000

authority that in England, one acre of land produced in one year 80 tons of grass. Grass most effectively removes [p. 184] the fertilizing material from the water made to pass over it. It also offers the best protection to the land to be irrigated, it entirely covers the land.

8) Winter Irrigation

As I have already said, the water of systems, when in a swollen condition, contains more plant food than at ordinary times. Now, streams are most likely to be in this condition especially in sub-tropical countries during the winter season; and hence, it is that irrigation in the winter is highly beneficial to grass lands, wherever it can be used. If the land is irrigated during the winter, a supply of plant food is deposited in and upon it for use during the following summer. Irrigation during the winter, however, can be used only where the frosts are neither severe nor long continuing. Whenever water is likely to freeze to any great depth [p. 185] and remain frozen for a long time, irrigation at this season would be injurious rather than beneficial. It will be injurious because the plants would be deprived of the necessary supply of oxygen. In this latitude, irrigation during the winter is, of course, impractical, but in the southern regions of this empire, grass may be made to grow much more abundantly by winter irrigation. Although irrigation in the winter will be impossible in this province Hokkaido, it will be advantageous to cover our grass lands with water in early spring, especially during frosty nights, to prevent the cold from injuring the grass. In this manner, the grass may be made to start much earlier in the spring than it otherwise would.

9) Use of Spring water for Irrigation

The water of springs, as we have seen, does not contain [p. 186] as much plant food as other kinds of water, but it is nevertheless much better than no water, and should be used wherever it is possible to do so. Springs, although they appear quite small since water that flows from them usually sinks into the earth near them, often yield quite large amounts of water. A spring yielding 2 quarts of water per second would, in 24 hours, supply 43,200 gallons which is nearly one quart per sq.-foot for 4 acres of land. This will be an abundant irrigation. It would be preferable, however to store more water and to irrigate more copiously at less frequent intervals. A reservoir containing 5,760 cu.-ft. will hold 43,200 gallons. The dimensions necessary to give this capacity might be 40 feet in length, 20 feet in breadth, and 7 feet in depth, but a reservoir which is not so deep will be preferable for two reasons:

(1) Because the temperature of spring water is usually cooler [p. 187] than that of the atmosphere. Now, all plants need for their growth a plentiful supply of warmth. Watering with this cold water, therefore, checks their growth somewhat. For this reason, it is essential that the temperature of the water to be used is raised before it is applied to crops. This can be done only by exposing the water to sunlight; hence, the larger the surface area of a reservoir of any given capacity, the more quickly or the more perfectly the water contained in it is warmed.

(2) Because a deep reservoir lessens the head, since it is usually necessary to take the water from the bottom of it. Now, if this bottom is lowered for the sake of getting a deep reservoir, we lose just as much head as we gain in depth. This very much lessens, in most cases, the amount of land that can be irrigated from the reservoir, since the water taken from the bottom [p. 188] will not be high enough to flow in open canals to the land in question. Figure 39 represents a cross-section through the middle of a reservoir and land. The line ABCD represents the former level of the ground. At the point E, the spring issues. The line AB represents the present bottom of the reservoir, HIJ shows the upper surface and sides of the dam, and the line KL its foundation. The portion at the center, represented by double shading, is composed of puddled clay. The other portions of the dam may be made of common earth the earth which has been removed from the bottom of the reservoir may be used for this purpose. The double line AF represents [p. 189] a pipe which discharges the water into the ditch at H. At V is a valve in the pipe which may be opened or closed at pleasure. The heavy shading represented by crossed lines shows the original earth.

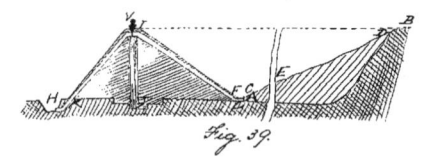

The dam, if a high one, must be twice as wide at the bottom as high, and the top should be 1/5 as wide as the bottom. It must be founded upon solid subsoil, for if this is not done, and there water will find its way under it, and there will be constant leakage and danger that the dam may be washed away. One-fourth of the width at the top should be filled with puddled clay, in order still to decrease the probability that water will still find its way through the dam. The inner slope of the dam should be what is technically stated: a slope of $1\frac{1}{2}$ to 1, that is, there should be $1\frac{1}{2}$ times as much horizontal distance as height. This slope should be covered with sods, because growing grass will most effectively prevent [p. 190] washing away by water erosion.

In the construction of the whole dam, care should be taken to make it as solid as possible, and to have the foundation a little lower at the center than at other points, in order to increase the strength of the dam to resist the pressure of the water. The outer slope of the dam should be such that ordinary rains will not be likely to wash the earth away. A slope of 1 to1 will, in most cases, be sufficient to ensure its stability.

The outlet pipe should be large enough to discharge at least 4 times as much water as enters. This must be so in order that it may irrigate quickly. For such a reservoir as we have been describing, a 3 inch diameter pipe will be sufficient. This would discharge 9 quarts of water per second, and this amount would furnish 2 inches of water to 4 acres of land in 12 hours. Figure 40 is intended to illustrate a balanced trap for the purpose of emptying [p. 191] a reservoir automatically. ABCD represents a dam, and OP a pipe running through the bottom of the dam and emptying into the canal L; EF is a pipe leading from a point near the top of the dam, and HR is an escape pipe leading from EF; I is a box which is hinged at J. At the other end of the rod is a weight K which is sufficient, when there is no water in the box, to cause its end to rest upon a support S′; "ab" represents a rod which is hinged at b. At the place on this rod which is opposite the mouth of the pipe OP is a leather or rubber cushion which, [p. 192] when the rod is in its present position, entirely closes the pipe.

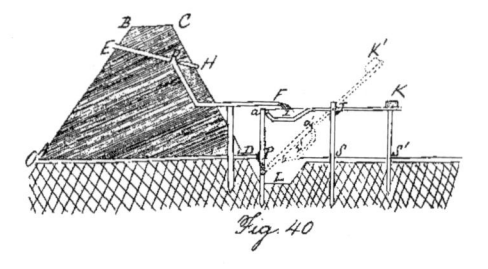

The method of operation is as follows: when the reservoir becomes filled to the point E, water flows through the pipe EF and empties into the box (I). The box, then becoming heavier than the weight, descends and the balanced rod takes the position represented by the dotted line, with the box at J′ and the weight at K′. At the same time, the pressure of the water through the pipe OP forces back the cushioned rod "ab" into the position shown by the dotted line "a′b". The force of water flowing out of the pipe OP keeps the balanced rod in this position. When the reservoir is emptied, however, the flow of water through the pipe OP ceases; the weight K descends from the position K′ to K. The cushioned rod is raised to its original position "ab", thus closing the pipe OP. It will remain closed until the reservoir is again filled and water once more flows from the pipe EF [p. 193] into the box, again causing it to descend and thus repeating the whole operation. This arrangement is entirely self-regulating, and is one of the most reliable automatic systems of emptying a reservoir. It is, as you see, extremely simple in construction and hence is not likely to get out of repair. In case it should not work, all that could happen would be that water will continue to flow through the pipe EF, doing no damage. This is quite extensively used in Switzerland and some other countries. The same rules as to the manner of building the dam, the size of pipes etcetera, apply as in the case where the water is emptied by a siphon. By using the water of springs for irrigation, we often gain two desirable objects, namely, the drainage of the land in which the spring is situated, and the capacity to use the water for irrigating purposes. [p. 194]

3-13. Water Meadows

A water meadow is a meadow, devoted to the raising of grass, which can be overflowed with water. Figure 41 shows the method of irrigating a valley through which a spring flows. The original course of the stream is represented by the dotted line AB. If the original water-course is too meandering, as is represented in Figure 41, it should be straightened somewhat. If it is entirely straight, that is better than any other position, but it is seldom possible to make the course of a stream perfectly straight. so it may be allowed to flow in a gently curving channel as represented by the double line AB. [p. 195] The water of the spring should be dammed at the point H, and turned into canals (ACD in Fig. 41), one on either side. These canals should follow a course which is nearly on a level with the point from which the water is taken, until they reach the higher land on the sides of the valley, where they should follow the foot of the slopes, but still as nearly level as possible. The double lines EF represent the positions of drains which it may be necessary to have at the foot of the slopes to prevent water from flowing down into the low land of the valley. The dotted lines CD represent the positions of drains which should lead from EF to the stream AB. The lines "ab" represent distributing furrows leading from the canals ACD to the stream. The lines 1 & 2 are distributing furrows leading from the main channels "ab". The course of the water in the different parts of the field is shown by the directions of the arrows. The land along the main channels "ab" [p. 196] must be slightly higher than adjacent parts of the field.

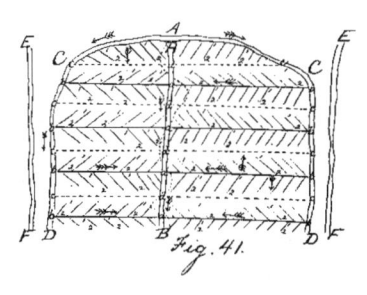

Figure 42 represents a water meadow in which AB is the stream or canal, "ab" an embankment along the edge of the stream, ACD a canal which takes water from the stream, "acd" an embankment along the edge of this canal, EF a drain for catching superfluous water from the portion of land above it, "ef" an embankment between successive sections of land, and HI a distribution channel leading from the canal ACD. This method of irrigation can be used only when the slope of the land is slight and uniform. The location of the canal ACD [p. 197] should be similar to that described in the last system. The land to be irrigated is divided into successive beds, the surfaces of which slope only slightly. Each bed may be 1 to 2 feet lower than the one which is next above it. Each field is enclosed between an embankment along the side of the stream and one along the side of the main supply canal on two sides; on the other sides, it is bounded on the upper part by a distribution channel, and on the lower part by a drain which carries off superfluous water.

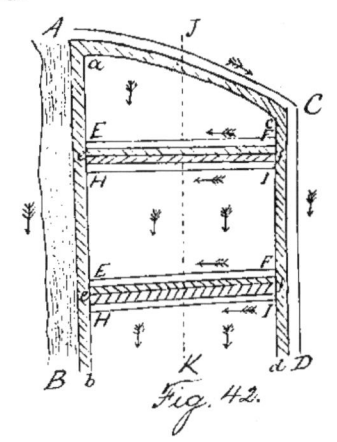

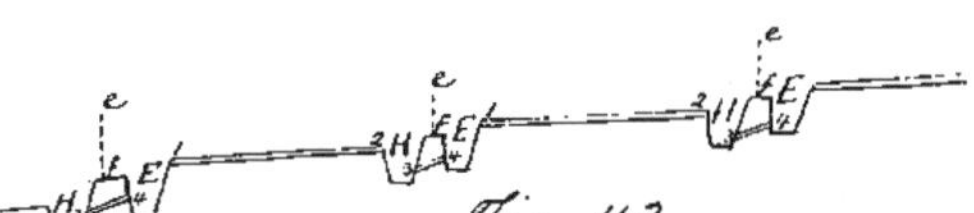

Figure 43 represents a cross-section through the plan of Figure 42 along the line JK. The drainage ditch EF is located above the highest canal bank "ef" in the plan. The cross-section of the distribution channel H is on the corresponds to HI in the plan. Lines 1-2 represent the line of the land surface from the distribution channel H to the open drainage ditch E, which is shown on the plan as ef. You will see from the section that each bed is lower than the one next above it and that the surface of each slopes slightly towards [p. 198] the one next below it. Each bed, however, may be made perfectly level, but it is actually better to have a very gentle slope. The double line 3-4 in the cross section represents a spout through the embankment for the purpose of carrying superfluous water from the drain at the foot of one bed into the distributing canal of the next. In the plan, the direction of the flow of water is shown by the arrow.

Figure 44 represents a spout AB through the bank CDE, with a hinged gate F which will yield to the pressure of water through the spout in the direction of the arrow, but which will close if the water into which it empties rises above it.

Figure 45 represents a dam which may be used in irrigation canals. ABCD shows the cross-section of the canal, and AHID that of the dam. The portions AHB and DIC are the banks of the canal to give [p. 199] stability to the dam. The middle portion of the dam EF is so arranged that it may be conveniently raised or lowered at pleasure. If the dam is a large one, its middle section should be raised by means of a rack and pinion. If the dam is a small one, it may be raised by hand and held in place by means of pins. Figure 46 shows a crude drawing of a rack and pinion. The larger the area that can be irrigated in this manner, the better, as far as expense is concerned. The canals, embankments, dams, etcetera must it is true if the amount of land [p. 200] to be irrigated is large be more substantial and expensive than if the amount of land is small; notwithstanding this, it will be cheaper in the end in proportion to the amount of crops raised than to irrigate smaller pieces of land. One reason for this is that the works, once properly constructed, are very permanent. All embankments should be made with a moderate slope of ample size, and should be covered with grass, since grass most effectively prevents injury from washing.

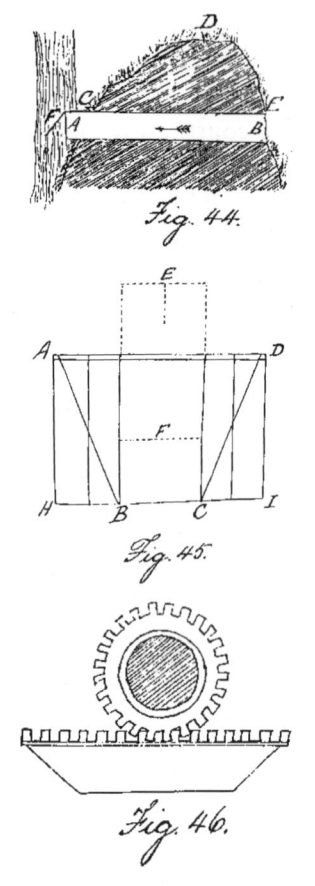

The amount of water used in this system of irrigation varies very greatly. It has been stated that in some instances, from 13 to 27 feet per annum have been used, but those are very large amounts, and will only be necessary in very dry climates on soil which is very sandy. On level land of clayey nature, the least amount is sufficient. When the land slopes regularly in one direction, the arrangement of canals and drains is very simple [p. 201] as shown in the figure. The maximum slope with which this system can be used is a fall of 1 foot in 100 feet (1% incline). If the fall is greater than this, another system must be adopted. If the surface of the land slopes irregularly, the arrangement of the canals and drains must be made to suit the slope.

The frequency and duration of irrigation must be done to cover the grass lands to a depth of 2 or 3 inches, and permit it to flow over the meadow in a gentle current; the amount of water admitted into the distribution channels and allowed to flow out through the drains, drainage ditches must be regulated so as to bring about this result. An irrigation of 10 or 15 days and an interval of 5 days without irrigation is a common practice. For about 2 weeks before you intend to cut the grass, you should not irrigate, because you wish the land to become hard so that you can work conveniently. [p. 202]

(1) Preparing the Surface of a Water Meadow; The surface of a water meadow must be entirely free from little inequalities, in order that the water may cover all portions of it to an equal depth. Surface inequalities must, therefore, be carefully removed before grass seed is sown. The larger inequalities may be smoothed by the use of a scraper. The land should then be ploughed without back furrows, and then be thoroughly harrowed and rolled. By the use of these implements, the land may usually be made smooth enough, but there may sometimes be places in the corners or even in the middle of the field, which may require a little hand labor. The distributing furrows may be made mostly by the use of a plow. Fig. 47 represents a cross-section of a furrow [p. 203] made by the use of a plow; In Fig. 47, AB shows the line of the surface, C is the place from which the furrow has been turned, and C' is the furrow slice which has been thrown out of the furrow. The slice should be thrown towards the upper side of the furrow. This will protect from the water of the bed next above it, which the water contained in it may flow over the other edge of the furrow. It will usually be necessary to perform some hand labor in smoothing the furrow. After the necessary embankments, canals, and furrows have been made, and after the surface has been properly smoothed, the seed may be sown.

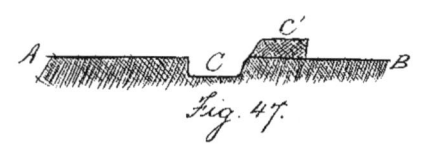

Figure 48 represents an arrangement of canals and furrows that may be resorted to [p. 204] where the slope is regular but too great for the system previously described. The principal slope of the land is in the direction of the arrow. AB represents a canal at the head of this slope and CD are branch canals which run diagonally down the slope. By running in this direction, they have the proper amount of fall, whereas if they run directly down the slope, the fall will be too great. The lines marked "a" show the positions of distributing furrows, which take the water from the canal CD and distribute it equally over the surface. These furrows "a" may be very shallow; they should be about 2 feet in width at the place where they leave the canals CD, and should gradually come to a point. The depth at the center may be 4 inches. At the other end, they should have no depth. The slope of the sides of these furrows is so gradual that grass may be grown in them. The distance between the CD may vary from 100 to 150 feet, according to the nature of the soil. [p. 205] In soil which is very porous, they must be placed nearer than one which contains a considerable amount of clay. The "a" may, therefore, be 50 to 75 feet in length. The distance between successive furrows should be 15 to 20 feet. If the slope of a field is very irregular, the arrangement of the canals must be made to suit the nature of the surface. It will be impossible to give rules which will cover every case. All that is necessary is that you arrange your canals so as to conduct the

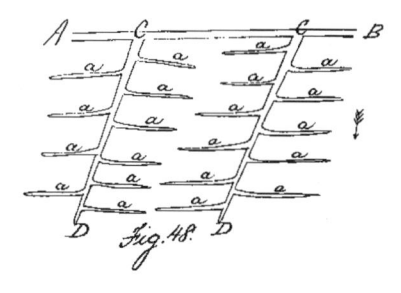

106 Chapter 3 Farm Drainage and Irrigation

water to all parts of the field conveniently, without having too great a fall in any of the watercourses.

Figure 49 is intended to represent a field in which the slope is [p. 206] very irregular. AB is a principal canal, and the lines CD are secondary canals which are so arranged as to have a little fall, to make it possible to spread the water evenly over the surface. The lines marked "a" represent distributing furrows. The rules with regard to the distance between the canals and furrows, and with regard to their size and form of the latter, are applicable as in the manner of the system last described.

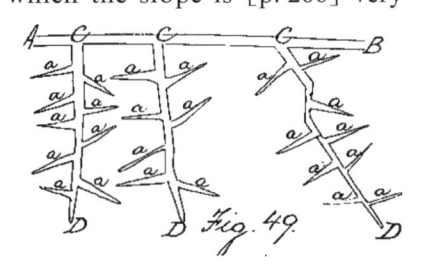

(2) Catch Water System; Where the slope of land is very great, it may be impossible to irrigate according to any of the systems yet described; and in such cases, the catch water system may be used. In this system, each canal should follow the same level throughout its course. [p. 207] Figure 50 represents a section of a hill-side arranged by irrigation according to this system. AB represents the line of the surface; at the head of the slope is the canal C, and a short distance below is the canal C', and they follow down the whole slope. Water is let into canal C, which it fills and overflows the edge and passes down to C'. After filling this canal, it again overflows and passes to C'', and so on down the slope. Since the slope of the land on which this system is used is quite great, there will be considerable water lost, unless the bottoms and lower sides of these canals are rendered somewhat impervious to water. If the soil contains considerable clay, the bottom of the canal may be made quite impervious by puddling clay, and the sides would probably be sufficiently impervious. If the soil, however, [p. 208] is loose or sandy in its nature, you must either bring clay and place it in the bottom of the ditch, or resort to some other method for preventing the water from soaking down. A very good way would be to use boards for the bottoms and lower sides. The bottom boards should be 14 inches wide, and the side boards 18 inches wide. The side boards will effectively prevent the edge of the canal from being injured by water.

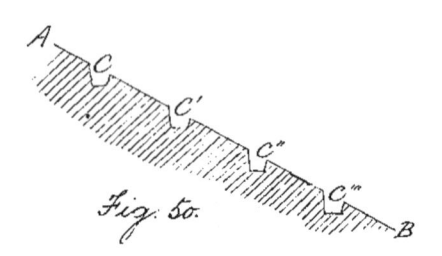

3-14. Management of Irrigated Fields

Irrigated fields are frequently pastured, but great care must be used not to pasture animals for too great a length of time on such fields, as so doing would be likely to injure their health. Sheep, especially, are liable to contract bad diseases if pastured for a long time upon an irrigated field. The land must, of course, be dry while the animals are upon it. [p. 209] It is best to confine the animals upon a small piece of land, and to keep them there until they have eaten all the grass, since if allowed to run at pleasure over a large field, they will cut down a great deal of grass. They should be allowed no more area than they will eat all the grass in one day. They may be confined by means of movable fences, and they should be given fresh pasture each day. You must not allow animals to enter an irrigated field until it has become quite hard, since they will sink into the soil and leave the surface rough. It will be found useful to frequently roll irrigated land which is pastured. Rolling will smooth the little inequalities of the surface caused by the trampling of the animals. The best time for irrigating pastures or mowing fields is either at night or on cloudy days, for the reasons already given. The quantity of water needed [p. 210] depends upon conditions previously mentioned, —i.e. such as the nature of the soil, climate, etcetera. But you should remember that grass will thrive under more abundant irrigation than most other farm crops. The irrigator must exercise constant watchfulness to keep all his works in good repair. Every autumn and spring, all ditches and canals should be inspected and put in perfect order. Rolling the surface of grass land in the springtime, after the soil has become dry, is imperative. Irrigated fields should generally be fertilized with substances which contain plant food in a soluble condition. Nitrate of soda, Peruvian guano, and fish guano will be found very valuable. Feeding animals pastured on irrigated fields a liberal allowance of such foods as corn or grain of any kind, oil-cake, or indeed any rich food, will furnish sufficient fertilizing [p. 211] material to the fields pastured.

1) Cost of Irrigation

In deciding whether we can irrigate or not, we must carefully estimate the cost of procuring a sufficient supply of water, and the increased value of crops which will result from it. The British government has spent an estimated $701,000,000 in India for irrigation, and the investment in almost every instance has been very profitable. In Spain, the charge for water is $7 per annum for each acre. This includes 12 watering in which 33 inches of water are supplied. In Italy, 1,600,000 acres of land are irrigated at an expense of about $20 per acre. The increased value of crops from this irrigation is such that the land rents for $4,500,000 more per year than before. This returns 15% interest on the money invested. The average cost of water to the farmer is $7.50 to $8.50 per cubic foot per second. [p. 212] This equals $2.50 per acre for Indian corn, $7.50 per acre for grass, and $20 per acre for rice. The irrigation works in Italy are very permanent. Some of the European works now in operation have been in use for 1500 years, and some which are 2000 years old, though not used at present, are still serviceable.

The figures which I have given you show the cost of irrigation in some of the countries where it is abundantly practiced, but this cost must vary, of course, according to the case with which a supply of water can be obtained, according to the nature of land to be irrigated, and the cost of labor. The first and last items are those which will be most likely to vary. These must differ according to the varying conditions of each individual case. Hence, no definite rule as to cost can be given. In Japan, since labor is cheap, the cost, wherever water can be [p. 213] conveniently obtained, must compare favorably with the cost in the countries which I have mentioned.

2) Consumption of Water

This must vary, of course, according to the climate and the nature of soil. The average amount is, however, stated to be 72 square inches per 100 acres, continually flowing at the rate of 4 miles per hour. This does not include loss by evaporation or soakage through the bottom of the canal, but is the amount which must be actually furnished to the field. The amount which will evaporate will depend largely upon the climate of the country. If the country is prevalently hot, a large amount of water will be evaporated. If the soil is loose and porous, much water will sink into it. In one canal in California, the loss by evaporation and soakage amounts to 40% of the quantity of water [p. 214] entering at the mouth of the canal. On account of this loss, which must be greater or less in all cases while the water is conducted through open canals, you must make provision for more water to enter the canal than will be needed in the field where it is to be used.

3-15. Crops most Benefited by Irrigation

As a general rule, broad leaved foliaceous plants are more benefited by irrigation than others. This is because the larger the leaf surface of a plant, the greater is the amount of water which can be thrown off by it. Long tap-rooted plants such as beets, turnips, and parsnips are also much benefited by irrigation. Plants belonging to the order Leguminousae are generally much benefited by a plentiful supply of water. Rice, as you well know, [p. 215] must have a plentiful supply of moisture. Cranberries must be similarly treated. Field crops and the methods to be pursued in irrigating them are as follows:-

(1) Wheat can be made to produce more abundant crops under irrigation, but caution must be used in supplying water. An excessive supply encourages the growth of grass rather than of grain. If the wheat is sown in drills, the water is allowed to flow between them. If it is sown broadcast, the land should be rolled with a corrugated roller, which will leave it thrown up in little ridges between which the water may be allowed to flow. Wheat should be irrigated at intervals of 7 to 14 days, the amount of water varying [p. 216] according to conditions already stated. After the grain is well formed, no water should be supplied, because the grain must be left to ripen.

(2) Corn thrives best in a moist warm soil; therefore, in a warm climate, irrigation is beneficial to this crop. It may be planted either in drills or hills, and water may be allowed to flow between these drills of hills. No water should be supplied after the corn is glazed.

(3) Flax flourishes best in cool moist soils. In such soils, the fiber is finer and of better quality than in dry soil. Hence, irrigation is beneficial to this crop, and it may be quite abundant.

(4) Hemp thrives under similar treatment to that which is adapted to flax. [p. 217] These two crops should be sown in drills, and water should be made to flow between them.

(5) Tobacco; the tobacco crop is very much benefited by irrigation. A leaf of fine texture and mild flavor, as well as good color, may be produced by judicious management. The land should be made up into ridges, and water allowed to flow between them, in order to thoroughly saturate the soil before the plants are set out. After the plants have been transplanted, the land should be watered again. After 2 or 3 days, it should be watered still again, the objective being to supply abundant moisture until such time as the plants become rooted. Then, an interval of 20 days without irrigation should follow, after which water is furnished once in every 8 to 14 days. [p. 218]

(6) Cotton; few crops need so little moisture as cotton, yet it has been grown

successfully under irrigation; however, caution must be used to keep the soil simply moist, and no more. One irrigation just before the seed is sown, in order to make the soil moist so that the seed may germinate quickly, and another during the period of growth, may be sufficient, except in very dry climates.

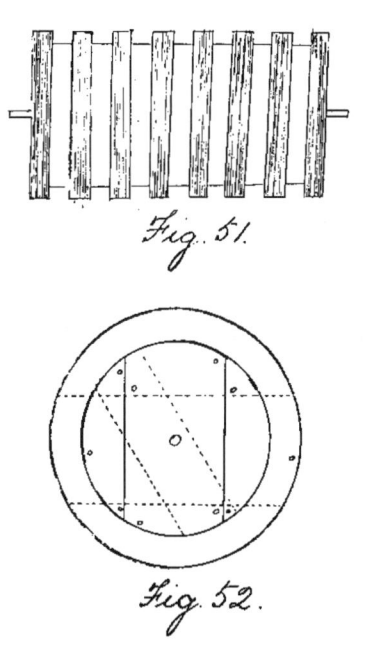

(7) Sorghum; In irrigating this crop, care must be used not to supply too much water, since doing so would make the juice, and the manufacture will cost much more than if the amount of sap is less, and the percentage of sugar greater. This crop should be irrigated only enough to promote a uniform, healthy growth. [p. 219]

The Section Roller (Fig. 51) is a very good roller for use in fields sown broadcast with grain crops, for preparing the surface for irrigation. It is made of circular sections of oak plank, 30 and 36 inches in diameter, the different sizes being placed alternately. These sections revolve upon a common axis which passes through their centers; each section may be left to move independently, but a better way is to join three sections together for convenience in fastening the planks of each section (Fig. 52). This roller may be provided with [p. 220] a framework and a seat for the driver, like an ordinary roller, and should be drawn by two horses.

3-16. *Means of Getting a Supply of Water*

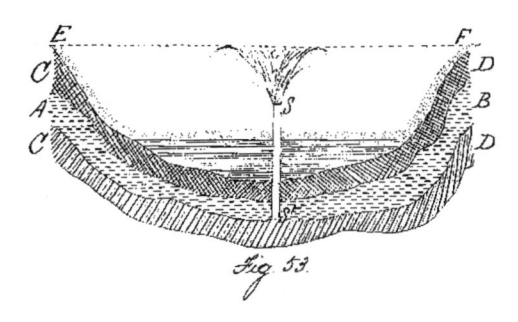

Artesian wells are recommended by many writers as a source of a supply of water for irrigation. Now, while they may be useful in cases where water cannot otherwise be obtained, they do not ordinarily furnish sufficient water to irrigate enough land to pay for the cost of their construction. Artesian wells are so called from the name of a province in France where they were extensively used at an early period, although they had been known in China at a much earlier time. They can be successfully made in soil of peculiar geological formation. The formation in which they may be made is illustrated [p. 221] by Figure 53, in which EF represents the surface of the land. CD are impervious strata, between which is the pervious stratum AB which comes to the surface at the points A and B. Water falling to the surface at these points will sink to the lowest part of this stratum S' and will, in most cases, almost if not entirely fill it. If a hole is bored in the valley until it strikes this pervious stratum which is full of water, the water will rise to the surface and often may be thrown a considerable distance above the surface. The height to which it is thrown [p. 222] depends upon the height of the points A and B. If unresisted by the pressure of the atmosphere, the water would rise as high as these points, which may be hundreds of miles distant, situated upon the tops of high mountains but it will make no difference, for the water will always seek its level. These wells are sometimes bored to a very great depth some are more than 2,000 feet deep. The supply of water is not infrequently very great. There is one well in France which yields 600,000 gallons of water per day. and another in Kentucky in the United States which yields 300,000 gallons. It is not often, however, that they furnish so large a supply of water. The amount usually furnished is not sufficient to irrigate land enough to repay the cost of making the wells, since they are very expensive. In ordinary cases, therefore, we cannot depend upon those wells to furnish enough water to irrigate a large amount of land. [p. 223] However, in countries where it is impossible to obtain water from any other source, it is of course a good policy to bore these wells.

(1) The Water of Streams; The most common sources of a supply of water for irrigation are the rivers and streams of a country. In many cases, this water is high enough to be taken directly from the stream and carried in canals to the land which you wish to irrigate. Not infrequently, however, it is cheaper to raise the level of the water in the stream by means of a dam to take the water from a point higher up a stream, and construct a canal to convey the water to the place where it is needed. These dams may be of various sorts, a few of which I will briefly describe. [p. 224] Where it is necessary to raise a water only a few feet, it may often be done by building what is called a wing dam. In Figure 54, CD represents a river or stream, EF a dam constructed so as to raise the water a few feet, and AB a canal to conduct the water to the field to be irrigated. At point F is a gate to admit or exclude water as desired. In building such a dam as this, the first step must be to survey the land to be irrigated very carefully. The objective of this survey is to determine how much the water in the stream must be raised. If you find that by carrying the dam a reasonable distance [p. 225] upstream the water can be raised sufficiently, you may proceed with the work.

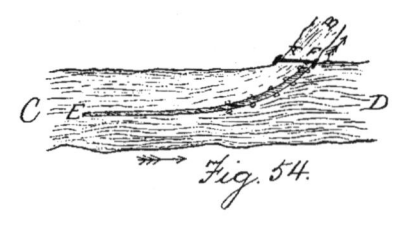

The next step will be excavation of the canal. The dam should then be commenced at the point F and carried up the stream until a point is reached where the level of the water's surface is as high as the land to be irrigated. This dam should be made of materials that will bind together well. Piles may be driven 3 feet apart in a double row in the first portion of the dam, and then be continued a greater or lesser distance upstream, according to the strength desired. These piles may be fastened together by means of girders extending from one to another. It is also good to make a network of brush extending from one pile to another. The distance between these rows of [p. 226] piles may vary according to the strength of the dam required. Their size and number must also vary according to the same consideration. This wing dam must be stronger at the end near F than at the other, since the water is deepest at the former end. It will usually be necessary to use piles throughout the whole length of the dam. The upper portion may be made of earth and stones, if you can find them.

In Figure 55, AB represents a stream, C is a canal leading from it, and a gate "d" which controls the flow of water into the canal, and D, D' are dams, each of which is built halfway across the stream. The distance necessary between the ends of dams D and D' depends upon the amount of water [p. 227] flowing in the stream. If the amount is great, the space must be greater than in cases where the amount of water is small. Water flowing down the stream striking against the dams is checked in its course and raised a few feet, with most of it finally escaping in a rapid current between the ends of the two dams. Such dams can raise the water but a few feet, and it used only in streams where there is a large volume of water, all of which cannot be stopped, and where it is desirable that vessels may be able to pass up the stream. The construction of the dams D and D' may vary according to the degree of strength required, but since they are subjected to the continual action of strong currents of water, they must be made of materials which will not be washed away. Stones held in place by crib work or piles [p. 228] would answer this purpose well. This system of damming water is called a Cross Wing Dam.

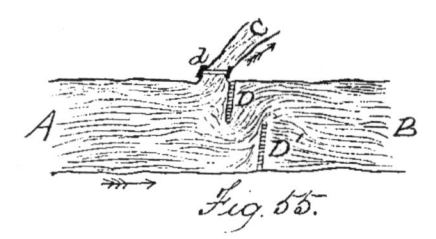

Figure 56 represents a cross-section of a cribwork dam. AB is a horizontal timber; at point E is mortised a timber EC; AC and CB are timbers placed diagonally. Several such frames are made and placed in the stream at distances varying according to the degree of strength required. After being placed in position, planks are nailed to them; after those on the upper side being put on, the interior space is filled with stones; lastly, planks are nailed to the lower side. Care must be taken, in making such a dam as this, to pack [p. 229] material such as clay solidly at the bottom, in order to prevent water from finding its way under it.

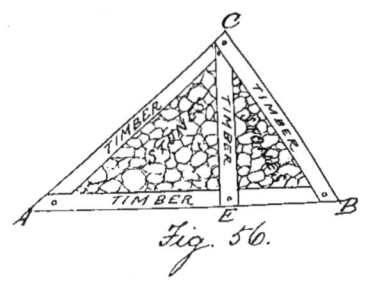

Figure 57 represents a cross-section of a dam made of piles and stones, which is called a dam of piles and rock. CD represents piles of different lengths driven far enough into the ground to be secure. The lengths should be graduated to give the proper slope to the upper side of the dam. AB is a timber connecting the ends of these piles. The interior spaces and the lower side of the dam should be filled with stones or gravel as shown. Several such frameworks may be placed in the stream, the number varying according to the strength required. Planks may be nailed from one diagonal timber to another [p. 230] as in the case of the cribwork dam. Such dams are best adapted to rivers whose bottoms so soft that piles may be easily driven.

(2) Method of Building Earth Dams; I have already given you some directions as to the manner of building dams of earth, but those were intended for small dams. The same care with regard to the foundation, selection of materials, proportion between height and width, top and the bottom, and the slope of the sides, must be exercised in the cases of dams of the largest size. If the

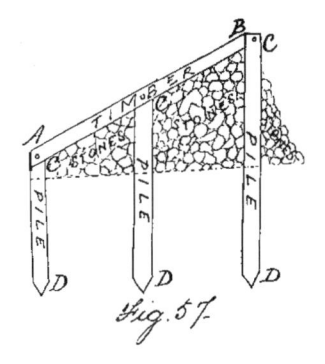

dam is a large one, the puddled work in the middle should increase 2 inches in thickness for every foot of descent from the top toward the bottom. In filling in the earth, care should be taken to place it in curved layers, [p. 231] with the lowest part of the curve being at the middle. Carefully selected materials should be placed next to the wall of puddled clay, that is, material which is known to bind together well. On the inner side of the dam, less carefully selected materials may be placed, but the surface should be covered with something which cannot easily be washed away by water. The outer side of the dam should also be covered with such material, so that in case water flows over the dam it will not break it down. The top should likewise be protected with similar material. Stones or boards, fastened securely in place, will be suitable for this purpose. Every reservoir for the storage of water should be provided with a waste way. This should be at least 4 feet lower than the top of [p. 232] the dam for one 25 feet high. For each additional 25 feet in the height of the dam, the waste way should be one foot lower. The size of the spillway must vary according to the circumstances of each case. If the water is liable to rise suddenly and with great rapidity, the spillway must be quite large in order to provide room for the water to flow out rapidly. The spillway, since it is intended that water should be allowed to flow over it, must be constructed of material which will resist the wearing action of currents of water. It is usual to place a few boards across the top of the spillway in order, in ordinary times, to raise the water in the reservoir nearly to the height of the dam. These boards are not securely fastened in place, so in time of freshet they will be carried away, thus allowing the excess water to pass out. [p. 233]

(3) The Pressure on Dams; The pressure on a dam depends upon the depth of the body of water confined by it. The pressure on any given area of an immersed surface is equal to the weight of a column of water whose base is the area of that surface, and whose height is equal to the distance of its center of gravity below the surface of the water. The weight of one cubic foot of water may be taken, in round numbers, to be $62\frac{1}{2}$ pounds. The total pressure upon an immersed rectangular surface is equal to the weight of a column of water whose base is the area of that surface, and whose height is equal to 1/2 the distance from the inferior to the superior bases of the rectangle. The center of pressure upon an immersed surface is a point to which a force, equal and opposite to the resultant of all pressure acting upon [p. 234] the surface must be applied, in order that it may maintain its stability. The center of pressure of a rectangular surface is on a line connecting the centers of its inferior and superior bases, at a point 1/3 of the distance from the former to the latter.

If a dam is built of masonry, all that can resist the pressure of water behind it is the weight of the dam; therefore, in computing how thick the dam must be, you must find what the total pressure upon it will be. Since in a rectangular dam the center of pressure is located at a point 1/3 of the distance from the inferior to superior base, you must multiply this total pressure by 1/3 of the height of the dam in feet, and divide the product thus obtained by 1/2, which is the weight of a cubic foot of masonry. The quotient will be the number of cubic feet that you must have in the dam in order to give the required weight; since you know the length and height, you can easily [p. 235] compute the thickness. The rules which I have given you for calculating the pressure upon a dam will enable you to form a correct estimate of the required strength, and you will be able, after a little practical experience with the strength of materials, to construct a dam of any sort. You should remember that is much better to err on the side of superfluous strength than to make a dam too weak.

(4) Pumping to Raise Water; Besides the methods already described of raising water for irrigation, pumps using wind or steam power may often be used economically. Pumps have recently been invented which are capable of raising a large quantity of water very cheaply. Shaws' Compound Propeller is a very powerful pump. [p. 236] One exhibited at the Centennial Exposition in 1876 had the capacity to throw 100,000 gallons of water per minute. In cases where water cannot otherwise be raised, such pumps may often be used to great advantage.

(5) Supply Canals: The average fall in water supply canals should not be more than one foot in 1,000. This gives a current of $1\frac{1}{2}$ miles per hour. Half of this fall or $2\frac{1}{2}$ feet per mile would be still better, giving a flow of one mile per hour. The reason why the flow must be small is that, otherwise, the current will be so rapid as to carry away portions of the bank. The degree of slope possible varies, therefore, in different soils because some are much more easily carried away than others. The following table gives the degree of slope necessary in different kinds of soil. [p. 237]

In cases where the slope is necessarily quite great, the sides of canals are frequently made of masonry or timber. The fall should be uniform throughout the length of the canal, otherwise injury would be likely to result to its banks from the unequal flow of water. It is sometimes necessary to make canals in places where the slope is so great that the bottom and sides would be washed away by water.

KIND OF SOIL	Degree of Slope per 100,000 feet
Fine mud	16
Soft Clay	45
Sand	136
Gravel	433
Solid Clay	570

In such cases, it is customary to make the canals with a slight slope for the greater part of the way, and to have occasional shoots made of timber having an abrupt slope, down which water can flow to the next level. [p. 238] The size of a supply canal depends somewhat upon the amount of fall and the form of the canal. The deeper a canal of a certain size, the more water will flow through it. The proper inclination of the banks differs depending on the kind of soil.

In deep canals, the slope should be broken (Fig. 58, a & a'), by a bank slightly above the level of the water. The object of this bank is to catch any earth which may be washed down the slope, [p. 239] and prevent it from being carried into the water. In canals extending around a hillside, the water impinges upon the bank at every point of the outward curve. Sometimes, therefore, it is necessary to build a wall of stone or brick upon this side. Such materials will be most durable, but very good protection can also be given with woodwork of various sorts.

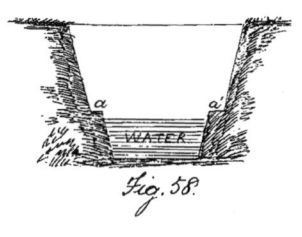

In finding the capacity of a canal, you must multiply the area of the cross-section in feet by the number of feet the water flows in a certain length of time. This will give you the number of cubic feet of water which would be furnished in that time, if there was no friction: however, there is friction both at the bottom and sides of the canal, and therefore this result should be reduced by 1/5. The usual estimate of the amount of water needed in common for ordinary irrigation is one cubic foot (about $7\frac{1}{2}$ gallons) per second for 100 acres of land. [p. 240]

(6) Dams with Gates; In the construction of dams with gates for the admission or exclusion of water in different canals, care must be taken to so make them that they will not allow water to leak through them. In order to effect this, the gate should slide in close-fitting grooves. At the bottom, an iron plate should be fastened, and to the bottom of the gate a plate of iron to fit the other should be fastened. The size and strength necessary will depend upon the size of the canal and the depth of water in them. The gate may be raised either by a screw or a lever. Hand gates, which consist simply of a board of proper size with a hole in it for convenience in handling, are very useful in smaller for controlling the flow of water. [p. 241] The number of gates and dams must be sufficient to enable you to perfectly control the flow of water. The material out of which gates are made should be the most durable oak timber. There are many other kinds of hand gates, some of which are represented below; explanations and modes of using them are left to the judgment and consideration of the reader. [p. 242]

KIND OF SOIL	The Slope Necessary from the Horizontal Plane
Wet Clay	16 degree
Dray Clay	45 degree
Coatse Gravel	40 degree
Ordinary Compact Earth	50 degree
Arable Loam	28 degree
Wet Sand	22 degree
Dry Sand	38 degree
Fine Gravel	40 degree

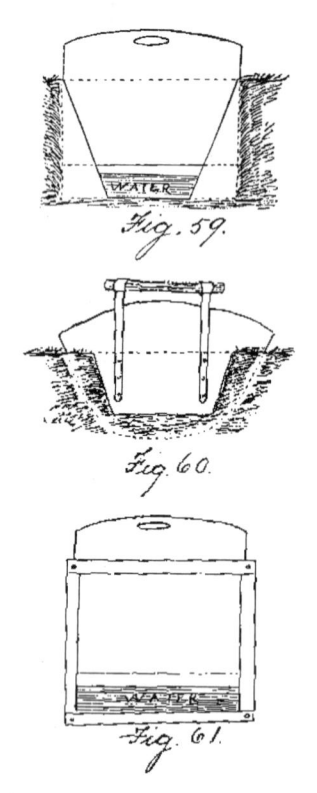

Fig. 59.

Fig. 60.

Fig. 61.

解　　題

1. ブルックスの排水

　ブルックスの講義ノートの目次を見ると，3-1 からの「排水」，3-9 以降の「灌漑」の順になっている。これはわが国において灌漑→排水としている一般的形式の逆である。明治の後半になって出版された上野英三郎・有働良夫の『土地改良論』が灌漑→排水の順である。これは水田農業を主体としているわが国では当然のこととして受け入れられた。しかし，その後に出版された札幌農学校出身の鈴木敬策の『農業土木学』では排水→灌漑となっていて本講義の影響かと考えられる。前者での灌漑→排水は水田への湛水，その水の排除を対象とし

ているのに対し，後者の排水→灌漑は冷湿地においては排水によって初めて農地が造成され，その農地への灌水を意味していることの差異を明確に表しているのであろう。これは今日でもカナダ，アメリカ北東部の大学では学科目として Drainage & Irrigation があり，アメリカ西部では Irrigation & Drainage としていることと符合することであろう。ブルックスの農学の第3章は，単なる作物対応ではなく農地基盤として位置づけられていたのである。今，排水関連の項の内容を理解するために小項目次をあげると次のごときものである。

　1) 排水の効果，2) 明渠排水と水路，3) 暗渠の種類，4) 土管暗渠，5) 暗渠管の設置法，6) 排水の必要性，7) 施工用具，8) 土管暗渠の配置

まず明渠に関してはきわめて簡単に触れている程度で，それもあまりよい評価を与えておらず，今日で言う承水路形式の利用に止めている。これは別に「土工学」の講義があり，「道路ならびに鉄道の建設，排水ならびに灌漑工事に必要なる範囲」を扱う内容のため，重複を避けたとも考えられる。一方，暗渠排水については，きわめて具体的な材料，施工技術などについて述べている。当時の技術としては世界(イギリスが先進国)レベルのものを講義していたと言えよう。ただ，それが当時の日本でどれだけ受け入れ可能であるかまでの十分な検討がなされていたかは不明である。しかし1851年のロンドン万国博覧会に初めて出展された土管製造機を輸入し，1880(明治13)年に農学校農場に埋設して，1891(明治24)年には明治天皇の御前で土管を製造したという記録もあり，その意欲は壮とするものがあったと見るべきであろう。

2. 明治以前の暗渠排水

わが国でも最古の暗渠排水は古墳に用いられたものであるとされている。水田を主体とするわが国の農業では水はその死命を制するものであるが，灌漑が主であって排水にはあまり留意されなかったようである。また，排水を必要とするようなところで，それが可能なところに適用されるようになったのは，土地利用・技術が進んでからであったのであろう。大阪府豊能郡の「ガマ」と呼ばれている特殊な水田排水溝は太閤検地(1593年)以前からあったことが知られている。これは構造から見て排水溝の構築後にその上部に水田が造成されたものと考えられている。いわゆる棚田の地下を通る排水溝で，形状的には中国大塞地区の排水溝と形状的にはきわめて類似しているもののようである。ただ，中国のは河川的排水溝であるのに，「ガマ」は用水路兼用の排水路のようである。

排水の効果として，①裏作の可能化，②乾田化による増収，③農作業の容易化，④冷水害の軽減，なども考えられたようであるが，逆に湛水をよしとする考え方もあり，必ずしも排水が勧められたわけではない。また卑湿を除く工法としても雑木類の投入，客土などがなされていて暗渠は用いられるに至っていない。『若林農書』(若林宗氏・若林利朝，1689年)では石礫暗渠と石垣を直結させると用水を得ることができるとしている。『農業全書』(宮崎安貞，1697年)では「手立てを以てその水を落し」とあり，排水のための手段を講ずることが述べられている。一方，『農具便利論』(大蔵永常，1817年)では明確に暗渠について述べられている。すなわち，石礫暗渠，丸太暗渠の工法について詳細に書かれている。なお，現場

では天保期(1830-44年)に耕地整理とともに実施され，暗渠の配置は組織的に計画され，大きな進歩をみたとされている。

3．明治初期の暗渠排水

　明治時代に入り，暗渠排水の技術は他の諸技術と同じように知識として導入されてきた。暗渠排水の技術は当時のヨーロッパでも最新の技術であったのである。それだけに，地域性の強い技術であるとともに実践しづらいものであったことが想像される。暗渠排水の技術の導入というよりも，多くの著書の翻訳がなされた。すなわち19世紀後半にRault, Stephensによって，暗渠間隔は土性に応ずること，材料としては石礫・土管がよいことを知り，Norton では，適切な暗渠とは上部の土に乾燥亀裂を生じ，それが全体に行きわたり土壌を膨軟にするものであることなどを知った。

　知識として導入された暗渠排水も現場ではほとんど進捗をみなかった。これはわが国が水稲耕作を主としているため，灌漑系統へその力を注いでいたためばかりではなく，東日本を主としてではあるが，排水可能な水田にも人為的に冬期間湛水・灌水するのがよいとする考え方，慣行が広く行きわたっていたためである。すなわち，貞享年間(約300年前)出版の『会津農書』にも次のごとく示されている。

　「山田，郷田ともに惣じて田え冬水かけて良し。何の河にも何の堀にも川ごみ有効にて，取分け町尻，村尻，そのほか深水の懸かるところは冬水かけて良し。そのうえ路辺より恵水入れて良し。卑泥田は春水かけても不苦。陸田は春水を干すべし。遅くては生田にて悪し」。この考え方は明治中期まで全体を支配していたと見てよいであろう。

　『改良日本米作法』(酒匂常明，1887年)も湛水反対の立場をとりながらも，湛水が実施されていることを認めている。また，乾田化を熱心に唱えた横井時敬も『稲作改良論』(1905年)では乾田に例外のあることをほのめかしている。すなわち，肥料の流亡防止，耕起の容易，害虫の凍死，用水不足の予防などのため，静岡県から新潟・福島・千葉・茨城・秋田県などの単作地方では，冬期間の湛水・灌水がよいとされていて湿田のまま放置されていたのである。

　それに対して二毛作の進んでいた西日本では乾田化が必然的に進められていた。『会津農書』より12年後に発刊された宮崎安貞の『農業全書』では，「泥田に麦を蒔くことはなりがたく，されども水を落とし干田になるべき痩せたる地は，手立てをもってその水を落し干田となし，犂きて其のまま播かず，其の塊をくだかずしておき，力次第段々に耕し，下まで良くほしたる時雨を得てよくかきこなしたるは，土よくとけて陽気を保ち，苗よくさかゆるものなり。これは塊のよく干せたるほど実りおほし。〔中略〕又山谷などにやせたる深田，或は冷水所，赤さび水の出づる地，常の作りやうにては稲の生長せざるところをも手立てを用いて水を落とし，干田となして山の若草をいれ，手立てよくして作りぬれば，甚だ利を得ることあり。〔中略〕冷水の出づるところには溝を立て，わきに其の水をぬきとり」としている。

　このような状況からして，暗渠排水は西日本の方から徐々に進められてきた。そのなかで北海道ではアメリカから暗渠排水が直輸入されたのである。すなわち，前述のように札幌農学

校で農園に埋設している。それにより，1887(明治20)年に江別に数軒の土管工場ができて江別周辺の湿地帯に利用した記録があり，広くは普及しなかったものの，江別煉瓦の発祥契機となっている。また町村牧場で実施したことは知られている。

4．その後の排水

　ブルックスの講義の成果としての暗集排水の実施は，その当時としては札幌農学校関係のみであり，外部で展開するまでには至らなかった。これは当時の開拓が条件のよいところから実施されていたため，いまだ暗渠排水を必要とする地域ではなかったこと，経済的に農地の整備レベルが暗渠排水を取り入れるまで達していなかったことによるものであろう。1893(明治26)年に新渡戸稲造によって排水を主とする泥炭地の圃場試験が対雁で実施された。またその弟子の時任一彦は，昭和初期の「土地改良学」の講義で「諸君のなかから土管暗渠排水を実施する者の出ることを望む」と述べていたという。

　ただ，北海道の開拓すなわち農地開発の進捗に伴い，農地造成には排水が絶対的条件であった。そのためまず河川工事によって基幹となるべき排水の整備が進められた。その結果，排水可能域が拡大した。また，暗渠排水の効果が認められだし，1932(昭和7)年に至って「暗渠排水奨励に関する件」なる通達が農林次官から各地方長官に出されて，暗渠排水が事業としてスタートしたと言えよう。しかし，土管を用いた本暗渠の本格的施工にはさらに十余年を要するのであった。

5．ブルックスの灌漑

　3-9は灌漑について述べているが，冒頭の定義のように畑作での灌漑について述べていて，日本の稲作灌漑とは異なる。したがって，この視点で見るとブルックスは灌漑について本来的にあまり詳しい知見を持っていなかったであろう。19世紀の欧米，特にアメリカの農業教育者としては，それは当然であり，また十分であったのであろう。

　灌漑の原始はナイル川の氾濫洪水で，バイブルにも示されているとしていると説き起こし，19世紀にあっては東アジアの国々や中国の降雨の多い地域のみでなく，降雨のない地域での農業も可能にしているなかで，イタリアでの大規模な灌漑やドイツ，フランス，ペルーなどをはじめ，アメリカのカリフォルニア，ユタ，コロラド各州などでも灌漑が実施されていることを紹介している。その上で灌漑用水の種類，必要性などの知識を若干示している。そして灌漑手法としての畑地圃場内の水配分の水路の配置，ポンプ揚水などを示すとともに，より土木技術的な水路，用水確保のためのダム，水路の水位調整の施設(ダム)などを列挙している。

　これはいずれも経験的事例の紹介的なものであるが，新開拓地における発展過程からすると，実用的な土木技術の伝達と評価すべきであろう。これは水田改良，新開田を意識していた受講学生を満足させたか否か，きわめて疑問である。水田水利の発達していた日本における灌漑用水の水理・水利システムの技術と直接的に対応させることは困難である。ただ，当時の学生のなかにこの種の日本での経験的知識に通暁する者もなく，また，強く知識欲を持

つ者もいなかったのではないかと推察される。本章の項目の順序がきわめて不整備である印象を与えるのがその一例であろう。灌漑に関しては，農業体系としての通過部分的評価であったのかもしれない。水に関する科学・技術が体系化してきたのが19世紀であることを考慮するならば，それが応用分野に取り込まれるまでには至っていなかったということであろう。

6. 北海道の水田禁止令

今，東北・北海道への水田水稲作の伝播を見る時，青森垂柳の水田遺構が紀元前後のものであることは大きな知見であろう。しかし，そこにはイヌビエのプラントオパールが多かったという。日本の古代水田の調査では，奈良・平安時代を境としてイネのプラントオパールの多様性が小さくなってきている。これは古代には危険分散のため意識的に多品種混作をしていたものが，国家体制を確立して租税を徴収するため，条里制の水田奨励とともにイネの系統の統一化が進められたためであろうとされている。江戸時代に入り収量の多い晩稲が要求され，それが冷害〜飢饉の一因となっていくのである。津軽の冷害について太宰治（『津軽』1951年）は「凶作年表」を引用して「現在までの330年の間に，約60回の凶作があったのである。まず，5年に1度ずつ凶作に見舞われているという勘定になる」としている。そして，その原因を強く指摘しているのが司馬遼太郎（『北のまほろば』1997年）である。すなわち「この〔津軽〕藩は，明治4年(1871)に終幕するまでの3世紀近いあいだ，世間のならいに従って――あるいは幕藩体制の原理どおりに――コメのみに頼った。コメというのは食糧と言う以上に通貨であり，その多寡は身分をあらわした」，「米は金にしなければならない。むろん，穫れ高の中から家臣の禄や士庶の食糧は残す。あとは廻米にした」，「飢饉の最中でも，しばしば藩では廻米を上方にむかって輸出した。すでに〝廻米〟は食うものではなくなっていた」，「領民が飢えようが死のうが，借金返済のための廻米を鰺ヶ沢から出さねば，藩財政そのものがつぶれてしまうのである」と言っている。東北地方での水田稲作の急速な東進北上のエネルギーは何であったのか。当時の全国的幕藩体制から振り落とされまいとする必死の体制へのしがみつきではなかったのか，また，そう仕掛けられていたのではないか，"identity traps"にはめられていたのではないだろうか。

それに対して，明治開拓初期の北海道の状況はどうであったか。北海道の拓殖史の研究をとりまとめたA. ベルクは，その論文の副題に「浮氷の中の水田」と付けて，日本人の勤勉と英知に敬意を表し，評価している。明治初期，北海道の開拓はいわゆる内地の人口圧などと言うよりも，明治維新という静かなる大革命の後処理のため，多くの人々の住むところを求め，その住む人々の食糧を得るための農業開発であった。基本的には文字通りの開拓殖民だった。そのために入植した人々の苦労は筆舌に尽くしがたいものがあったが，明治政府による多くの支援もあったのである。北海道土功組合法などが内地府県での耕地整理法とは別途に立法されている。そこで計画され，展開した農業は，そこの人々が生きていくための食糧を得るためのものであった。そして今日，日本の食糧基地と言うまでになっているのである。現実，その渦中にいる人間にとっては多くの問題を抱えているとしても，世界史的に見

ると大きな成功例，他に類例のない成功例と言ってよいであろう。

　飽食の時代にあっては想像もつかないが，食糧自給ということは容易ならざることなのである。中世以降ヨーロッパ貴族が一族郎党の食糧を確保するための策術が家政学の原型であり，その発展したものが国家統治の官房術であると言われている。その好例が明治開拓初期の北海道における稲作の排除であったと見られる。

　開拓使のお雇い外国人アンチセルは，1871(明治4)年に，石狩平野での米作は気温条件，さらに函館の気象観測から推測して不可能であるとした。すなわち「ここに稲田を作らんとするは，最もなるも知るべし。毎歳，北海道寒冷なれば，穂を出すに至らざるなり」と報告している。稲作の条件としてドイツ人地理学者ライマンは20°Cを下らない平均気温が6カ月間連続することが必要とし，ハーベルラントは積算温度が3500°C～4500°C，マックス・フェスカは同じく3000°C以上の高温を要するとしていた。しかし，その時すでに稲作前線は渡島地方にあった。そして後年に，酒匂常明は「2800°C以上で青米なし，2700°C以上では青米を含み，2600°Cでも熟田ならば収穫が望める。しかも夏の一時の高温は稲の成長を促進するので積算温度にばかりこだわることはない」として，1893(明治26)年に稲作奨励に入ったのである。そして，同年に札幌農学校教授であった新渡戸稲造は，北海道庁からの依頼で泥炭地の畑地利用の調査試験を開始していることに，北海道開拓の裏面史を見る思いもするのである。

　明治初年に，農業振興の開拓使顧問ケプロンは「農は百工の基なり」と，日本式アメリカ中心農業を改革して耕作が簡便な欧米農法の導入を建策している。それを受けて開拓使10カ年計画のなかで「北海道たる洹寒の故を以て水田に利ならず，必ず漸を以て変食すべきを化誘せざるべからず」としている。これに基づいた開拓使10カ年計画の政策は，屯田兵の営農指導，3県時代，北海道庁発足後までも影響を与え，米作の排斥までも起きているのである。しかし，北海道の開拓の推進官庁である開拓使は，本心から入植者の食生活を変え得ると考えていたのだろうか。日本のなかで北海道が異色の生活習慣を形成してよしとしていただろうか。さらに，北海道農業は稲作ではなくアメリカを範とする農業でよしとしていたかはきわめて疑問である。そして同時期に，明治以前に開田されていた渡島を起点とする稲作，石狩などにおける篤農家の稲作の試みなどが続けられていたのである。今，「稲作経験略記」という小冊子(近年農業土木古典選集に収録された)が北大附属図書館北方資料室に所蔵されている。1884(明治17)年4月に農商務省北海道事業管理局札幌農業事務所札幌育種場で出版されたものである。1876(明治9)年から開拓使が札幌村に水田を設けて，実際に水田稲作経営を行なった克明な記録である。前述の水田不利の政策は1874(明治7)年のものである。この相反する開拓使の動向の意味するところは何か。単なる内部不統一程度では説明できないであろう。この「稲作経験略記」の緒言に，「米は五穀の第一に位し，本邦常食中一日も欠く可らさる者なり，而し我北海道の地たる到る処平野広野に富み，地味饒沃にして水流其間に環繞し，水田を開設する最も容易なり。然るに農民の此に着手する者甚だ稀なるは何そや，豈其の気候風土の之れに適せざるか為乎，本場に於て水稲を試作するに8年其間豊凶なき能はすと雖も，其結果に至りては良好なる者と謂はさるを得ず。今や年々試作の実

況を略記し，併て栽培の方法を録し，以て農家の参考に供す．読者苟も此書に因て得る所あらは，豈特に本場の幸のみならんや」とある．開拓使が1882(明治15)年に廃止になり，その後継である農商務省の名称になっていればこそという感じすらする．いずれにしても水田稲作の小さな狼煙を官側から上げたというところであろうか．本格的奨励策は酒勾常明の道庁財務部長着任によって1893(明治26)年からということになる．この明治開拓使時代の米作を熱望する入植者を抑え込んでの水田不利政策時代の意味するものは何なのであろうか．

　北海道への入植，そのなかには士族も多くいた．当然のことながら米食を最高のものとする習慣を持った人々である．自給自足の生活を始めるにあたって米作を求め，水田を目標とするであろう．しかし，水田の造成は容易ならざるものがある．まず原始林の伐開に始まる開墾作業をした上に，造田作業である．均平化，畦畔の構築，さらには圃場土壌の水もち(減水深)などが問題である．さらに困難なこととして水利がある．水源を確保し，取水し，各水田に配水するための取水施設，水路の構築は考えただけでも気の遠くなるような膨大な事業量である．これだけの資本，労力，そしてその間の自給食糧をいかにして確保するか，ここにきわめて施策的な"identity traps"がちらつくのである．

7. 北海道の水田と灌漑

　日本の水田灌漑――アジアの水田灌漑――は，ヨーロッパを主とする農業施術のなかにおけるいわゆる世界の灌漑とは大きく異なる特性を有する．アジアの農地面積は世界の農地の約30%を占めるにすぎないが，そこには世界人口の約60%が依拠する．アジアが持つこの高い人口扶養力の背景には，言うまでもなく水田農業がある．今，試みに世界的食糧不足という極限的な状況下で，米に期待できる人口扶養力を試算してみると，単位収量(世界平均で3.61 t/ha)は，麦(同2.5 t/ha)の約1.4倍であり，米がほぼ全量直接食用に供されるのに対して，他の主要穀類の半分は飼料として消費されていること，ならびに主要穀類中，米だけが二期作(世界平均で1.3期作/年)可能であることなど種々の累乗効果から，米は他の主要穀類に比べて単位面積当たり3〜4倍の人口扶養力を持つ．このことを念頭に世界の水田適地を厳選し，重点的な技術・資本の投入を行なえば，1億haの水田(世界の永久農地の1/15)から，2025年の途上国全人口70億人におよそ100 kg/人/年の米を供給できる計算となる．

　しかも水田は，水を張ることによって土壌の汚染を浄化し，そのことが数百・数千年にわたる米の連作を可能にしてきた．また，モンスーンの集中的降水を水田テラスに貯え，順次下流の田で反復利用することによって，水資源を最大限に有効利用してきたし，肥沃な表土の流亡を防ぐ上でも不可欠であった．これの事実は，もし水田というクッション材がなかったとしたら，モンスーンアジアの農業と社会がどのような軌跡をたどったかを考えれば容易に理解できる．水田農業は，アジア社会のアイデンティティそのものであり，同時に21世紀における世界の人口・食糧・環境問題を解く鍵でもある．この重要なテーマが国際的な食糧環境問題の場において，なぜ中心的な課題とならないのか．

　"Aquatic Crop"(水生作物)としての米は，欧米的気候風土やそこから生起した畑作物と

出自を異にするため，国際的には十分に理解されていない．一例として，年間を通して安定的な降水が得られるヨーロッパ等の畑作灌漑では，「Etc(作物蒸発散位)だけが必要水量であり，その他の要素(水田の水張りや地下浸透等)はすべてロスである」等，基本的な認識の齟齬が見られる．畑作物における灌漑水が直截的な「消費水量」を意味するのに対して，稲作の灌漑水は栽培管理上の「使用水量」であって，消費水量はその一部分にすぎないことが理解されていないのである．

　戦後，世界の農業は，欧米型畑作における営農技術が，世界に伝播していった過程を物語る．在来的なアジアの灌漑が，雨期・乾期の存在という不可避的な自然に対処するための"Must Irrigation"であるのに対して，近々200年の歴史に立つ欧米的灌漑は，土地に付加価値をつけるための"Not must but advantageous Irrigation"と言える．最近では，世界的水不足が懸念されるなかで，単位水量当たりの作物生産量あるいは農業収入を最大にするための技術，すなわち"crop/drop"のコンセプトが国際的な中心課題となりつつある．しかし，一作ごとの生産性を追求するスケールからは，アジアの水田農業が果たしてきた歴史的事実は説明できないし，長期的な視点に立った地域環境資源の持続性を失わせしめる危険性も大きい．第2次大戦後，生産拡大を第一義とする農業コンセプトが世界に浸透する過程で，塩害や地下水低下などの環境劣化が顕在化していったことは，疑うべくもない事実である．

［梅田安治］

[*Lecture for the Second Semester*]
Chapter 4   Pulverization of Soils (Tillage)

It is a well-known fact that for the successful cultivation of most crops, the soil must be well pulverized. This is necessary that the roots of plants may be able to penetrate it in all directions, and that the various agencies of nature, such as frost, heat, water, carbonic acid and *etc.*, may act readily upon it. Soil, in its original state, consists mostly of pulverized rocks and the remains of organic life. ⟨See pp. 30-31⟩ Neither the pulverized rock nor the organic material is of use as plant food until it has been acted upon by those agencies of nature. It is very important, then, to have an accurate knowledge of the soil and the means of pulverizing it, in order to attain the greatest success in farming. [p. 243] As I have already stated, there are two great classes of matter, organic and inorganic. ⟨See p. 18⟩

Matter[1] may be defined as something which occupies a certain portion of space to the exclusion of everything else. Some of the properties of matter are divisibility (all matter can be divided), inertia, attraction of gravitation, cohesion, and chemical affinity. All matter is made up of 63 different elements, but of these only a small number are of importance to agriculture, that are carbon, hydrogen, nitrogen, oxygen, phosphorus, magnesium, iron, silicon, chlorine, manganese, and calcium. Some of the other elements, although occasionally found in plants, are not of any importance to the farmer. The different elements of matter, [p. 244] when brought into contact under favorable conditions, act upon each other chemically. Organic matter acts upon inorganic, as I pointed out early in my lectures. ⟨See pp. 28-29⟩ Inorganic elements sometimes act upon organic, as for example the alkalis, lime, soda, or potash, which have a caustic, burning action that tends to oxidize the organic substances with which it comes in contact. With regard to the chemical actions of elements upon each other, chemistry will teach you. For the farmer, it is only important to know in what form these various elements are used by plants as food, and how to treat the soil, both in the manner of cultivation and the application of fertilizer, to bring them into the ideal condition. It has been believed by [p. 245] some men in the past that by thoroughly pulverizing the soil, it will be possible to produce large crops many years in succession without the application of any manure. ⟨See p. 41⟩ One of the most celebrated believers in the possibility of that was Jethro Tull (1674-1741), an Englishman, who actually succeeded in raising good crops of wheat for about 20 years without manure, simply by stirring the soil frequently and pulverizing it very thoroughly.

Exposing the soil to the action of atmospheric influences, as would effectively be done by thorough cultivation, doubtless renders the elements more quickly available as plant food. But to rely wholly upon this means of sustaining fertility is unwise. We should seek to increase the fertility of the soil as much as possible by pulverization, but we should also apply fertilizers. [p. 246] It should be

124   Chapter 4   Pulverization of Soils (Tillage)

remembered that nothing is added to the soil by pulverization; therefore, while it is an important adjunct in the raising of crops, it cannot be relied upon to permanently sustain the fertility of the soil.

4-1. Methods of Pulverizing the Soil

Plowing the soil or inverting it, and at the same time crushing it as much as possible by an implement drawn through it, is and always has been the means most commonly resorted to for pulverization of the soil. It is the best method, all things considered, although there are other means of accomplishing this objective more perfectly. By the use of spades, forks, and the Japanese *kuwa*, the soil may be very well pulverized, but the use of these implements involves [p. 247] the expenditure of much more labor, and hence is more expensive.

1) Kinds of Plowing

There are three distinct kinds of plowing[2] namely flat furrows, furrows on edge, and lap furrows. The flat furrow, as its name indicates, leaves the land flat or level. Each furrow slice is completely inverted, and lies wholly in the preceding furrow (Fig. 1). The furrow on edge, as is also indicated by its name, leaves the land with each furrow slice standing on edge in the preceding furrow (Fig. 2). The lap furrow is [p. 251] intermediate between flat furrows and furrows on edge. In that method of plowing, each furrow slice is partially turned over and rests partly in the preceding furrow and upon the preceding furrow slice. You will notice that with both the flat furrows and furrows on edge, the surface of the land is left nearly flat. With lap furrows, however, it is left in ridges, and therefore more soil is exposed to the action of the atmosphere than in either of the other methods of plowing. The land should be plowed in lap furrows in autumn, if the objective is to expose the land most effectively to the action of the frost. [p. 252]

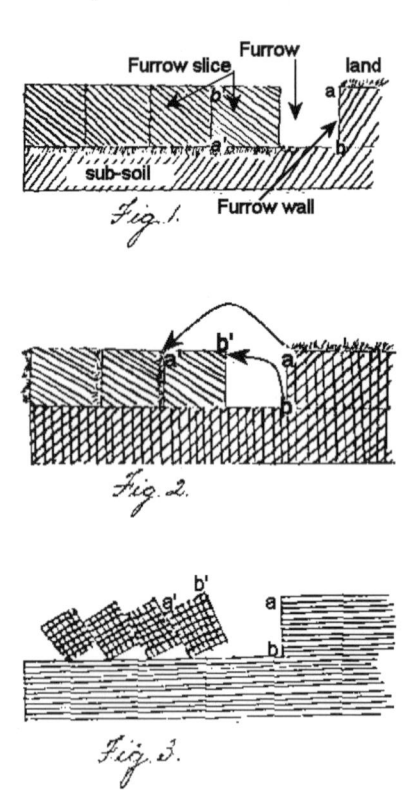

2) Time of Plowing

It is a good rule to plowing and cultivates both in and out of season, whenever you have the time. There is no danger of plowing too much. It is better, however, to have some definite purpose in view, and to be guided by some clearly known principle. One rule should always be remembered; never plowing when the soil is wet. So doing defeats the objective in view which is pulverization of the soil, the plowing wet soil is thrown up in clods, when they become dry, it goes hard and cannot be penetrated by the roots of plants. ⟨See p. 187⟩ Some farmers advocate plowing in spring, others   [p. 248] in the autumn, and it is amusing to hear their arguments. Some say that we should plowing in the autumn in order to destroy worms, to have the land dry more quickly in spring, to help along the spring work, *etc*. Others say spring plowing is best because the top of the soil gets warmed, and then, if turned over, the whole of the soil is warmed. Another argument is that in spring, growing grass may be plowed in to enrich the land; yet another is that plowing in this season prevents weeds and grass from starting up among the growing crops. These reasons are only secondary, our great objective being to pulverize the soil as a means of fertility, so we should plow in spring or autumn as best serves this purpose. [p. 249] (1) Autumn Plowing: We are plowing in the autumn for the sake of the beneficial effects of frost upon the soil. This is the principal reason for plowing at this season, those hitherto given being merely secondary. The action of the frost is to break up or pulverize the soil, hence all heavy, compact lands should be plowed in the autumn for the sake of this influence. Open, porous, alluvial, or indeed any light soil should not generally be plowed in autumn. If, however, the land is level and there is no danger that fine particles of soil will be washed away, any land may be plowed in autumn if it is more convenient to do so at that time. If the objective of autumn plowing is to get the beneficial effect [p. 250] of the pulverizing action of winter frost, it should be done in such a manner as to most expose the soil to this action. It is evident that the greater the surface exposed, the greater will be the effect of the frost. (2) Summer Plowing: The principal object of plowing in summer is usually to turn under weeds; therefore, the kind of furrow should be that which will most effectively cover the weeds. That is obviously the flat furrow. (3) Spring Plowing: By far the greatest amount of plowing is done in spring, the objective being to loosen and pulverize the soil in order to make it ready for the reception of seeds. That kind of plowing is best, then, which most effectively loosens and pulverizes the soil; this is usually the lap furrow. The furrow on edge has little or no merit. [p. 253]

3) Depth of Plowing

The rule in plowing should be to plow as deep as your agricultural soil extends, and if this is not as deep as you would like to have it, you should plowing a little deeper each year until you get the desired depth. ⟨See p. 18⟩ If too much soil is brought to the surface at once, it will interfere greatly with the cultivation of crops, since it does not contain elements of plant food in an available condition. ⟨See p. 41⟩ Different crops require different kinds of soils and depths. ⟨See p. 69⟩ Some thrive well with shallow culture, but others must have deep culture. Therefore, no absolute rule can be given as to the depth of plowing; but you should remember that soils are much [p. 254] more likely to be too shallow than too deep. Corn, onions, grass, and plants whose roots do not extend very deeply into the soil thrive well with shallow culture, but they would thrive equally well or probably even better with deep culture. Root crops, such as beets, turnips, carrots, and deep-rooted plants must generally have deep culture in order to attain the highest success. ⟨See p. 22⟩ From 5 to 8 inches is the most usual depth for plowing[3].

4) Subsoil Plowing

Subsoil plowing, as its name indicates, is the plowing of the subsoil. Unlike surface plowing, [p. 255] however, the soil is not inverted and neither is it brought to the surface in subsoil plowing, being simply broken up and loosened. The subsoil plow follows in the furrow made by the ordinary plow, simply breaking up and loosening subsoil to a greater or less depth as may be desired. Land in which the subsoil is more or less hard and impervious to water should be subsoil plowing. Land in which the subsoil is already loose and porous need not be subsoil plowing. It is desirable to subsoil plowing the land for those crops which send their roots down deep into the soil. Such plants are root crops —generally, clover and fruit trees. It is desirable to subsoil plowing the land for the same crops as those [p. 256] for which it is beneficial to plowing deeply. ⟨See p. 21⟩ It is an advantage also to subsoil plowing the land for almost any crops, provided the subsoil is impervious and the land in consequence is more or less wet, since the breaking up of the subsoil allows the water to pass down more freely.

4-2. The History of the Plow

The plow has been used from very ancient times. We find it mentioned in the Bible at a very early time, but those first plows were very crude implements consisting of little more than a sharpened stick (Fig. 4) drawn through the land by oxen. The first [p. 257] improvement was the selection of the forked branch of a tree, one part of which was sharpened with the other serving as a beam.

Another very crude form was made as shown in Figure 5, still entirely of wood, with the part used for stirring the ground being simply a sharpened stick; having two handles, it was an improvement upon the first form. Such plows as these were used in China, Palestine, and Egypt in very ancient times. In the time of the Roman Empire, the plow was greatly improved, assuming at that time a form somewhat similar to the present one, but it was still made almost entirely of wood. Such plows as these were used in Europe from the time of the Romans to the 18th century, in which [p. 258] the improvement of the plow in England was very remarkable. For a long time after this, however, plows were made mostly of wood. The first step in their improvement was the fastening of strips of iron upon those parts which were most subject to wear. The cast iron plow was invented by James Small, an Englishman; the point of this plow was of wrought iron.

The first iron plows were made in two pieces, one the point, and the other the remaining parts. In the year 1785, Robert Ransom, also an Englishman invented the cast iron point. A man by the name of Joseph Foljambe obtained a patent for a plow of peculiar shape, which was found to be better than any other form; plows of that kind were used for many years, not having gone out of use entirely until about 40 years ago. [p. 259] Early in the 19th century, the plow was greatly improved both in America and England, but people were at first much prejudiced against the use of iron plows. The iron plow, therefore, did not come into general use until about the year 1825. For many years after, plows were made entirely of cast iron, but in the year 1855 or thereabouts plow was first made of steel[4]. Steel is the best material from which plows can be made, for many reasons. (1) Since steel is stronger than any other kind of iron, the requisite strength in the plow can be gained by the use of less material than with cast or wrought iron; hence, plows can be made much lighter. (2) Steel will keep clean, [p. 260] and is not as likely to rust as other kinds of iron; since it is important that the plow be clean and bright in order to do the best work, this is an important consideration.

4-3. The Mechanical Principles Involved in the Plow
The mechanical principles involved in the plow are the wedge, the screw, and the lever. The wedge is used in the share and moldboard. The screw is also used in the moldboard. The lever is used in the beam and handles.
1) The Parts of the Plow
   The parts of the Plow are: [p. 261] the Beam (Fig. 6, a), the Share or Point (b), the Moldboard (c), the Landside (d), the Handle (e), the Clevis (f), and sometimes, though not always, a Beam Wheel (g) and a Coulter (h)[5].

The Beam is that part of the plow which is usually nearly horizontal and to which the other parts are attached, and by which the implement is drawn. [p. 262] Beams are made of various lengths and with different degrees of curvature according to the kind of plow. The beam is most frequently made of wood, although it is sometimes made of iron. A long beam gives the drawing power of plow a greater leverage, and hence we can more easily turn it. The beam should always be very strong, since it is subject to great strains.

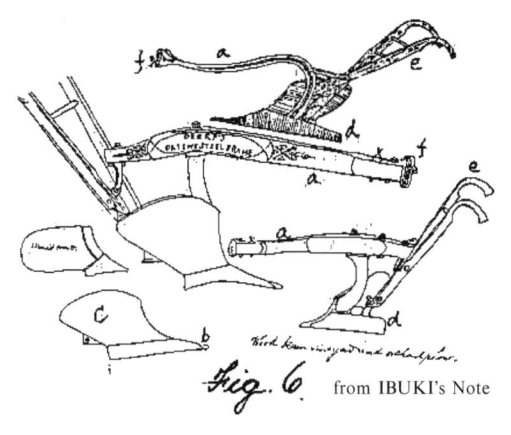

Fig. 6   from IBUKI's Note

The Landside of a plow is that metallic portion of the implement which, when the plow is in use, passes along the furrow wall of adjacent unplowed land. Its function is to give the plow steadiness, and to furnish a horizontal base upon which the plow rests. [p. 263]

The Share or Point of a plow is the metallic portion at the point of the implement. Its function is to cut beneath the furrow slice horizontally. Points are made in quite different shapes ands sizes. The longer the point, the plow can be turned the greater the width of the furrow. Besides cutting horizontally, the point acts as a wedge to raise the furrow slice somewhat. The point should always be made of very hard material, since it is subjected to the greatest amount of wear.

The Moldboard is that part of the plow whose function is to raise and turn over the furrow slice. [p. 264] Upon the degree of curvature, shape, form, size, and length of the moldboard depends the kind of work that can be done by the plow. A long moldboard with great curvature more completely inverts the furrow slice than one which has a slight curvature or which is very short. The form of the moldboard should be such that it will remain clean and bright when passing through the soil. The landside, point, and moldboard are all fastened in place in such a manner that they can easily be removed. This is essential in order that, should any part be broken, a new one can be put in place of it.

The Clevis is that part [p. 265] of the plow which is always made of iron, and by which the beam is attached to the implement. It is so arranged that, by different adjustments, the depth and width of the furrow can be varied. Three different holes in the horizontally arranged clevis, through any of which a pin can be fastened in place, permits variations to be made in the width of the furrow. Turning the clevis to the right permits the plowing of a wide furrow, and turning to the left, permits a narrow one. Holes also vertically arranged clevis, one above the other in the end of the clevis to permit variations in the depth of plowing. The arrangement of clevis hole attached to the upper one the plow will run deep, and attached to the lower ones, more shallow. The bolt on which [p. 266] the clevis swings should be made with a head of such shape or form and size that it may be used as a wrench, fitting the nuts used in fastening the plow together.

The Handles of the plow are those parts by taking hold the plow steady and by which the plowman guides. They are usually made of wood, and should be very strong. Variations in the length of the handles give additional or less leverage according to whether their length is increased or decreased.

The Beam Wheel, sometimes though not always used, is for the purpose of giving the plow additional steadiness and for more readily controlling its depth. [p. 267] It is so fastened near the end of the beam that it can be conveniently raised or lowered, thus its function make the furrow deeper or shallower.

The Coulter, also sometimes though not always used, is that part of the plow whose function is to cut the furrow slice vertically. It is sometimes a simple knife, and sometimes a revolving disc with sharpened edges. It is usually fastened to the beam of the plow just ahead of the point, and is so arranged that it can be raised or lowered at pleasure. In some plows, the knife coulter is attached to the point of the plow instead of the beam. [p. 268]

4-4. Varieties of Plows

Many different styles of plows are made for plowing different kinds of land; some of the more important are the sod, stubble, sedge or bog, breaking up, swivel or hillside, and subsoil plows[6].

1) Sod Plows

It is used for grass lands, and should always be provided with a beam wheel and a coulter. They should also be made quite strong, since the lands in which they are intended for use are usually quite hard and compact. They should turn either flat furrows or lap furrows. One of the best plows for turning over green sward [p. 269] is the double plow or, as it is sometimes called, the Michigan plow. In this plow [Fig. 7], the forward moldboard, which is smaller than the main one, cuts and turns over the upper portion of the sod, thus leaving the grass in the bottom of furrow, and the main moldboard covers upside of the furrow-slice by the lower portion of the sod.

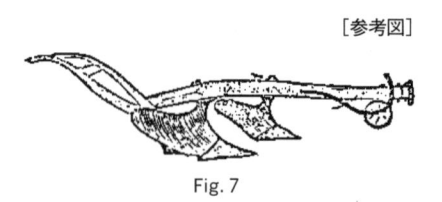

[参考図]

Fig. 7

2) Stubble Plows

Stubble plow is designed to plowing land which is covered with the stubbles of preceding crops. (Fig. 6) For this reason, such plows need not be [p. 270] as strong as the sod plow; neither is it necessary that a beam wheel or a coulter be present. These plows may turn flat furrows, furrows on edge, or lap furrows.

3) Sedge or Bog Plows

The roots of plants such as grow in bogs, sedges, rushes, *etc.* are very large and strong, so a plow for use in such land must be large and have great strength. It should be provided with a beam wheel and a coulter [Fig. 8], and should turn a flat furrow. [p. 271]

Fig. 8
from Brooks, *Agriculture*

4) Breaking-up Plow

Breaking-up plows are used for plowing land which has never been plowed before; since such land is often full of the roots of trees, bushes, and other large wild plants, these plows must be the largest and strongest of all. They should be provided with beam wheels and coulters, and should turn flat furrows. [Fig. 9]

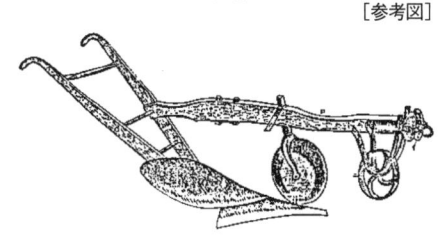

Fig. 9

5) Swivel or Hillside Plow

Swivel plows [p. 272] are one which is so constructed that the moldboard may be fastened on either side of the beam, thus enabling the plowman to turn all the furrows of the field in the same direction as his team travels back and forth. [Fig. 10] Although such plows are often very convenient in plowing fields where it is desirable to leave the land perfectly smooth, they have not yet been made in a form so perfect as to turn the furrow as well as the common plow. Still, in good smooth land they will work sufficiently well for ordinary purposes, and therefore they are quite extensively used. The turn-rest plow is designed to accomplish the same objective as that for which the swivel plow is used, but they are not now in general use. [p. 273]

Fig. 10
from Brooks, *Agriculture*

6) Subsoil Plows

It must be made very strong, but they are not usually furnished with either beam wheels or coulters; the moldboard and the point (especially the former) are small, being designed mainly to stir the soil and not to invert it. The curvature of the moldboard is very slight. The iron connecting the moldboard and the point with the beam is long, to permit deep plowing [Fig. 11]. [p. 274]

Fig. 11
from IBUKI's Note

7) Sulky Plow

Sulky plow is a comparatively recent invention, not having been used much, and not being employed within a few years; but they have been made to do their

130   Chapter 4 Pulverization of Soils (Tillage)

work so perfectly that they are undoubtedly better than ordinary plows. [Fig. 12] A sulky plow is one in which the plow is mounted upon wheels, the driver riding upon and governing the plow by means of a lever placed within convenient reach. Principal advantages over the ordinary plow, which it is clear that the sulky plow possesses are as follows: (1) Uniformity of work; since the sulky plow cannot be thrown out of the ground, it will turn a furrow of the same depth everywhere. (2) Lightness of draught; [p. 275] since the weight of the plow and the pressure of earth upon it is supported by the wheels and plow is drawn by the bottom of the framework, the team of drawing it will experience greater ease than with an ordinary plow. (3) Ease of operation; since the plow is controlled by means of a lever, which is easily moved, this contention is correct. In addition to this, the plowman rides in a seated position, which is of course easier work than walking, which he must do with the ordinary plow. On the other hand, the sulky plow costs more money than an ordinary one, and unless the farmer owns a large farm or labor is very expensive, it will not probably pay him to own a plow of this sort. [p. 276]

[参考図]

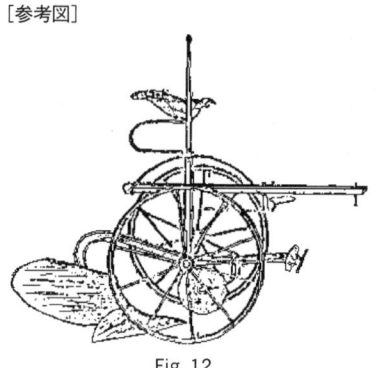

Fig. 12

8) Gang Plows

The gang plow [Fig. 13] is an implement by the use of which two or more furrows are turned at a time. The plow in this case, as in the sulky plow, is mounted upon wheels controlled by a lever so placed as to be within convenient reach of the man driving the horses. The merits of the gang plow are the same as those claimed for the sulky plow, with an additional one, that is, it economizes the labor of man, [p. 277] one man being able to plowing as much land as two men can with ordinary implements. This kind of plow, however, is quite expensive and it will be profitable to use it only on large farms; it cannot be used conveniently in small fields. On very large farms, such as those in California, plows are used which have three bottoms in the gang, but on ordinary farms, the most common number is two. [p. 278]

[参考図]

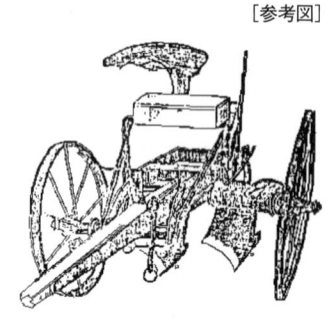

Fig. 13

9) One-horse Plows

Small plows designed to be drawn by one horse, are much used for stirring the soil among cultivated crops, but not for plowing the ground to prepare it for the planting of seed.

10) Steam Plowing

Ever since steam has been used as motive power, it has been the aim of inventors to apply it to the plowing of land, and various steam plows employed steam tractors have been invented. They have not yet been made so perfect, however, as to make their use in many cases preferable to that of plows drawn by horses or oxen. Inventors of steam plows have worked on two different principles, some designing [p. 279] to have the steam tractor travel back and forth over the field as a team of horses or oxen does, dragging the plow behind it, while another principles to use stationary tractor which drag the plows back and forth by means of a cable and windlass or a pair of steam tractors[7]. So far it has been found impracticable to use a locomotive engine in plowing. The reasons are as follows: (1) It is found difficult to make the wheels take hold of the ground with sufficient strength to go forward and pull the plows after them. (2) The nature of land to be plowed is usually such that it requires very great power over it, to say nothing of the power required in plowing.

These steam engine tractors must [p. 280] necessarily be very heavy, and hence the wheels sink deeply into the ground. Now, the power used in simply propelling the tractor is wasted, and since the amount of power so used must always be quite great, it is not probable that steam plows drawn in this manner will ever be extensively used. Because of these practical difficulties, therefore, plows drawn by locomotive tractors have not yet been successfully used.

11) Plowing with Stationary Steam Tractors

Various steam plowing systems have been invented which are worked by means of stationary steam tractors. Some systems have used two steam tractors, one at each end of the field, with one of them drawing the plow in one direction and the other [p. 281] in the other direction. Another steam plowing system uses just one engine, stationed at one end of the field, with an anchor being used at the other end. Using two tractors require the labor of two skillful engineers at each tractor. That system also requires the use of more fuel than when just one steam tractor is used. Plows drawn in that manner, therefore, have not been so much used, as those drawn by one tractor with the assistance of windlass enabling the use of plows in two directions. In the case of one stationary tractor, plows are drawn by means of a wire cable which is wound upon a revolving drum.

One of the most successful steam plows that have been used is Fowler's system. [Fig. 14] This uses eight plow bottoms in two gangs, [p. 282] one gang working in one direction and the other gang working in the other direction. In this system, the wire cable is kept from dragging by the use of little pulleys so mounted on iron anchor stakes that they can be conveniently moved from place to place. In the operation of this system, three men and two boys are necessary in addition to the men employed in bringing fuel and water to the tractor. The business of one man is to manage the tractor, of another man to ride upon the gangs of plows, and of the third man to take charge of the anchor stakes. The business of the boys is to attend to the little pulleys supporting the cable, taking them up in front of the plows and putting them down again behind them. By the use of this system, land can be plowed with great rapidity, but its initial cost is quite expensive. One farmer, unless he carries on an extensive business, [p. 283] cannot afford a steam plowing system by owning it entirely himself; however, several farmers may form a company and buy one steam tractor and plows to be shared by all. Steam plowing systems can be used to advantage only on large, smooth fields. On soil which

[参考図]

Fig. 14
from Brooks, *Agriculture*

132   Chapter 4   Pulverization of Soils (Tillage)

contains large rocks or which is very rough, such a system cannot be used.
12) Steam Digging Machines

If a implement could be devised that would dig over the land in somewhat the same manner as is done by the spade or the Japanese *kuwa*, which could be worked by steam tractor, it would be a very valuable machine. Land can be more thoroughly pulverized by the use of [p. 284] an implement of this nature than by plowing; therefore, practical farmers have long been desirous of having such a machine. Already, various digging machines worked by steam tractor power have been invented, but it may be said of them as has been said of the steam plow that they have not yet been so perfectly made as to come into general use.

4-5. Harrows and their Uses

The harrow is an implement which is extensively used in the pulverization of soil[8]. It is used mostly upon land which has been plowed for the purposes of breaking up clods and smoothing the surface. It is also used for covering seed sown broadcast, as well as for the cultivation of certain crops. [p. 285] The harrow is the implement which ranks next in importance to the plow as a means of pulverizing the soil, and no farmer should be without one. Many different kinds of harrows are made, the method of construction being varied according to the use for which they are designed.

1) Square Harrow [Fig. 15]

The most common harrow is the square harrow which has a square or rectangular frame, and straight teeth. This frame is sometimes made in sections joined together by hinges, though it is often made in one piece. The material of the frame is usually wood, and the teeth are of iron. The frame made in two or three sections is better than the single frame because the implement can adapt itself to the inequalities [p. 286] of the surface. The size of the common square or rectangular harrow may vary according to the purpose for which it is designed, and according to whether it is meant to be drawn by one or two horses. For use on very rough, heavy land, the harrow must be heavier than for use on comparatively light, smooth land. This end is sometimes attained by putting weights upon the harrow. The number and size of teeth must also vary with the land. The rougher the land, and the more large lumps and clods it contains, the larger and less numerous should be the teeth. For use on finer soil, more and smaller teeth are preferable.

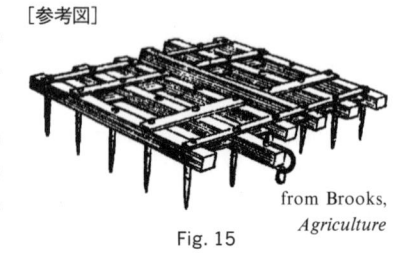

[参考図]

Fig. 15

from Brooks, *Agriculture*

The point of drawing to a square harrow should be a short distance from one corner of the implement, [p. 287] since this makes each tooth follow a separate path, and thus very thoroughly pulverizes the soil. If a rectangular harrow made in two sections joined together by a hinge is used, the point of drawing should be such as to ensure the same end. What has been said with regard to the weight of the frame, and the size and number of teeth, applies with equal force to all kinds of harrows.

2) The A-frame Harrow

The frame of a harrow is sometimes made somewhat in the shape of the letter

A. Each side of the implement usually has two rows of teeth. This form of harrow, when simple straight teeth are used, [p. 288] is not as good as the square or rectangular one. An implement of this sort is drawn from the apex of the angle formed by the two sides.

3) The Revolving Harrow

Various harrows have, from time to time, been invented which when drawn through the soil, revolve with greater or less rapidity. A harrow having such a motion would very thoroughly pulverize the soil, but harrows of this kind are somewhat expensive, and it is found to be difficult to construct them so that they have sufficient durability. For these reasons, they have not been extensively used. [p. 289]

4) The Harrow with Sets of Revolving Teeth

Harrows have also been made which have sets of teeth, each set of which was made to revolve independently when the implement was drawn through the soil. These implements also thoroughly pulverize the soil, but the same practical objections are made to them as to the revolving harrow; therefore, they have not been much used.

5) The Nishwitz Harrow

This harrow has a frame made in the shape of the letter A; its teeth, instead of being straight pieces of iron, are thin plates of iron, the edges of which are sharpened. These plates are so arranged that when the implement is drawn through the soil, they revolve. [p. 290] This implement is especially useful in sod land, where these cutting plates thoroughly break up and pulverize the sod. Implements of this sort were first known by the name "Nishwitz harrow", but lately many new ones have been made, and they are now called simply "Pulverizing Harrows". The frame is usually made of wood, and the implement is usually drawn from its apex, but recently one has been invented, the frame of which is made of iron, and which is drawn with the opening between the sides in advance. The driver usually rides upon a seat attached to the frame. It requires quite a strong team of horses to draw a harrow of this sort. [p. 291]

6) The Smoothing Harrow [Fig. 16]

Harrows have been made for the special purpose of smoothing land, but harrows of this sort are also very useful in covering small seeds which have been sown broadcast, and in the early cultivation of certain crops, such as corn, potatoes, and grains of different kinds. The frame of these harrows is made of wood, is rectangular in shape, and is usually made in three sections. Its teeth are more numerous than in the ordinary harrow, and are smaller and inclined to the rear. One of the best smoothing harrows is Thomas'.

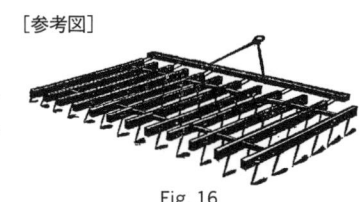

[参考図]

Fig. 16
from Brooks, *Agriculture*

7) Harrowing Work

In harrowing field the first time after it has been plowed, it is best to drive [p. 292] across the furrows. If land is harrowed more than once, it should alternately be harrowed in different directions. Land is more effectively pulverized by harrowing if the implement is drawn quickly. If the motion is slow, the clods of field will simply be pushed out of the way, and not be broken. This is especially true of the implement which has straight teeth. With the harrow having teeth on revolving plates, it does not make so much difference.

8) Drawbacks of Mentioned Harrows

One great drawback in the mentioned harrow is its lack of adjustability. Ordinary harrows cannot be changed to fit them for different kinds of soil. [p. 293] A harrow has been invented, however, which has A-type frame of iron so arranged that two frame meet at an acute angle can be expanded or contracted at pleasure. This implement was very expensive and has not come into general use. At present, every large farmer should own harrows of three kinds, the ordinary rectangular harrow, the pulverizing harrow, and the smoothing harrow.

9) The Wooden Harrow

For use on a small scale, a harrow which will do tolerably good work can be made entirely of wood. I would not advise the use of such harrow, however, if the farmer can afford to buy a harrow with iron teeth; however, in this newly settled country, Hokkaido where many [p. 294] of the farmers are quite poor, such an implement may be made very useful. Any farmer having an axe, a saw, and a chisel can make one for himself. The wood used should the strongest that can be obtained, such as oak, ash, or walnut will be suitable for this purpose. The size and weight of the frame, and the number of teeth may be varied for different purposes, as with ordinary harrows, but an implement having the following dimensions will be best suited for most uses-the framework may be 5 feet square, with 5 strips of 3 inches square running in one direction, and 5 strips of 1×3 inches intersecting them in the other direction. The teeth should be about 10 inches long and $1\frac{1}{2}$ inches square; 25 will be a very suitable number. The teeth should be evenly distributed throughout the frame; I have said that the teeth should be $1\frac{1}{2}$ inches square, but the point [p. 295] should be sharpened, not blunt. It is particularly important that the wood from which the teeth are made be well-seasoned; if made of green wood, they will be likely to bend.

4-6. Cultivators

The cultivator is an implement which is designed for stirring the soil and cutting weeds among growing crops[9]. Implements of this sort are exceedingly useful, for by their use it is possible to do nearly all the work of cultivating many crops. Many different kinds of cultivators are made, some being designed to be drawn by one horse, and some by two, or others being suited for use by manpower. [p. 296]

1) One-horse Cultivator

This cultivator is perhaps most commonly used, especially by farmers who do not carry on a very extensive business, because its initial cost is much less than that of the two-horse cultivator. The one-horse cultivator has a framework so assembled that the implement can be made to work at different widths. This framework usually assumes a form the outline of which is nearly triangular, consisting of different pieces which are jointed at the apex, this permitting expansion or contraction of the frame. The frame is sometimes made of wood, and sometimes of iron or steel. Handles somewhat similar to those of a plow are attached to it. Teeth of many different shapes are used, according to the kind of work for which [p. 297] the implement is designed. [Fig. 17] The teeth are attached beneath the frame. The team of drawing horses is attached to the clevis, which is similar to that of the plow, permitting variation of the depth to which the implement works. A wheel is also sometimes attached to the frame ahead of the teeth, its objective being to enable better control of the implement.

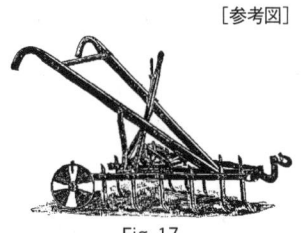

[参考図]

Fig. 17
from Brooks, *Agriculture*

2) Sulky Cultivators

They, like the sulky plows, are mounted upon wheels. It is usually designed to cultivate two rows at once, and is drawn by two horses; the driver rides on a seat attached to the axle of the wheels, and controls the cultivator by means of [p. 298] levers placed within convenient reach. With an ordinary cultivator, the driver walks behind, controlling the implement by means of handles like those of a plow. The advantages of the sulky cultivator over the ordinary implement are similar to those claimed for the sulky plow-ease of draft, regularity of work, and economy in the labor of man. One man is able, with an implement of this sort, to do as much or more work as two men can do with the one-horse implement. A small farmer cannot afford to own a sulky cultivator, but several farmers might associate in owning one implement in common, and each using it in turn. Farmers who raise large amounts of crops, on the other hand, will find it economical to use these implements. [p. 299] A very good sulky cultivator is one called the "Bertrand and Same's Sulky Cultivator", and another good one is called the "Advance Walking Cultivator". A cultivator designed to be propelled by a man who at the same time guides it, is useful among crops which are planted so close together as to make it impossible for a horse to pass between the rows. In no other cases would it be economical to use hand cultivators.

3) Horse Hoe

The hoe, like the cultivator, [p. 300] is used in stirring the ground and killing weeds among growing crops. It is very much similar to a cultivator. In fact, the difference is more in name than in reality, since many horse hoes very much resemble certain implements called "cultivators". A man by the name of Ross makes several kinds of horse hoes of several shapes, all of which are called by his name and are designed for the cultivation of various plants in the different stages of their growth. His implements are all very good.

4-7. Rollers

This is an implement of great use upon the farm. It is useful in breaking up clods and in pressing small stones and other small things [p. 301] lying upon the surface into the soil, thus making the land smooth and pulverizing the soil. The roller is used principally upon fields which have been sown with grain or grass, because it prepares the land very well for the use of harvesting implements such as mowing machines, reapers and *etc*. [Fig. 18]

The essential part of a roller is a cylinder so arranged that it revolves upon an axis.

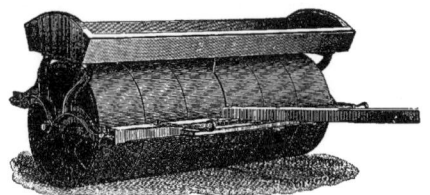

[参考図]

Fig. 18

The size and weight of this cylinder are varied [p. 302] according to the use for which the implement is designed, and according to whether it is intended to draw it with one or two horses, or by a man. The usual width of a farm roller is about 7 feet, the cylinder being made in two sections, each of which can revolve independently. There are sometimes more than two sections, and sometimes just one. It is better to have two sections, at least, because when the implement is turned around, each can revolve independently, so the surface of the soil will not be torn up as it would be, if there was only one section. The different sections of a roller all revolve upon a common axis, and are all of the same diameter. The cylinder diameter best suited [p. 303] for ordinary use is about 2 feet 8 inches. Both iron and wood are used in making these rollers. Cylinders made of iron are better than those of wood, but are also much more expensive; since wood answers the purpose well, it is most commonly used. The trunk of a tree may be used as a roller cylinder, but it is seldom that one can be found the trunk which is suitable for the purpose, and if one could found, it would be very heavy. The cylinder is, therefore, usually made of planks 2 inches in thickness nailed to circular end pieces. The axle passes through the centers of the end pieces, and is supported by a framework to which the pole, tongue, or thills by which the implement is drawn is attached. [p. 304] A seat for the driver is sometimes attached to this framework; sometimes there is also a box upon it, into which weights can be placed when it is desirable to make the implement heavier. The wood used in making an implement of this sort should be of the very strongest and most durable character. Oak or ash will probably be the best kinds of wood that can be used here.

(1) The Corrugated Roller; There are various forms have been made, some of them are quite useful on the farm. They have been made, as I told you when speaking of irrigation ⟨See p. 230⟩, with cylinders of such form as to leave the land in ridges. [p. 305] They have also been made with corrugations of an irregular nature running in the other direction, the object being to make the implement a more effective pulverizer of the soil. Such implements, however, are somewhat expensive and are not in very general use.

(2) Rolling; Rolling makes land more compact, therefore, those soils which are of a light nature are most benefited by it. Land in which the soil is very fine, and where the surface is quite level, may sometimes be damaged [p. 306] by rolling; it makes the surface so smooth that when the soil becomes dry, if violent winds occur, much of the finer soil will be blown away. Grass land, the surface of which has become rough from any cause, is benefited by rolling. Land which is already hard and compact should, as a rule, be rolled, because so doing will make them still more compact. If the weather is likely to be dry, after seed has been planted, it will be found advantageous to use a roller upon the land, since doing this has the tendency to make the soil more compact, thus giving it a greater power to retain moisture, so the seed will be likely to germinate. [p. 307]

4-8. Simple Tools
1) The Bush Harrow or simply Brush Harrow

A very cheap, and at the same time very useful, implement can be made by every farmer for himself at slight expense. It is the bush harrow, as it is sometimes called. It may be made of a post about 6 inches in diameter, the tops of a few young birch trees, and a few ropes or vines. The post should be about 5 feet in length; at points equidistant from each other should be made 5 holes, each 3×6 inches in size. The tops of the birch trees should have many small branches and be about 8 to 10 feet in length. The thicker ends of these tops are inserted through the holes in the post, being fastened in place by pins or strings in such a manner as to form [p. 308] a large broom, the undersurface of which should be smooth, so that when drawn over the ground it will smooth every inch of the surface. It may be necessary to connect the middle portions of these branches by means of vines twined from one to another, or of sticks laid across them, and fastened to each by vines or strings. This implement is especially useful in pulverizing and smoothing soil. It is also used for covering small seeds, such as grass or grain, which are sown broadcast. Its special use is in covering grass seed, which it does very nicely. Grass seed must not be covered deeply, for if it is, it will not germinate. A harrow would cover many seeds too deeply. [p. 309]

2) The Spade

It is an implement by the use of which land or soil may be thoroughly pulverized. The spade is also used in digging holes, ditches, *etc*. It is more useful for these latter purposes than for pulverizing the soil, as its use for this purpose is quite expensive. The spade is never used for pulverizing the soil except in small gardens or places where the plow cannot be used. Spades are made with both long and short handles. Spades with short handles are called "D-handle spades" [p. 310] Spades with long handles are simply called long-handled spades. The blades are made of different widths, lengths, and weights. The blade of the spade should always be made of the best steel, and should be sharp.

3) The Garden Fork

It is an implement which, like a spade, is used for pulverizing the soil in small gardens and places where it is impossible for any reason to plowing. By the use of this implement, very effective work can be done. The garden fork is also sometimes used in digging out the roots of trees and grass, because its tines can be forced into the ground below the roots of plants which it is desirable to dig out; the handle [p. 311] is then depressed, and a mass of earth containing the roots is raised; this is then shaken so that the earth falls between the tines, and the roots are held upon them. The tines of a garden fork are more numerous than those of an ordinary fork, usually being five or six in number. They are also broader than the tines of forks used for other purposes. Garden forks are made with both long and short handles.

4) The hand Rake

It is useful on the farm for pulverizing the soil and preparing it [p. 312] for the reception of seed. Rakes are used especially in gardens and in places where small seeds, such as those of onions, carrots, beets, *etc*, are to be planted. If such seeds are to be sown with a machine, it is especially important that the soil be raked, not only for the purpose of pulverizing it but for the sake of removing the roots of plants, or coarse material of any kind which would interfere with the working of the seed sower. Both fully wooden-made rakes and those with steel heads teeth are used. The latter are designed especially for garden use, while the former are used for raking hay. Their handles are always made of wood.

5) The Hoe

It is an implement [p. 313] of very extensive use upon the farm. It is used in digging holes and furrows, for the reception and covering of seed, in the cultivation of crops, and for manifold other uses. Its most important uses are for the covering of seeds and the cultivation of crops. Various kinds of hoes are made. Some have blades which forms are nearly rectangular; others have blades which are nearly triangular, the apex of the triangle being at the point where the handle is attached. Other blades are triangular with the apex [p. 314] at the bottom. Still other blades have a general outline which is rectangular but a cutting edge which, instead of being straight, is serrated, having teeth like those of a saw, except much larger pitch. The first kind is the common hoe, which is most extensively used. The second kind and the third having sharp points, are especially adapted to working among small plants which grow near each other. The fourth kind is useful in cutting up weeds and dragging their roots from the ground. The handles of each of these four kinds are similar. They should be of convenient size and of such a length that the implement may be used without inconvenient stooping. [p. 315]

6) The Scuffle Hoe

This hoe has been made with a blade which will cut on both edges; the blade is so fastened to the handle that when it is held at a convenient height, the blade rests horizontally upon the ground. The hoe is then moved back and forth with a rapid motion, cutting and stirring the soil in both directions.

7) The Wheel Hoe

It is made with little wooden wheels. The blade of this hoe [p. 316] is so arranged that when a man holds the handle of the implement at a convenient height, the blade rests horizontally upon the soil. The blade of this hoe is also made to cut on two edges. The implement is worked by moving it back and forth in much the same way as the scuffle hoe is used. It is especially useful for working among small crops, such as onions, carrots, *etc*.

8) Material to be used in making Forks, Spades, Rakes and Hoes

Since these implements are designed to be used by hand, it is important that they be light in weight but at the same time very strong. Particular care should therefore be taken in making these implements. The cutting parts, the blades of spades and hoes, the tines of forks, and the teeth [p. 317] of rakes, should be made of the best steel, since it enables us to get the required strength with the least weight. The handles should be made of wood which is both strong and light-weight. These points will be of special importance in introducing these implements among the farmers of Hokkaido. If for instance, when a Hokkaido farmer for the first time ventures to use a hoe or a spade, it breaks in his hands or soon shows the effects of wear, he will be disgusted with the implement and think that it is useless, when in reality the fault was in its manufacture and not in the implement itself. For the handles of these implements, [p. 318] ash will probably be the best wood that can be used, although oak will answer the purpose very well. [p. 319, p. 320 is blank]

9) The Pronged Hoe

It is an implement which is very useful in stirring soil among cultivated crops which are planted so closely together that a horse cannot be used. This hoe has a handle like that of an ordinary hoe, but instead of a blade it usually has prongs. These prongs are made of steel, are very sharp, and are attached to a steel head.

10) The Hand Roller

It is very useful in pulverizing the soil in gardens, and in making soil compact after the seed, especially of beets, carrots, onions, radishes, *etc.*, has been sown. A good hand roller may be made of a cylindrical piece [p. 321] of hard wood 9 inches in diameter and 5 or 6 feet long. A hole 2 inches in diameter should be bored longitudinally through the center along the axis. Through this hole should be passed an iron axle 2 inches in diameter. This axle should project 2 or 3 inches beyond each end of the cylinder; to these projecting ends, the handle should be fastened.

11) The Clod Crusher

It is an implement which has recently been quite extensively used in market gardens. It consists simply [p. 322] of a sort of drag platform made of planks. The forward end of this drag platform is bent upward somewhat, so that it will pass over the soil instead of pushing it up in front of itself, as it would if the planks were straight. Crosspieces are fastened to the bottom of the planks for the purpose of giving it greater crushing power. This is an implement of very simple construction, and can be made by any farmer for himself. It should be drawn by two horses. [p. 323]

解　　題

1. ブルックスの講義内容

　第4章のタイトルは，直訳すれば「砕土」となるが，その内容から「耕耘 Tillage」と置き換えたい。ブルックスが帰国後に出版した教科書 *Agriculture*（全3巻，1901年）では明らかに Tillage としながら，本章で Pulverization of soil を使った理由は，本文中に Tillage の語が登場しないことや「無施肥でも砕土を十分にすれば多年栽培ができると信じた人たちがいた」と述べているため，まだ Tillage という語が定着する前だったかもしれない。しかし，講義冒頭に出る Webster 辞典1864年版は，"Tillage"; The operation, Practice or art of tilling or preparing land for seed, and keeping the ground in a state favorable for the growth of crop. "Pulverize"; To reduce to fine powder as by beating, grinding or the like. と明確に区別しているため，その真意は明らかでない。

　この章は，播種前のすべての作業を一括した「耕耘」作業の意義と目的および時期と方法に始まり，それに用いる各種プラウやハローなどの構造機能，利用法までを細述していて，農機具を扱った章である。また，講義の全体から農機具を探すと，いずれも構造等の説明はないものの，第3章でサブソイラ(Mole Plow)など排水施設用の作業機があり，第6章では1節を設けて乾草収穫調製用の作業機 Implement Necessary for Harvesting Hay がある。さらに第7章では，作物ごとの栽培管理文中に中耕除草用のカルチベータ，収穫調製用のポテト

プラウ，開発間際のリーパやスレッシャなどを取り上げているが，すべての章で今日のように分類体系が整っていない．しかしながら，登場する作業機は，日本が受容しようとした泰西農業の農機具，すなわちアメリカにあった一連の農機具を網羅しており，当時のわが国の在来農機具が鍬と鎌を代表とする人力農具のみであったのと大差がある．ただし，注4に説明するように，日本にはプラウと同じ用途の犁（り，からすき，和犁）があり，すでに5世紀末に伝来して正倉院宝物にもなっているが，小規模経営と増収追求や牛崇拝から役畜がいないなどによって，大正時代までほとんど普及していない．そのように初体験の農機具であったから，ブルックスは毎日午後に行なわれる「農場実習」も担当し，講義の作業機をすべて輸入して構造の詳細，取り扱い作業法を教え，実体験させている．なお，これに登場した農機具は，本邦最古の洋式農業建築として重要文化財に指定された「札幌農学校第2農場」の模範家畜房と穀物庫などに収蔵され，今日でも講義内容に沿って現物を確認することができる．

わが国の在来農具を代表する鍬と鋤は，用途に応じて世界に類のない多様な形態に発展し，田畑を耕し，土塊を砕き，播種溝を作り，培土を行ない，草を削り，根菜類を掘る，さらには原野を開墾するという万能の作業機であった．これを講義の洋式農具で置き換えると，記載順にプラウ，ハロー，リッジャー，培土と除草にカルチベータや人力ホー，ポテトプラウやデガーとなる．また，鍬とともに代表農具である鎌は，稲麦や畑作物の収穫時に茎幹を刈り取るばかりか，畦草や雑草を刈り，牧草を刈る，薪などの灌木を切るという万能の刃物である．これ以外の主な在来農具と言えば，脱穀に使う千歯，汎用運搬車としての荷車，唯一機械と呼べる揚水用の足踏水車程度であり，これに箕と篩，筵や肥桶を加えれば，当時のわが国の農作業用装備が揃う．少し学問的になるが，厳密な意味での「機械」とは，複数の部品で構成され，それに外部からエネルギーを受けて，部品間に定まった運動を行なわせ有益な仕事をするものと定義されるから，講義に登場したリーパ，モーア，スレッシャ，西洋唐箕が農業用に用いる機械，すなわち「農業機械」である．一方，部品間の運動がないために機械としての条件を満たさないものは「道具」とされ，このうちで農業に用いる道具が「農具」と呼ばれて鍬や鎌が相当する．さらに，文頭で「農機具」と記載したのは，農業機械と農具を合わせた名称として使い分けている．したがって，講義を行なった頃の日本在来農具は，まったくの道具レベルであるのに，講義では高度な機械を含む一連の農機具を扱っていて，彼我の技術的な格差が明確になる．

2．北海道での農業機械化の推移

作物栽培は田畑を耕す耕耘に始まるが，その代表となるのがプラウである．第4章のプラウは，基本となる畜力2頭引の再墾・新墾・傾斜地プラウから，その頃実用化されたばかりのスチームトラクタを使ったプラウまで余すところなく説明している．しかし，ブルックスの出身地マサチューセッツは，移住開拓後すでに200年を経て農地が熟畑化しているから使えるのであり，開拓を始めたばかりの北海道では樹木の根が張り，灌木が生い茂っていて講義通りには進まなかった．したがって，北海道開拓の移住者は，入植時に給与される洋式農

機具，札幌農学校や官園で普及指導する農機具や農業技術に対して，稲作のない畑酪混同農業の受容と同様に，そのまま受け入れることが難しかった。特に洋式農具が畜力作業機であるため，例えば七重勧業試験場の第1号報告によると「2頭引プラウ作業に6頭の馬を付けても引けない」という資料も残されていて，農耕馬の輸入・増殖と調教や使役技術も問題になった。しかし，開拓初期に移住した武士団が早々に洋式農機具を受け入れたことと，入植後にただちに農業を始めなければ生活できないこともあって，いろいろの失敗を重ねながら洋式農機具の普及が進んだ。この普及状況を移住希望者向けに刊行された『北海道移民必携』(1896(明治29)年版)から関連事項を抜粋する。「農具の種類と数量を定むる前に，貸下ぐ地籍の5町歩〔ha〕に何々を仕付け，その割合が如何なる程かを定めざれば示し難し。10余年前に移住した伊達紋別，当別，白石の農家の作付け割合はほぼ一様にして，大麦1町歩，小麦5反歩〔50a〕，裸麦3乃至4反歩，大小豆5乃至6反歩，雑穀，蔬菜等4反歩内外，合計2町9反内外にして，残る2町歩は地方に適する作物を耕作せり。当別なれば麻，仁木なれば豇豆類，札幌近郊なれば亜麻，甜菜，藍または米を仕付けるが如し。家の口数4乃至5人にして農事に堪ゆる者3人とし，前掲の作物を作付けると仮定すれば，凡そ以下の農具を要すべし。2頭曳プラウ1挺，ハローは3戸に1挺，次には平鍬4挺，唐鍬2挺，鎌3挺，斧2挺，鋸大小各1枚，ホー4本，レーキ4本は入用。なるべく2戸に付き2輪車1車，橇2台を要す。播付けまたは収穫など田圃の業務多忙なる時に日雇を使役して使用せしむるを見込みしなり。以上の農具はいずれも和製にして札幌市中にて製造販売せらるる」。すなわち，入植農家の資金の問題もあって，推奨した洋式農具からまずプラウとハローが普及し，次いで明治時代後半に中耕除草用のカルチベータが普及して，ようやく洋式農具を使った「北海道農法」が確立されるにすぎない。

　この農機具の普及には，その国産化を図った先進的な鍛冶屋の活躍を見逃せない。開拓使は屯田兵制を実施に移した1875(明治8)年に農馬具製造所を設置し，アメリカ人技師2名を招いて鍛冶工12名を指導しながら，プラウやハローと馬具の製造を始めた。ここでの農具生産は年々拡大されて当時の需要に応えたが，1882(明治15)年に開拓使を廃して3県制になると，工場は民間に払い下げられて札幌農具製作所と変わった。しかし，間もなく官営工場払い下げ事件に発展して農具生産の停滞期を迎える。他方，農村では生産の停滞が許されないため，函館七重官園等では農具の補修や製造法を指導推奨したし，札幌には農具製作所の鍛冶工を招いて小さな鍛冶屋が生まれた。その動きのなかで移住者のなかにいた刀や武具鍛冶たちが農具の改造と生産に取り組み，地域の土性に合ったプラウを作り，鉄の表層に炭素を滲ませて強度を高める滲炭硬化法を発明するなどの成果があった。しかし，これら技術は職人の秘伝として広まらないばかりか，砂鉄中心の時代で均一な材質の鉄が得がたいため，わずかの鍛冶屋が自立したのみであった。

　日本の産業革命は，国営八幡製鉄の創業(1901(明治34)年)や日清(1894年)・日露戦争(1904年)を契機に始まったと言われるが，これによって農村にも新しい工業技術が伝わり，均質の鉄材が供給され始めると各地に鍛冶屋が生まれた。特に府県の先進的鍛冶屋は，1910年代から足踏脱穀機，籾摺機，和犂などを相次いで発明して日本の農機具工業の濫觴となっ

た。また北海道では，鉄道が延長されるなど交通網が整備されたことと稲作の普及によって移住者が急増したため，動力となる馬の売買をする馬喰とともに鍛冶屋が急増し，この鍛冶屋が今日も活躍する農機会社の源流になった。しかし，新しい鍛冶屋が生まれても，農機具の製作技術は秘伝のままであったから，熱心な鍛冶工たちは札幌農学校と発足したばかりの北海道農事試験場を訪れて指導を乞うようになった。さらに1914(大正3)年の第1次大戦はエンジン，自動車などの実用化を促して工業技術を急速に発展させ，日本農業に好況をもたらしたことから，大規模農場に石油発動機，耕耘機，ガソリントラクタ，コンバインなどが輸入され，農作業の手法を根本から変え始めたが，発動機を除く作業機類が普及するのは第2次大戦後を待たねばならなかった。加えて戦争は砂糖の国際価格を暴騰させたため，最低限の自給を図ろうとテンサイ栽培が国策で推奨されたが，それには深起こしが不可欠としてプラウの性能向上が叫ばれた。続いて間もなく農業が不況となって農家の耕作意欲が減衰すると，北海道はドイツの農家を招請して模範的営農を営ませたが，そこでも農機具の製作技術がクローズアップされた。そのような経過から，農学校から昇格した北海道帝国大学と農事試験場は農機具の研究指導体制を整え，まずプラウとカルチベータの研究に取り組んだ。これまで民間任せであった農機具研究が研究機関で始まったことになる。

3．農業機械研究と農業の近代化

　札幌農学校から北海道帝大への流れのなかで農機具講義を振り返ると，ブルックスの農学は，1889(明治22)年頃にいくつかの学科目に分けられ，最初の「農機具学」は南鷹次郎と佐藤昌介が洋式農機具と在来和農具に分けて分担担当した。両者は作物・農業経済学の専門家で農機具を扱うのも不思議に思われるが，その後も生物学の明峰正夫など専門外の教授が担当していることから，海外留学をした教授陣にとって農機具は基礎知識であったようである。職務の上で農機具分野を初めて担ったのは，1897(明治30)年に農学科を卒業した時任一彦である。今日の農業工学科の開祖となる時任は，農業と園芸分野のすべての物理学を農芸物理学と名づけ，専門は農業土木学ながら，明治後半から大正時代を支えた日本の農機具研究の数少ない先覚者となった。前述のようにプラウへの要求が高まると，時任は教え子から農業土木学に権平昌司，農業機械学に常松栄を指定し，それぞれの専門家として養成した。新しい農業機械の研究と勉学に在学年限いっぱいを費やした常松は，1929(昭和4)年に農学部に採用され，ただちに鍛冶場のある農機具実験室を農場に新設してもらってプラウとカルチベータの研究を始めた。そこでは鍛冶屋の指導をしつつ農村を巡り，数多くの標本から焼入れや深起こしができて抵抗の少ないよいプラウ曲面の作り方を平準化させたし，地域の土性ごとに形態の異なるプラウ曲面の解析を行ない，経験と勘に頼ってきた曲面の作り方を数値表現できるようにした。その活躍中に日本が戦争に突入したため，満州国開拓総局の求めによっていく度も現在の中国東北部を訪れ，北海道から半ば強制的に移住させられた鍛冶屋を指導し，開拓に適したプラウ，ハロー，カルチベータなどを製造させた。なお，当時の移駐工場は，数百人を雇用して需要に応じており，鍛冶屋と呼ぶような規模を超えていたが，これら工場は終戦とともに壊滅して裸一貫で引き揚げている。

第2次大戦が終わると，日本は海外植民地からの引揚者を加えて極度の食糧難となったため，駐留軍はアメリカの著名な農機研究者を代表者とする調査団を送り込んで食糧増産の手法について勧告を行なわせた。この勧告は，原野を緊急に開拓して農地の拡大と雇用の増大を図るとともに，既耕地も機械化によって生産性を高めることを基本とし，付随して農業機械技術者の早急な養成を求めていた。そこで農学部のある全国の旧帝大に農業機械講座が設置されたばかりか，北大には学生定員25名の農林専門部農業機械科も設置された。さらに民間では，終戦によって解体された軍需産業が農地開拓用建設機械と農機具製造こそ平和産業であるとして製造販売に取り組んだ。第2次大戦のもたらした影響のひとつにゴムタイヤの実用化があるが，旧軍需産業はゴムタイヤ耕耘機，ハンドトラクタなどを国産化し，今日のトラクタ機械化時代の幕開けとなった。特に耕耘機（歩行式トラクタ）は，全国の水田の耕耘に使われて大量生産が進み，1960（昭和35）年頃から輸出が増大して，世界を席捲するようになる。

　神武・いざなぎ景気などと叫んで日本経済が好況になると，農村の雇用労力が急減して一連の作業をすべて農業機械で行なう一貫機械化体系が求められ，その中核となる動力源として農用トラクタの時代がやってきた。それに応じるため，北大では農業機械講座に加え，農業原動機と農畜産加工機械学講座との3講座へと拡充し，研究スタッフの増加を図った。他方，行政は農地基盤を機械化できるようにする構造改善事業を推進したため，稲作の主作業となる耕耘，田植，稲刈りなどは，1967（昭和42）年頃から5〜7年でほぼ全国の90%を機械で作業できるようになった。一方，北海道の畑作と酪農は，輸入農業機械のまま使えるため，稲作機械に先駆けて導入が進んだが，昭和40年代に輸入機を模倣し，わが国の栽培基準に沿うように改造した国産の農作業機が開発されると，もはや機械なくして農業はできないと言われるまでに普及し，1977（昭和52）年頃，田畑の一般作物すべてについて一貫機械化体系が完成した。

　1876（明治9）年に導入を企画した機械化泰西農業は，時代の推移によって動力源を馬からトラクタに置き換えるが，極言すると，ようやく100年を経て実現できたと言ってよいであろう。

注
1) ここで講義は，物質の定義から物質を構成する元素までを説明する。現代の物質論は20世紀からの原子論的な定義であるが，これをブルックスが採用するのは無理のため，アメリカで絶大な人気を誇っていた1853年刊の化学教科書(B. Silliman, Jr., *First Principle of Chemistry*, この北大附属図書館蔵本はクラーク博士の署名がある)と想像する。この定義は，講義の通りに域性，碍性，分性の3性で定義し，鬆性，縮性，張性，弾性，引性，惰性を有するとしている。次いで，現在107種とされる元素数を講義で63と述べたのは，1869年にメンデレーエフが最初に工夫した(短周期型)周期表のもので古くないが，講義と同時期に刊行された『米欧回覧実記』では，「元素は，70余種ある」（岩波文庫第5巻，199頁）としており，講義時の最新データを使えば71種が正確であった。このうちで農業に必要として11元素をあげているが，今日では微量要素が加わり，講義のように限定できない。
2) 耕起用作業機のプラウは，英語でPlough，米語でPlowと表記する。また，日本の辞書ではPlowを「スキ(犂)，からすき」(同音の鋤はスコップのような手農具)と表記しているが，注4のように発達経路・過程，基本構造が異なるため，専門分野ではプラウ（西洋犂）とスキ（和犂）に区分する。Figs. 1-3に示す墢土(Furrow Slice，長さを加味すれば墢条)の反転図は，プラウが原則として右反転であるた

め，進行方向の反対側から見た図である。また，プラウは墢土を滑らかに反転させるため，矩形を転がすと考えればよいから，図の墢溝(Furrow)の幅が狭すぎる。Fig. 3 では，一般に墢土断面を $1:\sqrt{2}$ とし，それを 45° 倒した時に表面積が最大になり，風乾効果が大きくなる。

3) プラウの耕深は，土に直接作用する犂体 Bottom の大きさ，特に刃幅でほぼ決まる。講義の時点は，開墾初期で木の株間や根を避けるように利用するためと馬の体格が劣っていたため，刃幅 17-24 cm (7-9.5 in.，プラウは今でもインチサイズで規格化されているが，度量衡法でインチと言えないため型に置き換えている。また，前者を 1 頭引，後者を 1 頭 7 分引とも呼ぶ)を使用し，講義のように耕深 13-20 cm (5-8 in.，通常 16 cm)にすぎなかった。ブルックスの後を引き継いで「農機学」を担当した南鷹次郎(第 7 章参照)は，1895(明治 28)年頃の講義で「2 頭引の刃幅 10.5 型や 2 頭 5 分引の刃幅 12 型が普及し始め，20 cm 程度の耕深が普通」と述べており，馬格改良と熟畑化が進んだことを示している。また，第 2 次大戦後に畜力に代わってトラクタでプラウを引くようになると，当初こそ 12-14 型で耕深 25 cm 程度であったが，徐々に深くなって最近では 35 cm 以上が通常の耕深となっている。

4) プラウの原型を「アード，掘り棒」と呼び，今日では西南アジア，中近東あたりの乾燥地で発明されたことが定説となっており，講義のように古代エジプトやローマ帝国を経てドイツ，イギリスに伝わって，畑作に適したように改良されたのがプラウである。これがさらにアメリカを経て北海道に導入され，今も基本農具として活躍している。一方，逆に西南アジアから東方に伝わってインドや中国の水田とともに発達したのがスキであり，正倉院宝物にスキの原型があるが，古代日本に中国から直接か朝鮮半島を経るかで伝来して「からすき(唐犂)」と呼ばれている。このことから日本は，原型が 1 つの耕起用具が東西両方に別れてプラウとスキとなり，世界を一周して到達した終着点の国である。なお，日本に伝来したからすきは，解題の通り普及しなかったが，北海道開拓のプラウの成果や乾田思想の普及によって明治末期に改良され，大正時代に全国に普及する。

5) 本文中や Fig. 6 に示すプラウ部品は，大正年代の和犂の普及によって和名ができるが，英語名のままカタカナ書きした方が通用することが多い。しかし，北海道での普及過程で俗称も生まれているから，それらの名称を「英語；カナ(和訳名，俗称)；すき」での名称順に対比して示す。Beam；ビーム；ねり木，Share；シェア(刃板)，犂先，Point；ポイント(刃先)，Moldboard；モールドボード(発土板，羽根)；へら，Landside；ランドサイド(地側板)；床，Handle；ハンドル；かじとり，Clevis；クレビス；はずな，Beam Wheel；ビームホイール(定規車)，Coulter；ナイフまたは円板コルター(犂刃)。その後の文中で登場する用語では，Bottom；犂体，Standard；スタンダード(犂柱)；たたり，Heal；踵，Frog；結合板，Shin；脛などがある。

6) ここでプラウの種類を述べているが，今日のように犂体曲面や用途で分類されていないばかりか，典型的な形態を製造会社名で示す状況である。講義のなかで Sod Plows；草地プラウ，Stubble Plow；再墾プラウ，Sedge or Bog Plow；新墾プラウ，Breaking-up Plow；新墾プラウは曲面形態による分類であり，今日の All-purpose Plow；汎用型または兼用プラウが抜けている。Swivel or Hillside Plow；傾斜地プラウ，Subsoil Plow；心土プラウは用途別分類，Sulky Plow；乗用型プラウ，Gang Plow；多連プラウ，One-horse Plow；1 頭引プラウは形態による分類，Steam Plow；スチームトラクタ用プラウは畜力に代わる新動力での牽引を指して動力源による分類である。札幌農学校でプラウの分類法を教えるのは，残された湯地定彦(1898(明治 31)年卒業)の講義ノートから，ブルックス帰国後に農業を担当したブリガムにもなく，明治 20 年代後半から農機具学を担当した橋本左五郎(畜産食品製造学初代教授)を待たねばならない。この頃は，日本農学の始まりと言われる時であるが，農機具についてはアメリカでやっと整理されたのであろう。

7) 動力源に蒸気機関を用いるスチームトラクタは，今日の農業機械化を支えている内燃機関トラクタを生み出す先駆けとなった。18 世紀中葉のワットによる蒸気機関の発明は，イギリスの産業革命のなかでトピックとなっているが，これはまず製粉工場等の定置式動力源として普及し，構造が大きすぎて移動式には向かなかった。しかし，ワットは，将来を見込んで蒸気機関を積んだ汎用牽引車，すなわちスチームトラクタの特許を申請した第 1 号であるが，実際には 1900 年頃にようやく一般化されるにすぎない。講義では，開発・実用化の目処が立ったばかりのスチームトラクタを取り上げ，利用法としてプラウを牽引しながら畑を走る方法，圃場外に定置させてウインチでプラウを引く方法の 2 通りを紹介した，わが国に最初の紹介者である。なお，スチームトラクタは，わが国には輸入されていないという定説であったが，1908(明治 41)年に盛岡の小岩井農場でイギリス製蒸気トラクタを実演した記録が発見され，輸送費を考えれば 1 台購入した可能性がある。また，上富良野町の博物館「土の館」に 1904(明治 37)年製造のスチームトラクタが動態保存され，国内唯一の標本となっているが，これは三川町の三谷農場主がカナダから近年に購入したもので利用実績はない。

8) 土塊を崩し，地均しをする作業機をハロー，砕土整地作業機と言う。今日では，ロータリやデスク

ハローによる砕土が一般的であるが，後者ですら，バネ鋼を使ったスプリングツースハローとほぼ同時期，工業化が進んで均質な鋼板加工ができるようになった 1880 年代の発明である．さらに高い工作精度を求めるロータリは，所要動力が大きくてトラクタ化が進んだ 1940 年代に発明されるにすぎないから，講義時点では，土塊を十分に乾燥して破砕しやすい条件で作業するのがコツであった．本節の最後にある木製ハローは，北海道の資金力のない開拓農家向けに手作りのハローを紹介した部分で興味深い．

9) 作物の畦間の中耕除草，培土に用いる作業機をカルチベータ，中耕除草作業機と言い，作物の生育に合わせて数回の作業をするが，株間と根際の除草ができないために人力除草も不可欠である．作物栽培をすべて人力で行なった場合，耕起・砕土と中耕除草作業が重労働であるため，北海道開拓ではプラウ，ハローとカルチベータの 3 機種が明治末期までに普及し，播種や収穫は在来通りの人手頼りの「北海道農法」が実現した．

参考文献
1. 農業機械学会編，農業機械ハンドブック（初版(1957)：畜力作業機主体で本章向き；改訂版(1969)：トラクタ作業機中心），コロナ社
2. 髙井宗宏・南部 悟(1982)，発見された札幌農学校の畜力農機具（北の技術文化，4号，北海道産業考古学会，所収）
3. 髙井宗宏(1993)，明治維新における泰西農業の導入過程（田中 彰・髙田誠二編著，『米欧回覧実記』の学際的研究，北海道大学図書刊行会，所収）
4. 関連する教科書類：西村英十郎著，農用器具学(1900)博文館；明峰正夫著，最新農具論(1907，改訂版，1923)六盟館；稲垣乙丙著，農具学(1911)博文館；水野夏一著，農業動力及改良農具(1921)帝国農会；森周六著，農業機械学(1927)成美堂；広部達三著，農用機具学〈動力編，作業機編Ⅰ・Ⅱ〉(1929-30)西ヶ原刊行会；常松栄著，北方農機具解説(1943)北方文化社；庄司英信著，農業機械学概論(1953，改訂版，1956)養賢堂

［髙井宗宏］

Chapter 5   Manures (and Fertilizer)

Some agricultural writers define manure as a previous crop gone to decay. Webster dictionary definition is "anything which makes land productive", "fertilizing substance, as the contents of stables and barnyards, marl, ashes, *etc.*" The common idea with regard to manure is that it is always some form of animal excrement. None of these definitions which I have given you is complete. Manure is properly defined as being any substance that contains plant food[1], or which has a tendency, either by mechanical or chemical action, to develop plant food out of the elements already present in the soil. I need not dwell upon [p. 324] the importance of a full consideration of the nature and mode of action of manures. You all know that without manure, we cannot raise crop yield. Manure is raw material which, with the aid of Mother Nature (*zokuna hyogen*), the farmer wishes to incorporate into crops. His crops will generally be proportional to the amount of manure he uses.

5-1. Classification of Manures

Manures may be divided into two categories, mineral and organic. Organic manures may also be subdivided into two categories, atmospheric and compound. Atmospheric manures are the gases in the air. Compound manures are those which contain these gases [p. 325] together with organic substances.

1) Mineral Manures

Minerals are those substances which are of rock origin, the elements of which are taken up in a solvent state by plants, and which constitute ash when the plants are destroyed by combustion or decay. The ash left after quick combustion is itself a mineral manure. The action of mineral manures is twofold: first, they supply the necessary food for plants; and second, they act on the particles of the soil, producing mechanical or chemical results, thus converting the soil into plant food. We should always keep this latter point in mind, that some substances, [p. 326] though they may not contain plant food themselves, may be very beneficial in creating available from the elements as plant food in the soil from the elements already contained in it. Most farmers apply manure to land simply for the crop which they wish to produce, aiming to add to the soil such material as will supply such elements as the plants will need. We should also aim to add substances which, by their action upon the particles of the soil, will hasten their conversion into available nutrients[2]. The mineral manures are potash, lime, soda, magnesia, silica, iron oxide, phosphoric acid, sulfuric acid, nitric acid, and chlorine[3]. Among these, potash, lime, soda, and magnesia are called alkalis, and as bases unite with certain acids to form salts. All alkalis are of a caustic, [p. 327] burning nature and, in combination with organic substances, promote decomposition. The most important alkalis for the farmer are potash, lime, and magnesia. The most important acid is phosphoric acid.

2) Effect of Alkalis on Animal and Vegetable Matter

As I have just stated, the alkalis hasten the decomposition of organic substances; hence, they should never be mixed with animal excrement alone, nor with any pure organic substance, because they cause such rapid decomposition that much of the nitrogen in those substances is thrown off. If, for any reason, it is desirable to cause rapid decomposition of any organic substance, [p. 328] an alkali should be added, but the precaution should be taken to mix it with a large quantity of soil[4)], and also to cover the decomposing matter with soil. This soil will act as an absorbent, and no nitrogen will be lost. There is no danger of nitrogen loss from the application of alkalis directly to the soil, since any gases which may be generated will be absorbed and retained. Alkalis are, in fact, often extremely useful in promoting the conversion of substances contained in the soil into plant food.

3) Importance of Mineral Manures

There have been two theories with regard to the relative value of mineral and organic manures, some experts claim that only mineral manures are useful, [p. 329] while others make the same claim for organic manures. Neither theory is correct, because both are needed by the plant. One is equally as important as the other, as far as the plant is concerned. There may, however, be certain soils which contain an abundance of organic substances but not a sufficient quantity of mineral elements. In such soils, of course, it will be most important to apply mineral manures. In other soils, just the reverse will be true.

4) Potash

Potash is a very powerful alkali, and has a corresponding mode of action. It unites with all the different acids, forming salts, but its nature and value are very different in different salts. [p. 330] In some salts it is a very valuable manure, but in others a deadly poison. Nitrate of potash, commonly known as saltpeter, is one of the most valuable forms of potash. At present, however, in Japan, the principal source of potash must be wood ashes. Wood ashes, next to animal excrements, are the best manure that a farmer can use. They are very far from being pure potash, but contain all the mineral elements that were in the trees burned. The quality of the ash from different kinds of wood varies, but they all contain much more lime than potash.

5) Percentage of Ash in Different Woods

There is a great difference in the percentage of ash in wood [p. 331] from different parts of the same tree, the trunk wood contains a small amount, the branches more, and the smallest twigs with their bark the most. This is clearly seen from the following example; the trunk wood of the birch tree contains 0.66% ash, the branch wood 1.05%, and the twigs 1.45%. The figures

The Birch tree	0.31%
The Oak tree	2.80%
The Elm tree	1.50%
The Apple tree	1.25%
Twigs of Walnut tree	2.99%
The Pine tree	0.28%

which I am about to give you are, where not otherwise stated, the amounts of ash in wood as it is ordinarily burned.

6) Materials Contained in the Ash from Different Woods

The composition of ash [p. 332] also varies in different parts of the same tree, but there seems to be no general rule of variation, except for the element silicon, which increases in the smaller parts of the tree, probably owing to the relative

Name of Wood	Potash	Soda	Magnesia	Lime	Phosphoric Acid	Sulfuric Acid
Birch	11.6%	5.8%	8.9%	60.0%	0.3%	4.8%
Beech	16.1%	3.4%	10.8%	56.4%	1.0%	4.7%
Oak	10.0%	3.6%	4.8%	73.5%	1.4%	1.1%
Elm	24.1%	2.6%	10.0%	37.9%	5.4%	6.2%
Apple	12.0%	1.6%	5.7%	70.1%	2.9%	1.8%
Pine	15.3%	9.9%	5.9%	50.1%	3.0%	6.0%
Walnut	15.3%	0.0%	8.1%	55.9%	3.2%	2.9%

increase in the quantity of bark as compared with wood. From the table which I have just given you, you will see that the percentage of lime varies in the ash from different kinds of wood from 37 to 73%, being present in all kinds in much greater quantity than any other elements. Potash varies in quantity from 10 to 24%; those kinds of wood [p. 333] which contain the most being beech, pine, elm, and walnut. Those woods which contain the most lime are oak, apple, and birch.

7) Solubility of Ashes

To be effective as manure, ashes must be soluble in water. In hardwood ashes, a somewhat smaller percentage is soluble than in the ashes of softwoods, and therefore the latter act more quickly than the former. On most soils, ashes are chiefly valuable for the potash which they contain, although other compounds; such as lime, magnesia, and phosphoric acid, are also valuable. By being leached, ashes lose about 4/5 (80%) of the potash which they contain; therefore, if used [p. 334] as manure for the beneficial effects of potash alone, one bushel of unleached ashes is worth as much as 5 bushels of leached ashes. As a rule, by leaching, ashes lose little but potash. Nearly all lime, magnesia, iron oxide, soda, phosphoric acid, sulfuric acid, and silica remain; also, 20% of the potash is present in combination with silica as silicate of potash, this being quite insoluble. Ashes that are leached for the purpose of making soaps contain more lime than the original ashes, since lime has been added to them in order to make the potash more readily soluble. Taking into account all the elements which promote fertility, 5 bushels of leached ashes are worth as much as 4 of unleached. It should be remembered that ashes are rendered much more compact by leaching, and that [p. 335] a certain measure would not contain any greater quantity of leached ashes than unleached. However, as I have stated, since potash is the element that is usually of most value in ashes, every farmer should be careful to store what ashes he has, and in such a place as rain cannot fall upon them, because it would dissolve and carry away a large proportion of this valuable substance.

8) The Mode of Action of Ashes

When used as manure, the first and most important effect of ashes is to furnish plants with mineral food; but they also produce chemical changes in the soil, which result in making plant food available. Ashes also influence the physical condition of the soil, making it [p. 336] more retentive, compact, moist, and cold.

9) Soil Adaptation of Ashes

Sandy loam soil most benefits by the addition of ashes, and cultivated fields for long years benefit more than new fields because ashes supply to them the minerals and elements which have been taken away by years of successive crops. Ashes should not generally be used on wet soils because—though they may seem to be beneficial to the first crop—they will ultimately make such soils even wetter, and therefore injure them. Soils which contain much organic matter should receive an application of ashes, because their alkaline action upon that matter will hasten its decomposition. The proper mode of applying ashes is to sow them broadcast, and then harrow them into the soil. Ashes should not be placed [p. 337] in direct contact with seeds, since they will prevent germination. The best time to apply ashes is in the autumn, but they will be very beneficial even if applied in the springtime.

10) Crop Adaptation of Ashes

Plants are all special feeders, and are categorized as potash, lime, or phosphoric acid plants according to their most prominent mineral ingredients. All root crops —potatoes, tobacco, grain-crops, and grasses—are known as potash plants. Therefore, ashes which contain a large amount of potash are best suited to these plants.

11) Ashes in Compost

Ashes should never be [p. 338] composted with animal excrements, since they set free the nitrogen contained in them. It is desirable, however, to compost ashes with material containing a large amount of organic matter, such as muck. A combination of 20 bushels of ashes (1 Japanese *koku* = 5.3 bushels[5]) and one cord of muck (1 cord[6] = 128 cubic feet) composted are nearly as effective as animal excrements.

12) Value of Ashes as Fertilizer

In sandy loam soil, [p. 339] 15 bushels of unleached ashes will give one acre of land 15 bushels more corn than would grow without any fertilizer added, and the yield would increase at the rate of one bushel of corn per bushel of ashes, up to a certain point. On one acre of land of the same kind, 15 bushels of ashes will make a difference of 1,000 pounds[7] of hay; besides making this much difference in yield of crops, ashes will permanently[8] improve the condition of the soil. From these figures, you can form an estimate as to what you can afford to pay for ashes in any region in which you may be situated. The value of ashes will, of course, vary according to the value of the crops produced. Knowing the value of the crops, you can soon determine how much you can afford to pay for ashes.

13) Value of Coal Ashes as a Fertilizer

The value of coal ashes as fertilizer is very small. [p. 342] I would never buy them for the sake of the plant food they contain, yet if I had some I would apply them to wet, heavy soil, since they would make it lighter, and therefore more readily acted upon by the atmosphere. The reason that coal ashes are so valueless is that the plants from which coal originated did not contain the mineral elements which are necessary for the growth of plants at the present time. Coal contains mostly silica and silicates, which are comparatively valueless as fertilizer.

5-2. *Potash*

Many of the salts of potash are very valuable fertilizers; Of those, nitrate is the most valuable. For use as a fertilizer, the crude material is most commonly employed, since it is less expensive than that which has been purified. Nitrate of potash is commonly called saltpeter; it is valuable not only for the potash [p. 340] it contains, but also for its nitrogen. It is estimated that, in America, a farmer can afford to pay 2 to 5 cents per pound for saltpeter. Here in Japan, since the prices paid for crops are cheaper, the price paid for fertilizer should be proportionally cheaper. German potash salts have recently been extensively used by farmers, both in Europe and America. These salts are by no means pure; they usually contain some magnesia and other elements. One of the most valuable salts is sulfate, but that is more expensive than chloride, and for fertilizing purposes the latter is almost equally good. Crude chloride of potash, known by the name "kainite", is very valuable for many crops, but should not be used on crops which contain a large percentage of starch or sugar, [p. 341] since the hydrochloric acid in it is believed to prevent the formation of these substances. Kainite is found to decrease the quality of sugar beets and potatoes. The German crude potash salts vary broadly in the amount of potash which they contain, some containing 80% or more sulfate or chloride, while others contain much less. For use at a considerable distance from where those salts are mined, it is cheapest to use those which contain a high percentage of potash, since it is not profitable to transport as much valueless material as is contained in the salts of lower grade.

5-3. *Lime*

Lime enters more largely into all plants than any other mineral elements. In one part of the ash of most trees, 1/2 of the ash of cereal grains and about 1/3 of the ash of forage plants [p. 343] are lime. Lime is also a very powerful agent in the soil, in converting its elements into an available form for plant food. Since it is used so extensively by all plants, providence has very wisely furnished an abundance of it in nearly all soils. There are some soils, however, in which lime is deficient. Lime is very extensively distributed; it is found in most rocks and in nearly all soils. It is also found in limestone, chalk, and marble—each of these being a form of carbonate of lime. As silicate, lime is found in most soils, but in that form it is not directly available as plant food. There are no deposits of pure lime in Japan; it is always found in combination with some mineral or acid. It is most commonly found in [p. 344] combination with carbonic acid, as carbonate of lime. In that form it weighs so much, and is so difficult to pulverize, that it will not pay to use it as fertilizer. Nearly 1/2 of carbonate of lime by weight is carbonic acid—in one ton (2,000 pounds) of that substance, there are 875 pounds of carbonic acid and 1,125 pounds of lime. If carbonate of lime is heated, one ton of it will lose from 1,100 to 1,200 pounds of carbonic acid and water given off by the heat. In that modified form, it is known by the names of "quicklime" and "caustic lime". After carbonate of lime has been heated, it begins to absorb moisture from the air; after absorbing as much as it can, it is called air slaked lime. If we add a quantity of water to caustic lime, it will be slaked quickly, accompanied by the generation of a great deal of heat; the resulting substance, which is a compound of lime and water, is called "hydrate of lime" [p. 345] or "calcium hydrate". By hydration, lime is changed from a stone to a fine powder. After lime is slacked, if exposed to the atmosphere, it immediately begins to absorb carbonic acid from it and becomes, in a short time, chemically like its original form; but physically, it is very different, being now a fine powder. In this form, it is known as "mild lime", it having lost its caustic, burning quality. In some places, lime is obtained in large quantities from the shells of oysters or clams. This lime, although not quite as effective as stone lime, is often much used because it costs less. In America, one bushel of shell lime costs about 25 cents, while the same quantity of stone lime costs 75 cents. Those seashells are burned, in order to [p. 346] easily reduce them to a fine powder.

1) Action and Results of Lime

When I speak simply of lime, I am referring to caustic lime. Being caustic, it acts as a decomposing agent on organic matter, either in or out of the soil. It also has a strong affinity for various acids, and therefore often promotes chemical reactions. Applied to clayey soils, lime makes them more porous, lighter, and drier; it is, therefore, very beneficial. Sandy soils usually contain, in chemical combination with silica, some clay as silicate of alumina. Lime acts on this silicate, setting the alumina free, and forming silicate of lime. By means of the free alumina, the soil is made more retentive, and therefore better. [p. 347] In soils which contain much organic matter, lime is very beneficial, since its caustic action will hasten the decomposition of this matter, converting the elements in it into an available form of plant food. Lime is of more benefit to soils of this kind than to any other. Lime acts, or may act, in four different ways: (1) it acts as a neutralizer of injurious acids in soils which contain such acid; (2) it acts as a converter of both organic and inorganic matters into plant food; (3) by producing chemical changes, it often improves the physical condition of soils; and (4) it is itself plant food, and in soils deficient in it, it will be useful in furnishing this [p. 348] most important nutrient. As a carbonate, this form of lime is by no means useless; it will act as an alkali, and also furnish plant food.

Lime, by its caustic action, hastens the conversion of elements in the soil into plant food, acting upon both organic and inorganic substances. Thus, these elements are used up more quickly when lime is applied to the soil than they would otherwise be. Lime is therefore sometimes called an "exhauster". If the farmer depends on lime alone, he will soon exhaust all the organic material contained in the soil, and thus impoverish it.

2) Method of Application and Quantity to be Used

The method of applying lime, and the form in which it is be used, depend largely on the circumstances. [p. 349] If needed as plant food only, lime carbonate may be used. If needed for its action upon soil, it is necessary to have some other form. If the soil is cold or full of organic matter, shell lime (unslaked and ground fine) should be used. The granulated shells will become slaked in the soil, thus making it warmer. Lime should never be used as a top dressing for grass land, because its caustic action will kill the grass. It should be spread on ploughed land, then harrowed or cultivated in. It should never be ploughed in, because it has a tendency to absorb moisture and work downward; therefore, it should be applied near the surface, so that it is not beyond the reach of plants. The quantity of lime necessary per acre cannot be stated exactly, [p. 350] since that will vary in each individual case. It will vary from 30 to 100 bushels of slaked lime.

3) Sulfate of Lime

This substance, formed by the combination of sulfuric acid with [p. 357] lime in the chemical equivalents of 40 pounds of sulfuric acid and 28 pounds of lime, is very valuable as fertilizer. It is commonly known by the names "Gypsum", "Plaster of Paris", or simply "Plaster". Geographically, gypsum is very widely distributed, being found in nearly all parts of the world. That it was useful as fertilizer is said to have been discovered near Paris, it being noticed that where the workmen engaged in grinding it shook the dust from their clothes, the grass grew much better than it did on adjoining land of the same character. This salt may be a source of supply of sulfuric acid and lime, both of which are used as plant food; but aside from this, it is also of great value as an absorbent. [p. 358] We know that it has the power to fix ammonia, and that quality makes it valuable as a substance to mix with compost, or to use wherever decomposition is taking place. It will be wise to apply plaster in all cases where there is ongoing decomposition, such as on land which has just been covered with coarse barnyard manure, and on newly covered sod land. Plaster is also an absorber and retainer of moisture; for this reason, poor sandy soils—even when they have no organic matter to decompose—benefit from the application of it. Plaster may also be used in manure piles and in stables to absorb the ammonia always generated in such places.

4) Other Salts of Lime

Some of the other salts of lime [p. 359] are the nitrate and chloride, but these are not of much use as fertilizers; they are, however, disinfecting agents, and also have the power to absorb moisture. Chloride of lime has that power to a remarkable degree—it has been used in some places upon streets to suppress dust. It has the power of absorbing moisture to such an extent that even when the air is apparently dry; it absorbs sufficient moisture to keep the streets always moist. This salt must, therefore, be valuable for use on dry, sandy soils since it would make them much more retentive of moisture. Lime and its salts are special fertilizers for [p. 360] all grasses, foliage plants, grains, and seeds; but the most important use of caustic lime is for its action in the soil in converting its elements into available forms of plant food.

5-4. Soda

Soda is a mineral found in all plants in small quantities. Although not found in such large quantities as lime or potash, it is equally important for plants. It is, however, present in most soils in small quantities. Soda is found in nature, in the minerals of which rocks are composed, especially in mica, feldspar, and hornblende. It is found in South America in the form of nitrate, commonly known as "Chilean Saltpeter". It is also found in the form of chloride in the crystalline [p. 361] form of lime called "rock salt", in the waters of the ocean, and in saline springs.

1) Sodium Chloride or Salt

The deliquescent and absorbent power of salt is its value as fertilizer. For this reason, it is always beneficial to light, sandy inland soils. It is not as beneficial near the sea as in localities further inland, because in severe storms salt spray is often blown many miles inland from the coast. Salt should never be mixed with organic or nitrogenous matter, or in composts, because it hinders decomposition. Salt may often be used to advantage together with lime or sulfate of lime. [p. 362] The best way to mix it with lime is to dissolve it in water, and then to slake the lime with the brine. Another way is to mix them together and allow them to sit for a few months, when the same end will have been accomplished. In both methods, they should be mixed in the proportions of 1 part of salt to 3 parts of calcium.

2) Crop Adaptation of Salt

There are certain crops, known as "saline plants", for which salt has a special affinity in nearly all soils. One of the most important of these crops is Asparagus; this plant needs a great deal of salt, but it also needs other fertilizer. An application of [p. 363] 5 to 10 bushels of salt is needed on one acre of Mangold Wurzels (Sugar Beets). All grain crops benefit from salt—it makes the kernels fuller and the straw stouter. Since the straw is stouter, the grain will not lodge. Salt is also a good fertilizer for potatoes.

3) Methods and Quantities of Salt Application

Salt should always be mixed into thesoil, and never be applied directly to the hills, since it would be likely to prevent the germination of seed. The amount of salt [p. 364] necessary per acre varies from 5 to 25 bushels, but usually 10 to 20 bushels is sufficient. It should be sown broadcast, and mixed with the soil by harrowing or cultivating. In large quantities, salt is an agent of sterility, because of the chlorine it contains.

4) Nitrate of Soda

This salt, commonly known as "Chilean Saltpeter", is found in large quantities in certain countries which are almost rainless. It is found in South America (Chile and Bolivia), and in the western regions of the United States. It is valuable as a fertilizer because of the nitrogen it contains, of which it is a cheap source. It is not important because of its soda content; indeed, as I have stated before, it is seldom necessary [p. 365] to supply soda—though it is absorbed by all plants in small quantities, the soil usually contains a sufficient quantity.

5-5. Magnesia

Magnesia is an element found in the ash of all plants, in which there is more magnesia than any other mineral element except potash and lime. In the ash of wheat, there is 12% magnesia; in the ash of rye 10%; in the ash of corn 15%; and in the ash of buckwheat 13%. In root crops, there is more magnesia than lime. Although magnesia is abundantly found in the ash of plants, it is seldom necessary to apply it as a fertilizer, since it is present in sufficient quantities in most soils. It may sometimes be found [p. 366] necessary, or rather beneficial, to apply magnesia to certain crops. Magnesia is found in almost all rocks. In mica contains 30% magnesia. It is found in amphibious pyroxene and crysolite. At the present time, the most important source of magnesia for fertilizing purposes are the saline deposits in Germany. A material called "Kieserite", the most important constituent of which is sulfate of magnesia, is abundantly found in Germany. This is a cheap source of supply for this mineral, and Kieserite is now extensively used in Europe and the United States. [p. 367]

5-6. Phosphoric Acid

1) Phosphate of Lime

Lime has a strong chemical affinity for acids, forming with them salts. The most valuable of those to the farmer is phosphate of lime. There are a few principal reasons for its value: (1) it is a very important ingredient of all plants that produce seed, nearly 1/2 of the mineral elements in plant seeds is phosphate of lime; (2) it is deficient in all soils; (3) since it is always being taken from the soil into our bones and the bones of animals, this material does not return to the soil as manure in any considerable amount; and (4) it is so readily dissolved that a quantity of it is soon used up. [p. 351] Notwithstanding its great value, there are few natural deposits of phosphate of lime, though in some localities the rocks contain it. It is found in the mineral apatite, and also in the state of New York in the form of green crystals. It is also found in Canada, New Jersey, and Norway; but the richest and most extensive deposit in rock formations is in Spain. The most important source of phosphate of lime at the present time are the phosphate beds of South Carolina. There it is found in a nearly pure state, in nodular form, at the depth of 2 to 10 feet. The chief objection to the use of this material has been the difficulty of pulverizing it, but a comparatively simple method has now been discovered. The substance as taken from ground [p. 352] is first heated, then wet with cold water, whence it is easily pulverized.

2) Bone Phosphates

Bones were, until quite recently, the principal source of phosphate. In the United States, thousands of tons of bone phosphates are sold annually. Some are crude, some ineffective, and some actually harmful owing to improper or dishonest manufacture. The bones of animals are composed of 1/3 organic matter rich in nitrogen, and 2/3 phosphate of lime. In a raw state, bones are nearly useless because they decompose so slowly that their effects are scarcely perceptible. They are much better if ground up fine, and will act even more quickly if burned before being ground; but burning would result in the loss of the nitrogen which they contain. [p. 353] A good way to use bones is to pulverize them as finely as possible, then mix them with wood ashes. You should take a hogshead large wooden barrel for storage or box, and first put in a layer of ashes 5 or 6 inches thick, then a layer of bones, and so on until the receptacle is full. Then the entire contents should be thoroughly wet with water. After a short time, the ashes and bones should be well mixed. It will be good to cover the top with fine earth, in order to prevent the loss of ammonia. In a few weeks' time, the bones will be acted upon by the ashes, and the result will be a very fine fertilizer.

3) Super-phosphate of Lime

Super-phosphate of lime is made by adding sulfuric acid to phosphate of lime, [p. 354] the object being to reduce the bones to a fine powder to make the phosphate more soluble and therefore more quickly available to plants. The process is as follows: in bones, phosphoric acid and lime combine in the chemical equivalents of 1 part (72 pounds) of phosphoric acid with 3 parts (84 pounds) of lime. If you add 2 parts (80 pounds) of sulfuric acid to the bones, after they have been reduced to a fine powder, it will unite with two parts (56 pounds) of lime, producing 136 pounds of sulfate of lime, leaving 72 pounds of phosphoric acid and 28 pounds of lime as main products. Super phosphates are extensively manufactured in the United States, the best kinds being Bradley's XL, Russell Bowes, and Wilson's.

4) Directions for Making Super-phosphates

The bones should first be finely pulverized, [p. 355] after which they should be moistened with water. Sulfuric acid should then be added, in the proportion of 4 gallons[9] of acid to 1 barrel[10] of bone. Instead of moistening the bones with water, it is sometimes preferable to dilute the acid with an equal quantity of water and add it directly to the dry bones. It is best to add the diluted acid at two different times, but water should not be mixed with it until just before it is added to the bones. After adding half of the diluted acid, the whole mixture should be thoroughly stirred, in order that the acid may act upon the entire volume. After 24 hours, the remaining half of the acid should be diluted with water and added to the pile mixture which should be stirred as before. In the course of 48 [p. 356] hours more, chemical action will have ceased and the super-phosphate may be spread out and dried, or some fine powder such as dried muck may be added to it in sufficient quantity as to make it convenient to handle. Sulfuric acid has a very strong affinity with water; in mixing the two, great care should be taken, or else you will be likely to get severely burned. They should be mixed in a very strong vessel, and the acid should be poured into the water—not the water into the acid. A strong porcelain receptacle is the best kind to use, but if it is impossible to obtain one, a very strong hooped wooden vessel will answer the purpose.

5-7. Organic Manures

Organic manure is any substance which contains plant food of atmospheric origin, and which forms part of an organic body. All of those substances contain some minerals, but the greater part is carbonaceous material. Among organic manures, peat (or muck, as it is sometimes called) is of great importance. Peat is of vegetable origin, and is formed by successive years' growth of plants, which fall down and partially decompose underwater without full access of the air. Such deposits are formed in any place in temperate climates where plants grow partially under water; they are also often formed along the banks of streams which [p. 368] overflow from any cause, and also in shallow ponds, where plants falling down underwater become partially decomposed and carbonized. Peat usually contains 10% to 12% more carbon than the plants from which it is formed, and is deficient of approximately the same quantity of oxygen. The percentage of hydrogen and nitrogen in peat varies little from that contained in the original plants. Peat also contains most of the mineral elements that are found in the plants from which it is formed. What is lost in mineral elements—due to their having become soluble, and thus either used up by succeeding plants or washed away by water—is usually more than compensated for by the mineral elements contained in water which overflows the place of deposition, and [p. 369] by fine particles of soil from the surrounding land that are washed in by heavy rains. By the partial decomposition which the plants undergo beneath the water, various organic compounds are formed; among these are ulmin, humin, ulmic and humic acids, and also crenic and apocrenic. The organic compounds formed contain varying proportions of carbon, hydrogen, and oxygen; the acids are capable of entering into combination with various compounds such as FeO, Al_2O_3, MgO, MnO, CaO, K_2O, Na_2O and NH_3. Some of these compounds are soluble, but others are insoluble. All the ulmates and humates of the alkalis are readily soluble in water, as are the crenates and apocrenates. The compounds formed by ulmic and humic acids with other element which I have mentioned are not, in general, [p. 370] soluble, although compounds of the lower oxides of iron are soluble. If the compounds of these acids with Al_2O_3, MnO, MgO, and CaO are subjected to the influence of a strong alkali, they decompose, the acid combining with the alkali. Water containing a considerable amount of ulmates or humates of the alkalis in solution has the power to dissolve some of the ulmates and humates which are not soluble in ordinary water. These acids, upon being subjected to the action of the atmosphere, are oxidized and broken up into different compounds. For this reason, it is common practice to throw peat out of the place where it is deposited and expose it to the action of air before using it, since some of the soluble compounds often found in peat are harmful to plants, [p. 371] for example, $FeSO_4$ is very injurious. Compounds of the organic acids with the lower oxides of iron, on account of their solubility, are also injurious. By exposure to the atmosphere, the lower oxides of iron are changed into higher oxides; their compounds are insoluble, and therefore not injurious. For the purpose of more effectively neutralizing injurious substances, CaO or ashes are frequently composted with peat; that practice is highly recommended. There is yet another reason why peat must be thrown out and exposed to the air before being used—that is because when it is taken from the place of deposition, it always contains a very large amount of water (usually about 70%). [p. 372] Peat shrinks in volume by about half and in weight by about 70% if it is dried completely.

1) Peat

(1) Mode of Action; Peat may prove beneficial to soils in two different ways. (a) because of its physical properties, and (b) because of the elements of plant food which it contains. (2) Physical properties and influence; Peat may very properly be called "vegetable charcoal", and like charcoal it has great absorbent properties. It will absorb more than its dry weight, and nearly its bulk, of any liquid. Peat also [p. 373] has the power of absorbing very large quantities of NH_3. The following table shows the number of pounds of water which 100 pounds of different kinds of soils will absorb from the atmosphere in 12 hours of an average night. And retained water in table shows the amount of water retained by 100 pounds of different soils.

Peat is therefore an invaluable material to be used around any of the receptacles for manure, since it will completely absorb and retain all the water, as well as the gases, generated in such receptacles. [p. 374]

Kind of Soil	Absorbed Water/12 hour	Retained water/100 lb
Sand	2 lb	25 lb
Strong Sandy Loam	21 lb	40 lb
Clayey Loam	24 lb	50 lb
Peat	50 lb	100 lb

2) Muck

(1) Action of Muck upon Different Soils and Muck as a Source of Supply of Plant Food—On account of its absorbent and retentive powers, muck has a very beneficial physical influence upon all light, dry and sandy soils. Sand lacks absorbent and retentive powers, while muck possesses them in the highest degree; therefore, a liberal application of muck proves highly beneficial to all sandy soils, since it increases their retentive power, both for moisture and the gases of the atmosphere. On account of its color, which is usually black, muck promotes the

absorption of more heat, and this is beneficial to most soils. Due to the moisture which it absorbs, muck tends to [p. 375] equalize the temperature of sandy soils. Due to the large amount of carbon it contains, muck—when mixed with a dry soil, and thus subjected to the influence of the atmosphere—gives rise to a large quantity of CO_2. CO_2 is formed by oxidation of the carbon in the muck. Now, we have previously seen that CO_2, when present in water, exerts a very decided influence in the disintegration and dissolution of rocks. Therefore, since muck not only increases the amount of CO_2, but also of moisture in the soil, it must prove very beneficial in changing the mineral elements of the soil into such a condition that they will be available as plant food. It is also believed that the other acids found in muck [p. 376]—ulmic, humic, crenic, and apocrenic acids, have considerable solvent power; therefore, muck will prove beneficial to light soil, because of the action of the acids upon its mineral elements. Muck should not be applied to heavy or wet soils, since it would make those even heavier or wetter, and that deleterious physical influence would more than counterbalance its beneficial effects as plant food. In order to prepare muck for use upon light, dry soils, it is a very good practice to compost it with ashes or CaO, which—on account of their alkaline action—serve to neutralize any injurious acids which might be present in the muck. CaO may be added in varying quantities; there is little danger of adding too much, [p. 377] since both are themselves good fertilizers. About 10 to 20 bushels of either substance per cord of muck would be a very good mixture. Muck may be applied to soil as a top dressing either in spring or autumn. If applied to grass lands, autumn will be a better time; if applied to cultivated lands, the season will make little difference.

(2) Muck as a Source of Supply of Plant Food; its constituents show that muck contains the elements of plant food in considerable quantity. It contains, on the average, much more nitrogen than ordinary barnyard manure, and nitrogen is one of the most costly and valuable [p. 378] of the elements of plant food. Muck also contains small quantities of iron, CaO, MgO, K_2O, and Na_2O, all of which are useful as plant food; when muck is taken from a place of deposition, these elements are not usually in an available form, but the action of the atmosphere will in time make them available. This process may be hastened, as I stated before, by the addition of ashes or CaO.

3) Organic Manures Other than Muck

The principal organic manures other than muck are leaf-mould, and sediments found along roadsides, in fence corners, and on the outlines of the farm. Leaf-mould is better for fertilizer than peat, because the elements contained in leaves are much more valuable than those in the materials [p. 379] which form peat. Leaf-mould, however, should not be gathered in a forest many years in succession, since doing so will materially damage the forest. The material which collects along roadsides is very valuable, consisting of the excrements dropped by animals passing over it, and of the finely pulverized dust of the road. It is wise to frequently collect organic rubbish of all sorts which collects on different parts of the farm, such as in fence-corners, the weeds growing on the road and around buildings and *etc*. Two reasons may be given for so doing: (1) because it makes the farm look much better, and (2) because such material adds to the value of the manure pile. [p. 380] Such material may also be thrown into any of the yards where animals run about, into hog-sties, or directly onto the compost heap. [p. 381, p. 382 is blank]

4) Animal Excrements as Manures

Animal excrements contain all the elements of plant food, and for general purposes are the best known fertilizers. An animal can excrete only what it has eaten, so animal excrements are simply food which has been ground up and made partially solvent and fine by mastication and digestion, with wastes of the system added.

(1) Process and Results of Digestion: The food is first ground to paste, with an addition of the animal's saliva. In the stomach, that food is more or less dissolved by being mixed with gastric juices. From the stomach, it passes in a semi-liquid state [p. 383] into the large intestine; there the solvent part is taken up by vessels, called ducts, and carried to the blood, while the dissolved portion passes through the intestinal canal and is then excreted. This is the dung, or solid excrement; some of the solid portion has been added during the passage through the intestines, from some of the wastes of the system. Dung is simply the undigested portion of the food mixed with these wastes. The food has been ground up more or less thoroughly, and is readily decomposed. That part of the food that has been dissolved, called chyle, is taken up by the blood; on passing through the lungs, a portion of its carbon combines with oxygen in the air and forms CO_2, which passes off in the breath. A part of the oxygen and hydrogen also combine and [p. 384] pass off in the breath as waste. The nitrogen and solvent mineral portion which is not needed to build up the system pass into the blood and, by the action of the kidneys, thence into the bladder, whence they are passed out of the system as urine.

(2) Comparative Value of Solid and Liquid Excrements: Since the liquid portion of animal waste contains the most nitrogen and all the solvent mineral portion, it is of course the most valuable. If 100 pounds of cow urine are evaporated, approximately 8 pounds of solid material would remain, which would be one of the most powerful substances ever used as a fertilizer. The annual amount of liquid excrement from one cow will [p. 385] keep one acre of land in fertility. One cord of loam, muck, or leaf-mould saturated with cow urine is worth as much as one cord of solid excrement. Such a fertilizer will have a remarkable effect upon the growth of plants because of the nitrogen it contains.

(3) Quality of Manure as Affected by the Food of Animals: The process of digestion and its results prove that the better the food given to the animal, the better will be its manure. ⟨See p. 383⟩ This applies to entire classes of animals, as well as to individuals of the same class. If two animals of the same age are fed, one on coarse wild hay, and the other on fine hay and grain, the manure of the latter will be worth much more than that [p. 386] of the former.

(4) Quality of Manure as Affected by the Age and Condition of Animals: The manure from young growing animals is not as good as that from mature animals, because most of the $Ca_3(PO_4)_2$ in the food is taken up by young animals to form their bones, and the nitrogen to form their muscles. Older animals, fed for fattening, make the best manure because they need no $Ca_3(PO_4)_2$ or nitrogen, except to supply the wastes of the system. The element which they ingest in greatest quantity from their food is carbon, which as you know is supplied to plants

mostly by the air, and is not therefore very important as manure. [p. 387]

(5) Quality of Manure as Affected by the Milking of the Animals: A good dairy milch cow will, in one year, make use of 100 pounds of $Ca_3(PO_4)_2$ in her milk; therefore, manure from a good cow when she is giving milk is of very poor quality. An important question, which demands on answer from the farmer, is which product it is best for him to sell from the farm—milk, cheese, or butter. Careful consideration of the constituents of these products and of their relative prices will be necessary in order to determine the answer to this question. If milk is sold, everything in it is lost to the farm—all its nitrogen and H_3PO_4. If cheese is sold, all the nitrogen and H_3PO_4 [p. 388] also go with it. However, if the farmer sells butter, he would sell only the carbon; therefore, as far as the elements of plant food are concerned, it would be better for the farmer to sell butter than either milk or cheese.

(6) Quality of Manure as Affected by Work: In working animals, the blood circulates more rapidly than when they are at rest. When at work, therefore, an animal appropriates more of its food; in addition to this, a large portion of the soluble elements in its body pass out in the form of sweat. For these reasons, the manure of working horses or cattle is not nearly as valuable as that of animals which do not work.

(7) Comparative Value of the Excrements of Different Animals: [p. 389] As already stated, human excrement is richer than that of any other animals. This is because humans eat richer food than the lower animals. Taking cow manure (solid excrement) as the standard and giving it the value of 1.0, horse manure will be 1.5 and sheep manure 1.0. The reason why the manure of the horse is better than that of the cow or sheep is because the horse does not have such good digestive powers as they. The urine of the horse is better than that of the cow because it is not as diluted. A given quantity of food would, however, produce more fertilizing material in the urine of the cow than in that of a horse. The solid excrement [p. 390] of a cow contains are shown in the following table. And, in 100 pounds of that organic material, there is present 2 pounds and 2 ounces of $(NH_4)_2CO_3$. The solid excrement of a horse contains are also shown in the Table. And in 100 pounds of that organic material, there are 3 pounds and 3 ounces of $(NH_4)_2CO_3$. Horse manure, since it heats up

	Cow	horse
organic matter	15.45%	27.00%
mineral salts	0.95%	0.96%
water	83.60%	71.00%
Total	100.00%	100.00%

more easily than cow manure, is much better adapted to a cold soil. Sheep manure is usually better than cow or horse manure, because sheep are commonly yarded on it and because they are usually well bedded, and thus all the urine is saved. [p. 391] Sheep produce poor dung, but rich urine. The manure of swine, owing principally to the quality of their food, is richer than that of most other domestic animals. As a general rule, the manure of carnivorous animals is richer than that of herbivorous ones. The manure of swine is particularly well adapted to the production of corn, and is also good for grass lands. On account of its potency, it should always be composted. Muck and loam are excellent for such composts. The manure of domesticated fowl poultry is next to human excrement in value. This is so partly because their food is rich, and partly because solid and liquid excrements [p. 392] are voided together. The same conditions influence the quality of the excrements of other animals.

5) Modes of Utilizing Human Excrement

In most Western countries, there is a very great waste of human excrement, and the question of utilizing it has become one of great importance. In the United States, the products of more than 10,000,000 acres of land are annually used to support the inhabitants of the cities, and little of the plant food in these products is ever returned to the soil. Up to the present time, one of the principal methods of utilizing the sewerage of cities has been to conduct it to some convenient receptacle and there evaporate the water, but that method is usually too expensive. [p. 393] Another method of utilizing it has been to convert it into "poudrette", which is made by collecting "nightsoil" (human excrement) and deodorizing it with some substance such as finely ground charcoal. The principal objection to this method is that it is both disagreeable and unhealthy to the people engaged in the work. Poudrette, if properly made, is an excellent manure, since the charcoal absorbs and retains all the valuable liquid and gaseous elements. Another method for utilizing human excrement is an earth closet, which is so arranged that after each deposit of human excrement, a quantity of dry earth may be sprinkled upon it. This earth, if added in sufficient quantity, makes the closet entirely inodorous, and it can be kept in [p. 394] any part of the house. This mixture of human excrement and earth may be handled as easily as ashes, and it may be removed from the cities by the public authorities in the same manner as ashes are removed. Moules Patent Earth Closet is one of the best for this purpose. If we consider cow manure to have a value of 1.0, then nightsoil will be 3.0. Since nightsoil is so potent, it should either be composted or diluted with a large amount of water before use. Using an excessive quantity of nightsoil as fertilizer would cause excessive vegetative growth. For this reason, it must be used with caution, with plants valuable for their seeds or fruit. With plants which are valuable for their foliage—such as grass, cabbage, lettuce, *etc.,* and nightsoil may be used more freely. [p. 395]

6) Guano as a Fertilizer

There are several different kinds of guano, some of which are the excrement of wildfowl, and others which are the excrement of bats. Peruvian guano, which is one of the very best kinds ever discovered, is the excrement of wildfowl which has been accumulating for thousands of years. It is often found in beds 100 feet in thickness. Such deposits can only be formed in counties where the climate is hot and dry. Peruvian guano is found mostly on islands located off the coast of Peru; those islands are under the control of the Peruvian government. Peruvian guano principally consists of nitrogenous material and phosphates. In its pure state, it contains about 10% water, 59% nitrogenous and combustible material, [p. 396] 25% phosphates, and 2 to 3% each of K_2O, Na_2O, and SiO_2. In 100 pounds of guano are found 32 pounds of salts of NH_4. Guano is especially suitable for use on grain crops and tobacco. It is, moreover, an excellent manure for almost any crop. Since it contains a very large percentage of nitrogen, the same caution must be exercised in its use as with other nitrogenous manures. It is so potent that it must not be put in direct contact with seeds, since it will prevent germination. Due to the large amount of nitrogen it contains, guano is an excellent manure with which to force plants to make rapid and early growth.

7) Slaughterhouse Manure.

This consists of the excrements of swine, mixed with the blood and offal [p. 397] of the animals killed in the establishment. Since the blood and offal are rich in elements of plant food, as is the excrement of swine, this is very potent manure, so much so that it should always be composted. On account of its high nitrogen content, caution must be exercised in its use on certain plants. This is a potent forcing manure.

8) Dead Animals (Carcasses)

All animal carcasses should be regarded as fertilizer, because they are rich in all the elements of plant food. There is enough nitrogen in a dead horse to supply half an acre of land with a sufficient quantity. All carcasses should be cut up into small pieces, then composted with [p. 398] some material, such as horse manure, which will generate a large amount of heat. This will cause the flesh to thoroughly decay and the bones to become quite soft. Some other material, such as peat or loam, should be added to absorb the gases generated in decomposition.

9) Sink-drainage

This always contains some elements of plant food. From the soap used in the household, it collects a considerable amount of K_2O or Na_2O, and it will also contain nitrogen from scraps of food washed from dishes. This material should be saved for two reasons: (1) it contains plant food, and (2) if allowed to drain out and soak into the ground, as is often the case, it will generate foul gases which are deleterious to health. [p. 399]

10) Fish Guano

This is the refuse from the manufacture of fish oil. The fishes are boiled or steamed, then pressed to extract the oil, which is used for various purposes. This "pomace" contains a high percentage of nitrogen and some H_3PO_4, but little or no K_2O; therefore, it is not a complete fertilizer. It is too potent to be used in direct contact with seeds. If it is to be used in that manner, it should first be composted with some material, such as peat or loam. In order to obtain the greatest possible benefit in the least amount of time from the use of fish guano, [p. 400] it should first be ground up fine, then composted before it is applied to the soil. Composting, however, involves considerable labor, so it may sometimes be better to apply raw guano. The refuse from all sorts of fishes makes excellent fertilizer, and farmers will find it profitable to obtain and use such refuse whenever they can do so at reasonable expense.

11) Possible Waste of Manure

The best components of manure are those which are soluble and volatile; manure is therefore subject to waste, both by leaching and by exposure to the atmosphere. Loss of the volatile portion may be prevented, either by stopping [p. 401] decomposition or by the addition of some absorbent material. Manure in heaps always heats up when it is piled loosely. If that process is permitted to continue, the manure will become "fire fanged", as we say. In that condition, it contains only mineral elements, because all the nitrogen and carbon has been lost in the course of decomposition. To prevent this, loam or peat should be mixed with the manure, or decomposition may be prevented by treading down the pile of manure until it is so compact that air cannot readily gain access to its interior. If swine are allowed to run over the manure, they will make it so compact that it will not become "fire fanged". Manure can be lost through the leaching of all its soluble components. For this reason, water should not be allowed to run through it. [p. 402]

12) Composting Manures

In deciding whether or not to compost manure, the farmer must evaluate the quality of the manure and the use to which it is to be applied. Unless you desire some particular effect from the manure, and unless the application of raw manure would be deleterious to the physical condition of the soil, it is wise to apply raw manure and let it decompose in the soil. This course saves expense in transport, since you are not obliged to carry peat or loam (or whatever is used in composting) in addition to manure. If the manure is to be used for plants which need an early start, it should be thoroughly composted and decomposed. If the physical condition of [p. 403] the soil needs modification, that should be considered. If the soil is too hot or too cold, too dry or too moist, composted and thoroughly decomposed (or undecomposed) manure should be used accordingly. Coarse, undecomposed manure—on account of the heat which it will generate in decomposition—is best for cold, wet soils, while well-rotted manure is best for dry soils. It should be the farmer's aim, in the application of manure, not only to supply the elements of plant food, but to apply it in such a condition and in such a manner as to aid in the development of plant food from the elements already in the soil.

13) Changing Manure

Fertilizers should sometimes be rotated, that is, it is not [p. 404] wise to always use the same kind of manure upon the soil. Changes should be made from animal manure to mineral fertilizer, and from one kind of each to another. After an application of guano to the same piece of land many years in succession, it will not prove as beneficial as at first, so a change should be made, either to some kind of mineral fertilizer or to animal manure.

5-8. Application and Physical Properties of Manure

Foliaceous plants and seed-bearing plants require very different kinds of fertilizer; nitrogenous manures are best for foliaceous plants, while manure containing much H_3PO_4 and other mineral substances is best for cereals cereal grains. Some plants which need a great deal of warmth are often benefited by being fertilized with a warm manure, or one which in the course of decomposition will generate [p. 405] considerable warmth. One of the best fertilizers to apply to such plants is horse manure.

1) Method of Applying Manure

Knowledge of the nature of manure—its action and the way in which it is distributed throughout the soil—is necessary in order to know how to apply it in the most advantageous manner. There are a great many opinions in regard to this subject. Some farmers think manure should be applied at one depth, and other farmers, at another. Some say it should be composted, and others that composting does not pay. None of those men who advocate always following the same course are correct in their opinions, because the method of application [p. 406] which will produce the best results must vary in different cases. If manure is applied merely as plant food, it should always be in the form of well-rotted compost. If the objective is to change the mechanical or chemical condition of the soil,

164   Chapter 5   Manures (and Fertilizer)

manure should not be composted. Regarding the depth at which manure should be applied, if any general rule can be given, that is to mix it thoroughly with the surface soil. Upon light sandy soil, however, it should be ploughed in 7 inches deep. The lighter the soil, the deeper the manure should be inserted, and the heavier the soil, the nearer to the surface. Manure should never be spread on the surface of ploughed land and left there. In the case of permanent grass lands, manure cannot be mixed with the soil; therefore, we must leave it on the surface. The best season [p. 407] for the application of manure to grass lands is autumn; in order to prevent loss, manures applied to such lands should always be well-rotted composts.

2) Comparative Value of Nitrogenous and Mineral Manures

There has been a great deal of controversy about this subject, some experts advocating one side, and some the other. The fact is, however, that insofar as the plant is concerned, one is just as important as the other. No plant can grow without nitrogen, and neither can any plant germinate or thrive without mineral fertilizer. Thus, we sometimes apply nitrogenous and sometimes mineral manure, according to the condition and [p. 408] the nature of soils. [p. 409]

5-9. The Atmosphere as Related to Vegetation

A multitude of observations has demonstrated the fact that 95% to 99% of the entire mass of agricultural plants is derived, directly or indirectly, from the atmosphere. Since such a large proportion of plant matter comes from the air, a study of its relationship to plant growth must prove highly useful. The composition of the atmosphere is perhaps quite familiar to all of you, but in commencing to consider this subject, we will first devote a short time to the study of the composition of the atmosphere. You will see in the accompanying table the proportions of the two major gaseous elements in the atmosphere. [p. 410] Besides oxygen and nitrogen, the atmosphere

	by weight	by volume
Oxygen	23.17	20.95
Nitrogen	76.83	79.05
total	100.00	100.00

Water Vapor Average proportion by weight	1/100
CO_2 gas Average proportion by weight	6/10,000[11]
NH_3 Average proportion by weight	1/50,000,000

contains several other chemical compounds in small quantities; those are shown in the above table; Minute traces of ozone, HNO_3, N_2O_3, and CH_4 are also found. In the air of towns and cities, carbon monoxide, SO_2, and H_2S are sometimes found.

1) Relationship of Oxygen to Vegetable Nutrition

It has been determined by many careful experiments that oxygen is essential to all processes of vegetable growth. Seeds cannot germinate unless the atmosphere has access to them. The role played by oxygen in the germination of seeds is [p. 411] not thoroughly understood, but it is probable that it is necessary in order to convert certain ingredients of the seed into an available form for the nourishment of the young growing plant. Oxygen is essential to the growth of plants; this has been proven by many experiments. One Frenchman by the name of De Saussure experimented in the following manner:

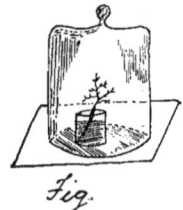

Early in the spring, he collected some twigs from a tree, upon which the buds were just about to expand; he placed these twigs into small vessels containing water, and then placed the vessels under bell-jars containing different gases. Into one bell-jar, he introduced ordinary air, into another hydrogen, [p. 412] and into another nitrogen. The buds on the twigs in ordinary air expanded their leaves and grew quite vigorously, but those on the twigs in hydrogen and nitrogen did not grow at all. That proved conclusively that the presence of oxygen is necessary for plant growth. Oxygen is also essential to the roots of plants. That was proved by the same experimenter in a somewhat similar manner. He took young chestnut seedlings, removing them from the earth carefully, in order not to injure the roots. He then carefully washed the roots to remove the adhering earth. He next carefully inserted the roots of the seedlings into glass jars containing water, fastening them in such a way that the ends of the roots were immersed in the water. [p. 413] He cemented the stems in the mouths of the jars in such a manner as to make them air-tight; into one jar was introduced carbon dioxide, into another nitrogen, into another hydrogen, and into another ordinary air. The result was that, after three weeks, those seedlings whose roots were placed in ordinary air were still perfectly healthy. The seedlings whose roots were in nitrogen or hydrogen died in 13 to 14 days. The seedlings whose roots were in carbon dioxide died in 7 to 8 days. Oxygen is also found to be essential to the blooming of flowers. Just around the time of blooming, flowers use a considerable quantity of oxygen. That oxygen is necessary to flowering plants is shown by the natural fact that all aquatic plants [p. 414] send their flowering stalks to the surface of the water; the flowers cannot open underwater. The oxygen used by various parts of plants is not retained within them to any considerable extent, since just as much oxygen is exhaled as is taken in. It is probable that oxygen is useful to plants as an agent of assimilation —that is, it enables plants to assimilate or make use of the various elements which they take in. Within the seed, oxygen is useful in changing insoluble constituents into soluble form for the use of young growing plants. In the various parts of growing plants, oxygen is useful in transforming crude material which the plants absorb into the various compounds which form a part of plant. [p. 415]
2) Hydrogen's Relationship to Plant Growth

The hydrogen in plants is taken in the form of water, and most of that water is drawn up by the roots of plants. With that portion of hydrogen taken up in the form of water, we will not be concerned at present. We shall simply consider the water vapor contained in the atmosphere, with reference to its possible use by plants. The amount of water vapor in the atmosphere varies much in different places, and even in the same place at different times. A question about which there has been much discussion is whether or not plants absorb water vapor through their leaves or other drawing parts. [p. 416] Some observers have noticed that on hot days, when the leaves of plants have become much wilted, a sudden rain shower will at once cause the leaves to become fresh. This change takes place before it would have been possible for rainwater to have reached the roots. From this fact, they have argued that the leaves of the plants must have absorbed rainwater. It is not necessary, however, to come to this conclusion in order to explain the sudden revival of the leaves of plants in such cases. The leaves wilt because evaporation takes away water faster than the roots can replenish it. That

166   Chapter 5   Manures (and Fertilizer)

evaporation, then, takes away a portion of the water which is a normal constituent of the leaves; [p. 417] having lost a portion of their water, the leaves thus lose their turgidity and freshness. In order to revive, it is only necessary to check evaporation, and soon the roots will supply enough water to restore the leaves' natural freshness. Evaporation would be checked just before and during the rain shower due to the sudden rise in humidity, because at that time the atmosphere is nearly saturated with moisture. Hence, it is probable that leaves revive not because they absorb water, but simply because evaporation is checked. It is now generally thought that most plants have the ability to absorb water from the soil through their roots. Some plants, however, such as lichens and mosses, probably absorb water through their foliaceous portions. Air plants "Epiphytes"; a group of plants that grow attaching to other plants [p. 418] have the power of absorbing water vapor through their roots. Plants, under certain conditions, exhale water from portions other than the foliage (leaves). If the air is very hot and dry, some water will doubtless be exhaled from the young growing portions of the plant. In general, however, leaves are the only parts of plants which exhale any considerable amount of moisture.

3) Relationship of Nitrogen to Vegetable Nutrition

It is now generally thought that most plants do not have the power to make use of free nitrogen in the atmosphere. If plants could do so, it would never be necessary for the farmer to use nitrogenous manures, since the atmosphere [p. 419] contains an inexhaustible amount of that element. Many experimenters have investigated this subject, with some coming to the conclusion that plants have the ability to make use of free nitrogen in the atmosphere, and others reaching the opposite conclusion. Among the most famous of the former was a French scientist by the name of Ville, while among the latter, Boussingoult, Lawes, Gilbert, and Pugh stand preeminent[12]. Careful experiments have proved that nitrogen is not emitted by living plants.

4) Relationship of CO_2 to Vegetable Nutrition

Carbon is found in all plants in large percentages; it comprises 44% of the dry substances [p. 420] in plants. Many experiments have determined that most, if not all, of that carbon is taken from the air by the leaves or growing parts of plants. Carbon dioxide is found in animal manure in immense quantities. It is one of the most important ingredients of the various minerals found on the earth. In limestone, marble, and chalk, it is found in very large quantities (all of those minerals consisting largely of $CaCO_3$). The atmosphere also contains much carbon dioxide in the form of gas—it has been estimated that the amount of CO_2 in the atmosphere is approximately 3,400 billion tons. This is sufficient to supply 28 tons to each acre of the earth's surface. You can readily prove that the air contains carbon dioxide by exposing CaO water to its influence. In a very short time, you will notice a white precipitate forming in the clear CaO water. [p. 421] This precipitate is $CaCO_3$, formed by carbon dioxide in the atmosphere with CaO in the water. Ordinary water absorbs approximately its own volume of carbon dioxide under normal conditions. When under pressure, water will absorb more of it, and at the freezing point, twice its own volume. The leaves of plants in sunlight

Fig.

absorb much carbon dioxide from the air. This has been proved by many experiments; one conducted by Boussingoult was as follows:

He took a glass jar having three orifices at the top, through one of which he introduced a branch of a plant full of green leaves, and cemented it air-tight. [p. 422] Through another orifice he forced air containing a known quantity of carbon dioxide. This air, after passing over the leaves of the plant, passed out of the vessel through the third orifice, whence it was made to pass through an apparatus for collecting and measuring the amount of CO_2 present. In one experiment, the air made to pass through the vessel contained 4/10,000[11] of carbon dioxide. It was also found that when the leaves were exposed to strong sunlight, they removed nearly all of CO_2. In the shade, little or no carbon dioxide was absorbed by the plant. Boussingoult also observed the effect of an increased quantity of carbon dioxide upon the growth of plants. He found that young pea plants lived for some time in an atmosphere containing 50% carbon dioxide. He also found that in an atmosphere containing 1/12th [p. 423] carbon dioxide, such plants grew better than in an atmosphere containing a normal amount. However, that was only true when the plants were placed in strong sunlight. In the shade, the presence of so much carbon dioxide proves detrimental. It has also been proved by experiment that plants cannot live long in direct sunlight in an atmosphere which contains no carbon dioxide. The amount of oxygen exhaled by plants is very nearly equal to the amount of carbon dioxide absorbed by them. The CO_2 absorbed by plants in sunlight is rapidly decomposed, with the carbon entering the organic constituents of the plant and the oxygen being exhaled. Since the volume of oxygen necessary to form a certain amount of the resultant gas, the fact that plants [p. 424] decompose the carbon dioxide which they take in is shown by the fact that the amount of oxygen exhaled is equal to the volume inhaled. Carbon dioxide is also exhaled by plants, and this process occurs all the time. There are, therefore, two opposite processes going on in plants at the same time when they are exposed to direct sunlight. The carbon dioxide exhaled must arise from the oxidation of a portion of the carbon in plants by oxygen taken in by them. This process is analogous to respiration in animals. The amount of CO_2 absorbed by plants is very much greater, however, than the amount exhaled. The composition of air contained in plants varies much at different times; the most marked difference is between its composition in darkness and in sunlight. In one experiment, [p. 425] the following results were obtained in next table.

As previously stated, the amount of carbon dioxide in the air is very great. If the entire surface of the

Conditions during Exhalation	Total quantity of gas collected cm^3	Nitrogen %	Oxygen %	Carbonic %
Dark	24.0	77.08	3.75	19.17
Sunlight	34.5	68.67	24.93	6.38

earth were covered with a rapidly growing forest of beech-trees, it is estimated that the amount of CO_2 in the air would be sufficient to last 25 years. Beech forest, however, requires much more carbon dioxide than most other kinds of vegetation; and besides this, only one-fourth of the earth's surface is land. Therefore, the amount of carbon dioxide present in the air, even if it were not renewed, would probably be sufficient to last 100 years. It is, however, being constantly renewed by various kinds of decomposition, combustion, and [p. 426] the respiration of animals. The amount of carbon dioxide contained in the air, therefore, varies little over time. Numerous experiments have proved that it is not necessary, for the best growth of plants, that carbon or CO_2 be present in the soil. Plants are able to obtain their whole supply of CO_2 from the atmosphere. Carbon dioxide is assimilated in the plant either in the chlorophyll grains or in close connection with them. It is a well-known fact, demonstrated by microscopic examination, that starch is formed only in the presence of chlorophyll grains, and carbon is the leading constituent of starch. That the presence of chlorophyll grains is necessary to the formation of starch has also been proved in another way. Chlorophyll grains can be formed only when the plant is supplied with iron; [p. 427] if iron is not supplied, no starch is formed. Many experiments have proved that carbon dioxide is assimilated by plants in darkness. Seeds have been made to germinate in darkness and allowed to grow for a considerable time. Then the amount of carbon contained in the young plant was measured; in every case, the amount of carbon has been found to be no greater than was contained in the original seeds.

5) Ammonia in the Atmosphere and Its Relationship to Vegetable Nutrition

NH_3 is usually present in the atmosphere in small quantities, and since it is such an important plant food, it will be interesting to see whether [p. 428] plants have the power to make use of this atmospheric NH_3. At its freezing point, water absorbs 150 times its volume of NH_3. While at normal summer temperatures only half as much is retained. Boiling the water expels all the NH_3. It was formerly thought that if nitrogen and hydrogen in a nascent state were brought together in the proper proportions to form NH_3, that they would unite, but this is now known to be untrue. In many cases of combustion some NH_3 is formed, but it is most abundantly and readily formed by the decay or dry distillation of organic nitrogenous bodies—that is, of albuminoids. "Ammonia" was so called because it was first made near the temple of Supiter Ammon from dried camel dung. When CO_2 and NH_3 are bought in contact under proper conditions, [p. 429] they at once unite to form $(NH_4)_2CO_3$. Now, since the amount of carbon dioxide in the atmosphere is much greater than the amount of NH_3, all the NH_4 in the atmosphere must exist in the form of $(NH_4)_2CO_3$. Both NH_3 and $(NH_4)_2CO_3$ are readily soluble in water; for that reason, the amount of those substances varies greatly at different times. After a protracted period of dry weather, they may be present in quite large quantities; but a rain will wash all the NH_3 and NH_4 compounds from the atmosphere. The salts of NH_3 are appropriated by plants by means of their roots, being taken in solution in water. It is probable, however, that some plants have the ability to appropriate NH_3 through their leaves. To how great [p. 430] a degree they have this ability is not yet well known. A German experimenter by the name of Sachs has proved that plants growing in an atmosphere containing NH_3 compounds appropriated more nitrogen than those growing under precisely similar conditions, with one exception: that the atmosphere surrounding them contained no NH or NH compounds. This experiment was done with bean plants. It is thought that clover, and leguminous plants in general, have a greater ability to appropriate atmospheric NH_3 than most other plants[12]. The appropriation of atmospheric NH_3 does not depend upon sunlight, but takes place equally well in darkness. The effect of NH_3 upon plants is to produce excessive vegetative growth if used [p. 431] in too large a quantity. Plants plentifully supplied with NH always have a very rich dark green color. It is possible, by supplying a large amount of NH₃, to increase the percentage of NH₃ natural to plants; thus, for example, wheat grown in ordinary air contains 2.09% nitrogen, but grown under the influence of NH₃ it contains 3.40%. A question which has long engaged the attention of scientific men is whether or not plants exhale NH₃.

One man tried an experiment with an aquatic plant (Typhalatifolia) in the following manner (Fig.): he took a long glass tube JK, and enclosed [p. 432] the foliage of a plant E within it, carrying the lower end of the tube a considerable distance beneath the surface of the water. The upper end of the tube was closed airtight with a rubber cap D, through which passed a small tube BC connected to an apparatus AB containing bits of glass M moistened with HCl. At the lower end of the large tube JK, he introduced a small U-shaped tube FOG, one end of which F was just above the surface of the water within the tube JK; and the other G outside it. At the mouth of the U-shaped tube was another apparatus for removing all traces of NH₃ from air which was made to pass through it. To prevent NH₃ from rising into the large tube from decaying substances present in the water, [p. 433] the surface of the water PT inside the tube was covered with a layer of oil. The experiment was then continued for about 48 hours, with a current of air being made to enter the large tube through the U-shaped tube and, after passing over the foliage of the plant, to pass out through the small tube BC and the apparatus AB for collecting NH₃ at the upper end. At the end of that time, the apparatus was carefully examined for NH₃, but no trace of it could be found. Since the means of detecting NH₃ are so sensitive as to enable us to notice very slight traces of it, it was therefore clearly proven that this plant, at least, did not exhale NH₃. It is probable, therefore, that no healthy plants exhale NH₃ from their foliage, but the matter [p. 434] is not yet fully resolved.

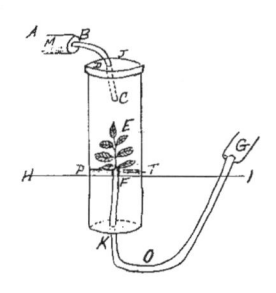

The proportion of NH₃ brought down by rainwater varies much at different times and in different places. In the various observations which have been made, the amount has been found to vary from 1.33 parts to 10,000,000 parts of rainwater. This, though quite small, is sufficient to have a considerable effect upon the growth of plants, since the amount of rainfall during a year is large. In an experiment conducted in England in the year 1855, it was calculated that 7 pounds of NH₃ were brought down upon an acre of land; in the year 1856, 9.5 pounds were brought down. In other places, the amount has been found to vary from 6.33 to 12 pounds per acre. Various other experiments have been conducted for the purpose of determining the effect of NH₃ of the atmospheric [p. 435] waters upon plant growth.

6) Ozone

This is a substance sometimes found in small quantities in the air. Electricity, whether artificially generated or the natural kind, seems to have the power of producing ozone, which may be considered to be an allotropic form of oxygen. Ozone is oxygen in a very active state; it has much greater oxidizing power than oxygen. Besides being produced by electricity, ozone is also formed in various chemical reactions. Ozone is probably also produced in small quantities by plants;

some experimenters have been convinced that they proved that ozone is sometimes exhaled by plants. The quantity of ozone contained in the air [p. 436] at any time is small, it being more abundant in winter. This is partly because there is more electrical activity in the atmosphere at that season, and also partly because, since snow usually covers the ground in temperate zones during winter, there are fewer substances exposed to the atmosphere upon which it can exert its oxidizing power. The amount of ozone present in the atmosphere has been calculated to be one part in 13,000,000 to 65,000,000 parts of air. It is certain that much more ozone than this is produced, but—even in winter—a large portion of it quickly expends itself in oxidizing substances with which it comes in contact. Ozone is probably important to the farmer simply because of its great oxidizing power. It is a powerful agent in the conversion of substances in the soil [p. 437] into a condition in which they are available for the use of plants.

7) Compounds of Nitrogen and Oxygen in the Atmosphere

Compounds of nitrogen and oxygen found in the atmosphere are NHO_3, anhydrous NHO_3, N_2O_3, N_2O_5, NO_2, and NHO_2[13]. Any of these compounds, upon being subjected to the proper conditions, is readily convertible into any of the others, and may also be converted into NH_3. Nitrogenous compounds are formed in the atmosphere in various ways: (1) From free nitrogen by electrical ozone—in the presence of water vapor, nitric peroxide may be converted into HNO_3 and HNO_2. (2) In the process of combustion and slow oxidation—it is probable that this is brought [p. 438] about by the influence of ozone, which is known to be generated in such cases, but this has not yet been conclusively proven. In the case of combustion or oxidation, NH_3 is sometimes formed, and sometimes HNO_3. Nitrogenous compounds are also formed by the evaporation of water. (3) Nitrogenous compounds are also formed from free nitrogen by the ozone accompanying the oxygen exhaled from the green foliage of plants in the sunlight. This latter theory is not yet fully proven, yet many known facts seem to prove that such is the case. (4) NHO_3 may be formed from NH_3 in the atmosphere by the action of oxygen. In the varying conditions of the atmosphere, the processes of oxidation and reduction of nitrogenous compounds must be constantly going on.

There is no rest in the great laboratory of the air. The natural occurrence of HNO_3 or NH_4NO_3 in the atmosphere [p. 439] has been abundantly proven. If a substance which has a very strong affinity for HNO_3 is exposed to the atmosphere, it will form a nitrate. The amount of NHO_3 in rainwater is quite variable, being especially abundant in the rain accompanying thunderstorms. The amount varies from 0 to 25 parts in 10,000,000 parts of water, the average being perhaps about 6 parts in 10,000,000. The amount brought down in one year upon one acre of land was calculated at Rothamstead in England, and found to be about 2.80 pounds. The amount of nitrogen brought down during a year upon an acre of land at that place varied, in the years in which the experiments were conducted, from 6.63 to 8.31 pounds, a portion of that being in the form [p. 440] of NH_3 and HNO_3. The amount of HNO_3 present in the atmosphere as is shown by the figures I have given you, is usually less than the amount of NH_3; hence, the HNO_3 must commonly exist in the air in combination with NH_3 as NH_4NO_3. There are, however, exceptions to this general rule. One of the most remarkable of these ever noticed was at Nimes, France where, upon a certain occasion, the rain of a hailstorm contained so much HNO_3 as to be sour to the taste. NH_4NO_3 is doubtless held in mechanical suspension; in that condition, it cannot be appropriated by the foliage of plants, but since it is readily soluble in water, all of it that the air contains is frequently washed out and carried to the ground by falling rain. It has been demonstrated [p. 441] experimentally that plants have the ability to take from the HNO_3 all the nitrogen necessary for their healthy growth. The various nitrates of the alkalies are extremely valuable fertilizers. HNO_3 in the atmosphere, therefore, when carried to the soil by rains, is doubtless used by plants; though its amount is small, it is not altogether unimportant.

8) Marsh Gas

This is a compound of carbon and hydrogen which, when mixed with air, is highly flammable. It is formed by the decay of substances containing much carbon without the full access of air. Though CH_4 might serve as a source of supply of both carbon and hydrogen, it is not probable that ordinary plants make use of it. Certain aquatic plants, however, have been [p. 442] found to live and grow quite well in an atmosphere consisting largely of this gas. Experiments also have been conducted for the purpose of determining whether CH_4 is exhaled by plants or not. Some experimenters have been convinced that they have proved that such is the case, but their experiments were conducted in such a manner as to leave room for error to creep in.

9) Carbon Monoxide

This is a gas formed by the combination of one equivalent of carbon and one of oxygen. It is a gas which is extremely poisonous to man and all animals. Certain plants have been made to live a long time in an atmosphere containing a large quantity of carbon monoxide, and during that time they assimilated some carbon and emitted oxygen. It is not thought, [p. 443] however, that as far as plant growth is concerned, carbon monoxide is of any importance.

10) Nitrous Acid

This gas consists of two equivalents of nitrogen and one of oxygen. Its effect upon animals is first to exhilarate them, and afterward to produce unconsciousness. It has never been proven that N_2O exists naturally in the atmosphere[13]. The means for detecting it are not so exact, so small quantities of it might be present but escape detection. Until it is proven that this gas is naturally present in the air, it is not important to consider its possible relationship to plant growth.

11) Hydrochloric Acid Gas

This gas is sometimes found in the air. [p. 444] It is especially abundant in the air around salt marshes, doubtless arising from the decomposition of a portion of the $MgCl_2$ deposited in the marsh by sea-water. This gas is extremely destructive to vegetable life, but it is not often present in sufficient quantities to do much harm. [p. 445, p. 446 is blank]

12) Recapitulation of the Atmospheric Supply of Food to Plants

(1) Oxygen; whether in a quiet, free state or in combination with carbon as CO_2, is abundantly supplied by the atmosphere.

(2) Carbonic Acid Gas (Carbon Dinoxide); This is also abundantly supplied by the atmosphere.

(3) Hydrogen; This is adequately supplied to crops by water from the soil, which enters plants chiefly through the roots.

172   Chapter 5   Manures (and Fertilizer)

(4) Nitrogen; Nitrogen exists in immense quantities in the atmosphere in a free state, which is doubtless the primary source of the nitrogen of the organic world, but in that form it is not directly available. Its assimilable compounds, NH_3 and HNO_3, are not present in sufficient quantity [p. 447] to supply vegetation with the needed amount. According to Dr. Anderson, the amount of nitrogen in a wheat crop with 28 bushels of grain and the normal amount of straw is 45.33 pounds. In 2.5 tons of clover hay, the amount is 108 pounds. The other ingredients of the atmosphere, with the exception of ozone, are—so far as we know—of little importance to vegetation; ozone is important simply because of its action in changing the substances in the soil into available forms.

13) Assimilation of Atmospheric Food

A question of great interest to scientific men is how it is that the food of the plant is assimilated. It has been suggested [p. 448] that this process may take place in the chlorophyll grains, according to the following sequence of chemical reactions:

(1) $CO_2 + H_2O = CO + H_2 + O_2$
(2) $CH_2O \times 12 = C_{12}H_{24}O_{12}$; Glucose
(3) $C_{12}H_{24}O_{12} - H_2O = C_{12}H_{22}O_{11}$; Saccharose
(4) $C_{12}H_{24}O_{12} - 2H_2O = C_{12}H_{20}O_{10}$; Cellulose

This process is, in other words: (1) the CO_2 taken in by the plant under the influence of sunlight loses its oxygen, leaving CH_2O, which is supposed to be retained in plants while O_2 is exhaled; (2) 12 times CH_2O gives glucose; (3) less one equivalent of water from the (2) step leaves saccharose; and (4) less two parts of water, we obtain cellulose.

1. Absorbed by Plants	O:	by roots, flowers, ripening fruits an all growing parts.
	CO_2:	by foliage and green parts, but only in the light.
	NH_4CO_3:	by foliage probably at all times.
	H_2O:	as liquid through the roots.
	N_2O	united to NH and devolved in H_2O through the roots.
	NHO_3	
	Ozone	uncertain
	CH_4	
2. Not Absorbed by Plants	N	in the state of vapor.
	H_2O	
3. Exhaled by Plants	O	by foliage and green parts, but only in the light.
	Ozone	
	CH_4:	in traces by aquatic plants.
	H_2O:	as vapor through the surface tissue of plants at all times.
	CO_2:	from the growing parts at all times. [p. 450]

14) The Physical Relationship of the Atmosphere to Plant Growth

That the atmosphere is a mixture of gases—and not of chemical compounds—I have already stated. The truth of this statement is evident from the fact that the composition of the atmosphere varies, though this variation is within a very limited range. That range is so small because of the remarkable balance maintained by nature between growth and decay, life and death. All gases have the remarkable property of diffusion. If two gases are placed in any confined space, they will soon become evenly distributed throughout that space; when gases pass through a membrane in which the pores are not discernible by optical means in order to become evenly distributed, the process is called *osmosis*. [p. 451] Gaseous osmosis, therefore, is simply diffusion modified by the influence of the membrane. Gases are absorbed by plants through their membranes by osmosis, and those portions of the atmosphere which are in contact with the plant are soon depleted of their content of these gases, which are available as plant food. Were it not for the property of diffusion possessed by all gases, the amount of atmospheric plant food obtainable by plants [would] be very small, since the percentage of such substances as CO_2 and NH_4 in the air is quite small. Since all gases, however, are governed by the laws of diffusion, the NH_4 and CO_2 in the atmosphere must be constantly flowing toward the plant in order to supply the air surrounding it with the proper amount. [p. 452]

解　　題

1. 肥料学の展開から見たブルックス講義録の時代背景

　植物が何を食べて生育しているのかという疑問は，古くアリストテレス(384-322 B.C.)の時代から強い関心を持たれていた課題であり，歴史的に見るとヨーロッパを中心として水説，土説，火説，硝石説，腐植説などの諸説が提案されてきた。19世紀に入って，植物は炭酸ガスを葉から吸収して光合成を行なうことが発見され(ド・ソシュール(1767-1845))，同時期に窒素，リン，カリウムなどの必須元素が根から吸収されて植物の生育を促進することも発見された。1840年にドイツの化学者であるリービッヒがこれらの知見を整理して無機栄養説を提唱し，1850年代には多くの実験によって無機栄養説が正しいことが立証され，多くの科学者がこの説を支持するに至った。無機栄養説に基づいてイギリスでローズとギルバートによって過リン酸石灰の生産が開始されたのが1843年，ハーバー゠ボッシュ法によってアンモニア合成が開始され，これを用いて窒素質肥料の合成が開始されたのが1913年である。一方，わが国においては作物を栽培してある程度の収量を得るためには人糞尿(こやし)や魚かすなどを水田や畑に与えなければならないことが古くから知られており，元禄時代には江戸の人糞尿を農村に送り，得られた生産物を農村から江戸に運び込む循環システムがすでに構築されていた。しかし，わが国において化学肥料製造の第1号である過リン酸石灰の製造が開始されたのが1887(明治20)年であり，作物の生産向上を目的とした肥料の施与法に関する研究活動の中心になる日本土壌肥料学会が発足したのが1927(昭和2)年であることを考えると，この講義は，世界的に見ても国内的に見ても肥料養分に関する研究のごく初期になされたものであると見ることができる。

2. ブルックス講義録の内容と北海道農業発展に対する貢献

本章は，肥料の分類，カリ，石灰，ソーダ，マグネシア，リン酸，有機質肥料，施肥法と土壌の物理性に及ぼす影響，作物の生育と深くかかわる大気組成，の9節から構成されている。

「肥料の分類」の節では，肥料を無機質肥料と有機質肥料に分類し，無機質肥料としてカリ，石灰，ソーダ，マグネシウム，珪素，酸化鉄，リン酸，硫酸，硝酸，塩素をあげている。さらに上記の肥料成分中，硝酸を除くすべての成分を含む樹木の灰が無機質肥料として大きな作物生育促進効果を持つことを，灰の分析データを示しながらその作用機作を含めて詳細に論じていることは興味深い。北海道は新たな開発地であり，新大陸であるアメリカにおける農業開発において樹木の灰が大きな役割を果たした経験をもとに，樹木の灰が持つ肥料効果が強調されたと理解される。この提案は北海道の農業開発において実際に採用され，一定の役割を果たしたであろうことは容易に想像できる。

「カリ」の節では，カリ質肥料としては硝酸を含む硝酸カリが作物の生育促進にとって最も効果があることを強調しているが，硝酸カリには作物の生育にとって必要不可欠な硝酸とカリの2大必須要素を含むために生育促進効果が特に大きいことを示唆しており，重要である。さらに，欧米でドイツ産の硫酸カリと塩化カリの消費が増加しつつあることを述べている。

「石灰」の節では，石灰質肥料として炭酸石灰，消石灰，生石灰，硫酸石灰(石膏)，硝酸石灰，塩化石灰があることを述べ，炭酸石灰，消石灰，生石灰については，それらの使用上の留意点について当時知られていた最新の情報が詳細に述べられている。また，硫酸石灰については製造法まで詳細に述べていることは重要である。

「ソーダ」の節では，ソーダ質肥料として塩化ナトリウムと硝酸ナトリウムが紹介され，それらの効果と使用する場合に留意すべき点を述べている。作物の種類によって上記の各種肥料に対する適応性が異なることがすでに講義されていることも注目される。例えば，ソーダ質肥料についてはテンサイが好む肥料であることが述べられている。テンサイに対して塩化ナトリウムを肥料として用いることには賛成できないが，その後，硝酸ナトリウムを主成分とするチリ硝石が北海道に輸入されてテンサイ用肥料として好んで用いられた。

「マグネシア」の節では，マグネシウムを肥料として施与しても効果がない土壌と効果がある土壌があることを述べ，その理由は土壌中に十分量のマグネシウムを含んでいる土壌と含量が少ない土壌が存在することによることを指摘していることは，当時の農業研究レベルとしては重要である。北海道では，その後，農耕地におけるマグネシウムその他の肥料養分の可給態含量とマグネシウム肥料を施与する必要性の有無に関する詳細な調査がなされたが，その際にこの講義の内容は大いに参考になったと考えられる。マグネシウム質肥料としてドイツ産のキーゼライト(硫酸マグネシウム)が紹介され，ヨーロッパとアメリカにおいて作物栽培のための施与が増加しつつあることが記載されている。

「リン酸」の節では，リン酸質肥料としてリン酸石灰，骨粉リン酸，過リン酸石灰があげられており，リン酸石灰資源であるアパタイト(リン鉱石)がニューヨーク州，ニュージャー

ジー州，サウスカロライナ州，カナダ，スペイン等に埋蔵されていると述べている。アメリカではフロリダ州が最大のリン鉱石埋蔵州であるが，この講義の時点ではまだ発見されていなかったと思われる。当時はアメリカにおいても肥料として使用されていたリン酸の主体は骨粉リン酸であったが，骨粉中に含まれるリン酸は難溶性であるので，これを肥料として使用する場合にリン酸を溶解しやすい形態にする方法と骨粉から過リン酸石灰を製造する方法を詳細に説明していることは注目される。先に記載したように，わが国における過リン酸石灰の製造はこの講義がなされた10年後の1887(明治20)年に開始されており，何らかの影響を与えたかもしれない。

「有機質肥料」の節では，有機質肥料として泥炭，muck，落ち葉堆積物，道路側溝の集積物，家畜糞尿，人糞尿，グアノ質肥料，屠殺場で発生する廃棄物，動物死骸，流し排出物，魚かす，および堆肥をあげ，それらの特性と使用法を詳細に論じている。わが国においては古くから人糞尿，魚かすや堆肥を作物の生産のために有効利用してきた歴史を持っていたが，その他の有機質肥料の利用法についてはこの講義が大いに役立ったものと考えられる。

「施肥法と土壌の物理性に及ぼす影響」の節では，有機質肥料を施与する場合に留意すべき点，例えば，対象とする作物種による違い，混合する土層深度と土性による違い，施与時期などが説明されている。さらに無機質肥料と有機質肥料の効果を比較して，どちらも作物の生育にとって役立つことを述べ，土壌の性質や養分状態によってどちらに重点をおいて施肥するかを決めるべきであると論じている。

「作物の生育と深くかかわる大気組成」の節では，光合成に関する研究のごく初期のデータが紹介されている。このような欧米の研究成果の情報をもとに，その後，わが国においても光合成に関する研究が活発に行なわれ，作物栄養学(植物栄養学)の発展と農業生産の向上，技術の発展に大いに貢献した。なお，13) Assimilation of Atmospheric Food の項に記載されている光合成の各プロセスは，当時としては止むを得ないことではあるが正しくない。各プロセスについての正しい反応の解明は，生化学や植物栄養学の進歩を待たなければならなかった。しかし，この当時に光合成の機構がこの段階まで理解されていたことは興味深い。この講義を受講した俊才たちは大きな学問的刺激を受けたに違いないと思われる。

ブルックス以後の肥料学に関する講義は，札幌農学校時代にはブルックスの後任として招聘されたブリガム，南鷹次郎，吉井豊造，1907(明治40)年に東北帝国大学農科大学になって農芸化学科が設置されてからは，大島金太郎，吉井豊造，鈴木重礼，三宅康次，石塚喜明，田中明に引き継がれ，各種作物に対する施肥法に関する研究が活発に展開されて，北海道農業の発展に対して大きな貢献をした。特に，石塚喜明と田中明によってイネの生育相と要素要求の研究が水耕法の開発をもとに展開されたことは，肥料の分追肥法に理論的根拠を与え，肥料学から作物栄養学(植物栄養学)を生み出す原動力にもなった。

3. 講義録の詳細評価

5-1. Classification of Manures

(1) Manure の作用には2種類あることを指摘した上で，「第1の作用は植物養分の供給，

第2の作用は土壌粒子に対する物理的・化学的作用であり，第2の作用の結果，土壌を植物の養分に変換する」としていることは興味深い。その後の研究によって「第1の作用は植物養分の供給，第2の作用は土壌粒子に対する物理的・化学的作用であり，manure の施与による微生物作用を主体とする生物作用の増加がこれらに深くかかわっていること」が証明された。しかし，「第2の作用の結果，土壌を植物の養分に変換する」は誤りであり，「第2の作用の結果，土壌からの植物養分の供給を増加する」ことが証明されている。このような誤りは当時の研究レベルを考えると非難されるべきものではなく，逆に当時すでに manure が保持する機能について，ブルックスがここまで明確に理解していたことを高く評価すべきである。

(2) 第7項で，カリウムが溶脱しやすいことと，ケイ酸カリウムが難溶性でこの形態で土壌中に残存することを指摘しているのは高く評価できる。しかし，土壌粒子が保持する陽イオン交換機能によって土壌粒子に K^+，Na^+，Ca^{2+}，Mg^{2+} などのカチオンが吸着保持されることについては，その後に解明されたためにまだ記載されていない。また，大正時代以降の研究により，土壌水に溶存する，リン酸，硝酸，硫酸などや土壌粒子表面に吸着されて存在するアルカリ元素の形態はイオン形態であることが明らかにされた。

(3) 第8項の内容は，大正時代以降の研究により，ash を土壌に施与した場合に起こる pH 上昇とカルシウムやマグネシウムによる効果であることが証明された。

(4) 第9項で ash の施与時期や施与をする場合に種子と接触しないようにしなければならないことなどについて論じている。これらの記載は，大正時代以降にアルカリ性の強い石灰窒素の施肥時期に関する研究が実施されて，播種7～10日前の施肥が好ましいことが明らかにされたことや，化学肥料を施肥する場合には種子と直接接触しないようにしなければならない，等の技術開発につながった。

(5) 第10項に記載されているように，この時代から「好カリウム植物」，「好石灰植物」，「好リン酸植物」などの用語が用いられ，栽培植物が分類されていたことは大変重要である。その後，植物の養分要求性に関する研究が進められ，好石灰植物，好ケイ酸植物，好カリウム植物などの用語が現在でも用いられている。

5-2．Potash

(1) ドイツ産の硫酸カリと塩化カリの特性が詳細に説明されている。塩化カリがデンプンや砂糖を生産する作物に用いるのは好ましくないということまで記載されているのは，この時代を考えると高く評価される。

5-3．Lime

(1) 緒論の節で，カルシウムは炭酸カルシウムとして存在することが多いが，粉末化が困難なために肥料として利用されていないと記載されている。その後，粉末化する方法が導入され，酸性土壌の酸性を矯正するために施与されるようになった。

(2) 第1項で石灰の施与効果として，①土壌中に存在する有害な酸を中和する，②有機物および無機物を植物の栄養物に変換する，③化学的変化をもたらすことにより，土壌の物理性を改善する，④カルシウム自身が植物の養分であり，これをカルシウムに欠乏した土壌に与

える場合には植物の生育に大変有益な効果をもたらす，という4つをあげているのは高く評価される。しかし，石灰を施与することによって土壌の酸性が矯正され，作物の生育にとって有害なアルミニウムの溶解性が低下するために，作物の生育が改善されることについてはまだ記載されていない。石灰施与がもたらすこの効果については，明治時代以後の研究の展開を待たなければならなかったことを示している。

(3) 第2項に記載されている石灰の施与法に関しては，わが国においても明治以後に詳細に研究がなされ，得られた結果に基づく施与法が実際に利用されてきている。石灰の施与量については条件によって異なると述べるに止めているが，この時代の記載としては最も正しい。その後の研究により，石灰の施与量は土壌のpH，有機物含有率などによって異なることが明らかにされた。

(4) 第4項で塩化カルシウムが吸湿剤として利用されること，砂土に施与することによって空気中の水分を吸収して土壌水分を高めるので有効であることが記載されている。土壌水分を高めるほど多量の塩化カルシウムを施与する場合には，土壌が塩類土壌になってしまうので，この記載は実験を伴わない想像によるものであろう。

5-4. Soda

(1) ナトリウム(Soda)は，カルシウムやカリウムと同様に植物の生育にとって重要な元素であると述べているが，その後の研究により，植物にはナトリウムによって生育が促進されるものとされないものがあることが明らかにされている。

(2) 第1項でNaClの施与効果は$CaCO_3$や$CaSO_4$と混ぜて施与する場合に大きいと述べているが，現在ではそのような効果は否定されており，施肥技術としては採用されていない。

(3) 第2項でアスパラガスが多量のNaClを要求すると記載しているが，この記載は正しいとは言えない。現在では，アスパラガスは耐塩性の強い作物であることが明らかにされているが，NaClによって生育が促進されるわけではない。おそらく当時のNaClに混入していた硝酸やカリウムが，NaClを多量施与した場合にアスパラガスの生育を促進したのではないかと推測される。

5-6. Phosphoric Acid

(1) 第1項の土壌に石灰を施与した場合にリン酸カルシウムが生成することが重要である理由を述べたなかに，(1) nearly 1/2 of the mineral elements in plant seeds is phosphate of lime と記載されているが，これは当時の分析法が未発達であったことによる誤解である。その後の研究により，子実中に含まれるリンの主体は有機態のリン化合物であり，リン酸カルシウムという形態では含有されていないことが明らかにされている。次いで (3) since it is always being taken from the soils into our bones and the bones of animals, this material does not return to the soil as manure in any considerable amount とあるが，manureの材料になる植物茎葉や家畜の糞尿のなかにかなりの量のリン酸とカルシウムが含まれているので，この説明文は正しくない。しかし，当時はそのように考えられていたという事実は興味深い。さらに (4) it is so readily dissolved that a quantity of it is soon used up と説明されているが，石灰を施与した場合に生成されるリン酸カルシウム$Ca_3(PO_4)_2$は土壌中では難溶性であり，植

物によって短期間で吸収し尽くされるようなことはないことが，その後証明された。ここでの記載は，リン酸は土壌によって急速に固定されて不可給態化する性質を持っているので，それを植物によって吸収されたと誤解して書いたものであろう。

(2) 前項に続いて there are few natural deposits of phosphate of lime, though in some localities the rocks contain it と講義している。わが国では明治時代の末期までにリン鉱石鉱山の存在に関する調査を終了し，日本にはリン酸埋蔵量が多いリン鉱石鉱山はほとんど存在しないという調査結果であった。

(3) 第3項に過リン酸石灰の製造法が記載されているが，わが国においてはこのような情報をもとに 1887(明治 20)年に過リン酸石灰の製造が開始された。これがわが国における化学肥料製造の第1号である。

5-7. Organic Manures

(1) 緒論の項で泥炭中に含まれる有機化合物の分類を行なっている。このような研究はその後，土壌腐植の研究として展開された。また，泥炭土壌における還元の進行に伴う二価鉄(Fe^{2+})が作物の生育に障害を与えることを示唆する記載があるのは高く評価できる。このことはその後，より明確に立証された。

(2) 第3項で廃棄されているあらゆる有機物を，肥料の原料あるいは堆肥の原料として用いるように勧めているのは非常に現実的である。その後，有害な重金属による土壌汚染が問題視されて，そのような重金属を含む有機物は有機質肥料の原料として用いることを禁止したが，この時代では有害重金属問題に関する意識がなかったので止むを得ないことと理解される。肥料養分が手に入りにくい当時としては，まずは作物の生産をあげるために何をしなければならないかを示す考え方として高く評価されよう。

(3) 第4項で家畜から糞尿が排泄されるまでの体内プロセス，糞と尿の肥料価値の比較，家畜に与えられる飼料の栄養価による糞尿の肥料価値の差異，家畜の Age による糞尿の肥料価値の差異，ミルク用家畜からの糞尿の肥料価値の評価，労働用家畜と非労働用家畜からの糞尿の肥料価値，糞尿の肥料価値の家畜種間比較が詳細になされていることは，高く評価される。

(4) 第5項で人間の糞尿を肥料として利用する場合の方法を述べていることも重要である。人間の糞尿の肥料としての利用については，わが国では江戸時代以前から実施され，江戸時代には都市域で発生する人間の糞尿を農村域に送って直接農地に施与する循環システムが構築されていた。その後，下水道の整備が開始された後，ここで発生した下水汚泥の有機質肥料としての農地利用は，1930(昭和 5)年にわが国初の活性汚泥法による本格的な下水処理と汚泥処理の運転を開始した名古屋の堀留，熱田両処理場で発生した消化汚泥を天日乾燥して，農耕地に利用したのが最初である。その後，下水道から発生する汚泥の大部分は埋設処理あるいは焼却処理されてきたが，近年に至って埋設処理や焼却処理によって発生する環境問題や資源の有効利用のために，堆肥化をして農緑地に施与する方策が進められてきている。しかし，下水道が設置されていない地域は，世界的に見ると途上国はもとより先進国にもきわめて多く，これらの地域における人間の糞尿の有効利用のために，この項に記載されている

(5) 第6項に記載されているグアノは,窒素質肥料およびリン酸質肥料としてわが国においても輸入して利用された。

(6) 第9項で流し廃液を有機質肥料として使用することを勧めているとともに,その理由として流し廃液には作物の肥料養分が含まれているばかりでなく,環境に放出した場合には人間の健康にとって有害なガスを発生することをあげていることは高く評価される。現在,生ゴミ起源の環境問題が問題になっており,この問題の解消と資源の有効利用のために生ゴミの堆肥化,肥料化,熱エネルギー化などが推進されつつあるからである。

(7) 第11項の冒頭に The best components of manure are those which are soluble and volatile という記載がある。生物系有機性廃棄物を堆肥などの有機質肥料として利用する場合に,原料中に含まれる可溶性化合物や揮発性化合物に,しばしば植物の生育を阻害する有機化合物が含まれているので,原料を堆積して定期的な攪拌などにより通気しつつ,酸化条件の下でこれらの可溶性化合物や揮発性化合物を分解することの必要性がその後の研究によって明らかにされている。

(8) 第12項のなかに Coarse, undecomposed manure—on account of the heat which it will generate in decomposition—is best for cold, wet soils という一文があるが,未分解有機資材を湿潤土壌に施与した場合には,しばしば植物の生育を阻害する低分子量の有機化合物を集積することがその後の研究で明らかにされているので,この文章は当時の見解を述べたものであると思われるが,正しくない。また,単位圃場面積に施与される有機物の量はそれほど多くはないので,未分解有機物が分解する場合に放出する熱量を過大評価していると思われる。

(9) 第13項に記載されている事項は,理由が記載されていないが,その後の研究で堆肥をはじめとする有機質肥料や化学肥料を連年施与する場合には,土壌の pH が低下して作物の生育を阻害するので,4～5年ごとに適量の炭酸カルシウムを施与して土壌の酸性を矯正することが必要であることが明らかにされている。有機質肥料の形態であれ,化学肥料の形態であれ,土壌に養分を十分に施与した場合には次年度の作物に対する残効があるので,施肥量が多い場合には毎年等量の肥料を与える必要はないこともその後明らかにされた。さらに,堆肥を施与せずに化学肥料のみを連年施与する場合には地力が低下するので,圃場に堆肥を時々施与するか,少量でも毎年施与する必要があることが明らかにされている。

5-9. The Atmosphere as Related to Vegetation

(1) 第1項で酸素が植物の生育にとって必要不可欠であることを,それまでになされてきた植物生育実験の結果をもとに詳細に記載しているのは評価できる。その後,酸素は体内に含まれるすべての有機化合物の構成成分となっているとともに,植物の呼吸基質や酸化反応の基質として必要不可欠であることが明らかにされている。

(2) 第5項の文中に「クローバやマメ科植物は大気中の NH_3 を吸収する能力が他の植物より高い」という記載と,そのことを示すマメ(Bean)の実験結果が示されているが,この記載および実験結果はマメ科の植物が根粒菌と共生して根粒を形成し,根粒が大気中の N_2 を

吸収固定することによることを示すものである。生物的窒素固定の存在が解明されていなかった当時としては，根粒による窒素固定機能が，この講義録に記載されているように理解されていたことはきわめて興味深い。

(3) 第7項における大気中のNOxおよびNH₃濃度に関する記載は，当時は意識されてはいなかったが，現在問題になっている地球温暖化，酸性雨などをもたらす原因ガスであるので興味深い。

(4) 第8項でCH₄が植物から放出されるか否かについて当時から研究されており，ある研究者は放出されるという結論を出しているのは興味深い。近年，湛水条件で生育するイネは還元的な土壌中で生成されるCH₄を根から吸収し，吸収されたCH₄は通導組織を経由して葉の気孔から大気に放出されることが立証されている。

(5) 第13項の表中に記載されている植物による吸収形態には間違いが多い。葉から吸収されるNH₄CO₃はあったとしてもごく微量である。NH₄⁺は主に土壌から根によって吸収される。すなわち，土壌中に存在するNH₄NO₃はNH₄⁺とNO₃⁻のイオンに解離された後に根から吸収される。HNO₃も同様である。

注
1) plant food：現在ではplant nutrientを使う。
2) この文章は，「土壌に施与したある物質が土壌粒子に作用して土壌粒子を植物が吸収利用できる養分に変換する」という意味にとれる。その後，このような考え方の実態について研究が進められ，「土壌に施与したある物質が土壌粒子に作用して，土壌粒子中に含まれている不可給態の養分を植物が吸収利用し得る形態に変換する」ことの発見につながった。例えば，炭酸カルシウムを酸性土壌に施与した場合には，植物の生育にとって有害なアルミニウムの溶解度を低下して沈澱させることによりアルミニウム毒性を除去するとともに，土壌粒子に吸着しているリン酸やカリウムの植物による吸収を増加させることが明らかにされている。
3) その後の研究により，植物の生育にとって必須な養分元素は，炭素，酸素，水素，窒素，リン，カリウム，カルシウム，マグネシウム，硫黄，鉄，マンガン，亜鉛，銅，ホウ素，モリブデン，塩素，ニッケルの17元素であることが明らかにされている。これらのなかで無機肥料元素として位置づけられるのは，炭素，酸素，水素を除く14元素である。ナトリウムと珪素は特定の植物で生育促進作用を示すことが明らかにされているが，すべての植物にとって必須であるか否かについては疑問視されている。
4) soil：原文にはearthとあったが，現在ではsoilを用いるため，置き換えている。
5) 1 bushel＝34 L.
6) 1 cord＝11.9 m³
7) 1,000 pounds＝454 kg
8) permanentlyと記載されているが，実際には「永久に」土壌改良効果が持続されるのではなく，「長期間にわたって」持続されるの意味である。
9) 1 gallon＝3.8 L.（アメリカ）
10) 1 barrel＝31.5 gallon：120 L.（アメリカ）
11) 当時の大気中CO₂濃度はほぼ3/10,000であったことが現在では解明されている。
12) N₂は植物単独では同化されないが，植物がN₂固定微生物と共生関係にある時には，根粒で吸収同化されることがその後明らかにされている。例としてダイズの根粒をあげることができる。
13) N₂Oが大気中に含まれていることは現在では立証されており，地球温暖化や酸性雨の原因ガスとして注目されている。

参考文献
1. 熊沢喜久雄(1974)，植物栄養学大要，養賢堂

2. 栗原　淳・越野正義(1986)，肥料製造学，養賢堂
3. 安田　環・越野正義編(2001)，環境保全と新しい施肥技術，養賢堂
4. 但野利秋他編(2002)，植物栄養・肥料の事典，朝倉書店

［但野利秋］

Chapter 6 Farm Economy (Management)

Webster's definition of the word economy is "management without loss or waste". Another good definition is "a system by which business is carried on without loss or waste." In no other branch of business is economical management more important than in farm, for in no other business are there many ways in which it is easy to in cut loss.

6-1. System Necessary to Prevent Loss or Waste
In order to succeed in any business, it is absolutely necessary to prevent loss and waste, and in order to do this, the business must be managed according to some regulate system. From the nature of agriculture, [p. 453] it is more important to manage economically than in almost any other business, and yet owing to circumstances, it is more difficult to lays down and adheres to a system of farm operations than it is in any other business. (1) In all farming operations we are greatly dependent on the weather over which we have no control. That man, indeed, is wise who can with reasonable accuracy foretell what the weather is likely to be. By the weather, the daily plans of farmers are often broken and their men are left without employment unless storms and interruptions are included in their system when they have merely to change the kind of work. (2) Upon the farm there is so little division of labor. For these reasons [p. 454] a system of farming should embrace all possible contingencies. A well-contrived system firmly adhered to, is essential to success and is the first step towards securing farm economy. Farmer's system should embrace general plans of farm work and contemplated improvement for several years in advance as well as more minute plans of the work from year to year, from month to month, and from day to day.
1) Economy in Regard to Soils
The man who killed the goose which laid the golden egg, is on a par with the man who, owning land, allows it to deteriorate. Is the farmer allows his farm to grow poorer, he is losing his capital. Land is not valuable [p. 455] because of the number of acres, but because of the plant food it contains. If the farmer cannot keep his entire farm fertile he had better sell a portion of it and use the money thus obtained to fertilize the remaining.
2) Economy in Regard to Manures
Manure may be considered to be the raw material which the farmer wishes to manufacture into crops, and since his income depends upon the quantity of crops raised, if he wastes his manure he is losing its income. All wastes from improper management or application should be prevented.
3) Economy in Regard to Crops
It is even worse, [p. 456] if possible, to waste crops than to waste manure, since the manufactured article is worth more than the raw material. It is estimated that individual carelessness or want of foresight or intelligence annually wastes

millions of dollars work of crops in America. Crops are wasted in the field by improper modes of harvesting and in the barn by improper feeding or storage. Much money is yearly loss by harvesting crops at the wrong time. If for example, the harvesting of a grain crop is deferred a few days beyond the proper time, much grain will rattle out and be lost. By careless handing, too, much grain is shaken out and lost. It requires but little additional labor to so manage as to avoid such loss. Much value is also [p. 457] frequently loss by harvesting hay crop too late. To allow grass to stand until the seed is ripe, is poor economy for two reasons. (1) Because it is not as good for hay not containing as much digestible matter. (2) Because a crop of grass the seed of which is allowed to ripen, much more rapidly exhausts the soil than one cut before the seed is formed. Crop improperly stored very often decays or moulds thus causing very great loss. Careless feeding of crop is also a fruitful source of loss. It should be your aim to put before animals no more food than they will eat, since if you give them a larger quantity than this, they will trample it under their feet and destroy it. [p. 458]

4) Economy with Respect to the Growth of Needs

This topic should well be remembered by every practical farmer. Until the seeds of needs become more valuable than they are now, we cannot let them grow on our farms. If one crop of weed-seeds is allowed to ripen, it will not all germinate for a great number of years, and so, the man who lets weeds grow to seed is laying trouble for himself for many years to come. Most kinds of weeds produce an immense number of seeds, and many of the latter are provided with contrivances which make their distribution by wind or other means very easy. For this reason, no weeds should be allowed to grow along by fences, ditches or the borders of the farm. [p. 459] If such seeds are not destroyed early, they should at least be mown before they ripen their seeds.

5) The Performance of Work at the Proper Time, as Affecting Farm Economy

All kinds of farm work, especially those connected with the planting, cultivating and harvesting of crops, must be done at a certain time or else great loss will result. If, for example, a crop is planted a little too late, it may not have sufficient time to ripen, and the result will be a great loss. If a crop is not cultivated at the proper time there is a great loss in two ways, [p. 460] (1) because it is much more work to cultivate it after the weeds have become large than it is when they are small and (2) because the growth of large weeds will seriously injure the growing crop. One week's delay beyond the proper time for cultivating a crop often necessitates the expenditure of ten times as much labor to clear the land of weeds as would be required had the work been done at the proper time. I have already called the attention to the fact that much loss would often result from not harvesting crops at the proper time.

6) Economy in Regard to Labor

At almost all times and in almost all places, labor is the greatest expense in raising crops, and therefore it should never be wasted. [p. 461] The aim of the farmer should be to so manage his working force as to make a given amount of exertion produce the greatest possible result. The following language will more clearly express the meaning I wish to convey. —"Economic labor loads both ways. It kills two birds with one stone. It goes to mill, store, blacksmith's shop and post office at the same time. On the other hand, uneconomic labor milks one teat at a time and ploughs the land the short way". Men, working for a poor manager never know in the morning what they are going to do and thus they lose much time. An uneconomic man will sometimes start in the morning to now the grass [p. 462] in a distant field, and getting there, he will find that he must go back for his whetstone. By the time he reaches his field with the whetstone, he will have become thirsty and he must get back for his water, and he may work hard half of the morning and yet not cut much grass. A man, to succeed in farming, must utilize all his working force.

7) Economy in Regard to Teams and Stock

Many of the animals used on a farm, form a part of farm stock that is they are valuable not only because of their work, but also because of their growth. Such animals as oxen, for example, should be regarded as valuable both for their labor and growth. For this reason, such animals should be [p. 463] so fed and cared for that they will pay for their food in improvement, in which case, their work will be a clear gain. It is also very poor economy to so feed and overwork animals which are kept simply for their labor, that they will deteriorate in value. Will fed animals can perform much more work than those poorly fed, but they should never be obliged to perform so much labor as to because very poor. A certain amount of work will keep them in a better condition than they would be without it. Animals which are poorly fed or which are obliged to do so much work as to become little else than skin and bones, very rapidly deteriorate in value and will last but very few years. Is should be the farmer's aim, then to feed his working animals [p. 464] well and to oblige them only to do so mush work as they can do and still keep in good condition. It is poor economy to so manage any stock that will barely hold own and not gain. By so doing, you lose nearly all the food you gave them, all you have to show for the food they have eaten, is a pile of very poor manure whereas had you given them a little more food, they would have continued to grow, and by this growth, would have paid for the food you, have given them. Animals which by neglect or poor feeding have at any part of their lives, been stunted, can never, though fed ever so well be made to grow as well as they would had they never been stunted. It should be the aim of the farmer to so feed all his animals as to keep growing thriftily all the time. There is a great difference in different breeds of animals as to the amount of growth which they will make in a certain time, and that farmer who is truly economical, will keep only such as grow rapidly and mature early. It costs nearly as much to raise an animal of a poor breed as it dose to raise a good one, while at a certain age, two years for example, the animal of good breed would doubtless be work two or there times as much as the other.

8) Economy in Regard to Farm Implements

Good tools and machines and plenty of them are necessary on the farm. Good implements and machines are consistent with the most rigid economy, [p. 466] since with their assistance, a man can accomplish much more labor than he can perform with clumsy implements. A man cannot afford, in general, to borrow especially smaller tools and implements, since it often takes more time to go and borrow an implement than it would take with properly directed labor to earn money enough to buy it. Economy demands that all tools and implements should be in perfect order and that there should be place for everything, and that everything should be kept in its place when not in use. Economy also demands that everything should be protected from the weather both in summer and in winter, since implements exposed to the influence of weather are very much injured by it. The wooden parts of machines and [p. 467] tools will decay, and the iron or steel parts will rust so that those articles exposed to the influence of weather will not last nearly as long as those which are always protected from it. A very good plan to be pursued by the farmer who employs several men to work for him, would be to have a complete set of farm tools for each man. Then, When the farmer hires a man, he should charge him with these tools and when a laborer is discharged, he should be credited with what tools he has in good condition at the same price he was charged for them. The price of those tools which he has lost or broken, should be de-ducted from his wages. The laborer should be told when he is first employed that such would [p. 468] be the case. This would lead him to take excellent care of all his tools and therefore though this course would oblige the farmer to have more tools than the course ordinarily pursued, it would doubtless prove the best economy in the end. In order to conveniently make repairs on tools, buildings and machines, the farmer should have a workshop provided with both an iron and a wooden vice and a good supply of carpenter's tools. He should keep in his workshop plenty of well seasoned timber as well as nails, bolts etc. If he has these things, he can easily repair his wagons and tools much more cheaply than he could do it if he had to take them to a carpenter or blacksmith. [p. 469]

9) Economy in Regard to Buildings

Farm buildings are a part of the farmer's capital and they should be well suited for the purpose they were designed for. They should be to built that crops can be stored and fed conveniently and the stables can be easily cleaned out. Every barn should have a root room so placed as to make the business of feeding roots as easy as possible. Is should also have a completely convenient place in connection with it for the reception of manure from the stables. This place should be one into which the manure can be easily put and out of which it can be conveniently carried. The place for the storage of hay should be such [p. 470] that the hay can be rapidly and easily put into it, and such that the hay can be conveniently carried out to the place where it is to be fed to the cattle. The room in which the animals are to be kept, should be well lighted and well ventilated, and the place where the animals are to lie should be so arranged that they can easily be kept clean, and at the same time, be comfortable. [p. 471]

6-2. Farm Management

It was formerly believed that a man who can do nothing might become a successful farmer, but this old idea is now about obsolete, and as present, it is a general opinion that a man to be a successful farmer must have a particular education both practically and theoretically to fit him for his work. To be capable of managing a farm well, a man must be familiar with the great natural powers which have reduced the rock to soil and which still influence it. He must know a difference between an agricultural soil and subsoil, and what influence air and gases in it have upon the soil and also how frosts heat and waters affect it. He must know the chemical and [p. 472] physical changes of the soil, influence on the soil of cultivation, influence on the soil of producing crops both when cultivated and, removed, and by nature's process. And, he must also know the principles of adaptation of crops of crops to soil, and soils to crops. He should know—how and upon what different plants feed, how they perpetuate themselves and the specific requirements, structure and adaptation. He must know the nature, character and changes and mode of action of all fertilizers on soils and on plants. He must have an accurate knowledge of the structure, wants and adaptation of animals and the principles of health and breeding. He must be an adept in every possible phase of farm economy. And finally he must possess sufficient knowledge of the rule and system of business to enable him [p. 473] to execute promptly all business operations. Farm management is simply the application and use of all this knowledge in the business of managing the soil, crops and stock amid the varied circumstances of each individual case.

1) The Proper Size of Farms

You will often hear men making the statement "Ten acres is enough for a farm". Others have even said that three acres is enough. The proper size of farms, however, cannot be stated in absolute figures. It depends largely on the branch of business you intend to pursue. In makes gardening ten acres would be a large farm, and even four acres would require [p. 474] the labor of about four on five men and teams to manage it properly. On the other hand, stock and grain farming each requires a great many acres. If the work upon the farm be done thoroughly, the larger it is, the more profitable it will be for the following reasons. (1) It cost less proportionally to raise crops whatever may be the branch of business. (2) It is costs less for oversight on a large farm, because one man of ordinary capacity can oversee and direct and direct the labor on a thousand acres of land just as easily on a small farm. (3) It costs less in proportion to the crops raised for labor of both men and teams upon the large farm, [p. 475] because if only one or two men are at work on a farm, they work at disadvantage as many kinds of farm work. (4) If a man has only twenty-five acres of land, he has to keep just as large a team as a man who has two thousand acres, because at certain times of the year, he cannot do his work unless he has such teams. (5) It also costs more, proportionally, for tools and machines on a small farm, for the small farmer must have about as many as the large farmer must have. (6) A man, having a large quantity of anything for sale, can usually is more readily than a man who has but little, and in addition to this, the large farmer has the power to govern the market more or less. [p. 476]

2) Capital in Farming

Both permanent or fixed and floating capital is of great importance to farm operations. Capital invested in land, which is in a growing community, is the safest possible investment. It is commonly most profitable in the end thought at first is may not pay a large percent. Money invested in stock or general business, is called floating capital. It is wise to so invest some of our money that it can be readily obtained for use as any time. Money invested in implements and machinery, is called perishable capital. In order to keep a sufficient number of tools and machines, it is usually necessary to spend [p. 477] considerable money each year. As capital thus invested, is so perishable, it ought to pay, while in existence a large percent. If a man has $10,000 with which he intends to buy a farm and carry on the business of farming, he should invest only $6,000-7,000 in land and buildings, and keep the rest as floating capital. The kinds of capital enumerated, do not embrace all the capital that the farmer may have. Cost of education, whatever kind it may be, is capital and a man should be paid for the use of it. Character is also capital. A good character may be work many thousand dollars to a man in business, and no man can succeed without it. [p. 478]

6-3. Kind of Farming

There are two kinds of farming, namely general and special. General farming is understood to be the business of raising most of the crops and kinds of stock which the section in which the farm is located, will produce. Special is the business of raising a few special crops or kinds of stock. Sometimes, a special farmer confines himself to one or two crops, sometimes to a single kind of stock, raising only those crops necessary to feed it. Other men who may be called special farmers, raise perhaps three or four different crops and two or three kinds of stock. Since the general farmer raises so many crops and so many kinds of stock, [p. 479] he will always be comparatively sure of having some crops or some kinds of stock that will prove profitable. One or two crops may fail, but others will give a good result. If one or two crops raised by the special farmer fail, he has no other upon which he can rely. The general farmer raises his own food and much of the material necessary for the manufacture of his clothing, so that, he is perfectly independent for means of subsistence. Since the general farmer produces so many crops and cares for so many kinds of stock, he cannot become very skillful in their production while the special farmer, since he confines his attention to a few special crops, learns how to produce them [p. 480] in the most skillful and economical manner. The special farmer dose not need so many implements in proportion to the amounts of work he dose, as the general farmer requires and, therefore, this item of expense is less in proportion so the crops raised than in general farming. It notwithstanding the many advantages of special farming, it is impossible in a new country for many to be special farmers. Special farming can be successful only in a place where large markets are accessible. If in Hokkaido for example, a farmer produces a very large crop of potatoes, he can find no profitable market for them. The facilities for transportation must be good or else, special farming cannot prove profitable. There must [p. 481] be good roads, canals or railroads. Though the market is distant, if the means of transportation is such that goods may be cheaply carried to it, special farming may prove successful. Those districts where there is a large thriving population engaged in other pursuits, are best adapted to special farming, since the people engaged in these other pursuits, will require large quantities of farm products.

1) General Farming

The location of a farm for general farming, is of the utmost importance. The soil, climate, and markets should all be carefully considered. In a region as far north as Hokkaido, we should select the farm, [p. 482] the soil of which shall be as warm as possible, and we should strive to get a farm in such a location that frosts will not be likely to affect crops either late in spring or early in autumn. In a region where the rivers run east and west, select a farm on the north side of it, where they run north and south, select one on the west side. In selecting a farm in a region where there are mountain ranges or single mountains, select on the southern or eastern slope rather than on the northern or western for the reason that the general inclination of the farm in such locations will be such as to insure the greatest possible absorption of heat. There are of course, exceptions to these general rules. If the soil in that location would naturally [p. 483] be the warmer than that on the other side of the rivers or mountains, I would take the fertile soil in preference of the colder. Other conditions being the same, however, the rules which I have given you are good ones to follow. A general farmer should have various kinds of soil so that he may have soils suitable for all kinds of crops. In selecting a farm in a new country, much can be told with regard to the fertility of the soil, by the character of vegetation growing upon it. You should always select a farm in the place where the natural vegetation is very luxurious. A growth of large trees such as elms and oaks, indicates a richer soil than a growth of small stunted trees as the birch or pine. [p. 484] A growth of trees generally indicates a richer soil than a growth of grass or herbaceous plants, but, if the latter be very luxurious and vigorous, the soil may be a very excellent one. In selecting a farm in an old country, the buyer should be able to determine from its appearance as to the quality of the soil. Much can always be told by the character of the natural vegetation growing upon it. If the weeds are large and vigorous and grow close together, you may decide that the soil is rich. If on the other hand, there are but few and small weeds, you may conclude that the soil is poor. By a close inspection, you can determine whether the farm is simply neglected or exhausted. It may often be good policy [p. 485] to buy a neglected farm where it would not be to buy a worn out one, unless we could buy it very cheap. We must be careful to select a farm with plenty of good spring or running water upon it, if it is possible to do so. The supply of such water should be sufficient for the use of the family and farm stock and also for irrigation, if possible. Such a supply of water will be a great saving of expense and what is still better, it is far healthier. We should have a regard to markets in selecting a general farm, for, a convenient market is even more important to the general farmer than to the special. He usually has but small quantities of any one thing to sell, and, therefore, [p. 486] he must have a convenient market. A farmer should be particular to locate his farm on a good public high way, if it is possible to do so. By this means, he saves the expense of building roads, and keeping them in repair. He should locate so as to be near schools, and where his family can have social intercourse with other people. Good railroads are also highly important as a means of transportation. In a new country, it will not usually be possible to locate upon a railroad. But if a farm can be so located, it will be work much more than one located distant from the railroad. [p. 487]

2) Ownership of Land to be Farmed

It is often a question with a young man whether it will be wise for him to run in debt for a farm. Some hold the opinion that it will be wiser for him to work for some one else until he accumulates sufficient money to pay for a farm. It is a general fact that a borrower is the slave of a lender, and if a man runs in debt, he should do so only after mature deliberation. So one should ever run in debt for any personal expense, for whatever he buys, has no market value after he has used it himself. On the other hand, if a man does run in debt for a farm, he can, if the purchase has been judiciously made, sell it as any time for as much as, if not more than it costs. [p. 488] I would advice a young man to run in debt for a farm, if he has the following qualifications; (1) If he has fully decided to follow farming as a life pursuit. (2) If he has confidence in his own intelligence and skill to grapple with and manage the business and (3) If he has any experience as a managing director of the business of farming. The advantages of owning a farm, if you are qualified to manage it, are many. If you are a man such as, I have described, you will be capable of improving the farm, and if you own it, you will get all the benefit of these improvements. The influence on a man personally of owning a farm, is very great. It improves him for, [p. 489] it makes him independent, self-reliant and manly. It gives him a home and an object and incentive to labor. The power and productive capacity of a country, the land of which is divided among many of its inhabitants will be greater than the power and productive capacity of a country otherwise similarly situated but the land of which is owned by a few individuals. For this reason, it is good policy for nations to give public lands to industrious men who will settle on them. If an immigrant comes to Hokkaido to commence farming and has enough money to start in that business, the government is really a gainer if it gives him land to settle on. [p. 490]

3) Location of Buildings on the Farm

In selecting a farm which has no buildings on it, you should be careful to see that there is a good location for them. Every farmer should live on his farm, since if he dose not, there is a great loss of time in going to and from work, and in hauling crop and manure. There are two fundamental rules which should govern the location of farm buildings. (1) They should be so located as to be central to the field to and from which there is to be the most cartage of manure and products. (2) They should be so located as to be on or contiguous to a good public highway. [p. 491] The distance of the house from the road should never be less than from 75 to 100 feet. The best location on your building, therefore, since by such a location, both of the rules. I have given you are complied with is upon the two sides of a good public road, the amounts of land on each side of it being about equal and the general shape of the farm are square. The buildings may then be located on the road and at the same time, be centrally located with reference to the field. If your farm is of such a shape and so located that if the buildings lie in the center of it, they will not be near the public road. I will place them near it than in the center. [p. 492]

The climate of Hokkaido is such that shelter for farm buildings, is of considerable importance. A good natural shelter as one which can be afforded by a hill or dense forest often makes a change in temperature equivalent to several degrees of latitude. The temperature is greatly influenced by winds. The coldest winds in this region are those which blow from north, north-west and north-east. Therefore, if it is possible to do so, locate your buildings upon the south side of a mountain, hill or forest. A good artificial shelter may be obtained either by building a high board fence or by planting thick rows of trees. If natural shelter can be obtained in connection with the other advantages of which, [p. 493] I have spoken, it is wise to so locate your buildings as to take advantage of it, but since artificial shelter can be provided cheaply, the other advantages should not be sacrificed for the sake of natural shelter.

Farm buildings should also be so located that those living in them, may not be subjected to unhealthy influences. For this reason, they should not be located where the shade of trees falls upon them or on low or marshy ground or near a swamp unless it can be thoroughly drained or some shelter put between it and the house. A girdle of trees or sunflowers set out between the swamp and the house will thoroughly purify [p. 494] the air which passes through them, and thus mal-aria generated in this swamp, will be harmless to animals or persons living in farm buildings. It would be better, of course, to thoroughly drain the swamp than to depend upon the protection afforded by trees or sunflowers. But in place where, from any cause, it is impracticable to drain the swamp, it will be good policy to set out trees or sun flowers between it and the farm buildings. There is another point with reference to the location of farm buildings, which, though not so important as those already spoken of, should not be overlooked. The world is full of beautiful scenes, and to look upon them exerts the beneficial influence upon the mind, for this reason, [p. 495] if it is possible to secure a location for farm buildings in a place from which an extensive view of the surrounding country can be seen, it is wise to do so.

4) Farm Buildings

Every farmer should have a house of his own. He cannot hire a house. He must also have certain other buildings adapted to the kinds of business he intends to pursue. More buildings than he actually needs himself are usually superfluous, especially in a country. It should be the aim of a farmer in building a house to make it as beautiful as possible. In elevation and style it should comport with the peculiarity [p. 496] of its location. Is should be so constructed as to be convenient in its interior arrangement for the performance of the work of his wife. The portion of the house where the family are to do their work, the dining and the sitting room should receive most care and attention and, if need be, the most expense.

With reference to the road, I would locate a house on the north side of the road running east and west. With one running north and south it would not make so much difference, but since the afternoon sun is that which is usually most thought of, I would locate on the east in preference to the west side of it. With those running in directions other than the cardinal points, the house should be located on that side which would insure the greatest amount of sunlight [p. 497] in those parts of the house most used.

5) The Location and Structure of the Barn

The location and form of the farmer's barn must be varied according to the circumstance of each individual case. Men, not infrequently, build the barn at a little distance from the house and connect the two by means of a long shed, which is used for the storage of wood, wagons, *etc*. This arrangement is very convenient on some account. It is convenient, because it makes it possible for him to go from the house to the barn in all sorts of weather without exposure. On the other hand, it has some serious disadvantages, the principal one of which [p. 498] is that in case of the burning of either building, the other will be very likely to burn also. Whenever it is possible to do so, the barn should be located in this country on the north or northern side of the house, since the prevailing winds are from south southeast or southwest. During the summer season, if the farm were located on the

192   Chapter 6   Farm Economy (Management)

windward side of the house disagreeable odors would doubtless be carried by the wind to the house. The distance of the house from the barn necessary in order to make it unlikely that one will catch fire in case other burns, must vary according to the height of the buildings, the

[参考図]

Barn and Cow stable, Massachusetts Agricultural College
from Brooks, *Agriculture*

higher they are, the greater must be the distance. From one hundred to one hundred and twenty-five feet will ordinarily be found sufficient. [p. 499] Before commencing to build a barn, the farmer should give the subject careful consideration. He should decide exactly for what purposes he wishes to use it. So particular rule can be laid down other than this. Barns should be so constructed as to be convenient for housing stock, storing crops and manure and feeding the animals which are to be kept in it. The manure made by cattle should be either stored under some shed near the barn or in a cellar under it. Some think that the effluvia rising from the manure in the cellar, is hurtful to the stock and hay, but if the cellar is well ventilated and the floor above it, tight, this will not be the case. A cellar saves [p. 500] a great deal of expense in the handling of manure, and the first thing, then, in constructing the barn is to make a well ventilated room under it. A barn should be so constructed if possible, that you can drive in near the roof, as it is easier to pitch hay down than to pitch it up. The cattle stalls, if they occupy but one side of the barn, should be placed on the south side rather than on the north, since they will be much warmer. The stall should be so constructed that cattle can be got out and in quickly. Young cattle and cows, I would tie with stanchions, but large or fat cattle or cows

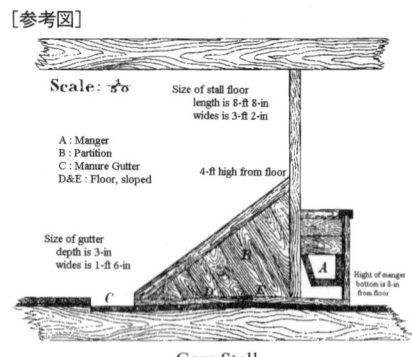

[参考図]

Cow Stall

heavy with calf, should be tied with ropes or chains. There should be a gutter behind the cattle stall [p. 501] of from four to six inches in depth and eighteen or twenty-four inches in width, for the reception of manure. The platform for cattle should incline to the rear a very little one inch for a platform six feet in length being sufficient. The length of the platform should be from $3\frac{1}{2}$ to 6ft. according to the size of an animal to be kept on it. The width allowed to each animal should be from 3-4 ft. It is not good economy to have too much room in the barn floor, since this space is vacant most of the time. A width of from 12-14 ft is ordinarily sufficient. The main floor should run the full length of the barn, so that you may drive through it and not be obliged to back out. [p. 502]
6) The Balancing or Division of a Farm for Different purposes

By the balancing or division of a farm is means a suitable division of it into mowing, pasture, tillage, *etc*. The proper division is determined by the character of the soil and the business pursued. A good division is one that is self-supporting up to the capacity of the farm. A farmer should have mowing land enough to furnish sufficient hay to keep what cattle he can pasture in summer through the winter. He should have tillage land enough to furnish what grain, roots and vegetable his stock and family can consume, and he should have [p. 503] woodland enough to furnish sufficient timber to keep his fences and buildings in repair and to supply the necessary fuel.

6-4. Management of Mowing Land

A strong retentive soil and one which has great absorbing power with a tendency to be moist, is best adapted for mowing land. Such a soil should be moist, not because of the overflow of water from surrounding high land nor because of the presence of springs, but because of its absorbing and retaining power. A soil with some clay or a large amount of organic matter in it, is, therefore, the best soil for grass. [p. 504] If a soil is absolutely wet, it should be under drained. The importance of the grass crop for the general farmer is very great. The number of animals he can keep, depends upon the amount of grass he can produce. The amount of manure he can make, depends upon the number of animals he can keep, and the amount of crops he can raise, bears a direct relation to the quantity of manure at his disposal. The first thing to be considered with reference to the proper management of mowing land, is the season best suited for sowing grass seed. We may adopt, as a general rule the following. Grass seed should be sown at a season when the weather is likely to be and to continue [p. 505] for some time, quite moist. This is so, because grass seed germinate much better under such conditions.

The climate of Hokkaido is so humid that grass seed may be sown with a certainly of success at almost any season. But the time most suitable for the performance of this work is the early spring or autumn. If sown in spring, the earlier the work is done, the better since the grass will have sent its roots deep into the soil before the hot suns of summer. Before sending its roots deep into the soil, grass subjected to hot and somewhat dry weather, is very likely to perish. If sown in autumn, this should be remembered; *viz.*, the crop for the following year. [p. 506] In a place, however, where the snows of winter are likely to be very deep, there is some danger that the young grass will be smothered. However, this may sometimes be prevented by cutting the young grass just before the approach of cold weather.

Whether I would sow grass seed in the spring or autumn, then would be decided by circumstances. If I wish to raise some crop which could be harvested early upon the land which I propose to convert into mowing land, I would do so and sow the grass seed after the crop had been harvested. The month be suited for autumn sowing is September. Success would be quite certain from the grass seed sown at any time during the twentieth of August and tenth of October. [p. 507] If I wish to raise some crop such as oats for fodder or some grain such as barley or oats on the land which I am going to convert into grass land, I would do so, sowing the grass seed with the other crop in early spring.

If sown in spring, I would always sow some grain with grass seed, since if grain is not sown, there will be a large growth of weeds which will take as much plant food from the soil as would the crop of grain. If you are so situated that you can make use a crop of fodder, it will be much better to cut grain just as it comes into blossom, as at this time it will not have removed nearly as much plant food from the soil as it would, were it allowed to ripen. In this locality grass seed, if sown in spring, [p. 508] should be sown as soon after the snow disappears as possible.

The method of doing the work, if a grain crop is sown with, it is as follows: (1) The land should be ploughed. If this is done the preceding autumn, it will be better as it will enable you to sow the grass seed earlier. (2) Harrowing the land until it becomes quite smooth; and if there is any abrupt depressions or mounds upon the surface, which cannot be sufficiently leveled by the harrow, they should be leveled by the use of scraper or by hand. When the land has been brought into a smooth condition. (3) Sow the grain, after which. (4) The land should be again harrowed for the purpose [p. 509] of covering the grain ; and (5) Brushed after brushing. (6) The grass seed should be sown, care being taken of course, in sowing both grain and grass seeds to scatter them evenly over the entire surface. This work of sowing may be done either by hand or by the use of machines.

A cheap and simple machine is—Cahoon's Broadcast Seed Sower. This does very excellent work. There are also machines worked by horse-power which sow grass seed broadcast in a very excellent manner. One of the best is Buckeye's Seed Drill. In sowing grain with grass seed, care should be taken to sow it rather thinly, since if sown too thickly, the growing grain will choke out the grass seed. [p. 510] If oats are sown, two bushels per acre will be sufficient. (7) After sowing grass seed, the land should be again brushed, (8) and Rolled. The rolling will leave the soil in good condition and the surface quite smooth. (9) If however, there remain any protruding stones, roots or materials of any sort which can be easily removed they should be picked up and carried away after the rolling is completed.

It is of the first importance as you know, that mowing lands be smooth in order that the machines used in harvesting hay may work to perfection, and, therefore, I have been thus particular in recommending the complete leveling and smoothing of the surface. [p. 511] Grass seed should not be buried very deeply, and it is for that reason that I have recommended that you cover it simply by the use of a brush. The harrow will bury it too deeply. In moist weather, most grass seeds will germinate upon the surface of the land, but a very slight covering of earth insures more perfect germination, and the brush gives a slight covering to most of the seeds.

If grass seed is sown in the autumn, the manner of doing the work, is almost precisely the same. (1) The land should be ploughed, (2) leveled, (3) harrowed until it becomes quite smooth. (4) Brushed or harrowed with a fine toothed harrow. (5) Grass seed should be sown. (6) The land should be brushed again. (7) It should be rolled. (8) If anything is left upon the surface, picked up and carried away.

In sowing permanent grass land, a variety of seeds should be used. Never confine yourself to one variety. Several varieties cover the land much more perfectly than one or even two or three. In selecting varieties for mowing, you should exercise care to plant upon the same field those varieties which blossom at about the same time. If you have varieties growing together which blossom at different time, some of them will not have become full grown when those which

Tab. 1	Number of POUNDS in an bushel	Number of SEEDS. In an ounce	Time of BLOOMING
Redtop (*Agrostis vulgaris*)	13	425,000	Jun., Jul.
Sweet Scented Vernal (*Anthoxanthum odratum*)	6	71,000	Apr., May
Tall Oat Grass (*Arrhenatherum elatius*)	7	21,000	May, Jul.
Orchard Grass (*Dactylis glomerata*)	12	21,000	May, Jun.
Tall Fescue (*Festuca elatior*)	14	20,500	Jun., Jul.
Sheep's Fescue (*Festuca ovina*)	14	64,000	Jun., Jul.
Meadow Fescue (*Festuca pratens*)	14	26,000	May, Jun.
Italian Rye Grass (*Lolium multiflorum*)	15	27,000	Jun.
Perenial Rye Grass (*Lolium perenne*)	18-30	15,000	Jun.
Timothy (*Phleum pratense*)	44	74,000	Jun., Jul.
Kentucky Blue Grass, June Grass (*Poa pratensis*)	13	243,000	May, Jun.
Rough-stalked Meadow (*Poa trivials*)	15	217,000	Jun., Jul.
Red Clover (*Trifolium pratense*) [p. 514]	64	16,000	May to Sep.
White Clover (*Trifolium repens*)	65	32,000	May to Sep.
Lucern, Alfalfa (*Medicago sativa*)	5	76,000	May, Jun.

blossom the earliest will be in proper condition to cut. [p. 513]

The time of blossoming which I have given you in the Table 1 is the time at which these varieties come into blossom in the Northern states of the United States. Some of them, you will notice, continue to blossom during only one month, [p. 515] the majority blossom during two months, while the clovers continue for four or five months. By consulting this table, you will be able to determine what varieties it is best to select for sowing together that they may blossom at the same time. You will also notice from the table that the number of seeds in an ounce by weigh varies very greatly being all weighing between 15,000 and 425,000. Of course, the greater the number of seeds in an ounce, the fewer pounds of seeds it will be necessary to sow per acre. A very common mixture sown in America is 13 lbs. of redtop, 11 lbs of timothy and 8 lbs of red clover. Where this mixture is sown, the prevailing plant for the first one or two years will be clover. [p. 516] After this, the timothy and redtop will take possession of the ground, the timothy being at first much more abundant than the redtop, but the latter will finally take possession of the greater part of the ground. This is a very good mixture since all of these varieties are excellent fodder crops, but a greater number of varieties will be preferable. Table 2 will be a very good mixture for mowing.

The total mixture is 43 lbs which is an ample allowance. For the first few years clover and orchard grass would occupy most of the land; [p. 517] but the quantity of the other varieties will gradually increase. One variety of grass seed should seldom be sown alone except for the purpose of raising seed. Table 3 gives

Tab. 2			
Redtop	3 lbs	Timothy	11 lbs
Italian Rye Grass	4 〃	Meadow Fescue	2 〃
Perenial Rye Grass	3 〃	Red Clover	8 〃
Orchard Grass	10 〃	White Clover	2 〃

Tab. 3			
Kentucky Blue Grass	10 lbs	Rye Grass	25-35 lbs
Sheep's Fescue	10-14 〃	Timothy	12-25 〃
Orchard Grass	14-20 〃	Sweet-Scented Vernal	6-10 〃
Redtop	8-14 〃		

the quantity per acre of some of the principal varieties. You will notice that in this table, the quantity recommended per acre is not stated definitely. It should vary according to the quality of the seed, condition of the land when it is sown as to fertility and moisture. If the seed is of the best quality and the land is rich, [p. 518] the smallest quantity mentioned if carefully and evenly sown, will be sufficient. If, on the other hand, you are rather doubtful about the seed or the weather or the soil is somewhat dry, you must sow more. [p. 519]

1) Timothy

Timothy is a very excellent variety of grass which perhaps deserves to be mentioned first. It requires a rather rich soil, but on such soil, it grows very luxuriantly, yielding large crops of most excellent hay. Timothy is somewhat coarse and for that reason, it is especially valuable as hay for horses. It is not as well suited for cattle especially young one and sheep although it is very good for them also. When once mown, timothy does not spring up again very quickly, and therefore unless the soil is very rich and the season moist, there is not usually much aftermath. [p. 520]

2) Redtop

Redtop is a very excellent grass which will grow upon the soils on which timothy would not thrive well. It, of course, grows more luxuriantly on rich soils. This grass does not as quickly gain possession of ground as many other varieties, neither will it yield as large as some other kinds, but it is a grass which makes hay of the best quality. It is finer than timothy and therefore better adapted to young cattle and sheep. Like timothy, it does not spring up quickly after being mown and hence does not produce much aftermath.

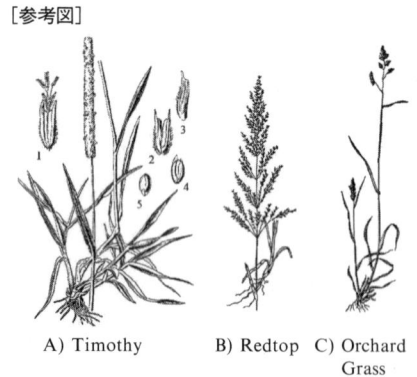

［参考図］

A) Timothy　　B) Redtop　C) Orchard Grass

Both A) and B), from Brooks, *Agriculture*

3) Orchard Grass

Orchard grass is a variety which quickly gains possession [p. 521] of the ground and which grows with considerable luxuriance producing on good soils very large crops. It is somewhat coarse in its nature and has the habit of growing in turfs. It should, therefore, never be sown alone except for the purpose of raising seed. The hay made from this grass is not generally regarded as being of as good quality as timothy or redtop, but if cut early and well cured, it is very good. After being mown, it springs up quite quickly, and hence furnishes a large amount of aftermath. As this grass blossoms early from May to June, it is particularly well suited for sowing with clover, and also since it blossoms early reaching a proper condition for cutting from the middle to the last of June—it is well suited for growth in this locality since June seems [p. 522] to be a very favorable month for the making hay.

4) Kentucky Blue Grass (June Grass)

Kentucky blue grass is a variety which was found growing wild in the western sections of the United States, particularly in Kentucky and in those regions, it is regarded as being the best of all grasses. It is a grass somewhat finer than timothy, but it produces on suitable soils very large crops. It takes some time to gain full possession of the land. It has the peculiar property of continuing to send up an abundance of green leaves throughout the entire season, and it therefore a grass particularly adapted to pastures. It is as a pasture grass chiefly that it is so much in the west of the United States. [p. 523] It also makes hay of very good quality.

5) Italian Rye Grass

Italian rye grass is a rather coarse variety which quickly takes possession of the ground, and grows with very great luxuriance. It allowed to remain standing until the seed is formed, it becomes very hard and wiry in its nature, not making a good hay. If, however, it is cut early, it makes very good hay. After being cut, it springs up quickly, thus furnishing a large amount of aftermath. It has many green leaves in proportion to the number of flowering stalks, and on this account and because it springs up early and grows with great luxuriance, it is a valuable pasture grass. [p. 524]

6) Perennial Rye Grass

Perennial rye grass is a variety very similar in its nature to the one last described, and it is valuable for about the same purposes. The quantity of the nutritive material in the rye grasses when green is not nearly so great as that in timothy. When made into hay, the difference is not so great, but particular care must be taken to cut the rye grass before the seed is formed. After at this stage, they very quickly pass into a condition in which they are hard and wiry and very difficult of digestion.

7) Tall Oat Grass

Tall oat grass is a variety having a habit of growing quite similar to that of the rye grasses. [p. 525] Like them, it is very vigorous and like them, it furnishes a very large quantity of green leaves. It is therefore, usually regarded as of more value for pasture than for mowing.

8) Meadow Foxtail

Meadow foxtail is a variety of grass which is very similar in its appearance to timothy. The head is, however, often softer, has a more wooly appearance than that of timothy. It also blossoms very much earlier, namely in May while timothy does not blossom until the last of June or July. There is, therefore, no difficulty in distinguishing the one from the other. The nutritive qualities of meadow foxtail are not so great as those of timothy. It sends up numerous green leaves [p. 526] and for this reason and also because it grows so early in spring, it is valuable as a pasture grass, being esteemed very highly in England. It takes in several years to gain full possession of the ground.

9) Rough-stalked Meadow

Rough-stalked meadow is a variety of grass having considerable value both for pastures and mowing. It is somewhat like Kentucky blue grass in general appearance and nature and like it is more valuable for pasture than for mowing. [p. 527]

10) Tall Fescue

Tall fescue is a variety of grass which is somewhat fine and is valuable as a mixture especially in pastures.

11) Sheep's Fescue

Sheep's fescue is a variety of grass somewhat like tall fescue, but which does not grow with so great luxuriance. It has the habit of growing in turfs and for this reason and also because it is rather a small grass, it should not constitute any considerable portion of a mixture for mowing land. It is exceedingly relished by sheep and it has an abundance of fine green leaves, and it may, therefore, be used abundantly in a mixture of pasture grasses for sheep.

12) Meadow Fescue

Meadow fescue resembles tall fescue considerably both in its nature and uses. It is, however, probably of a greater value [p. 528] in a mixture for mowing land than the tall fescue.

13) Sweet-Scented Vernal

This variety of grass is fine, begins to grow very early in spring and continues to furnish a liberal supply of green leaves throughout the season. Its nutritive qualities are not very great, but it is the only kind of grasses which has a decided odor. It is believed that dairy products from cows fed upon this grass have a finer aroma and flavor than those from cows fed upon any other variety of grass. Though it constitutes but a small portion of a mixture of grasses, yet it will import its peculiarly agreeable odor to the whole when made into hay and therefore it may well be used in small quantity in a mixture of seeds [p. 529] for permanent mowing. It should also be put in a mixture of seeds for pasture.

14) Red Clover

Red clover is a plant belonging to the natural order *Leguminous*, and as a fodder crop, it perhaps ranks second to none. It quickly takes possession of the ground, sends its root deep down into the soil and grows on rich soils furnishing large amount of fodder per acre. On poor soils, it cannot be grown with profit.

It blossoms early and after being cut it springs up very rapidly almost always furnishing two good crops per year and very frequently three even in this latitude. Its nutritive value [p. 530] is great and it is very much liked by all kinds of animals. It is believed to have the power of taking much nitrogen from the air. Clover, however, on account of the large size and succulence of its stems is cured only with considerable difficulty. The climate of Hokkaido at the time when clover must be cut is usually quite moist, so much so that is cured with great

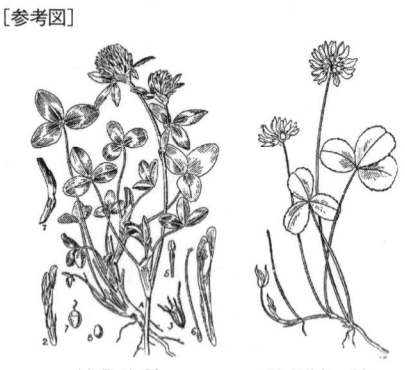

[参考図]

A) Red Clover   B) White Clover
Both A) and B), from Brooks, *Agriculture*

difficulty. Owing to the great value of clover as a fodder crop, it would doubtless be well to add some clover seed to a mixture of seeds for permanent mowing as well as for pastures.

15) White Clover

White clover is one which never grows as large as red clover. Like red clover, [p. 531] however, it blossoms early and springs up quickly after being cut or eaten off. The quantity produced per acre will never be so large as the quantity produced by many grasses, but while clover is very nutritious and is exceedingly well liked by all animals. It should, therefore, be sown in permanent grass fields and in pastures, much more so than red clover.

16) Alfalfa

Alfalfa is a plant also belonging to the *Leguminous* which is extensively cultivated in some regions as a fodder crop. It is somewhat like clover in appearance. Like clover, it sends its root to a great depth. Like it, it springs up quickly after being cut or eaten off. [p. 532] But unlike it, it is not perfectly hardy. It is doubtful whether it can be successfully cultivated in Hokkaido. As the alfalfa sends its root to such a great depth, it has the greater power to withstand the injurious effects of long continued dry weather than most other fodder crops.

17) Alsike Clover

This variety of clover is perhaps a hybrid between the white and red clover as it partakes of the characteristics of each. The shape of the blossoms is like the white and the color of the blossoms intermediate between the white and red. This clover is not so extensively [p. 533] cultivated as either the white or the red. This may be due to the fact that it is comparatively new though perhaps also to the fact that farmers have not found it so well adapted for their use. I consider it at least worthy of trial in Hokkaido. [p. 534]

6-5. Management of Permanent Mowing Land

So land except that which is occasionally overflowed by the water of some river or stream can long continue to bear large crops of grass without manure. The quantity of manure necessary, depends upon the character of the soil and the time at which the grass is cut. Those soils which are naturally adapted to grass; that is, which are absorptive and retentive of moisture, require less manure than those which are of dried nature. Your rule should be to apply manure whenever the crop of grass is so small as not to produce at least two tons of hay per acre. The manner of application should either be well rotted composts or some mineral manure specially adapted [p. 535] to the needs of the soil. Ashes naturally give very good results if the soil is not too wet. If under-composed manure is applied, there will be a loss in decomposition since from the nature of the case, it is impossible to mix manure, applied to grass lands, with the soil, it must lie upon the surface of the land. Manure applied to grass lands should always be fine, as coarse manure lying upon the surface of the ground will interfere with the working of harvesting implements and also be likely to be raked with the hay. The time manure should be applied to grass land is a matter of considerable importance.

If your farm is situated in a place, where the winters are very cold, and where it is not likely [p. 536] to be much snow, the early fall will be the best possible time for the application of manure to grass lands, since if applied at this time it will stimulate the grass to make a growth, which, remaining upon the land during the winter, will protect the roots of grass.

If situated in a country where there is always an abundance of snow, this protection will not be as important and therefore, you may apply manure to the best advantage whenever you can do it most cheaply, and with the least injury to the surface of the land.

In a region like Sapporo any time during the autumn or winter months will be suitable for top-dressing mowing lands. Manure should never be applied to grass lands late in spring, if it is possible to do the work at any other time. [p. 537] The ground is not infrequently rather soft at that season and driving upon it with heavily loaded wagons will make the surface uneven. Then, also, manure applied at this season does not, unless it is very fine, become well incorporated with the soil. Much of it is likely to be dried up and great lump of dry manure will be mixed with the hay. Manure should never be drawn upon grass lands and left in heaps for any great length of time.

Permanent mowing fields some times become "turf-bound" as it is called; that is, they become so full of grass roots and so solid that grass cannot grow well. A soil having clay in it will get into this condition much more quickly than those destitute of it. Mowing fields, when they become "turf bound" should be ploughed and reseeded; [p. 538] and if the object in view is to make them produce grass again as quickly as possible, the best time for the performance of the work is the autumn. Lands which can be irrigated, will produce large crops of grass without other manure, and whenever possible, mowing fields should be irrigated.

1) Time and Manner of Harvesting the Hay Crop

There are different opinions in regard to the proper time of cutting grass for hay. At different periods of their growth the grasses have very different elements in them; and we wish to cut them when they contain the greatest amount of digestible, nutritive material. As the seeds of grasses are not ordinarily digested, the best time for cutting grasses [p. 539] is just as they are coming into blossom. The time of cutting, however, should depend somewhat upon the kind of stock to which the hay is to be fed. For milk cows and young or fattening animals, it should be cut early, but for working cattle and horses, later. Good hay should have as much as possible of the sweet odor and oil of the grass in it. To insure this, it must have only just as much sunshine to keep it from moulding, and after it is cut, no rain or dew should be allowed to fall upon it. The amount of drying necessary depends much upon the condition of the grass when it is cut. If it is green and succulent, it will need more drying than if it is older. The number of days necessary will depend much upon the amount of sunshine, the humidity of the air and the direction of the wind. In this climate, it is ordinarily necessary [p. 540] to dry hay two days at least and not infrequently three days are necessary.

The manner and time of doing the work should be as follows: (1) Grass should be cut in the morning of a good day after the dew is off. The best method of mowing is, of course, by the use of a mowing machine. After the grass has been dried about two hours, it should be. (2) Turned with a hay tedder and if the day is a remarkably good one, or if the grass did not need much drying, it will dry sufficiently to be carried into the barn by two or three o'clock in the afternoon.

In far the greater number of cases, however, in Hokkaido, it will need another day's drying. At about four o'clock, therefore, it should be raked with the horse rake and made into cocks, which, if the weather looks as though it might rain soon should be [p. 541] carefully made and covered with hay caps. The following day, as soon as the dew has dried off, the hay should be spread more or less carefully according as it needs more or less drying. After about an hour and a half or two, it should be turned with a hay tedder, and if it needs a great deal of drying, it may be turned again after about another hour. In most cases, if the weather is favorable and the grass not too green, the hay will be dry enough by two or three o'clock on the second day, when it should be carried into the barn as quickly as possible. After about five o'clock, the air usually becomes somewhat damp, therefore it is better to get in hay early in the afternoon. If the weather is somewhat unfavorable and the grass rather green when cut, it may be necessary to dry another day, in which case, it should be raked up and made into cocks as described for the first day and [p. 542] spread and treated on the third as has been described for the second day. In almost all cases, three days' drying will be enough. The directions I have given you will, of course, be modified according to circumstances. We are entirely dependent upon the weather in making hay. It should be your rule never to allow any hay to get wet, and you should remember that hay which has been dried considerably is much more injured by wetting than that which is comparatively green.

You must also remember that hay which has been wet, always requires a great deal of subsequent drying. Mowing grass late in the afternoon, is a practice which has some advantages. Mown at this time, it will not wilt sufficiently before night to be injured by the dew, and this, falling upon the surface of it, quickly [p. 543] dries off the next morning, and if the grass is well tended the day after it is cut, it will have dried much more than grass mown after the dew is off.

2) Curing Clover

Well cured clover is the most excellent hay; but if cured in ordinary way; that is by exposure to the sunshine with frequent turnings, many of the leaves and blossoms will be broken off and lost and you will have left little but dry hard stalks. Clover, after it is cut, is injured more by the rain than most of the grass, and if it is cut green as it should be, it will need more drying than most grasses. The method of drying, however, [p. 544] should be different. It should be cut and turned the first day as directed for the grasses. It should then be made into cocks which should be covered with the hay caps. If the weather is fine, it should be allowed to stand in the cocks until cured, these cocks being simply turned over and spread very slightly on the day in which you are going to carry the hay to the barn. This method of curing is the best for clover, wherever there is not too much rainy weather, but if the climate is very moist and rains, frequent, the clover will be likely to heat very much in the cock and would, therefore, it may be the best policy to manage clover as directed for the grasses with this exception that after it becomes nearly dry, it should not be turned with the hay tedder nor handled roughly as the leaves and blossoms [p. 545] will be broken off by such that treatment.

6-6. Implement Necessary for Harvesting Hay

Every hay maker who has any considerable amount of grass to cut, should have a mowing machine, horse-rake and hay-tedder. Mowing machines are made for use with either one or two horses. The best mowing machine, in my opinion, is the Buckeye. Wood's mowers are also very good. Horse rakes are always drawn by one horse, and the best are those which are mounted upon wheels and on which the driver rides and works the implement. Among these, Taylor's is one of the best. The Lock Joints and [p. 546] the Bay State are also good rakes. The revolving horse rake is a cheap wooden implement which does very good work and since such rakes can be easily made here they may prove well adapted for use by common farmers. The American Hay Tedder is one of the best which have been made. It is usually operated by one horse. In a country where labor is high, if the farmer has much hay to secure, it may be profitable to have a hay loader.

It is an implement which will take hay from the windrows, and carry it on to a wagon. A horse fork is used in unloading hay in the barn, and is exceedingly useful when the hay is to be lifted to a great height. There are many good forks among which I know Gardener's to be one which does good work. I regard the machines which [p. 547] I have mentioned as ranking in importance in the following order:— Mowing machine, Horse Rake, Hay Tedder, Horse Fork, Hay Loader. Besides these machines, every farmer must, of course, have hand forks of various sizes, hand rakes, convenient wagons or carts for the transportation of the hay. A very useful hand rake is one which is called the "Loafer" or "Hand Drag Rake". The wagons or carts used in carrying the hay to the barn, should be such that you can make a long wide load, since it is much more convenient to put a large amount of hay into such a load than to a high one. [p. 548]

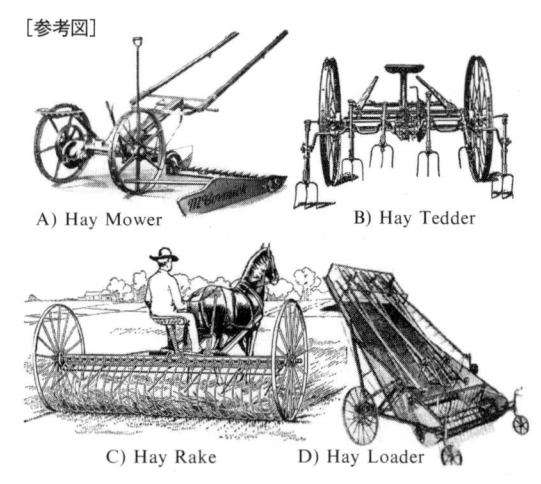

[参考図]
A) Hay Mower   B) Hay Tedder
C) Hay Rake   D) Hay Loader

1) Height at which Grasses should be Cut

It is not an uncommon custom to cut grass as near to the ground as possible. This is wrong, as a certain equilibrium should be kept up between the roots and the leaves. If all the leaves are cut off, it injures the plants, and therefore, it is for the general interest of the farmer to cut his grass as a little height from the ground. About three inches is usually the best height.

2) Storing Hay

There are two general ways in which hay is stored up, viz., in the barn and in stacks. [p. 549] All hay should be put into barns if possible, but hay, in a well-made stack will keep very well, although there is always some hay upon the surface of the stack which will be much injured. A circular stack is the best form, since in it, less surface is exposed in proportion to the amount of hay than in stacks of any other form. In building a stack, it should be made rather small at the bottom and should gradually grow larger as it rises from the ground until about the center when it should be made smaller to the top. As the stack is made, the hay should be trod down hard and should always be the highest in the center. After the stack is completed it should be well raked down, and if you wish to keep the hay [p. 550] in the best possible manner, the stack may be thatched with straw or some kind of coarse grass. In storing hay, in the barn, you should get in as much as possible at one time and tread it down solid.

3) Salting Hay

It is a common practice to salt hay when it is put into the mows, especially if it is a little too green. Doubtless, this is sometimes a good practice, but hay is often much injured by too much salting. As a general rule, therefore, do not salt hay. But, if at some time, you have a small quantity of hay, which you think is not dried quite enough, it may be well to use a little salt. A peck (1/4 bu.) of salt to one ton of hay would be sufficient, [p. 551] and it should be scattered uniformly throughout the mass. After hay has been put into the barn, it is usually found best to keep the doors of the barn closed as much as possible, since the access of the external air to the outside of the mows, will prove more injurious than beneficial. It is, of course, more important that the doors be kept closed during rainy weather and at night than during pleasant sunny days. [p. 552]

6-7. Pasture Land and its Management

On a general farm, pasture is of very great importance, since as the practice of nearly all farmers indicates, only in a pasture can farm stock be profitably kept through the summer. Pasture should therefore be well cared for, and should have a warm, rich and strong soil. The soil for pasture should be retentive of moisture, but not wet and should be one which will produce sweet nutrition grass. Such grass will not grow upon a soil which is cold and wet. In short, for pasture, we need a soil similar to that described as being well adapted for permanent mowing. It is essential that in a pasture, there shall be plenty of good, pure running water, since animals will thrive much better when they have access to such water than in cases where the water is poor or the amount insufficient. [p. 553] Indeed, for milk cows, it is important in order to insure the health of the people using milk that the water be pure. Very malignant fevers and in some cases other diseases have been known to originate from the use of milk from cows that had been drinking impure water. The surface of a pasture may be rough, rocky or very hilly, provided it produces sweet grass. In fact, we should select that portion of our land for a pasture which is not very well adapted for cultivation, if its soil is such as I have described. If the soil of a pasture is comparatively new and rich it should be the aim in its management it so. A good pasture should be able to support at least one full grown ox and a sheep per acre. The two leading substances taken from the pasture are potash and phosphate of lime, there being also, a large amount of nitrogen taken. [p. 554] If the animals are allowed to remain on the land both day and night, and if their droppings are frequently carefully spread, a pasture will remain fertile for a considerable length of time. The length of this time, however, would depend upon the kind of animals pastured on it. Full grown animals would remove less from the soil than young growing one. Horned cattle and horses would exhaust the pastures poorer than sheep. The droppings of sheep being fine and usually spread over the land, are of much greater use than those of horses or cattle, which are deposited in bunches. Whatever is carried away from pastures by the animals fed on them, should be returned in quantity and kind.

1) Pasture of Hokkaido

In a pasture for young growing [p. 555] animals or horses, it is not so important that the grasses be sweet and succulent, as it is in a pasture for milk cows or for sheep. Most of the natural grasses of Hokkaido, at least, so far as I have seen, are coarse and rank in their growth. In order to the attainment of the highest success, therefore, in the production of butter, cheese, milk, wool or mutton, it will be necessary in most cases to introduce foreign grasses. Hence, it is important to consider the best method of attaining this end. There are in this province several different classes of land which you may wish to convert into pastures. I will speak only of three different sorts, viz., —Woodlands, Open lands (which are now producing wild grasses) and Hills. We may consider the hills as belonging to two classes, viz., —those, covered with wood, and those, the greater portion of which is covered with a thick growth of bamboo. [p. 556]

2) Conversion of Woodland into Pasture

The first step will, of course, be the cutting down of most of the trees, a few only being left for shade. After cutting the trees and carrying away the timber and wood, all the bush remaining upon the land should be carefully burned. In burning this bush, it should be your object to so manage the fire as to burn over nearly all the land, as by this means, you will be able to kill much of the under-grows. In some forests, the ground under the trees is covered with a thick growth of shrubs, while in others the under-growth is very little. In the latter case, it may not be necessary to plough the land. It may simply be pulverized as much as possible by the use of the harrow and then grass seed sown. In far the greater number of cases, [p. 557] however, there will be so many wild plants growing under the trees that it will be best to plough the land and cultivate it two or three (or if long enough one) years for the purpose of killing these plants and then sowing the grass seeds. With regard to the stumps of trees, it will be found quite expensive to dig up the large ones immediately after cutting trees, and therefore, since you do not wish to cultivate the land for any great length of time, I would not advice the expenditure of a large amount of money for the purpose of removing them. All the smaller ones, and those which can be dug up without much labor should be removed when you first break up the land. It may be wiser in most to cases, to allow the larger ones to remain until they are partially decomposed when [p. 558] they can be taken up with much less labor. The time for sowing grass seeds, and the manner of doing the work in pasture land, should be the same as those recommended for permanent mowing with this exception that it is not necessary nor indeed profitable to pay so much attention to make the land smooth.

For pastures, it is more important than for mowing that a large variety of seeds be sown, and while in mowing, the varieties selected should be such as blossom at the same time, for pastures, on the other hand, you should select such varieties as blossom at different times in order to secure a plentiful supply of green leaves throughout the season. A very good mixture of grass seeds for a pasture would shown in Table 4. This makes in all forty seven pounds which is a very liberal allowance, and if any kinds are to be left out, they should be the Wood Meadow Grass, a portion of the oat grass and a portion of the Perenial Rye Grass. Wood Meadow Grass is a variety which is particularly [p. 560] adapted to moist soils: and if the soil of a pasture were wet, I would sow more of this variety. It blossoms from June to July. The Yellow Oat Grass is somewhat similar to the kind of oat grass which I have described. The time of blossoming is July. Some of the varieties of

Tab. 4

Meadow Foxtail	2 lbs	Redtop	2 lbs
Tall Fescue	1 lbs	Rough Stalked Meadow	2 lbs
Wood Meadow (Poa memoralis)	3 lbs	Yellow Oat Grass (Avena flavescens)	2 lbs
Italian Rye Grass	3 lbs	Tall Oat Grass	3 lbs
Timothy	3 lbs	Red Clover	2 lbs
Orchard Grass	6 lbs	White Clover	5 lbs
Meadow Fescue	2 lbs	Sweet Scented Vernal	3 lbs
June Grass	4 lbs		
Perenial Rye Grass	4 lbs	Total	47 lbs

grass are such as grow well in shaded places, and as pastures should always have some trees in them, it is important that some of the grass sown, be such as will grow well under trees.

3) Open Lands on which Wild Plants are Growing

To change such lands into good pastures, it will only be necessary to plough and cultivate [p. 561] the land one or two years, and then sow grass seeds as recommended in the case of woodlands.

4) Hills covered with Forests

In converting such lands into pastures, the course pursued, must be similar to that recommended in the case of ordinary woodlands, although since it is more difficult to harrow and cultivate such lands, I will try to avoid doing so if possible and would sow grass seeds after having simply cleared the land, burned the bush and harrowed it in all cases where I thought the seeds will take root and grow. It may be possible in some cases, by mowing and burning whatever grows on the land for several years in succession, to entirely kill the wild plants [p. 562] growing upon it, after which the surface of the land may be harrowed, and seeds sown.

5) Hills covered with Bamboo

With regard to the best treatment of hills covered with bamboo, I do not feel that I am qualified to positively, since I have had no actual experience with the plant, but it is my opinion that successive cutting down and burnings, would, in the course of a few years, very much weaken, if not entirely kill bamboo, after which the land may be immediately ploughed, cultivated a short time if necessary, and seeds sown, or the roots may be allowed to decay, the land harrowed and seeds sown without ploughing. [p. 563]

6) Management of Pastures

Animals should not be put into a pasture in the spring until the grass has got a vigorous start, since if they commence to eat the grass when it is small, they will check the growth, and the grass in the pasture will be short throughout the season, in which case you will not be able to pasture as many animals upon it as you could have done, had you waited until the grass had got a vigorous start. In Massachusetts, the legal time of turning animals into pastures is the twentieth of May, and here the time should be about the same as the season is similar. So many animals should not be kept in a pasture at any time as eat the grass down close to the roots, because if this is done, the vigor of growth will be much lessened, [p. 564] and in extreme cases, much of the grass killed. Animals should be taken from a pasture early enough in the autumn that there may be a good growth of grass left upon the land to protect the roots during the winter. The legal time for removing animals from the pasture in Massachusetts is the twentieth of November, but this is much later than is usually wise to keep them there. It is well to top-dress pastures from time to time with such manures as plaster, phosphate of lime, these substances being found very useful in maintaining the fertility of the soil. In deciding at what season to top-dress pastures, you must consider the same points as those spoken of in the management of permanent mowing. [p. 565]

6-8. Management of Tillage Lands

Warm sandy loams, if their surface is adapted to it, should be selected for tillage. If the surface slopes so much that heavy rains will be likely to wash the soil away, such lands should not be tilled, but kept permanently in grass, since grass will prevent washing. Clayey loams are also suitable for tillage. In the management of tillage lands, it will usually be found wisest to raise grass upon them as a rotation crop. Having selected our tillage land, we should plan a regular system of cultivation and rotation which are adapted to our means and which may last a life-time. In laying down this system, [p. 566] we must carefully consider the principles of crop rotation and the kind of business which will be found most profitable to us. The particular system pursued must, of course, vary with individual cases. With regard to the importance rotating crops, I have perhaps already said enough, and you understand the principles which should guide you in deciding about a system of rotation. Knowing these principles you can decide in each individual case what system of rotation will be best adapted for the conditions under which you may be placed. You should remember, however, that rotation alone cannot sustain the fertility of the land, and that plants will consume plant food faster than the forces of nature will in most cases, render it available. You must remember also that land will certainly [p. 567] deteriorate if all crops raised upon it are carried away and nothing returned. With regard to the folly of allowing the land to grow poorer, I have already spoken. You should aim to so manage your tillage lands that they will constantly be growing richer rather than poorer. This is the truest wisdom in the end, although it may at first seem to you as if you might gain more money by carrying away crops and returning nothing. You must remember that a farm is not valuable simply because it contains a certain number of acres, or a certain amount of earths, but only because of the plant food which it contains. In any system of rotation, some crop should be introduced which will give the land comparative rest and shade it. [p. 568] Such a crop is called an "Ameliorating" crop. A good one for this purpose is grass or clover, which may be either removed or pastured.

6-9. Forests on a General Farm

As already stated, a general farm should have sufficient woodland to furnish fuel and material to keep the farm buildings and implements in repair. In this latitude, a forest is also useful in furnishing protection against cold winds. The first point to be considered in selecting a place for a forest is the nature of the land. Wood-land should be located on the roughest and most inaccessible portion of the farm. In this vicinity, there is at present in most cases, [p. 569] sufficient woodland growing. In clearing up woodland for tillage, mowing and pasture, you should keep in view the two points which I have mentioned, namely, to leave the forest on that side of the farm where it will be most useful in furnishing protection against cold winds, and second, to leave it on the roughest and most inaccessible portion of the farm. If both of these points can be secured, it will be well to do so. If, however, that portion of which is the roughest and most unfit for cultivation, is not located on that side of the farm which will most effectually oppose the cold winds, I will leave the woodland in this rough place rather than on the north side of the farm. If all the land of the farm is of about equal value [p. 570] for cultivation I would, of course, leave the woodland on the northern side.

If your farm is located in a region where there are no woodlands or where forests are present in insufficient quantity, trees should be set out. In selecting a place for setting out trees, you should keep the same principles in view as those spoken of as being important in clearing up land. The varieties of trees most valuable for woodlands for furnishing a large supply of timber or wood, are such as grow rapidly and straight without many branches. Those most highly prized in America, are the White Ash, White Pine, Scotch Larch and European Pine. Of these, the white ash thrives best on rather moist soil and furnishes, when quite young, a good quantity of the very best timber. [p. 571] The timber of the ash is very strong and is useful for the manufacture of wheels, wagons, handles of implements, the wooden parts of machines and in fact in any place where strength and lightness are desirable. The pine will thrive quite well on both dry and moist soils, and it grows with very great rapidity, furnishing a large amount of valuable timber. The timber is used most extensively in the construction of houses, it being very light, quite strong, durable and easily worked. Pine timber is also extensively used in making boxes, trunks of all sorts, and it is also used for the masts of ships. The wood of pine is easily kindled when dry, because of the pitch which it contains, and it is therefore used extensively as kindling wood. [p. 572] It is also used whenever a quick and hot fire is desirable. The scotch larch grows well upon the soils which are dry and as it is extensively hardy it will thrive on hills of considerable elevation. Its timber is valuable for building purposes, being somewhat durable than pine, but not so strong, and tough as ash. Larch is also especially valuable for use incases where it will be subjected to alternate moisture and dryness. It is, therefore, extensively used for piles in the construction of wharves. European pine will thrive in many different kinds of soils and in very exposed situations such as the tops of high hills. It grows rapidly and furnishes timber which is a little harder and more compact than the white pine and also more durable. [p. 573] It is used for much the same purposes as the white pine, but as it is more compact it can resist a greater pressure, and is therefore more valuable for framing purposes than the former.

If a forest is to be set out, the trees should first be raised in a nursery. In a nursery the seed is planted quite thickly in rows which should be about three feet apart. When the trees have reached the age of two or three years, they should be transplanted to the place where they are to stand. In the performance of this work, it is best to exercise considerable care. The roots should be left as entire as possible, and the holes in which the trees are to be set should be made sufficiently large that when transplanted, the roots may be placed in their natural position. The trees may be planted quite closely together, [p. 574] and as they increase in size, a portion of them may be cut and removed to make room for the rest. These trees will furnish a supply of wood which will be valuable and will partially repay the expense of setting out trees. It is characteristic of all trees that when growing close together, they send out comparatively few branches and grow taller than they otherwise would. Trees sending out few branches are more valuable for timber than those having many branches and therefore, it is best to leave the trees quite close together.

It will not generally be profitable to cultivate forest land. The only care necessary is to prevent animals from getting among the young trees. [p. 575] Such animals as oxen, sheep, or horses, would do a great deal of damage by eating the young shoots and leaves. Horned cattle, also, by rubbing against the trees would be likely to break down a great many of them. After the forest trees have attained considerable size, it will be found profitable to go over the land at least once a year, and cut all those trees which are beginning to die. Such trees if allowed to remain soon fall down and decay, and the wood is lost. It may also be sometimes profitable, if the trees are very much branched, to cut off those branches which can be readily reached for fuel. There are a few additional varieties of trees which I wish to mention as being adapted to special uses, although they are not of as much general value as those kinds [p. 576] which I have already spoken of.

(1) White Cedar: This tree thrives in wet swampy places and is very valuable in making fences, shingles, tubs and pails, and boats. The tree has few branches and grows of quite uniform size to considerable height. It splits very easily and is durable, These characteristics making it valuable for fences.

(2) Red Cedar: This variety is of very slow and straggling growth and it is valuable only on account of its great durability. When used in any place where it will be subjected to alternate moisture and dryness, it will last for very many years. It is therefore, of great value for posts which are to be set into ground.

(3) Locust: This is another slow growing variety and its chief value is because of its great durability, it being useful for the same [p. 577] purpose as the kind last mentioned. [p. 578]

6-10. Orchard on a General Farm

A general farmer should raise at least fruit enough of all the varieties that the latitude in which his farm is located will produce to supply his own wants, and if there is a market for fruit in his vicinity, he will also find it profitable to raise fruit for sale. Ripe fruit is one of the most healthy articles of food, and it is much cheaper as well as more agreeable for the farmer to raise fruit than to pay doctor's bills. Fruit tree should be set in one place devoted exclusively to their use. It is not advisable to scatter them along by roadsides or by the boundaries of the farm, or to plant them in fields where other crops are to be raised. One reason why fruit trees should not be [p. 579] raised in fields devoted to other crops is that these crops will make use of the plant food which the trees themselves need. Another reason is that the trees will be the likely to be injured by the teams and implements used in cultivating other crops. A location for an orchard should be sheltered and on well drained land. The soil should be rich and in fine mechanical condition. It should not contain too large an amount of nitrogen; but there should be an abundance of mineral manures. A large proportion of nitrogen causes an excessively rapid growth which does not become well ripened before winter and is therefore likely to be killed. If a location is suitable in other ways, but is not sheltered, it should not be rejected; for an artificial shelter [p. 580] can be made at small cost.

1) Apples

In all western temperate climates, the apple is regarded as one of the nicest and most valuable fruits, and it is very extensively cultivated. Its uses are various. It is cooked in many different ways. It is eaten uncooked, and it is manufactured into cider, cider brandy and vinegar. The apple tree is more hardly than most fruit trees, and in this latitude can withstand almost any exposure to weather. For this reason, shelter for apple-trees is not as important as for other trees. Hence they should be set out on the north or west side of the piece of land which you propose to devote to fruit culture. [p. 581] While the trees of an orchard are young, it is good economy to raise crops upon the land-care being taken to return to it the elements of fertility, removed in the crops, and not to injured the trees by the implements used in cultivation. Low growing crops should be selected for cultivation in an orchard since they will not shade the trees. After the trees begin to furnish fruit, the soil should be thoroughly cultivated but no crops should be raised and removed. The trees for an orchard must be raised from seeds in a nursery and when one or two years old, should be grafted or budded. It is necessary to graft or bud in order that you may be certain as to the variety of fruit the trees will produce. In two or three years after having been grafted or budded, the trees will be ready for transplanting to the orchard [p. 582] where they should be set at least twenty-five or thirty feet apart each way. Each tree of one row should be so placed as to be opposite a space in the next row. In digging up tree, care should be taken to dig as much of the roots as possible; some of the larger roots must usually be cut off. But so far as possible all the fibrous roots should be left uninjured, they being more important in gathering nutrient than large roots. When transplanted, the end of all roots which have been cut off in digging up should be cut off with a smooth oblique cut, since such a wound will heal much more quickly than a lacerated irregular one. When transplanted the trees should be well pruned in order to maintain the balance between the amount of water that can be evaporated from the leaves and the amount [p. 583] that can be collected by the roots. This pruning should be down with a sharp knife in such a manner as to give the head of the tree an open symmetrical shape. Trees may be transplanted either in early spring before the buds expand or in autumn after the leaves have fallen. The hole in which you intend to plant the tree should be made large and good earth should be placed in the bottom of it. The roots should not be all crowded together, but should be placed as nearly as possible in their natural position. After having been transplanted, the trees should be carefully attended to and any branches growing in improper places should be removed when very small. It is much better to give young trees this constant attention and cut off branches growing in wrong places [p. 584] or those taking a wrong direction when they are small than to allow them to grow large and then to cut them off. The soil of the orchard should be carefully cultivated as in the case of ordinary crops.

The varieties of apple are very numerous and no definite rule can be given as to the kinds which should be cultivated in any given place before trial has been made, since those apples which thrive well in one locality may not do so in another. I will, however, give you a list which has been decided to embrace the best varieties for culture in Massachusetts; (1) For summer use: Early Harvest, Red Astranchan, William's Favorite. (2) For autumn use: Foundling, Porter, Golden Pipin, Mother Apple, Hubbardson Nonsuch. (3) For winter use: Rhode Island Greening, Baldwin, Roxbuly Russet. The list of apples recommended for the State of New York is as follows: (1) For winter use: New Town Pipin, Tompkin's Country King, Spitzenburg, Swaar, Pecks Peasant, Lady Apple, Bell Flower, Northern Spy. (2) For summer use: Early Strawberry, [p. 586] Sweet Harvest,

Sops of Wine, White Astrachan, Red Astrachan, Early Harvest. (3) For autumn use: Pipin, Maiden's Brush, Alexander, Beauty of the West, Beauty of Kent, Graverstein, Porter, Golden Pipin, Foundling.

2) Pears

The pear is, as you know, a very excellent fruit and it is usually better liked by most people for eating than the apple. It is not much used in cookery and is therefore not so useful a fruit as the apple. [p. 587] The Tree is not so hardy as the apple, although it will withstand with a fair degree of safety any amount of cold to which it is likely to be subjected in this latitude (Hokkaido or Sapporo). The pair tree needs more shelter than the apple tree, and should, therefore, be set on the south or east side of the apple trees. The best soil is a warm dry sandy loam which is very rich; but the pear will do well on clayey loam if it is well under-drained. Pear-trees should be supplied with a plentiful amount of mineral and nitrogenous manures. Pear-trees may be raised from the seeds and grafted in the same manner as appletrees; but it requires somewhat more skill to do this successfully than in the case of the latter. Pear trees are of two sorts; Standard and Dwarf[1]. [p. 588] Standard pear trees are those which have been grafted upon pear stocks. A dwarf pear tree is one which has been grafted upon a quince stock. Standard pear trees grown to larger size than dwarfs do not begin to produce fruit so early and live much longer. Dwarf pear trees, as their names indicate, are small and begin to produce fruit when quite young, and yield (if well cared for) very abundantly in proportion to their size; but they live comparatively few years. It is a good policy in setting out a pear orchard to plant both standard and dwarf trees, the latter being placed between the former. The dwarf tree will then produce fruit when young, and will be about through bearing by the time the standards begin to produce fruit when the dwarfs should be removed. The standard trees should be set [p. 589] from ten to fifteen feet apart according to size, and the same arrangement of trees in the rows should be followed as in the case of apples. The same rules in regard to digging up, setting out, pruning, and cultivation apply in the case of pears as those given in the case of apples. The varieties of pears are very numerous and are such that good fruit may be had through out the entire year. The varieties well adapted to Massachusetts are as follows: Bartlette, Flemish Beauty, Clapp's Favorite, Brandy Wine, Buffon, Urbanitc, Merriam, Onandaga, Lawrence. For the state of New York, the following varieties are recommended in addition to the list given above; Beurrediel, Dichess, Sheldon, Howel, Maria Louisa, Superfine.

3) Peaches

The peach is a fruit universally liked, but more are eaten uncooked than cooked. The peach tree is less hardy than either the apple or pear; but if grown upon the proper kind of soil and not forced to make too rapid a growth, it may be made to thrive in this province. As the peach is not entirely hardy, a warm sheltered location should be selected for it. The best soil is a warm sandy loam which must be thoroughly drained. [p. 591] The first buds of the peach tree are likely to commence to expand during the warm days in early spring, and as it is usual for severe frost to come after these warm days, the fruit buds will be killed. It is, therefore, considered best to plant peachtrees upon the northern rather than the southern hillsides, because in the former case, the trees will not feel the influence of the warm sun of the early spring. The soil should be manured with mineral manures among which ashes are perhaps the best. Peach-trees can be easily raised. The stones should be planted in autumn and the frosts of winter are likely to crack them. They should, however, be examined in the spring, and if not already cracked, the operation should be carefully performed with a hammer. [p. 592] Peach trees are usually budded, as grafting will not be likely to prove successful. If the trees are planted in suitable soil, they will be large enough for budding in August of the first year. The operations of digging up, transplanting, pruning and so forth should be performed with the same care as in the case of other fruit trees. The distance apart should be about from twelve to fifteen feet. The varieties are numerous, but not as much as those of apples or pears; Among the best are Hale's Early, Early York, Erawford's Late and Yellow Rare Ripe.

4) Quinces

The quince, [p. 593] although not so extensively raised as the fruits of which I have been speaking, is a valuable fruit especially in cookery. The quince is very hardy and thrives best in a gravelly soil. If you have not a lot of such a soil, you must at least select one which is warm and well drained. The quince may be cultivated as a tree or in bush form. They are raised from seeds and may be, although quite frequently they are not grafted. The varieties are not as nearly numerous as in the case of pears, apples and peaches. I will mention only two, the common quince and the orange. Of these, the first has the better flavor; but the second is the better cooking. The rows of quince bushes should be ten feet apart, and the plants should be five feet apart in the rows. [p. 594] Cultivation and management of the quince should be similar to those of other trees.

5) Cherries

The cherry is a fruit valuable both for cooking and for eating uncooked. The trees are very hardy as well as fine shaped and large, and therefore they may be used both as shade trees and fruit producers. If planted in an orchard, the distance apart should be the same as that recommended for apple trees. The best soil is that best adapted for pears. The trees are raised from the stones and usually budded, although they may be grafted. [p. 595] The best varieties are the Common English, Black Tartaria, Black Eagle, Ox Heart, Beauty and Kirkland's. The management necessary to produce the best results is the same as that the previously recommended for other trees.

6) Plums

The plum, although not so easily raised as some other fruits is excellent, and the farmer should endeavor to produce enough for the use of his own family. It is not raised so easily as some other fruits in America for the reason that an insect called the *Curculio* is very injurious to this fruit. This is the especially true of the eastern sections of the United States. Whether the same difficulty will be met with here[2] can be told only after trial. [p. 596] The plumtree in America is also peculiarly subject to a diseased condition which causes the formation of warty excrescences. These are believed by some to be caused by the sting of an insect, but others believe that they are owing to a diseased condition of the juices of the tree. If a tree is affected by these warts, it is best to cut off branches at considerable distance below the affected portion. Unless the spread of the disease is checked by this severe pruning, the whole tree will be affected and will finally die. The portions cut off should be destroyed by burning.

Plum trees are raised from stones and budded in the same manner as cherry or peach trees. The varieties of plums are not as numerous as those of some other fruits. [p. 597] There are two principal classes; namely the Purple and the Green. The Green Gage is one of the best varieties. All that I have said with reference to the management of other trees, applies with equal force in the case of plumtrees.

7) Grapes

The grape is a fruit extensively used both for eating and for manufacture into wine. It is of the production of grapes simply for eating purposes that I shall speak. Many varieties of grapes are cultivated; some of them being exceedingly hardy while others can be cultivated only in tropical climates. There are many varieties which will grow further north than Hokkaido, but these varieties [p. 598] are not usually of so good quality as those which grow further south. In order to cultivate the best varieties which it is possible to produce in this latitude, it is necessary to select as warm a place as can be found. A southern hillside is the best, and if the locality is not already sheltered, artificial shelter should be provided. The best soil is a gravel and one which is rather light and thoroughly drained. Having such a soil, it must be well manured with mineral manures. Nitrogenous manures are to be avoided for the reason that they will cause a large growth of wood which will be likely to be killed by the frosts of winter. Bone phosphates, ashes, potash and magnesium salts, chlorides are excepted, and some times lime are among the best manures. [p. 599] In order to prepare a piece of land for grapes, it should be well manured if not already fertile, and some other crops should be raised upon it for one or two years for the sake of bringing it into good mechanical conditions and killing the seeds of weeds that may be present. Until the vines begin to produce fruit, other crops may be raised upon the land, but you should return those elements of fertility removed in them. Every farmer should raise his own grape vines, as it is a very easy matter to do so. Vines may be propagated either from seed, or by layers or cuttings. They are raised from the seed only when the object is to secure new varieties.

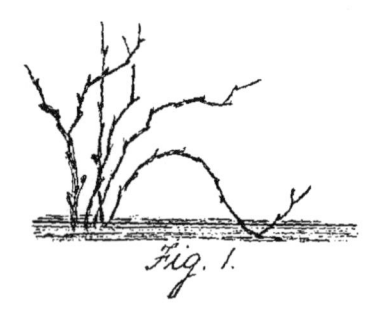

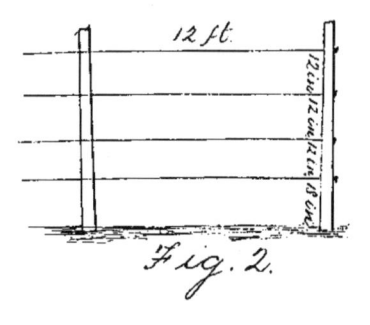

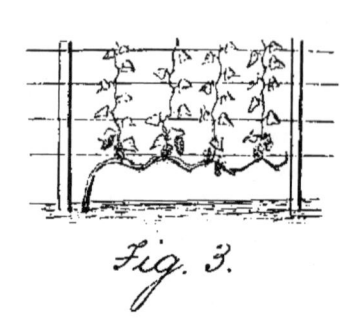

Vines may be propagated by layers in the following manner: [p. 600] A shallow hole is dug near the vine and a portion of its canes are bent down and covered with earth, the end of the cane extending beyond the hole. In a short time, the buried portion will send out roots and as soon as these are well established, the cane may be separated from the parent vine and planted in another place. This is an easy and certain method of propagation, but a far more rapid and almost certain method is propagation by cuttings. The scions should be cut in the autumn and kept in a cool moist place where they will not wither. Wrapped in a piece of moist flannel [p. 601] and kept in a cool cellar, they will suffer no injury. In March, they should be cut into pieces four or five inches long. Upon each piece, there should be one bud at the upper end, and the lower end will send out roots more quickly if it is cut at a place where there is a bud. The top end should be cut square and the bottom oblique. They should then be set obliquely in a box filled with sand, about one inch being left above the sand. These boxes may be kept in any warm light room where the atmosphere is moist; but the sun should not be allowed to shine directly upon them. A green-house is best; but any ordinary room, provided it is kept in a proper condition, will answer. These cuttings should be watered just enough to keep the sand constantly moist but not wet. [p. 602] After the roots are well started, the cuttings should be transplanted into beds out-of-doors. The length of time required to start them will be from four to six weeks. When first put out-of-doors, vines may be planted in rows about two and a half feet apart and eight inches apart in the rows. In the autumn, the young plants should be cut down nearly to the ground, only one or two buds being left. The following spring, the vines may be transplanted to the vineyard. The rows should run north and south; because with such rows, the soil will get more sun than with rows running in any other direction. The vines should be six feet apart in rows which are eight feet apart. In transplanting grape vines, [p. 603] they should be set three or four inches lower than they originally grew, and the same precautions in regard to placing the roots in their natural position *etc.* should be observed as in the case of other fruit-trees. Some cultivators train the vines altogether upon stakes, some using only one, others three; but a better way is to have a trellis. A good one may be made from posts eight feet long set once in six feet with slats of wood or wires fastened to them. It is well either to charr[3] the bottom of these posts or to paint them with coal tar in order to preserve them. These posts should be large enough to give considerable strength; but may be made smaller at the top than at the bottom. Four wires or slats are enough, the first of which should be eighteen inches above the ground and the others each about twelve inches apart. [p. 604] Grape vines should be pruned in the autumn after the leaves have fallen. It is customary the first year after transplanting to cut the vines down nearly to the ground, and unless they are vigorous, this operation may be repeated the following year.

It must be remembered that canes which are two years old never produce fruit; hence all old canes which are not needed to send out new ones, are each year cut away. While the fruit is growing, no new wood should be allowed to grow on the same cane. [p. 605] After the fruit is set, the end of the cane should be pinched off, leaving about two or three buds beyond the last bunch of fruit. At the same time, canes may be allowed to grow from other parts of the plant for the sake of having new wood for the next year.

In picking grapes for market, special care should be taken not to destroy the bloom as this adds very much to the beauty of the fruit, and it will command a better price. In very severe climates, it is good policy to take the vines down from the trellis late in the autumn, and lay it on the ground throwing a little earth on them to keep them down. So doing much lessens the danger of winter killing.

The varieties of grapes are numerous; but I shall mention a few of them which are found to be best [p. 606] for the cultivation in New England —that climate being about the same as this as regards severity. (1) *Concord Grape*: This variety is, everything considered, the best grape for New England, of good quality, quite early, hardy and prolific. There are other varieties which are of better quality, but few of equally good quality are as certain to produce good crops of fruit. (2) *Hartfold Prolific*: This is a good grape which is a little earlier than the concord; but not of quite as good quality. (3) *Isabella*: This is a grape of the very best quality; but in the latitude of Massachusetts, the fruit is likely to be injured by frosts before it is ripe. Unless in very favorable locations it would doubtless fail to ripen here. (4) *Delaware*: [p. 607] This is a grape of the very finest quality; but the vine is a slow grower and poor bearer, and for these reasons this variety is not ordinarily profitable as some others. (5) *Iona*: This grape is of exceedingly fine quality; but the vine is not as hardy and productive as some other sorts. (6) *Israella* and *Agawam*: These varieties were originated by a gentleman by the name of Rogers. They are sometimes called Roger's Hybrids and are designated by numbers. These grapes are of fair quality; but not generally so well liked as the other varieties I have mentioned. They are hardy and quite productive.

8) Strawberries

The strawberry is a desirable fruit [p. 608] on account of both its excellent flavor and being one of the earliest in ripening. A man establishing a home in a new place can get a return for his labor in good fruit in a shorter time by the cultivation of strawberries than that of any other fruit. The best soil for strawberries is a rich warm loam. They do not need very deep culture after they are set out, but the ground should be deeply cultivated before the vines are transplanted. The best manure is well rotted barnyard manure although mineral manures which contain the proper elements may be used with success. The application of undecomposed manures rich in nitrogen would result in a large growth of vines with little fruit. It is very important that lands designed for strawberries [p. 609] are placed in a good condition both physically and chemically, and that the roots of weeds and weed-seeds are killed before the vines are set out. Vines may be transplanted either spring or autumn, but the spring is generally regarded as being the best time. There are two methods of cultivation, namely in hills and in beds. In the first system, the vines are not allowed to spread all the runners that start from them being cut off. If they are to be cultivated in this manner, they should be set about one feet apart in rows about $1\frac{1}{2}$ foot apart. In this case, most of the cultivation must be done by hand. In the second method of culture, the vines are allowed to spread more or less and they may be put one foot apart in rows and three feet apart in furrows. [p. 610] During the early part of the season much of the labor of cultivation may be performed with a horse. The vines, if well cared for in suitable soil will before the end of the season, have spread entirely over the ground. During the latter part of the season, therefore, it may be necessary to do most of the work of cultivation by hand. This culture need not be deep, but should be sufficient in amount to prevent any weeds from growing. In either method of culture, the vines should not be allowed to produce fruit the first year. The second year, they will produce abundantly. If transplanted in the autumn, the work should be done during August or the early part of September, and in this case the vines will produce some fruit the following season. [p. 611] Strawberry vines should be covered late in autumn with some light material such as straw, pine or hemlock boughs or leaves. The object of this covering in cold climates with little snowfall is to protect the vines from severe cold. In countries where snow is abundant, less covering will be needed. Care must be taken in covering the vines not to put so much material upon them as to smother them. A heavy fall of damp snow is likely to smother the vines, and for this reason even in countries where the covering is not needed as a protection from cold, it may be useful in preventing the snow from becoming too compact about the vines. For this purpose, the branches of evergreen trees will be best adapted. [p. 612] This covering should be taken off in early spring, and before the ripening of the fruit, the land should be mulched with such material as sawdust, hay or straw. This keeps the soil moist and also prevents berries from becoming covered with dirt. This mulch should be removed after the berries are harvested and ground thoroughly cultivated. It is a practice with some cultivators not to give the vines any cultivation after the first year and to allow them to produce only one or two crops. Each year, an old bed is ploughed and a new one planted. Such cultivators usually have three beds; one which has just been set out; one which is producing fruit for the first year and one which is producing fruit for the second and last year. Vines for making a new bed [p. 613] should be only one year old and they may be taken in the spring from the bed which is to produce fruit for the first time that year. These vines, you will remember, have entirely covered the ground and in taking up plants it is best to take them all from strips about a foot in width. These strips should be about six feet apart and they serve as paths in which those picking the berries walk.

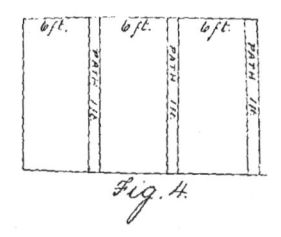

The varieties of strawberries are numerous, and the following are among the best suited for general culture. (1) *Wilson's Albany*: This variety, though not of quite as good quality as some others, is much more productive than most of them. [p. 614] The vines are hardy, the berries very fine looking, and they bear transportation well. For these reasons, this variety is more extensively cultivated in America than any other. It is perhaps the best market variety. (2) *Downer's Prolific* and *Hovey's Seedling*: These are also fine berries. The latter does not develop the proper proportion of male and female blossoms; therefore some other variety berries must be planted with them. A variety that produces an abundance of staminate flowers should be selected. (3) *President Wilder*: This is of very fine quality. (4) *Jucunda* and *Brighton Pine*: These are also good berries.

Strawberries are usually picked by children. [p. 615] If they are to be carried to a distant market, the best way is to leave the hulls on as the berries keep better with them on. If they are to be carried a long distance they should be put in baskets each holding about a quart, and these baskets should be packed in boxes of convenient size which are open at the sides.

9) Blackberries and Raspberries

These berries are very hardy and productive; consequently they can be easily raised. The bines are most easily propagated by dividing the bushes. The vines should be planted according to variety from $2\frac{1}{2}$ to $3\frac{1}{2}$ feet in rows which are from 3 to 5 feet apart. They will very soon begin to produce fruit [p. 616] and it will not usually be best to attempt to raise other crops among them. Good cultivation should be given, but most of the work can be done by means of a one-horse cultivator. The vines have a habit of forming very much wood and it is found to be conducive to fruitfulness to prune them severally. Every autumn, the canes should be cut back, leaving the vines from $2\frac{1}{2}$ to 4 feet in height according to variety, and all the old wood should be cut off. The best soil for these fruits is a warm sandy loam.

Some of the best varieties of blackberries are the Lowton, Imperial and Kitatimny. Raspberries may be divided into three classes, namely the Black, the Red and the White. Of these the Black is perhaps most easily raised, but the Red or the White are by most regarded as of better flavor. [p. 617] The same rules as those given in speaking of strawberries as to marketing are applicable in the case of these berries, since they are soft and somewhat similar in nature.

10) Currants and Gooseberries

These fruits are of much value in Western Countries and they can be easily cultivated. The vines are propagated both from cuttings and by dividing as in the case of blackberries. Almost any good loam is suitable for these berries. They should be planted at distances about the same as those given for the smaller varieties of raspberries. The second year after planting, the vines will produce fruit. The method of pruning [p. 618] is quite similar, except in degree to that of blackberries or raspberries. These vines do not grow as rapidly as blackberries; and therefore, the degree of pruning must be less. There are both red and white currants commonly cultivated; but the red varieties are most liked. There is a variety of black currants also sometimes cultivated, but not so extensively as the others. Gooseberries, when ripe, are of a dark red color and some of the deferent varieties vary very much in size. The English gooseberry is one of the largest and best. [p. 619]

6-11. Farm Road and Fence

1) Farm Road

The farmer should have such road as will enable him to pass readily to all parts of his farm at all seasons of the year. He should make these road as good as his circumstances will allow, but of course, those roads which will be little traveled need not receive as much attention as those which are much traveled. It is good economy to construct good roads. Teams can carry much larger loads upon a good road than across a field where there is no good road. If there is no road, persons driving across the farm will sometimes travel in one place and sometimes in another, and thus the crops will more or less be injured. It is therefore better economy, as far as crops are concerned, to devote a portion of a farm exclusively to road. [p. 620]

(1) Construction of farm roads and bridge: The construction of farm roads and bridge, since they are not generally much used, need not be expensive. Small stream may be bridged in the following way: Large tiles or wooden boxes of sufficient size to carry the water, are placed in the stream and then covered with earth. Lager stream may be bridged as follows: Three pieces of timber of sufficient length to reach across the stream and rest securely upon its banks, are stretched

across at equal distance from each other. Planks are them nailed to these cross timber and the bridge done. [p. 621] Such bridge must afford a channel large enough to carry the maximum amount of water that were flows in the stream. Where stones are abundant, good permanent bridges may be easily made from them. You have only to build abutments of stone and lay across the stream large flat stones if you can find those of suitable size. If such stones cannot be had, the superstructure of the bridge may be of planks.

(2) Location and direction of farm roads: Farm roads should as far as possible lie upon the boundary lines of the divisions of the farm into mowing, tillage and pasture, [p. 622] and they should all center at the farm buildings. They should run either at right angles or parallel with the boundaries of the farm.

Fences by the side of farm roads are unnecessary and worse than useless except by the side of those fields which are to be pastured. If the road is one over which animals are to be driven frequently to and from pasture, it must be fenced. The width of farm roads must vary according to the amount of business which is to be transacted upon them. For such roads as are little used, and where it is not likely that teams going in opposite directions will be traveling at the same time, the width need be but very little grater than the distance between the wheels of the carts and carriages used upon the farm. A width of eight feet will be sufficient for such road. If the road is one upon which there is to be considerable travel and where teams are likely to meet, the width must be sufficient to allow the passage of two vehicles at the same time. From twelve to fourteen feet will be a suitable width for such roads.

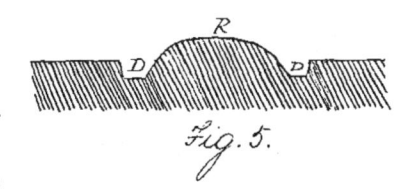

2) Details of Road Construction

For convenience, we may consider the manner of constructing farm roads under three heads according to the kind of soil upon which they are to be built.

(1) On well drained soil; On well drained soils, all that is necessary to make a good road for farm purposes, is to turnpike the soil, leaving shallow ditches at the sides. [p. 624] The earth taken from the ditches will be sufficient to make the bed of the road of proper height. The slope of the sides of the ditches may on these soils be very gradual, and ditches of slight depth will be sufficient. The ditches are so made that teams may cross them at any point. Such a road may be almost entirely constructed by the use of the plough and scraper. The plough is used to loosen the earth in the ditches and the scraper to drug this earth to the center of the road. After the use of these implements, the surface must be finished by the use of hand implements and after this, a heavy roller may be advantageously used to make the earth compact.

The ditches by the side of such a road, although they need not be deep, must be so constructed that the water which enters them will find a free outlet. [p. 625] If this is not so, the road will become soft. Such a road as I have described will not need a coating of gravel unless there is much business upon it, in which case it will be better to cover a space of about six feet in width in the center of the road with three or four inches of gravel evenly spread.

(2) In poorly drained soil: In soil which is poorly drained, the only essentials difference in the mode of construction is that better provision must be made for

drainage. This end may be effected by the use of deep ditches by the side of the road or still more effectually by the use of under drains in connection with shallow side ditches.

(3) In swamp; [p. 626] The mode of construction in swamps which it is impossible to drain, may be as follows; —Lay upon the natural surface of the ground two layers of poles about $2\frac{1}{2}$ or 3 inches in diameter. One layer should be parallel with the road and should have a length equal to its width. The other should be placed at right angles to the first and may have any convenient length. Deep ditches should then be dug at the sides of the road and the earth from them thrown upon it. It may be necessary in very swampy places to put more earth upon it than could be taken out of ditches of ordinary width in which case sufficient earth may be obtained by digging very wide ditches. It have very greatly improve roads [p. 627] through swamps or any poorly drained soil to give them a thick coating of gravel as the soil in such cases is naturally soft. If the road through the swamp is much elevated, the earth may be prevented from spreading by driving stakes on the side of road next the ditches.

3) Farm Fences

In America, fences are a considerable and often useless expense. The American farmers, as a rule, especially in the eastern sections of the U.S. have too many fences. It is estimated that farmers of the U.S. annually expend $250,000,000 for fences.

The farmer should have only so many fences as will enable him to control his animals upon his farm. His pasture, of course, must be surrounded by a strong fence. Lanes or road upon which animals are frequently driven must be fenced. Any piece of land which the farmer may sometimes wish to pasture must be fenced. If different kind of animals such as cattle, horses and sheep are kept upon the same pasture, each may be provided with a separate pasture. Milch cows should not be pastured with other cattle. It is reason mended by some that a pasture for milch cows be divided into two pieces so that they may one week be pastured upon one piece and the next upon another. This I do not regard as necessary.

There are many objections to having numerous fences on the earth. (a) They are expensive. (b) They occupy much land. (c) They obstruct cultivation. (d) They harbor vermin and weeds. (e) They injure the appearance of the farm.

(1) Road side fence: It is not necessary in Massachusetts to build a fence by the side of the road; because according to the Low in Massachusetts. The Low in Massachusetts regarding the roadside fence: No landowner is obliged to build a fence between his land and the public road. Any person may drive animals along the road and [p. 630] if he exercises due care and in spite of this care, the animals from some unseen circumstances break into the fields of the adjoining owner and there do damage, the land-owner can recover no pay for such damage. If on the other hand, animals straying about the roads or driven along without due care, go into the adjoining fields and there do damage, the owner can recover pay for all injury done. If fence is built by the side of a high-way, it must be placed entirely upon the land of the adjoining owner and not half on the high-way.

The Low in Mass. regarding fences between neighboring landowner: The common low of Mass. [p. 631] provides that every man must keep his cattle upon his own land at his peril. As it is expensive and useless to maintain parallel fences, it has been provided by the statute low that adjoining landowners must each build and keep in repair one half of a division fence in case either party desires such fence. Each year an officer called a Fence Viewer is elected in every town in the State. It is the duty of this officer to settle all the questions with regard to fences as provided by Low. If for example, A wishes to have a fence between his land and B's, he must notify the fence viewer, who will meet then upon the spot, divide the line and decide which portion of the fence each man must build. If B refuses to build the fence within a reasonable length of time, A can give notice to the fence viewer who will them give him authority to build B's portion of the fence also; and A can get pay from B for doing this work by a suit at low.

If the land of either party is unimproved, that party cannot be forced by low to build half of a fence between himself and a neighboring landowner. In that case, the party improving his land must either build the entire fence himself or keep his animals upon his own land in some other way. A fence between neighboring landowners is placed half upon the land of one owner and half upon that of the other. If animals belonging to one party break through the fence belonging to the neighboring landowner and do damage in the land of the man owning this broken down fence, [p. 633] he can called no pay for the injury done. If, however, these animals stray through the land of the adjusting landowner and break into the land of the third party, even through they go through the broken down fence of the third party, he can collect pay for the damage done.

Railroads in Massachusetts are obliged to build a fence on both sides of their line. If animals belonging to an adjoining landowner break through the fence and get upon the railroad and are there killed, the owner can recover damages from the railroad company. If animals belonging to another party stray through the lands adjoining the railroad and thence get upon the road and are there killed or injured the owner cannot get pay for them. [p. 634] A legal fence in Massachusetts is one which is four feet high and sufficiently strong and close to prevent the passage of horses, cattle, sheep and hogs. Animals escaping into the road through clearance of hence, doing damage upon lands of a third party owner of cattle.

(2) Different kind of Fence: The rule in making fences should be to use a durable material as possible. If may sometimes be cheaper to use perishable material and to renew the fence frequently; but if the cost is not too great, it is better to use durable material.

(a) Stone Walls: Wherever stones of suitable size [p. 635] and shape abundant one of the best fences can be made by their use. A good stonewall should be $2\frac{1}{2}$ feet at the bottom, $1\frac{1}{2}$ feet at the top and $4\frac{1}{2}$ high. Stones must be laid in such a manner that each will have a good bearing, otherwise the wall will tumble down. A stonewall is particularly good for cattle and horses. It is not well suited for sheep, as they will often run over it. One pole placed at the top of a stone wall will prevent sheep from going over it.

(b) Post and Rail Fences: A good fence may be made by the use of post and rails. The post should be 6 or 7 feet long and sufficiently large that holes $2\frac{1}{2}$ inches wide [p. 636] and 5 inches long can be made in them. A stick of timber from 7 to 9 inches in diameter can be sprit into two good posts. In these posts either 4 or 5 holes should be made. The first should be about 6 inches from the top and they should be placed more or less thickly according to the kind of animals which is intended to fence against. For large cattle or horses, four holes placed at intervals of 5 or 6 inches will be sufficient. For small cattle, sheep or hogs, it may be necessary to have 5 or 6 holes. These posts should be firmly set into the ground about twelve feet apart. The rails may be either of round or split timber and should be from 4 to 6 inch face if split; and if round should be from 3 to 5 inches in diameter. [p. 637] If the material out of which a post and rail fence is to be built is such that it cannot be readily split, it may be necessary to place posts nearer together and to use shorter rails.

(c) Post and Board Fences: In making a fence of this sort, the posts should be from 7 to 8 feet in length and from 5 to 6 in. in diameter, or if the posts are of the sawed timber, they may be 3×5 in. in size. The board should be 3 in. wide and 1 in. thick with the exception of the top board which should be 6 in. wide. Sometimes a cap board is used which should be of a width equal to the diameter of the posts so that it may cover their tops. These posts should be set once in 6 or 8 feet. [p. 638] The depth at which they should be set should be from $2\frac{1}{2}$ to 3 feet. The softer the ground the deeper they must be set in order that they may have sufficient firmness. The boards should be nailed upon them in such a way as to break joints. The post in the two kinds of fences described may be made more durable by charring that portion which is to be below the ground.

4) Hokkaido Fences

(1) Stakes Fence: A good fence may be made by the use of stakes in the following manner; Stakes six and one half feet in length for well drained soil [p. 639] and 8 feet for swampy land and 3 in. in diameter in either case are driven into the ground from 8 to 12 in. apart or even closer than this according to the kind of stock against which the fence is built. These stakes are driven into the ground from $2\frac{1}{2}$ to 3 feet so as to make a fence 4 or $4\frac{1}{2}$ feet in height. It is not necessary that they be round. If of sprit timber, they should be of a size equivalent to a stick 3 in. in diameter. These stakes are connected near the top by horizontal rails of the some size and any convenient length, which are fastened by mean of vines wound about the rail and each stake. This horizontal rail presents the stakes from being displaced. [p. 640] The part of such a fence which will be the soonest to fail is the vine which is used for fastening the rails. This will soon decompose and allow the rail to fall to the ground when animals are able to pass through the fence. These vines are made to bend easily by placing them for a short time over a fire where they are steamed in their own juices. In a country where iron is not expensive, iron wires might be substituted for these vines with great advantage. This kind of fence is best suited for swampy place.

(2) Virginia Fences: A very excellent fence for a cheap one can be built of stakes and rails in the following manner; [p. 641] Stakes 7 to 9 feet length according to the nature of the land and from 3 to 5 in. in diameter or of equivalent size, if split are drown in to the ground in pairs crossing each other diagonally at a distance of 3 feet between successive pairs. Rails 12 feet long and from 4 to 6 in. in diameter or of equivalent size if split are laid diagonally with one end resting upon the ground and a point about 3 feet from the other end resting in a crotch formed by a pair of stakes. The stakes should be driven into the ground to such a distance as to have a vertical height of about 4 or $4\frac{1}{2}$ feet, and the long rails should be

so laid that their upper ends will be of the same height from the ground as the ends of the stakes. [p. 642] These rails should be about a foot apart. Twinning of vines or wires about the crotches is useful in swampy land.

The two kinds of fences last described may be cheaply made and are well adapted for temporary fences in new countries where timber is cheap. The latter I consider to be much superior both of its formidability and durability to the fence.

(3) Worm Fence: A fence, which for want of a better name, we will call a Worm Fence, may be made of rails in the following manner; The rails should be 20 feet long and should be grown 6 to 7 in. in diameter or an equivalent in any other shape square or rectangular is the best [p. 643] since the rail will remain in place better. These rails are laid diagonally across the line of the center of the fence, the ends of successive length of the rails being placed alternately across each other. The angles formed between two lengths of rails may vary somewhat. The most usual is a little grater than a right angle. A stone or block of wood should be placed under the bottom rail at evenly angle of the fence and also between each rail at each end of the fence. The number of rails necessary varies according to their size; the longer they are the fewer will be necessary. Seven or eight is a common number. The fence may be strengthened by driving stakes at the points of intersection of different lengths of rails. A great objection to this fence is that it occupies too much land. [p. 644] A width of 8 or 9 feet is rendered unfit for cultivation, since it cannot be conveniently ploughed. And another objection is that it requires a great deal of timber. Of the three fences last described, I consider the Virginia fence as the last.

(4) Wire fence: In some portion of the U.S. wire has of late years been extensively used in the making of fences. Wherever wire is cheap and timber expensive, wire may be used with advantage.

There are two kinds of wire fences, Plane and Barbed. The latter make a fence which no animal will try to get through a second time. Some who have tried this fence object to it upon the score that it is likely to injure animals confined by it; but I think there need be no trouble in this respect. The animals will soon learn to respect such a fence and will give it a wide birth. Where plane wires are used, animals have sometimes been injured by unintentionally running against the fence at night as at that time they cannot distinctly see them. The post for a wire fence should be of about the same size as those for a post and board fence. The wire are tightly stretched between them. The number of wires necessary varies according to the kind of animals to be confined by the fence. From 4 to 6 is usually sufficient.

5) Hedge as Fences

In some countries, [p. 646] hedges are very extensively used as fences, in no country, perhaps more extensively than in England where the hedges are the pride of farmers. The principal good qualities of hedges are that they are durable and strong, that they afford considerable protection against wind and that they add much to the beauty of the landscape. Some of the hedges of England are more than a hundred of years of age.

Some of the demerits of hedge fences are that they are expensive, since they need constant care, cultivation and pruning, and that they occupy considerable space.

(1) Hawthorn (*Crataegues Oxyacantha*)[4]: This plant grows quickly, is very long

lived and makes a stiff and formidable fence [p. 647] on account of its habit of grows, and the thorns which it produces. It is regarded in England as the vest hedge plant. The general management in making a hedge of this plant is as follows; The seeds are in England gathered in October. They are kept for one year mixed with soil containing a large quantity of organic material. The following October they are sown. This plant is suited to variety of soils; but will not succeed well in any poorly drained land. Four years from the time of gathering the seed, the plants should be large enough for transplanting to the hedgerow. If a soil is good and the plants well cared for seven years after transplanting, the hedge should be strong enough to retain any kind of stock. [p. 648] The method of pruning should be such as to induce the free growth of lower branches. The more numerous and stocky the lower branches, the stronger will be the hedge.

(2) Blackthorn (*Prunus spinosa*)[5]: This plant is not so rigid nor does it produce so many branches as the hawthorn. If however, grows rapidly and will thrive in more exposed situation and on damper soil. The seed in England is gathered in October, rotted in a heap containing much organic matter until spring and sown early. The general treatment thereafter is much the same as that for the hawthorn.

(3) Beech (*Fagus sylvatica*)[6]: This tree has served several qualities which recommend it as a hedge plant. It may be trained from fifteen to thirty feet in height [p. 649] and yet have a width of only three feet at the bottom. Since the tree retains its withered leaves throughout the winter it afford much shelter. It is grown from seed which in England is ripe in November. The seed should be kept in a room till the time of sowing which is early the next spring.

(4) Hornbeam (*Carpinus betulus*)[7]: This plant does not make so good a fence as the beech; but it is able to withstand great exposure and will bear severe pruning. On these accounts it may sometimes be used with advantage. In England, the seed ripens in October and must be rotted in a heap for twelve months and then sown.

(5) Maple (*Acer campestre*)[8]: This species of maple is sometimes used as a hedge; but it is inferior in strength to those previously spoken of, and affords but little shelter. It has few good qualities to recommend it as a hedge plants. [p. 650]

6) Plants suitable for Evergreen Hedges

In ornamental grounds, evergreen hedges are sometimes desirable though they are seldom used as farm fences. Among the plants suitable for making such fences are as follows:

(1) Holly (*Ilex aquifolium*)[9]: This is the one of best. The holly ripens its seed in December and they must be rotted for twelve months. It is customary to sow the seed one year from the March or April following the time of its ripening. Great care is requisite for the successful cultivation of this plant. The method of pruning [p. 651] differs from that suitable for hedges of deciduous plants. The holly naturally sends out sufficiently side shoots; and the top must not be cut until the plant has reached the desired height. Pruning should be given as in the case of most evergreen plants soon after the frost disappears in spring.

(2) Osage Orange[10]: This is a hedge plant which has recently come to prominence in America. Different opinions as to its value; but the most reliable testimony is that under suitable cultivation, the plant is well suited for making hedge.

(3) Plant suitable for Hedges in Swamps: Among the plants best for hedges upon low ground, [p. 652] are the poplar, elder, willow, brich and alder belonging

respectively to the genera Populus, Sambucus, Salix, Betula and Alnus. The poplar, elder and willow are readily propagated by cuttings which are inserted in the ground in moist soil. The brich and the alder must be raised from seeds.

7) Ditches as Fences

Broad open ditches may sometimes be used as fences; but in order that they may effectually prevent animals from crossing the banks should be steep, and the ditches deep.

8) Farm Gates

Gates should be constructed at all points [p. 653] where it is necessary to pass through the fence with teams or cattle. Such gates will be more economical than bars which can be removed, since the length of time required to open them is so much shorter. Such gates should be strongly and cheaply made and furnished with strong hinges and convenient fastenings. The fastenings should be so arranged that the gates may be opened from either side.

9) Timber for Fence

(1) Chestnut (*Castanea vesca*)[11]: The timber of this tree is very useful for fence building. It splits very easily, and it is not difficult to split out long, straight rails. This tree grows with considerable rapidity and the timber is very durable when used for rails. [p. 654]

(2) White Cedar (*Cupressus thyoides*)[12]: This is perhaps the most valuable tree for furnishing fence timber of any that grows in Massachusetts. It sprits very readily and is exceedingly durable both for posts and rails. The tree grows in swamps with medium rapidly.

(3) Spruce (*Abies nigra*)[13]: The timber of this tree has considerable value for fence building. The tree grows rapidly and will flourish in very exposed situations as upon hilltops. The timber is most value for posts, being quite durable.

(4) Locust (*Robinia Pseudoacacia*)[14]: These are several species of locust all of which have some value for fence; but *Pseudacasia* is the best. The timber of this tree is exceedingly durable and has great value for posts. [p. 655]

(5) Oak (*Quercus* ...)[15]: Oak timber is considerably used for posts being very durable. This timber is not well suited for rails as it can seldom be found of sufficient length and straightness except when it is of too great size.

(6) Red Cedar (*Juniperus Virginiana*)[16]: The timber of this tree furnishes posts of greatest durability. It is no value for rails.

(7) Pine[17]: Pine furnishes the best boards which can be used for the construction of post and board fences or gates. Spruce also makes good boards. [p. 656]

6-12. Winter Work upon the General Farm

Winter is the time when not much out-of-door work can be performed. There are some branches of farm work, however, which can best be attended to as this season. During winter, sufficient wood should be cut, drawn from the forest and prepared for fuel to last the entire year. At this season, also any timber, which may be needed for repairing farm buildings or implements, should be cut and drawn to the farm. If the supply of timber or wood upon the farm is more than sufficient for home needs, winter is the time for cutting and marketing the surplus. The amount of time devoted to such work should not, however, be so great as to prevent the proper caring [p. 657] for the farm stock which is after all. The

Chapter 6 Farm Economy (Management)

principal work of the winter months much depends upon the careful and proper management of stock at this season. The farmer should devote a large share of his personal attention to this business. He should see that food is not wasted, and that the animals are kindly and judiciously managed.

Another kind of work which should be done in winter, is the composting of manures and pitching them over to put them into good mechanical condition. Farmers should strive in all possible ways to increase the size and improve the quality of their manure piles. Manure is the raw material which the farmer with the aid of nature wishes to manufacture into crops; [p. 658] and the more raw material, he can produce the larger crops. Farmers are often quite ignorant as to the elements of plant food which their manure contains. Not knowing this, they cannot tell with any degree of accuracy what crops they can raise with it. To enable you to form an intelligent idea as to the crop producing power of any manure which you may have, I will give you a few figures. Average barnyard manure when fresh contains the following constituents in one thousand pounds (Table 5). [p. 659]

Tab. 5 barnyard manure	
Winter	75.0 lbs.
Organic substances	24.6 lbs.
Total ash	44.1 lbs.
Nitrogen of the ash	4.5 lbs.
potash	5.2 lbs.
Lime	5.7 lbs.
Phosphoric Acid	2.1 lbs.
Soda	1.5 lbs.
Magnesia	1.4 lbs.
Silicic Acid	12.5 lbs.
Chlorine	1.5 lbs.

The composition of your barnyard manure will be better or poorer than this according as you feed better or poorer than the average farmer. It will be useful to know how to determine what elements of plant food should be contained in the manure of any animals which have been fed a certain amount of food. Of these elements nitrogen, potash and phosphoric acid are the only ones believed to be of general importance, and therefore we will consider only these. The manure, as you know, can contain only those amounts of these elements [p. 660] which are present in the undigested of the food and the wastes of the system. In order to determine what quantities of nitrogen, potash and phosphoric acid the manure of animals contains, we must first know the quantity in the food fed to these animals. Table 6 shows the number of pounds of nitrogen, potash, and phosphoric acid contained in some of the principal crops, [p. 661] and Table 7 shows the number of pounds of amounts of nitrogen, potash, phosphoric acid in various cattle products. [p. 662]

In the year 1876 these substances were worth in America as follows; Nitrogen: \$0.25/lb, Potash: \$0.06/lb, Phosphoric Acid: \$0.12/lb. These prices vary from time to time, but we will adopt them as a basis for calculation. From the figures showing the amount of nitrogen, potash and phosphoric acid contained in various products you see that it makes much difference what crops are sold from the farm some contain large amounts of substances valuable for plant food while others contain but little.

Let us suppose that a man sells 50 tons of English hay. This would contain of nitrogen 1,310 lbs of potash 1,710 lbs, and of phosphoric acid 410 lbs, worth \$327.50, \$102.60 and \$49.20 respectively. [p. 663] The same amount of clover would contain of nitrogen 2,130 lbs, of potash 2,000 lbs, and of phosphoric acid 560 lbs, worth \$526, \$120 and \$67.20 respectively. You see form these figures that the elements in the English hay are not worth nearly as much as those in the clover, —those in the former amounting to $478.80 and those in the latter to $713.20. In 50 tons of timothy there would be of nitrogen 540 lbs, of potash 610 lbs, and phosphoric acid 230 lbs, worth $135, $36.60 and $27.60 respectively. Therefore it will be less injurious to the farm to sell timothy than to sell English hay. Suppose a farmer sell the milk from fifteen cows each averaging 15 quarts of milk per day, for a year. He will have sold in that time 16,416 gallons of milk [p. 664] which would contain of nitrogen 919 lbs, of potash 246 lbs, and of phosphoric acid 262 lbs worth respectively $229.75, $ 14.76 and $31.44 in all $275.95. If he had sold simply the butter which he could have made from this milk, almost the

Tab. 6			Phosphoric acid
Crop	nitrogen	Potash	
English hay/ton	26.2 lbs.	34.2 lbs.	8.2
Red clover 〃	42.6	40.0	11.2
White clover 〃	50.6	21.2	17.0
Timothy 〃	10.8	12.2	4.6
Corn stalks 〃	9.6	32.2	3.8
Rye Straw 〃	4.8	15.2	3.8
Wheat Straw 〃	6.4	9.8	4.6
Buckwheat Straw 〃	2.6	48.2	12.2
Oat Straw 〃	8.0	19.4	36.0
Tobacco Leaf 〃	120.0	108.2	14.2
Wheat 100 bu	120.0	32.2	48.2
Rye 100 bu	97.7	30.5	48.2
Corn 〃	88.8	10.3	48.2
Oats 〃	64.0	14.0	33.3
Buckwheat 〃	68.6	10.0	21.0
Flax Seed 〃	59.0	61.0	76.0
Field beans 〃	244.0	72.0	97.0
Potatoes 〃	18.8	33.0	10.5
English Turnips 〃	10.5	17.6	5.8
Carrots 〃	12.3	18.8	6.4
Common Beets 〃	10.6	25.3	4.7
Barley 100 lbs	1.5	0.6	0.8

whole quantity of these valuable elements would have been left upon the form. Every farmer should, therefore, carefully estimate what amounts of these three elements are contained in the crops which he intends to sell, and provided he can sell at a fair price, he should aim to sell those things which contain as little nitrogen, potash and phosphoric acid as possible. If his circumstances are such that he must sell crops which contain a large amount of these substances, he must return them to his soil in some way in order to keep up its fertility. [p. 665] In order to know how much a certain amount of manure is worth, it is only necessary for the farmer to know what the food he gave to the stock contain and what portion of it the stock used. We will suppose a farmer to have fed 50 tons of English hay, 5 tons of red clover, 1 ton of buckwheat straw, 8 tons of corn stalks, 400 bushels of corn, 100 bushels of oats, 50 bushels of flax seed meal, 100 bushels of buckwheat, 400 bushels of turnips and 100 bushels of potatoes. Suppose he has used 6 tons of rye straw and 5 tons of oat straw for bedding. All these would contain in pounds nitrogen 2,292.2, potash 2,642.5 and phosphoric acid 797.4. We will suppose that his cattle have gained 1,000 lbs.

This will contain of nitrogen 26.0 lbs, of potash 1.7 lbs and of phosphoric acid 18.6 lbs. [p. 666] Suppose he has also kept eight cows each averaging 12 quarts of milk per day for five months in the year. In this milk there has been carried off of nitrogen 201.6 lbs, of potash 54 lbs, of phosphoric acid 57.6 lbs. This makes in all nitrogen 227.6 lbs, potash 55.6 lbs, and phosphoric acid 76.2 lbs. These amounts of nitrogen, potash and phosphoric acid are what the cattle have taken from their food and if there has been no waste there will therefore be in the manure 2,064.6 lbs of nitrogen, 2,586.8 lbs of potash and 721.2 lbs of phosphoric acid. The value of these elements would be as follows: nitrogen $266, potash $155.21 and phosphoric acid

Tab. 7	Nitrogen	Potash	phosphoric acid.
Milk 100 gal	5.6	1.5	1.6
Cheese 100 lbs	4.5	0.2	1.1
Cattle (Live weight) 1,000 lbs	26.0	1.7	18.6
Sheep 〃 〃	22.4	1.5	12.3
Swine 〃 〃	20.0	1.8	8.8
Washed wool 100 lbs	9.4	0.19	0.03

$86.54 in all $507.75. The value given for this manure is what it would be, had there been no lose, and if the elements which it contains were in such condition as to be directly available. [p. 667] Barnyard manure also contains other elements which are of value and its physical influence upon the soil is also sometimes beneficial. Therefore we can not decide exactly as to the money value of a certain pile of manure from such calculation as this. In this supposed enough for 233 bushels of corn, potash enough for 941 bushels and phosphoric acid enough for 202 bushels. These figures take into account only the elements contained in the grain.

The stalks contain a far greater proportion of potash than the grain and therefore the disproportion between the amounts of nitrogen, potash and phosphoric acid in this manure when applied to the corn crop is really less than a first glance at the figures I have given you might lead you to suppose. [p. 668] Still even when the stalks are taken into account, there is more than sufficient potash and therefore it will be wise, if this manure were to be used for the corn crop to use in conjunction with it some manure containing principally nitrogen and phosphoric acid, such for example as fish guano.

1) Relative Quantity of Nitrogen, Potash and Phosphoric acid Found in Crops, Furnished by the Processes of Nature and by Manure

It is stated by a French professor Ville that if the number of pounds of nitrogen, potash and phosphoric acid found in a crop be applied to the soil. [p. 669] You can raise as much as you choose within certain limits. For instance, if you apply sufficient nitrogen, potash and phosphoric acid for 25 bushels of wheat you can raise that amount of wheat. If you apply as much as is used in a certain quantity of corn, you would raise that quantity of corn.

Experiments to ascertain whether or not this conclusion is true have been tried by Prof. Stockbridge of the Massachusetts Agricultural College, and his results have thus far shown that the statement of Ville is correct. Prof Stockbridge has tried the experiments with all the leading crops and with all has been successful. Fertilizers made according to his formulae have been extensively used throughout New England and to a less extent in other sections of the United States, [p. 670] and the farmers using then have almost invariably met with good success. These fertilizers are however quite expensive and it is believed that the processes of nature will supply a portion of the necessary plant food. Prof Stockbridge is now engaged in trying experiments for the purpose of determining what proportions nature will supply. This proportion must of course vary in different soils and with different crops. Some crops are able to draw a portion of their nitrogen from the air. Some soil may contain sufficient of element but be deficient in others. In such cases it would be unwise to apply all the elements necessary to the crop since the soil or air would furnish a part of them. Farmers in different localities should, therefore, determine for themselves [p. 671] what elements are lacking in their soils. The result of the experiments thus far tried in Massachusetts indicates that in that state it is necessary to supply only nitrogen, potash, and phosphoric acid. The same may

Tab. 8	Nitrogen	potash	phosphoric acid
Potatoes 100 bu	21	34	11
Shelled corn 50 bu	64	77	31
Cured Corn Fodder, 2 t	20	66	16
English Hay 1 t	36	31	12
Wheat 25 bu	41	24	20
Oats 25 bu	23	20	12
Rye 20 bu	25	24	16
Buckwheat 25 bu	37	50	15
Rye Straw 2 t	10	31	8
Cabbages 1 t	28	12	4
Beets 100 bu	11	25	6
Peas or Beans 20 bu	53	33	28
Swedish turnips 100 bu	11	3	8
Onions 100 bu	11	18	4

or may not be true here. We can only tell by experiment.

Stockbridge Formula for 1876 as Published by W. H. Bowker & Co.: [p. 672] The quantities of potatoes, corn, corn fodder, English hay, wheat, oats, rye, rye straw and buckwheat given are about the quantities that can be economically raised per acre in addition to the crop naturally produced by the land. Larger quantities than those stated may however be applied and some farmers have attained very satisfactory results by the use of about a half more than the quantities given in the Table 8. All of these formula include sufficient nitrogen, potash and phosphoric acid to produce the crops stated with there natural proportion of straw stalks or tops *etc.* as the case may be. [p. 673] In the years 1869, '70, '71 and '72, the average crop of wheat in Massachusetts was 17 bushels per acre, of rye 17 bushels, of oats 29 bushels, of potatoes 109 bushels, of tobacco 1,325 pounds and of hay 2,500 pounds. These quantities may all be easily doubled by the application of sufficient fertilizer, and the farmer should not be satisfied with a yield less than double these quantities. Prof. Ville has decided from the experiments that he has tried in France that it is sometimes necessary to add lime in addition to the elements which I have told you are considered sufficient in Massachusetts. A manure containing nitrogen, potash and phosphoric acid in the proper proportions and in sufficient quantity to produce a given crop may in Massachusetts be called "a complete" manure. [p. 674] A complete manure in France must in addition contain lime. What a complete manure for Hokkaido must contain, can be told only after experiments. Actual experiments in the field are a more reliable means of ascertaining the needs of the soil than chemical analysis. By chemical analysis, it is true, we can tell what elements are in the soil and can determine also something in regard to their availability as plant food but plants are themselves the best judges as to whether elements are contained in the soil in a suitable form. Their growth, therefore, will furnish a reliable indication as to what a soil lacks. To determine what elements are wanting in a soil for the production of a certain crop, you should take two more plots of land than the total number of elements [p. 675] with which you are going to experiment. We will suppose that a man wishes to experiment with nitrogen, potash, phosphoric acid, magnesia and lime. He must then take seven small plots of land for each crop with which he proposes to experiment. These plots should be of precisely similar character as to the chemical and physical characteristics. The time of ploughing, of harrowing, of applying fertilizers, of planting seed, and the methods of culture should be alike. To the first plot for each crop, he should apply no fertilizer to the second a fertilizer containing the requisite amount of nitrogen, potash, phosphoric acid magnesia and lime, to the third the requisite amount of all except nitrogen, to the forth all except potash, [p. 676] to the fifth all except phosphoric acid, to the sixth all except magnesia and to the last all except lime. Provided the plants are not injured by insects and provided no unforeseen accident happens to any of the plots the former from the resulting harvest can tell what elements have been of use and what have been of no use.

2) The Power of Plant to Take Nitrogen from the Air

From carefully made experiments Prof. Ville has reached the conclusion that certain plants have the power to take a large proportion of their nitrogen from the air. Among those which possess this power in the highest degree are peas, beans and clover, the following table shows [p. 677] the percents of nitrogen which, according to Prof. Ville, are taken from the air and form the soil by the crops named in Table 9. In the case of clover, Ville asserts that he has proved that it will grow precisely as will without nitrogen in the soil as with it. He states that he has taken earth and thoroughly calcined it that there could be no nitrogen compounds in it, and then planting clover seed, if grew as well as seed planted in the soil which was known to contain nitrogen.

Tab. 9	From the Air	From the Soil
Clover	100	0
Barley	80	20
Rye	80	20
Wheat	50	50
Beets	60	40
Rape Seed	70	30

The additional precaution was taken to water the plants with distilled water. [p. 678] Winter is also the time when the farmer should make all his plans for the ensuing year. As I have already stated, the farmer should have a general plan of operations extending over several years, but more minute plans should be made at least as early as the winter previous to their execution. These plans should embrace the kinds of crops and quantities of each which are to be cultivated the land upon which they are to be raised, and the fertilizers to be applied to them. These plans should also embrace the improvements which you propose to make the kind and amount of stock that you intend to keep and in short, all the details of farm management. Winter is also a good time for the farmer to read and study about his business. During other seasons of the year, [p. 679] he will not have so much leisure, and therefore he should improve the winter hours to the utmost. It will be found possible to make most improvement by following a regular course in reading, having some definite object in view. Besides this general reading, the farmer should keep himself posted in the periodical literature of the times. He should take one or more good agricultural papers and there he will find most time to read and study during the winter. While reading, the farmer should not accept as absolute truth everything he reads, for many persons that write for paper may man not be any wiser than himself. He should therefore think, about what he reads and form an independent opinion of his own. Brains are found to be a very effective and [p. 680] at the same time an economical substitute for fertilizer. The exercise of brains in thinking costs nothing, but on the other hand, the more a man thinks, the more and the better he can think. Here than is one kind of fertilizer, if we may use this expression. From which the more we take, the more we have left. The man who uses his brains in forming may often produce better crops with less fertilizer than those not thinking much about their business. It should, therefore be the aim of every man to make use of this brain manure as much as possible. [p. 681]

6-13. The Routine and Time of Doing Spring Work

One of the first things which will demand attention in spring is the work of putting all the farm fences in perfect order. As winter draws near its close you should prepare whatever material you will need in repairing your fences. As soon as the land becomes hard enough to make it possible to pass over if with teams, you should go entirely round your farm and put all your fences in perfect order. Another kind of work which may be done very early in spring is the repairing of roads. If new fences or roads are to be built, this work can often be best done at this season, but the rule should be to do such work [p. 682] whenever you have most leisure.

The ploughing[18] of land for early garden crops will be one of the first kinds of work demanding attention. This should be done as early as the lands get into such condition that plowing will pulverize it. The crops which may safely be planted as soon as the land get into this condition are potatoes, peas, radishes beets, onions, lettuce and cabbages. The method of planting potatoes in the garden is in drills[19] about three feet apart with the plants at intervals of $1\frac{1}{2}$ to 2 feet. It is better to cut the potatoes into quite small prices and if it is desirable to have them very early, they may be sprouted on a hot bed before being planted. [p. 683] Peas may be planted in drills which should be about 3 feet apart. Some of the larger varieties require $3\frac{1}{2}$ feet. After peas have grown to the height of 2 or 3 inches, they should be provided with sticks upon which they may climb. There are some very dwarf varieties of peas which never grow to any great height and these need not be provided with sticks. The best kinds of sticks are the branches of birch trees. These should be from 3 to 4 feet in height according to the variety. Peas should be covered, when planted to the depth of 1/2 to 2 inches. The garden crops which must not be planted while there is danger of frost, are corn, beans, cucumbers, melons, and squashes. Any of these crops would be totally destroyed [p. 684] by a comparatively slight frost. Nothing is gained by planting them very early in the season; for even if uninjured by frost, they will fail to make much growth before the coming of hot weather.

[参考図]

A) Cahoon's seed sower

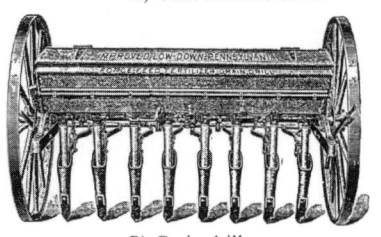

B) Grain drill
Both A) and B), from Brooks, *Agriculture*

After planting early garden crops, the work which will next demand attention is the sowing of spring grains such as oat, barley, wheat and rye. This work should be done as early as it is possible to bring the land into suitable physical condition. All these grains are natives of cold climates, and if sown early, they will thrive much better than if sown late. The quantity of seed necessary per acre varies according to the richness of the land. The richer it is, the less seed is needed. The quantity of wheat per acre necessary for broadcast sowing is about 2 bushels, [p. 685] of oats and barley from 2 to 2½ bushels, of rye about 2 bushels. If sown in drills from 1/4 to 1/3 of the above quantities will be sufficient. The best method of sowing these crops is in drills[20]. Many good machines are made for drilling in grains, but if the quantity to be sown is not very great, the work may be done quit cheaply in this country of cheap labor by hand. The seed should be covered to a depth of from 1½ to 2 inches. Seed is sow broadcast either by hand or by the use of machines. Machine sowing insures a more even and sure distribution than hand unless the workman is particularly careful, and it is also more quickly done. One of the best grain drill sower is the "Buckeye", and for broadcast sowing "Cahoon". [p. 686] Whether the grain is to be sown in drills or broadcast, the land should first be thoroughly harrowed. If drilled in by machines, the grain will be covered sufficiently. If sown broadcast the land must afterwards be harrowed. It is a common practice to sow grass seed in spring with certain grains. Of the grains which I have mentioned, barley is the one with which grass succeeds best; but it also does will with oats, particularly if they are cut early for fodder.

Such crops as beets, cabbages and onions may be planted quite early in spring. Onions should be planted about as soon as the soil can be put into good physical condition. The same is also true of beets. In this vicinity, there crops should be planted about the tenth or fifteenth of May. Carrots should be planted a little later than beets. [p. 687] About the twentieth of May will be found the best time. The work of planting corn, beans, squashes and melons must be deferred until the ground is thoroughly warm. The rule of the American Indians was to plant corn when the leaves of the oak tree became as large as the ears of mouse, and this is a very safe rule to follow. Of nearly all garden vegetables there are those varieties which ripen very early and those which ripen late. In all such cases the early varieties are usually planted first, thus early cabbages may be planted as early as anything can be planted, while the late ones should not be planted until a little later than the time of the planting of corn. Some turnips may be planted early in the season; but the late varieties should not be planted before the middle of June, [p. 688] and from that time to the first of July will be most suitable. These are some small varieties of turnips which will mature even if planted in August. It should be your rule to plant those crops which you intend to keep through the winter as late as possible, and have them mature. Buckwheat and millet should be sown after such crops as corn, and squashes have been planted. From about the tenth of June to the first of July for millet and a little later for buckwheat is the time most suitable. Hemp must be planted early in season, flax somewhat later. [p. 689]

6-14. *Summer Work on the General Farm*

The principal work of the summer is the cultivation of the various crops and the harvesting of hay and grain.

1) The Objects of Cultivation

We cultivate our crops for the sake of killing weeds, for the sake of pulverizing the soil so that the plant food may be made available, for the sake of making the soil more retentive of moisture, and in certain cases for the sake of pruning the roots of plants. Most farmers if asked why they cultivate their crops would answer "For the purpose of killing weeds" [p. 690] and this is perhaps the most important object. The cultivation of all crops should at least be sufficient to prevent the excessive growth of weeds. Many crops, however, grow so rapidly and shade the ground as completely that often they have reached considerable size, weeds will not grow to any great extent. Some farmers doubtless cultivate the soil more than is necessary not allowing any weeds to grow. This, however, is a very uncommon foiling but the farmer should remember that if it casts more to remove weeds than the amount of injury which they will do to his crops, it will be unprofitable to remove them. Weeds should never be allowed to ripen seeds, but sometimes it may be wisest to allow a few weeds to grow after the crops have reached [p. 691] such a size as not to be much injured by them.

The second object of cultivation, the pulverization of the soil, is exceedingly important and one which is often overlooked. Most farmers not seeing any weeds among their crops would say that they do not need cultivation. This, however, is not always the case. Cultivation may be very necessary although weeds are not present in large quantities. There is much difference in different crops as to the amount of pulverization which is requisite for the highest success, but we may state as a general rule that crops should be cultivated sufficiently often to keep the soil loose and friable. Soil in this condition is freely acted upon by the influences of nature and admits of the free extension of the roots of plants. [p. 692] The roots of plants, it is true, are able to find their way through very hard soil, but they are not so largely developed as in soil which is well pulverized, and therefore cannot collect as much plant food.

The third object of culture is in certain cases to make the soil more retentive of moisture and in others to cause a more rapid evaporation of moisture. As the first glance it may seem singular that we should cultivate sometimes to make the soil more moist and sometimes to make it more dry, but the results of recent experiments make it evident that culture does subserve these ends. Prof. Stockbridge of the Mass. Agricultural College has during the past year carried out an elaborate system of experiments [p. 693] for the purpose of ascertaining the effects of culture upon the retention of moisture by soils. His method of experiment was as follows: Boxes containing one cubic foot were filled with various kinds of soil, care being taken not to disturb the soil, but to leave it in its natural condition. Two boxes were filled with clayey soil, two with light sand and two with heavy loam. This experiment commenced just after a fall of 3/4 in. of rain and continued seven days. Those boxes filled with the undisturbed soil were placed in a trench in the open field so that the surface of the earth in the boxes was on a level with the surface of the surrounding field. They were weighed at the beginning of the experiment and again at its close. One box of each kind of soil [p. 694] was thoroughly cultivated each morning, the earth being stirred to the 4 in. depth and the moist soil each time brought to the surface. The other box of each kind of soil was left entirely undisturbed. At the close of the experiment, the different kinds of soil were found to have lost the following amounts:

Clayey soil, tilled;	5 lbs. 5 oz	Light sand, untilled;	7 lbs. 8 oz
Clayey soil, untilled;	6 lbs. 14 oz	Heavy loam, tilled;	6 lbs. 13 oz
Light sand, tilled;	3 lbs. 3 oz	Heavy loam, untilled;	7 lbs. 13 oz

You will notice that in every case, the untilled soil lost considerably more than the tilled. This amount through seems small for this small quantity of soil is very great for an acre of land. It therefore appears that in the case of soil moderately moist, [p. 695] culture makes it more retentive of moisture. These kinds of soil were afterward thoroughly saturated with water and then subject to the same treatment as previously. The experiment continued five days when the different boxes were found to have lost moisture as follows:

Clayey soil, tilled;	4 lbs. 4 oz	Light sand, untilled;	4 lbs. 0 oz
Clayey soil, untilled;	5 lbs. 15 oz	Heavy loam, tilled;	6 lbs. 15 oz
Light sand, tilled;	5 lbs. 13 oz	Heavy loam, untilled;	4 lbs. 11 oz

In but one case, that of the clayey soil, was the loss greater in the untilled soil than in the tilled. In all other soils, the tilled soil lost the most moisture. These boxes were weighed each day and for the first few days the lose of moisture in the tilled clay was greater than in the untilled. [p. 696] With each of the soils experimented with, it was found that the moisture evaporated by the tilled soil gradually grew less while the amount evaporated by the untilled soil was proportionally greater; i.e. while at first each of the tilled soils lost more moisture than the untilled, at the close of the experiment, the amount evaporated by the latter was about equal to the amount evaporated by the tilled soil in the case of light sand and heavy loam, and greater in the case of clay. The results of these experiments indicate, then that culture is needful in time of draught, because it makes the soil more retentive of moisture and also that it is useful when the soil is too wet, since it hastens the evaporation of the superabundant moisture. [p. 697] In the case of wet soils, however, you must remember that culture must not be given when the soil is so wet as to make the soil more compact by the pressure incident to culture.

Another object of culture to which attention has been directed very recently is the pruning of the roots of plants. It is now believed by same that root pruning at certain stages in the growth of most plants is quite useful. It is a well-known fact that when rootlets are quickly formed at the end. It is these small rootlets, which are useful in collecting plant food from the soil and an old roots has but little power to do this. If by root pruning, therefore, we can increase the amount of active roots, [p. 698] we shall doubtless improve the plant. Within the last two years, Dr. Sturtevant, the editor of the "Scientific farmer" has written much upon a subject which he calls "The New theory of Tillage". In his articles upon this subject, he has endeavored to show that root pruning is particularly beneficial to most crops. He has tried some experiments for the purpose of ascertaining whether or not this practice is beneficial. By judicious root pruning in connection with abundant manuring, he has been able to obtain some remarkable results. From a single kernel of corn he has raised in one season from fourteen to twenty good ears. [p. 699]

2) Method of Culture

The methods of culture one very numerous and must be varied according to the crop cultivated and the circumstances of each individual case. Of course, it is important that the work of cultivating crops be done as cheaply as possible since upon this largely depends the amount of profit. Since the labor of horses and animals is usually cheaper than that of man, it should be the general principle of the farmer to do as much of the work of cultivating his crops by the use of horses as possible. The implements drawn by horses which may be used are the plow, harrow, cultivator and the horse-hoe[21]. Those plows used in the culture of crops are generally much smaller than ordinary plows, [p. 700] and those with both double and single moldboards, are used. The plow may be used with advantage wherever it is desirable to stir the ground very deeply or to throw earth either toward or away from the growing crop. The double moldboard plow is especially useful in throwing earth towards the drills between which it is used. The use of the plow in the cultivation of crops is not so extensive as the use of other implements.

The harrow has of late years been much more extensively used in the work of cultivating crops than formerly. That harrow best suited for this work is the one with small round teeth sloping to the rear. The harrow is particularly useful in the early cultivation of such crops as potatoes, corn, [p. 701] and grains such as wheat and rye. The harrow is also used before weeds have reached any considerable size. Used at this time, it will up root and kill most of the small weeds just starting from seed without materially injuring the growing crop since its roots are more deeply established. Should cultivation be neglected until weeds have got well started, it will be useless to attempt to kill them with a harrow without seriously injuring the crop. The ordinary straight-toothed harrow is frequently used in the cultivation of wheat. Used in spring, the wheat is so well established that not much of it will be torn out by the harrow while most of the weeds will be destroyed and the ground well stirred. [p. 702]

There are many different styles of cultivators manufactured, this implement being adapted, therefore, to a great variety of uses. It is more extensively used in the cultivation of crops than any other implement. There are cultivators designed for use by horse power and those for use by hand. The former are of far more general application than the latter which can be profitably used only in the cultivation of such small crops as onions, carrot, *etc.* Horse cultivators may be divided into two classes, viz. the ordinary one horse implements and sulky cultivators. For use on large fields, the latter are preferable to the former, but ordinary small farmers will probably find it more economical [p. 703] and therefore wiser to use the former. Cultivators are made with teeth of many different shapes, some stirring the earth deeply others throwing it outwards, still others throwing most of it inwards and still others simply cutting the weeds near the weeds near the surface. Horse hoes differ from cultivators more in name than anything else, as there are implements called "horse hoes" which are exactly like others called "cultivators" For the cultivation of grains sown in drill, there are many grain hoes which cultivate several rows at a time. Wherever much grain is cultivated the use of these implements is recommended. For the cultivation of root crops such as beets, [p. 704] cultivators have been made which stir the soil of several rows at a time. Such implements are profitably used wherever these crops are extensively cultivated.

The various hand implements used in the cultivation of crops such as the hoe, rake, *etc.,* I have already sufficiently spoken of. The use of each of these implements is sometimes necessary and preferable, but the crops are more expensively cultivated by their use than by the use of implements drawn by horse power and therefore we should dispense with them as far s possible.

Besides the work of cultivation and harvesting hay and grains during the summer, the farmer will often find time to make permanent improvements such as the building of roads, the mowing in pastures, the drainage of land *etc.* [p. 705]

Drainage is a work which must be done when the land is driest, and in this climate the month of June seems to be the most favorable time. Land may be drained most profitably when it is quit free from water. The farmer will often planting his crops, he will have unoccupied time before the season of harvesting hay and before his crops need much cultivation. In this otherwise unoccupied time, he will find his best opportunity to drain land.

The farmer will also find that during the later summer months, there will be many days unsuitable for making hay which he can devote to the work of improvement. Improvements thus made during comparatively leisure time [p. 706] will cost but little, since if not engaged in this work, the farmer teams and men would remain comparatively idle. The mowing of bushes and wild plants in pastures can best be attended to during the summer months. Mown at this time, these plants will be more likely to be killed than if mown at any other season. Although the summer is not the most favorable season for making roads, but if there is no other work of improvement demanding attention, we may profitably occupy our leisure time in making or repairing roads.

During the summer months, also, manures intended for topdressing lands in the autumn may be drawn to the fields where they are to be used, and composted. [p. 707] The summer also is the best season for throwing out peat for use in the compost heap or any of the receptacles for manure. [p. 708]

6-15. Fall Work upon the General Farm

The principal work of the autumn months will be the harvesting of various crops, plowing land, sawing winter grains such as wheat and rye, sowing grass seed and top-dressing grass lands. During the autumn months as during the summer, the farmer will find considerable opportunity for making improvements such as those I have been just speaking of. Although the autumn is not so favorable a time for draining land as the summer, yet some of the drier fields needing drainage may be profitably drained at this time.

The autumn is a favorable season for clearing up and plowing new lands. [p. 709] During the fall months a plentiful supply of absorbents such as peat and loam should be drawn to the manure yards. During the later fall months preparation must be made for winter. Those crops needing protection must be covered.

All the buildings should be put in good repair and made as warm as possible for the winter. The autumn is a favorable time for transplanting trees. Most fruit trees may be successfully transplanted at any time after shedding their leaves.

The judicious farmer will find plenty to do during all seasons of the year, but it is not wise to confine yourself always to work upon your own farm. [p. 710] You may profitably spend some time in visiting other farmers and talking with them about their methods of work and their results. There is no time more favorable for this than the autumn when the farmer can see the crops of his neighbors in the field. Agricultural exhibitions should also receive a share of the farmer's attention. By attending such exhibitions, he will often learn what will be of great advantage to him. [p. 711]

解　題

1．農業経営学
1）第6章の要約
　ノートに書かれた原題は Farm Economy のみであるが，本書では内容から Farm Management も付け加えた。後に見るように，この時代は農業経済学と農業経営学とはまだ区別されておらず，両者とも広い意味での農学から分離されていない状態にあった。したがってこの題名は，農学の一分野としての「農場管理」と訳すのが適当であろう。全体は次の15の節からなっている。
　　1．ムダを防ぐためのシステム，2．農場管理，3．農場経営の種類，4．採草地の管理，
　　5．永久草地の管理，6．牧草収穫の農機具，7．放牧地の管理，8．耕地の管理，9．森林，10．果樹園，11．農場の道路と垣根，12．冬の農作業，13．春の農作業，14．夏の農作業，15．秋の農作業
　第1節は，Farm Economy についてのブルックスの定義である。ここでは土壌，肥料，作物，雑草，適期作業，労働，家畜，農機具，建物など農場経営を構成する諸要素につき，いかにしてムダを省いて効果的な要素投入を行なうかが考察されている。第2節では農場の適正規模および資本が経営形態によって異なることが説かれており，固定資本，流動資本という概念も登場する。第3節では農場経営の種類が General Farming と Special Farming に分けて説明されている。General Farming は何種類かの穀物と牧草との輪作と家畜飼養とを組み合わせた経営で，その後の北海道で「混同経営」と呼ばれた経営形態を指している。Special Farming は1種類か2種類の作物または家畜に特化した専門経営である。ブルックスは専門経営が成り立つのは市場条件に恵まれた特別の場合であり，北海道のような立地条件にあっては General Farming を行なうべきだとして，以下ではその前提の下に講述している。
　第4節から第7節までは採草地，永久草地，放牧地についての説明であり，牧草の種類，栽培管理法，収穫用の農機具，収穫後の貯蔵法などが詳しく述べられている。ここに登場する草種は，チモシー，オーチャード，ケンタッキーブルーグラス，イタリアンライグラス，レッドクローバ，ホワイトクローバ，アルファルファなど17種類に及んでいるが，日本人にとってはおそらく初めて耳にするものばかりであったと思われる。
　第8節は作物の種類や栽培法については第7章にゆずって，畑作地の管理について述べて

いる。第9節では農場経営における林地の意義について詳しい説明がなされ，立木に乏しい土地での植林の必要が説かれている。これは北海道における耕地防風林の意義の最初の言及として注目されよう。第10節ではリンゴ，モモ，ブドウ，サクランボ，プラムなど，今日北海道で最もポピュラーになっている果樹12種類について述べられ，第11節では農場の道路とフェンスについての詳しい説明がある。第12節から第15節までは，春から冬まで季節ごとの農作業が詳細に述べられている。

　以上の内容は，他の章と同じようにマサチューセッツ農科大学での教授内容にほぼ一致していると思われるが，ブルックスはそれを北海道にどう応用するかという点についても随所で工夫を凝らしている。例えば，草地のところでは北海道の野草についても調べ，それらは牧草に適さないとして外来種を導入することを勧め，垣根の作り方についても北海道に適したフェンスを何種類か紹介している。放牧地の造成の項では，北海道の山地を覆うクマイザサや根曲がり竹に困惑したようで，「私には経験がないのでうまく説明する自信がない」とこぼしている。

2) 農業経営学の「農学段階」

　ここで講述されている「農場管理」の内容は，今日では農業経営学が扱う問題である。しかし，今日の農業経営学の重要な構成要素となっている簿記会計についてはまったく触れられていないし，全体の叙述は「管理法」あるいは「管理術」と言うべき農場管理の技術，ノウハウを教えていると見るべきで，きわめて実践的ではあるが経営学とは言えない。それは当然であって，当時は世界的に見ても農業経営学が農学から分化して独自の学問になる以前の時期だったのである。

　農業経営学，およびその隣接科学としての農業経済学が，農学一般から分離して独自の学問領域となるのは19世紀中葉に現れたテーアの研究からとされるが，それもしばらくはリービッヒの新農学による農業技術学の目覚ましい発展のなかに埋もれていて，ようやくその重要性が認識されるのは19世紀末の農業恐慌期からであった。矢島武は19世紀の末葉以降を農業経営学の「経済段階」，それ以前を農学一般と未分化の「農学段階」としている。1877(明治10)年に札幌農学校で講述された農場管理論が「農学段階」に属するものであったことは明らかであろう。しかし，そのことはこの章の価値を低めるものでは決してない。個々の農業技術を総括するものとしての農場管理の重要性についてブルックスは随所で強調しており，論述も生き生きと熱のこもったものである。そしてこの熱気は，アメリカの南北戦争の終了後，新たに開放された西部のフロンティアに向かう人々のためにマサチューセッツ農科大学において講義されたものとして，この内容を読むことによってよく理解されるように思われる。

　南北戦争は周知のように奴隷解放を重要な争点として戦われたが，農業面では，メキシコとの戦争によってアメリカ領となった広大な西部地区の開拓を，奴隷農場方式で進めるか，独立自営農民によって進めるかの対決であった。北部の勝利は，ピルグリムファーザーズ以来のアメリカ民主主義の基礎である「独立自営の理念」の勝利でもあった。ホームステッド法によって分割された新天地を目指し，希望に燃えて幌馬車を進めていく開拓者たち，彼ら

のための必須の学問として伝授されたのが農場管理法なのである。

　そのようにして読むと，ブルックスの講義には時代の特徴が明瞭に刻印されていることが分かる。ここで対象とされている農場は，家族労働力を基本として経営される独立自営のファミリーファームであり，経営形態，作物，家畜，建物，農機具，そして四季の作業まで自由に選択することができる完全な「営業の自由」が前提にされている。これをこの時期なお封建時代の遺制を引きずるヨーロッパ大陸や日本の内地の農業と比べれば，そのアメリカ的特徴は明らかであろう。

　こうした独立自営志向が最もよく現れているのは，第3節「農場経営の種類」において所有形態による分類が示され，たとえ借金をしてでも若いうちから自営形態をとることを強く勧めているところである。自営は知識と技術を確かなものにしてくれるだけでなく，「独立心と誇りを持ち，男らしい」農業者を育てる；It makes him independents, self-reliant and manly。マサチューセッツに育っていたのはこうした強烈な独立自営の精神なのであり，それが南北戦争の勝利によってさらに高揚されて，湯気の立つような熱気を伴って札幌に伝えられていたことを，ブルックスの講述は私たちに教えてくれるのである。

3）農業経営学，経済学のその後の発展

　先述のように19世紀末の世界的農業恐慌は経営学，経済学の重要性を認識させ，農学から農業経営学，農業経済学を独立させることになった。ドイツではゴルツの『農業経営学入門』，アメリカではウォレンの *Farm Management* がその画期を示す業績とされている。その後の展開を見ると，アメリカでは純粋に私的経済の最大化を目指す経営学が発展し，ドイツでは国民経済における農業の位置づけや国家の役割を重視する農業経済学や農政学に比重が移るという特徴が見られる。

　このような農業経営学，農業経済学の発展は，札幌農学校にどのように伝えられたであろうか。ブルックスら外国人教師が札幌を去り，札幌農学校卒業生が教鞭をとる時代になると，早くも農業経済，農業簿記，農政学，経済学，農業史などの科目がカリキュラムに現れ，佐藤昌介(1期生)と新渡戸稲造(2期生)がこれらを担当している。経営学や簿記は佐藤，農政や経済は新渡戸という分担が見られ，佐藤を農業経営学の，新渡戸を農業経済学の日本人としての元祖としてよいであろう。

　佐藤も新渡戸も卒業後アメリカに留学し，ともにジョンズ・ホプキンス大学でイリーから農業経済学の指導を受けている。佐藤はここでアメリカ土地制度史の研究で学位を授けられて帰国し，日本人として初めての札幌農学校教授として『威氏〔イリー〕経済学』『世界農業史論』などを著す。他方で北海道開発，北海道農業のあり方について政府や北海道庁に多くの提言をなし，北海道農会の会長として農業団体の育成にあたるなど実学の道に徹している。新渡戸はアメリカからドイツに留学し，ドイツ歴史学派の経済学，農政学を本格的に学んでいる。農学校に着任してからは農政学や経済学のほかに，ブルックスの講義の後継と思われる農学総論を担当しているが，彼の病気療養中にこの講義ノートをもとにして書かれたのが有名な『農業本論』(1898(明治31)年。1976年に農山漁村文化協会から明治大正農政経済名著集第7巻として復刻された)である。これは農業を社会科学的に論じた本邦初の労作であ

り，名著として今日まで読み継がれている。佐藤，新渡戸の学問は，すでに「経済段階」に入った経営学，経済学としてブルックス講義を一歩進めているが，その中心的な主張にはブルックス講義のスピリットが濃厚に息づいている。佐藤の有名な「大農論」は，資本家的大経営ではなく，内地的零細規模から脱した家族経営としての適正規模を言っているのであり，経営組織としては「農牧混同」の有畜農業を勧めるなどブルックスの弟子としての面目躍如の観がある。新渡戸の『農業本論』は地主制の弊害を説いた点でも本邦初であり，独立自営の精神が生きている。

佐藤が総長となり，新渡戸が去った後の札幌農学校からは，日本で初めて『農業経営学』のタイトルで本を書いた伊藤清蔵(1900(明治33)年農業経済学科卒，盛岡高等農林学校教授時代の著書，前出名著集第8巻に復刻)，同じく最初の『農政学』を書いた高岡熊雄(1895(明治28)年農業経済学科卒，新渡戸の後任教授となり，1933-37年北大第3代総長，1950年北海道総合開発委員会委員長を歴任した)らが出た。特に高岡は後に日本社会政策学会会長となり，わが国の社会科学全体の指導者として活躍することになる。しかし彼らの学問は農学校初期のアメリカ的特徴を失い，著しくドイツ的な性格を帯びてくる。そしてそのことが当時の国家の大学に対する要求だったのである。

しかし，ブルックスが説いた独立自営農民の精神が，やがて日本の社会科学全体を覆うことになる国家主義，全体主義の堅い岩盤を破って随所に噴出することを私たちは知っている。東北帝国大学農科大学時代の教授有島武郎(1901(明治34)年農業経済学科卒)の農場解放は，その最も鮮烈な例であろう。そうした動きがやがて地主制批判の農民運動となって戦後の農地改革につながっていく。そうした日本における農業民主主義の流れをさかのぼっていく時，私たちはその源泉にブルックスを見出すのである。

〈追記〉　この稿をまとめるにあたって，北海道大学農学研究科黒河功教授の貴重なアドバイスを得ました。また高岡熊雄以降の農業経済学については，拙稿「農業経済学の形成と川村琢の位置」(太田原高昭『北海道農業の思想像』北海道大学図書刊行会，1990年，所収)を参照してください。

［太田原高昭］

2．北海道の牧草

6-4〜6は，本邦に初めて導入した牧草について述べている。北海道開拓使は北海道に大規模な畑作・畜産経営を構築することを想定し，1871(明治4)年に東京青山の第三官園にアメリカからの輸入牧草を試作した。次いで1874(明治7)年に，函館の開拓使七重官園にアメリカからチモシーほか17草種を導入した。この開拓使の記録が，これまで北海道における牧草栽培の始まりとされていた。しかし，田辺安一の最近の研究によると，幕末から明治初めに騒ぎとなったガルトネル事件の当事者であるプロシャ人のR.ガルトネルが撤退するに際して残した，「日本政府に引き渡すべき開墾場の附属財産目録」と元会津藩士の箕輪醇が残した同農場での「伝習雑記」を精査すると，すでに1869(明治2)年にガルトネル農場で牧草が栽培されていたことが分かってきた。その後の普及を考慮すれば定説のままでよいが，単なる最初の栽培では後者に軍配が上がることになる。その後，酪農普及の祖と言われるエドウィン・ダンが札幌に引っ越した1875(明治8)年には札幌，根室官園でも作付けられた。

1876(明治9)年に札幌農学校の開校に伴って農場用地とされた札幌官園の一部は，輸入牧草が播かれていたかどうか明らかでないが，排水が悪くて明渠を掘らなければならなかったし，収穫した「日本草」は繊維が多くて栄養価が低いため，セイロで蒸して牛の増体重を高めたとクラークが『札幌農黌年報』(第1年報)に報告している。そこでブルックスは1877(明治10)年に数種の牧草を導入した。

　従来の家畜飼養法は，野草や稲藁，麦稈を飼料としていた。1876(明治9)年にダンは「野草は栄養的に劣り，特に冬期間の貯蔵粗飼料には外国からの牧草が必要である」と牧草の必要性を提言し，さらに「試作結果から，イネ科牧草ではチモシー，オーチャードグラス，レッドトップ，ライグラス，ブルーグラスなどが適しており，混播するマメ科草種としてアカクローバが北海道に適する」と述べている。当時の北海道の内陸は原始林で，いわゆる草原には海岸でミヤコザサ，シバ，湿地でイワノガリヤス，スゲ，ハマニンニク等が生育していたと思われる。

　以上がブルックスの講義の始まった頃(1877年)の日本の状況である。一方，この頃のアメリカは，1849年にカリフォルニアで金鉱が見つかり，1865年に南北戦争も終わり，西部開拓を進めていた。草地農業は，大鎌による刈り取りとヘイフォークによる集草形態から，ようやく畜力機械化が始まったばかりで，有刺鉄線もこの頃実用化されている。

　ブルックスの講義は，上述のイネ科5草種，マメ科のアカクローバを含むイネ科13とマメ科4の計17草種を扱っている。これらはアメリカの北部州で栽培されていたものである。これらの草種には学名が添えられているが，1867(慶応3)年にパリで第1回国際植物学会が開かれ，植物命名規約が定められたのを早々に引用したものである。一部に今日の学名と異なるものもあるが，植物名の検索には非常に有効であった。講義ではレッドトップ，ブルーグラス類は優れたグラスとされているが，現在では，乳牛に対する嗜好性や施肥反応が劣るので低級牧草とされている。当時は，牧草の成分，品質，嗜好性や施肥反応を考慮せず，収量と良質の乾草ができるかどうかを重点的に捉えていたから，上述の地下茎型のイネ科牧草は草勢が優れていたので，適応するとされたものと思われる。またマメ科牧草の4草種のうちアカクローバは北海道の気候によく適応するが，刈り取り時期の関係でよい乾草を得るのは困難である。しかし，窒素固定をする能力があるのでクローバの混播を勧めている。一方，アルファルファは耐寒性が十分でないので，北海道での栽培に適するかどうか疑問視しながら，根が深く入るので乾燥の続く地方では有用であるとしている。今日では，北海道の土壌が酸性でアルファルファの栽培にはあまり適していないとされるが，当時は土壌の化学的性質について考慮されていなかったと思われる。また，アルサイククローバは，おそらくシロクローバとアカクローバの雑種であると述べているが，これは誤りである。メンデルの法則が1865(慶応元)年に発見されながら，1900(明治33)年まで周知されなかった時代の止むを得ない結論である。

　その後，ブルックスの後任のブリガムの指導を受けた1893(明治26)年卒業の小川二郎は，北海道農業の振興に強い意志を抱いて札幌興農園を開業し，輸入種子や農具を販売するとともに，数千haの農地を得て牧草栽培を行ない，率先して社名の通りに興農事業を進めた市

井の人であるが，1902(明治35)年には『家畜改良・牧草論』を出版した。札幌農学校や著者の牧草導入試験の結果やアメリカでの最新の研究成果が，寒地型牧草と暖地型牧草について記述されている。これはわが国における牧草についての最初の本格的な指導書である。そこでは「牧草とは，その茎葉を家畜の飼料とする目的をもって，とくに耕作される舶来の草」と定義している。

今日アメリカ合衆国の牧草・飼料作物の教科書 *Forages*(1995年の第5版, Iowa State University Press)に寒地型イネ科牧草として取り上げられているのは，チモシー，スムーズブロムグラス，オーチャードグラス，リードカナリグラス，フェスク類，ライグラス類，ホイートグラス類などであるし，マメ科ではアルファルファ，アカクローバ，シロクローバ，バーズフットトレフォイル，その他のクローバ類である。

ブルックスの講義で取り扱われていない草種で *Forages* の各論で取り上げられているのは，スムーズブロムグラス，リードカナリグラス，バーズフットトレフォイルである。これらの草種は1884，1885，1900年以降にそれぞれアメリカに導入されている。小川の『家畜改良・牧草論』にはスムーズブロムグラスがオーンレッスブロムグラスとして取り上げられている。

当時としては最新の知識，技術を講義したと思われる。従来の飼養方法から，牧草，トウモロコシ，根菜類を栽培し，給与し，さらに冬期間の貯蔵と発展し，今日の畜産の基礎となった粗飼料栽培の最初の講義である。その後，西洋式農業の浸透と発展は戦後まで時を要した。21世紀の環境保全型農業には，土—植物—動物の関係を維持していく必要が今後も続くだろう。

なお，6-6は，乾草を収穫する作業機を扱っているが，これらほぼすべては，開拓使が輸入して農場と官園で利用した畜力作業機であり，それらが今日も重要文化財「札幌農学校第2農場」に保管されていて一見の価値がある。また，これら乾草収穫機は，明治30年代にサイロが紹介された際にエンシレージカッタ類が加わったことを除いて，そのまま第2次大戦後のトラクタ化時代まで継続されていて参考資料も多いため，説明を要しないと思う。

［中嶋　博］

3．輪作講義の補足

ブルックスは，6-8で「輪作」について別に詳細説明しているとして簡略に述べている。その説明は，おそらく毎日午後に行なった「農場実習」の時に，輪作が畑作の重要事項であると位置づけ，目的や効果，実施法など克明に述べたと思うものの，本文からは内容が理解できない。特に日本人には，連作のできる稲作と異なって理解に苦しんだであろうし，今日の視点でも，泰西農業の重要事項としてどのように説明されたか知りたいところである。そこで20年ほど後に発刊されたブルックス著の *Agriculture* (1901)から，輪作部分(Vol. 2 Manure, Fertilizers and Crop, pp. 374-379)を引用して解題に代える。なお，第7章の作物各論には，前作物に何が適するかということを書いて輪作の順序に言及していることを付記する。

CROP ROTATION

What rotation is and its objects; The term crop rotation is used to designate the system whereby different crops are made to succeed each other in certain regular or definite order. A rotation is completed when the last one of the crops which are included in it has been harvested. The number of years that the rotation lasts may vary widely. It is designated by prefixing the figure to the word course. A four-course rotation is one including four crops and lasting four years. The object in view in crop rotation is to secure a larger aggregate product from a soil than would be secured should a proper system of arrangement of crops be neglected. It was formerly thought that the roots of our various crops each excreted certain substances harmful to that crop, but harmless or at any rate less harmful to other crops. It was believed that as a result of this root excretion the soil was gradually poisoned, as it were, for a crop continually grown, and that therefore a change was necessary. It is now well understood that this theory is incorrect. Roots do not excrete substances necessarily harmful. We are now able to give numerous other and important reasons to explain the well-known benefits which follow a proper system of rotation.

1. Reasons why crop rotation is beneficial

(a) Food requirements: To some extent the different crops consume the various elements in different proportions. They also have varying capacity to extract the different food elements from one and the same soil: one crop being able to obtain, for example, a sufficiency of phosphoric acid and potash in a soil where another, which may require no more of these elements, is unable to extract the quantity it needs. Some of our most important crops have been classified in accordance with their food requirements in paragraph "classification of crop". It is evidently an advantage, after producing on a soil a crop consuming a large proportion of one element, to follow with another whose demand upon the soil will be different. It is in this connection that one of the chief advantages of rotation is found.

(b) The depth of the root system: Our crops vary widely in respect to the depth to which they send their feeding roots. It is the object of the farmer to draw upon the soil for the food which his crops need as largely as possible. Clearly, then, a rotation should include deep rooting as well as shallow rooting crops. The former will take from the soil food the latter cannot reach, and thus the farmer more perfectly utilizes the resources which nature places at his disposal. Among the more important deep feeding crops may be mentioned all the root crops, cabbages, clovers, alfalfa, and hops. Among rather shallow feeding crops may be included most of the grasses, cereal grains, Indian corn, all the melons, squashes, cucumbers, onions; and strawberries. Beans and peas, potatoes, and tobacco send their roots to medium depth.

(c) All crops belong to one of two classes, nitrogen gatherers or nitrogen conservers; Under green manuring, the important distinction which exists between different crops in respect to the sources of the nitrogen they require has been pointed out. It will be remembered that all of our common crops, except the legumes, must take the nitrogen they need from the soil, while the legumes can draw upon the air. It will be remembered that when the legume is harvested it leaves the soil richer in nitrogen than before its growth, if it has been produced under the right conditions; a considerable surplus of nitrogen remaining in the stubble and roots, after the decay of which this nitrogen becomes available to the following crop. It will be remembered that nitrogen is the most costly of the elements of plant food. It must be clear, therefore, that an important economy may be effected in crop production if in every rotation at least one nitrogen-gathering crop can be included.

(d) Liability to disease, insect injury and weeds.

1) Disease: Most of the diseases which affect our field and garden crops are due primarily to the growth upon or in some part of the plant of a parasitic fungus. The parasitic fungi are of microscopic dimensions but none the less real. They are propagated in most cases by means of spores, in accordance with the same general rules as those which govern the propagation of the larger plants of the fields and woods. Each disease from which our various crops suffer is produced by a specific fungus. Potato rot by one fungus, potato scab by another, corn smut by another, and so on throughout the entire list of fungoid diseases. The fungus which can grow in or upon one plant, causing it to become diseased, is usually powerless for injury on other plants, though in some cases it may affect a few very closely related plants. It seldom happens that a crop is produced in which there are not some diseased plants. The amount of injury may be so slight as not to attract attention, but the fungi which cause plant diseases are most of them capable of multiplying with extraordinary rapidity, and thus, since we practically always hive some disease, if we continue to grow the same crop for a number of successive years on a given soil, the amount of disease tends to increase, and often at a very rapid rate. There is, it is true, a wide difference between different diseases in respect to the rapidity with which they increase. Some increase so slowly that it is safe to cultivate a crop for many years in succession, while in other cases disease tends to increase so rapidly that to cultivate a crop for even two successive years upon the same ground is unwise. Among diseases which do not appear to increase very rapidly, may be mentioned corn smut, ergot in rye, onion smut; and among diseases which increase very rapidly maybe named potato rot, potato scab, and club root in cabbages, turnips, *etc*. The advantage of rotation, as a means of lessening the ravages of plant diseases, must be evident.

2) Insects: The principles underlying the advantages connected with insect injury which may be derived from rotation are identical with those which have just been stated under disease. To a considerable extent, each crop has its own insect enemies, and these, in many cases, are entirety harmless to other crops. It will be understood that there are wide differences between insects: some have a large number of food plants, while many others prey only upon one or a few closely related plants. It must not be expected that by rotation we shall succeed in entirely escaping the ravages of insects, because many species at some stage in their life travel freely from field to field. There are many, however, which are quite sluggish in their movements and by rotation we, to some extent, lessen the probability of injury.

3) Weeds: The reasons connected with the growth of weeds which oftentimes render rotation an advantage are in part the same as those considered under the last two heads, for, to some extent, each crop has its own specially injurious weeds. There is, however, another important consideration. With certain crops,

there is much difficulty in keeping down the growth of weeds. Fields and gardens tend to become increasingly stocked with weeds of certain kinds. This is especially true of most of the grains as they are grown in this country. Other crops which receive more careful culture, or which come up quickly, grow rapidly from the start, and quickly shade the ground, are much more favorable to the keeping down of weeds; and, in order that the spread of these pests may be lessened as far as possible, it is expedient to include at least one crop of this character in every rotation.

2. Planning the rotation

The particular arrangement of crops in rotation which will prove best must vary with individual conditions. A system which meets all the scientific requirements may prove ill adapted to the needs of a particular farmer gardener. Crops which can be profitably produced are determined by local conditions of soil and market, and these vary widely. The planning of a rotation must, therefore, be largely left to the individual. Each for himself should aim to adopt a system satisfying at the same time the scientific principles which have been pointed out, and the economic requirements of the individual case. A few examples of systems of rotation, which have been found widely useful, will be given largely as a means of illustrating the application of the principles which have been laid down.

(a) The Norfolk system.

The Norfolk system of rotation is in extensive use in England and, with slight modifications, in many other countries. It takes its name from the county in England where it was first extensively followed. This is a four-course system, and the crops and the order in which they are placed is as follows: turnips, barley, clover, wheat, each for one year. The turnip may be either the English or the Swedish. This is a rank feeding crop and to it a large part of the manure made upon the farm is applied. It is planted late in the season so that the ground can be cleansed from weeds before the seed is put in; it then grows rapidly, shades the ground, and keeps down weeds. It is the cleansing crop in the rotation. It is a deep feeding crop and is especially benefited by potash. This is followed by barley, a more shallow feeding clop, a grain which does best on the old fertility of the soil. It is one of the crops especially requiring available nitrogen. The clover which follows the barley is a deep feeding crop, a nitrogen gatherer, especially benefited by potash. This leaves the soil in good condition for wheat, a nitrogen consuming crop. The principles which underlie rotation, as will be seen, are therefore very perfectly met by this system. No one of these crops is likely to be affected by the diseases or insect enemies which attack the preceding crop.

(b) Modifications of the Norfolk system.

1) On farms where a large amount of live stock is kept, it is desirable to produce more forage than the Norfolk system would give, since under that system only one-fourth of the land would be kept yearly in clover. On such farms, therefore, the following rotation is often followed: turnips, barley, mixed grass and clover, mixed grass and clover, mixed grass and clover, wheat. This makes a six-course rotation and one-half the land, as will be seen, would yearly produce hay.

2) In districts where sugar beets and cereal grains are largely grown, and where the quantity of live stock kept is not very large, the rotation is often as follows: sugar beets, barley, clover, wheat. This, as will be seen, is but very slightly

different from the Norfolk rotation.

3) Other possible modifications of the Norfolk system, which may render it better adapted to special conditions, will readily suggest themselves to any intelligent farmer. A few of them only will be named: mangel-wurzels may be grown instead of turnips, if desired; oats might be substituted for the barley; or rye for wheat. Modified in any of these ways, the rotation would still satisfy all the scientific requirements.

(c) Rotations followed in corn growing states.

In the chief corn growing states of the prairie regions, the following rotations are common: grass and clover for three years, corn two years, wheat or oats one year. With either of these grains would be sown mixed grass and clover seeds. This makes a six-course rotation and one-half the land would each year produce a hay crop. Under this system of rotation, most of the manure of home production is applied to the grass. In localities where the soil is especially rich, the corn is often grown more than two years, and on the other hand if it is poorer than the average or ill suited to corn, that crop may be grown for but one year.

Another rotation followed in many parts of the West is as follows: corn, oats, wheat, grass and clover, the seeds of the latter being sown in the wheat. In this rotation the manure is applied to the oat stubble in preparation for wheat.

(d) Terry's rotation.

Mr. T. B. Terry, the skillful farmer and prominent agricultural writer of Ohio, practices the following three-course rotation: clover, potatoes, winter wheat. Under this system, he finds that the productive capacity of his farm has been greatly increased. The frequent introduction of the clover crop is, in Terry's opinion, the chief reason for this improvement.

(e) A dairy farm rotation.

A rotation which is common in many parts of Massachusetts where dairying is prominent, and where the potato is usually a profitable money crop, is as follows: potatoes, corn for two years (the second year for ensilage), grass and clover three years. Under this system one-half the ground is annually in grass. The potatoes are raised chiefly on fertilizer. Homemade manures are applied to the corn crop, together with some material furnishing potash. The grass and clover seeds are sown in the standing corn, at the time of the last cultivation, and the third year the grass is often profitably top-dressed with nitrate of soda. This system is susceptible of several slight modifications to fit it better for local conditions. Corn as a grain crop may not be required or may be unprofitable, in which case corn may occur in the rotation, but one year. Or it may be that three years in grass and clover will give too large a proportion of hay ground, in which case these crops may occupy the land two years instead of three.

［髙井宗宏］

4．果樹について

 6-10 の果樹に関する事項は第 7 章の野菜と併せて解説しているため，第 7 章解題「3．果樹および野菜について」を参照してください。

5. 農道と牧柵

6-11 は，農場内の道（農道）とフェンス（柵，もしくは牧柵）について解説している。実際にこれから農場を作ろうとする者にとって，実にていねいにまた具体的な説明がなされている。

考えてみると，例えばヨーロッパやわが国の内地府県では，圃場内の農道や柵などの設置方法については数百年から千年近い歴史を持ち，伝統的に最も効率的もしくは集約的な方法があり，現在残っているかどうかは別にして，現場では農道や柵の構造や設置に考えを巡らせる必要がなかったものと思われる。北米の，特にブルックスが引き合いに出している中西部では，こうしたヨーロッパの伝統を中西部の大平原開拓に生かしつつ，ここで述べている技術がマニュアル化されたものだろう。北海道開拓当初，内地府県とはまったく異なる風土のなかで，新たに農場を拓く際はこうした講義は非常に有用であったことは想像に難くない。

農道については，その構造から配置，土壌との関係，湿地の場合や沼地に道を開く方法，さらには農道の橋のかけ方まで講義されている。こうした「農場を新たに拓く際」の基礎的な技術について，現在では農業工学の分野で詳しく述べた資料があるだろうが，畜産学分野では，こうした知見は現場に出てから実地に体得していくことがはるかに多く，講義としてはあまり聞いた覚えはない。今になってこの講義録を読むと，その内容はなるほど，その通りだと思う記述が多い。

柵についての記述も興味深い。ブルックスは農場における柵の難点について，①経費がかかる，②土地を占有してしまう，③耕作の邪魔になる，④害虫や雑草の隠れ場所になる，⑤農場の景観を損なう，の5点をあげている。家畜生産にとっては柵は必要不可欠なものであるので，こうした観点は斬新であるし，100年以上前に農場景観についても考慮に入れるという観点は目を見開かれる思いである。

牧柵については，特にマサチューセッツにおける当時の法律を例にとって説明している。家畜が脱柵した場合，ランドオーナーがその被害について弁償の義務がある場合とない場合があるが，それぞれプラクティカルな理由があるのであろう。Fence Viewer という役職があったという点も興味深い。

いくつかの牧柵の種類と構造が説明されている。まず，Stone Wall, Post and Rail Fence および Post and Board Fence が説明され，材料が揃うなら石壁が最もよいとしている。わが国でも東北の軍馬放牧場跡地で，こちらは石ではなく土壁だが，同じような構造物を見ることができる。Post and Rail Fence は一見いわゆる材木を組み合わせた普通の牧柵のように思えるが，実は柱に穴をうがち，これらに横木を挿入する方式で，以前は北大構内の研究農場でも一部見られたものである。現在，このタイプの牧柵は，静内にある北大研究牧場に，学生に対する教材の意味もあって100 m ほどが残されている。材はミズナラで，後述される柵の材料としてもブルックスはオークとして記述している。ただし，ミズナラで作る場合は非常に大きな原木がいるとされており，研究農場のそれは御料牧場時代に切り出した巨大なミズナラから，斧，手斧（ちょうな）とくさび（矢）で切り出したものと聞いている。Post and Board Fence は字面から板柵を想像するが記述は角材を横木とした柵であり，この構造は現在の北

米の家畜管理施設の教科書である *MidWest Plan Service* に示されている牧柵とほぼ等しい。こうした技術の伝統は北米ではまだ生き続けていると言える。

　北海道に適した柵としては4つをあげている。Stakes Fence, Virginia Fence, Worm Fence および Wire Fence である。Stakes Fence は丸太を地面に打ち込んで上部を横木で固定する柵であり，湿地や沼地に適切であるとしている。Virginia Fence は柱をX字型にクロスさせて立てて12 ft ほどの横材で支え，これに横木を渡していくものである。また Worm Fence は基本的に縦の柱を使わず，横木を地面に90°よりやや広い角度で互いに重なるようにジグザグにおいて柵を形成する。ブルックスはご本人も，Worm Fence が「奇妙な名称」だと紹介しているが，カナダのオンタリオ州で Snake Fence と言われる柵と同じ構造である。どちらもくねくねと地面にのたうって延びていくことからつけられた名称であろう。このフェンスは多量の材が必要で，また柵自体が占める面積も大きいという欠点がある。Wire Fence については鉄線をそのまま張る場合とバラ線，いわゆる鉄条網を使用する方法を述べており，両者の長所短所が記述されている。北米やわが国では，鉄線をそのまま張る牧柵は電気牧柵以外にほとんど見かけないが，アルゼンチンでは現在も一般的で，たて杭に穴をあけて鉄線を通し，7段張りで張り巡らしている。1930（昭和5）年に書かれた『アルゼンチンの農業』という書物にも当地の牧柵としてまったく同じ構造が記述されており，北米で使わなくなったものが，南米でまだ技術として伝承されている点が興味深い。

　Virginia Fence および Worm Fence は，どちらも杭を地面に打ち込む必要がないという特徴がある。ただし Stakes Fence も含めて多量の材が必要である。現代の家畜生産現場で，この技術を生かす状況としては2つあるだろう。1つは東南アジアなどの開発途上国で，特にインドネシアなどの山火事跡地での利用である。こうした地帯で家畜生産を行なうには牧柵が必要であるが，多量に残存する燃え残りの材や立ち枯れた材を利用し，大規模な作業機を用いないで牧柵を作り得る。いま1つは現在のわが国の林間放牧などでの間伐材の利用であろう。

　生垣はイギリスで一般的で，うまく作られたこの柵は恒久的であり，また風をも防ぐとしている。生垣にすべき木類の種類を落葉樹と常緑樹に分けて，学名とともに記述し，その長所短所，さらにどうやって育てるかに至るまで記述している。ちなみにセイヨウサンザシ（Hawthorn）の場合，種を採取してから苗圃で育てて柵として役に立つまで，ざっと12年かかるとしている。

　最後に牧柵用の材として適切な木材について述べている。ここで言っている chestnut はヨーロッパグリらしい。またニセアカシアも優れた柵材であるとしており興味深い。ミズナラについてはすでに述べた通りであり，トウヒ，杉や松はあまり評判がよろしくない。

[近藤誠司]

6．農作業の観点から

　第7章の作物各論，第8章の畜産では，種類，特性や管理法を述べるが，6-12〜15 でそれらの実際の作業法や手順を述べている。これは，講義が数種類の作物を栽培し，さらに家

畜を飼育するという混同農業を前提にし，それらを満遍なく見る農作業を行なう必要から，実際面に沿って整理したと考えられる。すなわち，春夏秋冬の季節変化に合わせて，何を行ない，何に注意すべきかなどを整理し，農家の思考と作業を容易にしようとしている。このような扱い方は，当時，泰西農業の実用書に多く見られたし，江戸時代の農書でも多いから，理解しやすい書き方と思われていたのであろう。

　江戸時代の農書は，著者の経験や見聞を記録して広く公開するというよりも，一族への秘伝として書き残したものも少なくないため，最初に「年中行事」や「耕稼春秋」などとして年間の作業手順を記し，次いで作物や作業ごとの詳細を述べるという形態が多く見られる。また，随所に作業に際しての精神論を追記していることも特色にあげられる。

　明治維新政府は殖産興業政策に則り，進んだ西欧の農学と農業技術を紹介するため，組織的に泰西農業の翻訳本を刊行させているが，その代表書に『斯氏農書』という1000頁を超える大冊があるが，これも作業手順に従って書かれた農業指導書である。本書は，Henry Stephens, *The Book of the farm* を内務省勧業寮で1875(明治8)年に刊行し，1886(明治19)年に再版していて，当時のベストセラーにあげられる本である。この巻頭に「斯氏農書は農書の大成全備せるものなり，故に独り英国に行わるるのみならず，欧米各国亦咸之を重んせさるものなし，苟も泰西農事の全貌を窺はんと欲せは宜しく一本を蔵せさるへからず」と翻訳の趣旨を述べている。本文は「実地農業に通暁する最良法」など6項の総論を書いた後，「実務・冬の部」から春夏秋の順に作業や作物，生産資材等のあらゆる農業関連事項について述べている。例えば冬の部は，「天気及び田野事務の概略」に2頁，「事業〔農作業〕の準備及び着手」に2頁を書いた後，「犂及び其附属器械」では耕耘用の犂(プラウ)の種類から馬具と牽引方法を細説して19頁，「諸種の犂の作用」では犂ごとの「れき土」反転状況を述べて7頁，「犂耕法及び競犂」では作業法から作業能率，さらに競技会の方法まで扱って25頁，「犂鋤法各種あること」ではいろいろの耕起法を扱って44頁を費やすなど，実に細かく説明している。しかしながら，本書第4章耕耘で述べている通り，日本の犂(からすき)は明治40年代にようやく実用化されるにすぎないから，本農書の初版や再版が出された明治前半から中期は，まだ人力耕耘の時代であり，海外の耕耘技術に驚異の目をもって単に読まれたにすぎないとも言える。

　一方，北海道は開拓の初めから洋式農作業機を採用し，日本人に初めての畑酪混同農業を実現しようと努力してきたし，多くの北海道史書も発展経過を裏づける記事が多い。しかしながら，それを積極的に受け入れられたのは，移住経過や入植当初の悲惨さを考慮外として，伊達，八雲や静内の武士移住集団，それに華族農場系列のみであり，一般農家には農具と馬の支給など行政援助がなされるまで普及していない。武士集団は組織としての団結力によって，大農場は資本量によって，それぞれ新技術研修生を官園に派遣できたが，個人では自らの生活に精一杯で，新技術の受容まで時間的，資本的に手が回らないのが本音であった。具体例として官が推進した屯田兵制度を見ても，支給された農具等は人力作業主体であった。すなわち，農具は，鍬：大小各1，唐鍬：大2・小1，鎌：柴刈と草刈各1，山刀と鋸が各1，その管理用の砥石：荒中各1，鑢(やすり)，それに筵(むしろ)10枚，蚕籠20枚，麻扱函1，培養桶1である。

これを用いて，札幌近郊の琴似・山鼻に入った1877(明治10)年頃が5000坪(1.6 ha)，江別や根室に入った明治10年代後半が1万坪(3.3 ha)，その後，滝川から上川地区一帯に入った明治20年代から最後の屯田兵入植の1899(明治32)年までが1万5000坪(5 ha)の支給面積を耕すのであるから，自らの開墾と生産作業および農業経営に精一杯であり，プラウなどの新技術導入が遅れている。まして個人の入植者は屯田兵並みの入植者支援があるものの，すでに住宅の用意された屯田兵と違って自宅を作りながら開拓を進めること，年貢を納める小作人として入植する者が多かったこと，情報不足から仲買人や馬喰に搾取を余儀なくされることなどから，厳しさは一層大きかったと思わざるを得ない。すなわち，生活や経営の厳しさに差がなくとも，集団作業によって余剰人員を生み出して洋式農法の研修生を派遣できる集団の力が，北海道の洋式農法普及の基礎を作ったと言えよう。そのために北海道は，1923(大正12)年にドイツから，U. グラバウ，F. コッホの2農家を招請して模範農家として実際に経営をさせたほどである。

　ブルックスの講義は，6-12冬の作業から13春，14夏，15秋の順に作業を扱うが，毎日の午後に農場を使った実習があるため，講義では意義を中心に述べているところが特色である。例えば，前述の『斯氏農書』で扱っている馬具と牽引方法，犂耕法などが登場しないし，第8章で扱う家畜の飼養技術についても触れていない。これらは農学講義として必須の事項であるため，本書に農場実習での内容を補充しなければ欠陥があると言わざるを得ない。ともかくも6-12冬の項では，本来の薪や住宅等の整備や堆肥の扱いに簡潔に触れた後，窒素・リン酸・カリ成分の農業における物質循環に多くを割いている。肥料・土地・雨から供給された窒素・リン酸・カリ成分が農作物に蓄積され，その一部が直接に人間の食料となるし，残りが家畜の飼料となって乳肉に変わり，これが人間の食料となる。しかし，人と家畜が消化してエネルギーに用いる物質成分はごくわずか(例えば窒素は乳肉合わせて約10%)であり，多くが糞尿と敷きわらになる。これを堆肥として畑に戻すと，肥料のいくらに相当するかを述べ，物質循環に則った農業経営が賢い農民のあり方だと説いている。現在の物質循環学は，海外の食料と飼料を輸入して環境汚染をもたらす環境観念に主題があるが，ブルックスの経済観念からの指摘は，当時では大変に斬新であったし，今日の「地球環境維持や自然調和」と対比して意義深いものがある。6-13春の項では作物別の播種法を作業順に要旨を述べており，通常の農書の領域に止まる。しかし，6-14夏の項に入ると，栽培管理の意義を列記して本来あるべき作業法は出てこない。同様に6-15秋の項は収穫について述べるが，これは簡潔すぎて内容がない。これらは前述の農場実習とのかかわりで，学生が理解済みとして講義したため，本書の内容のみで完璧性を論ずべきでない。

［髙井宗宏］

注
1) dwarf：「矮(わい)性の」または「矮性のもの」の意味。現在では，リンゴ等の果樹で矮化栽培が実用化されており，他の果樹でも試みられている．この場合，矮性台木や矮化剤が用いられる。
2) ここでは札幌を指す。
3) charr は char と同語で「軽く焼く」，「焦がす」という意味である。

4) Hawthorn (*Crataegus Oxyacantha*)：セイヨウサンザシ
 5) Blackthorn (*Prunus Spinosa*)：リンボクの一種，→リンボクの場合は，*Prunus Spinulosa*
 6) Beech (*Fagus Sylvatica*)：ヨーロッパブナ
 7) Hornbeam (*Carpinus Betulus*)：セイヨウシデ
 8) Maple (*Acer*)：カエデ科の総称，*Acer Campestre* ならコブカエデ
 9) Holly (*Ilex Aquifolium*)：セイヨウヒイラギ
10) Osage Orange (*Maclura Pomifera*)：アメリカハリグワ
11) Chestnut (*Castanea setiva*)：ヨーロッパグリ
12) White Cedar (*Chamaecyparis obtusa*)：ヒノキ
13) Spruce (*Picea Abies*)：トウヒ
14) Locust (*Robinia Pseudoacacia*)：ニセアカシア（ハリエンジュ）
15) Oak (*Quercus*)：コナラ属の総称
16) Red Cedar (*Juniperus Virginiana*)：シダーウッド（エンピツビャクシン）
17) Pine (*Pinus*)：マツ属の総称
18) 今日は，plough（英），plow（米）とされている。本書では作業機名に plow，作業名に ploughing と使い分けているようだが，この2行後に plowing もあって混乱している。
19) 今日では in rows, row space がよい。drilling は条播となる。
20) ここでは密条播(dense drilling)を指し，6行後の grain drill で播く。散播(broadcast seeding)は，人の手でばらまくか，手回しの回転羽根で種子を飛ばす農具(7行後の sower)を用いる。
21) plow：培土プラウ，harrow：除草ハロー，cultivator：中耕除草機，horse-hoe：畜力カルチベータ

Chapter 7   Crop Cultivation

7-1. Wheat (Triticum aestivum)
 Wheat is the grain crop which is perhaps more extensively cultivated for food than any other crop. It is especially suited to the needs of mankind. Wheat ground and cooked without sifting from it any of its bran is a complete food. None of the other grains such as corn, barley, or oats can fill its place. Corn contains too much fat, barley makes those who eat it too phlegmatic, and oats when used extensively as food causes certain skin diseases. Wheat is native of temperate zones, and is found in latitudes 25-60° North. It will also thrive in the tropical zone and as far north [p. 712] as the northern countries of Europe. It is exceedingly hardy, and is able to adapt itself to a great variety of soils and climates. Wheat has in comparatively recent years been developed from the wild plant *Aegylops ovata*, but whether the varieties of wheat which have been cultivated for thousands of years originated from that same wild plant is not known[1]. In the year 1838, M. Fabre (the French agronomist) planted the seeds of that plant, and after 8 years of cultivation succeeded in modifying it to such an extent that the seed produced was as good as most wheat being cultivated at that time.

 There are many varieties of wheat, but those now being cultivated are generally considered to belong to one species. The different varieties have doubtless [p. 713] been developed through the cultivation of the plant over such a wide range of country and in so many different kinds of soil. The different varieties of wheat are so distinctive that some agronomist consider those belonging to several species, but all probably have a common origin. There are two quite distinct kinds namely winter wheat and spring wheat. Some writers give to the former the name *Triticum Hybernum* and to the latter *Triticum Aestivum*[2]. It does not appear, however, that winter and spring wheat belong to different species, since either one may be readily changed into the other. In order to change winter wheat into spring wheat, it is only necessary to sow it a month later each year[3]. Indeed, it can be changed [p. 714] by simply sprouting it late in autumn, checking its growth and then sowing the sprouted seed very early in spring.

 There are four quite distinct varieties of winter and summer wheat, Beardless White and Beardless Red, and Bearded White and Bearded Red[4]. There are also some intermediate varieties which, although tinted or slightly colored, cannot be called Red Wheat. The red varieties are generally hardier than white, and upon inferior soil are rather more productive; however, they do not make as much, nor as white flour as do the white wheat. On account of its productiveness, red wheat is as profitable for the farmer as white wheat. Upon soil especially suited to wheat, [p. 715] in a good state of fertility, and in a climate well-suited for this crop, white wheat will prove equally productive and will be more profitable than red wheat. All other things being equal, those varieties of wheat which are bearded are less profitable, since the beards make the wheat disagreeable to handle.

1) Characteristics of an Ideal Variety

(1) Early maturity: This quality is regarded as being of special importance in America, since those varieties which mature early are much less likely to be injured by insects. Whether this quality will be of equal importance hence in Japan, [p. 716] where the same insects may not be found, we cannot say. Even if those insects do not prove injurious, all other things being equal, we prefer the varieties of wheat which ripen the earliest.

(2) Prolificacy: There is a great difference among varieties as to the amount produced under similar circumstances. It is estimated that one quart of wheat brought from North Carolina in 1845 benefited the farmers in one county of Ohio $100,000 because of its greater productiveness. It is possible, by judicious selection of the best wheat in each crop, to greatly increase the productiveness of any variety. In 1857, an English gentleman by the name of Hallett planted wheat from a head which was 4¾ inches long, [p. 717] contained 44 grains of wheat, and produced 10 ears from the best stool. In 1862, the best ear of wheat which he produced was 9¼ inches long and contained 132 grains; in one stool, there were 90 heads. From year to year, he selected and planted only the very best wheat, and you see he made a great improvement.

(3) Midge-proof: An ideal variety of wheat in America should be midge-proof (resistant). The midge is an insect which often does great damage. Those varieties of wheat in which the kernel is well covered by the husk are much less liable to be injured by midges than those in which the grain is comparatively naked.

(4) Thinness of skin: Wheat should have a thin skin, since a certain quantity of grain will make much more flour than [p. 718] an equal quantity of thick-skinned wheat. In the manufacture of flour, the exterior portions of the grain are sifted out; hence, there is much more waste with wheat which has a thin cuticle.

(5) Hardiness in Winter: Some varieties of wheat seem much better able to withstand the cold of winter than others; all other things being equal, the best variety is of course that which is the hardiest.

(6) Regularity of the Rows of Grain: This is an indication of purity in wheat, and we should therefore select those varieties which have straight, regular rows of grain.

(7) Stiffness of Straw: Cultivated wheat has a tendency to lodge, and indeed upon almost any soil there is a greater or less tendency [p. 719] in this direction. Those varieties of wheat which naturally produce stiff straw are less likely to lodge than others, and therefore we should select such varieties.

2) Habits of Growth

Right figure represents a young wheat plant, showing the grain from which it has grown "S" with the primary roots starting from that part of the plant. "C" is a ring or bulb in the stem from which secondary roots will afterwards start; it is also from this ring that the branches of the stems start. The process of sending out additional stems [p. 720] is called "tillering"[5]. In rich soil, tillering occurs to a large extent. There is one instance on record where 500 stalks were, by division and transplanting, developed in one year from a single grain.

3) Varieties

We may divide all varieties of wheat into 3 principal classes;

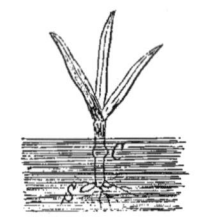

Hard Wheat, Soft Wheat, and Polish Wheat. Hard wheat is cultivated in [p. 721] warm climates, soft wheat in temperate climates, and Polish wheat principally in Poland. The Polish varieties are also hard.

(1) Pedigree Wheat: This variety was originated by careful selection of seed over many years. It is a white wheat of excellent quality, and the variety is productive. However, it does have one defect—it shells too easily, and in the handing necessary in harvesting, a considerable grain is lost.

(2) Red Blue Stem: This is a variety of red wheat originated in the states of Pennsylvania, and was for many years very extensively cultivated. Its quality is very fine.

(3) Diehl Wheat: This is a variety of white wheat which is very early. Its grain does not shell too freely, the straw is stiff, and the chaff is close [p. 722] to the kernels, so this grain is little liable to be injured by insects. This variety is fairly productive and of good quality; hence, all things considered, it is very desirable.

(4) Weeks Wheat: This is a variety of white wheat which has awns; it is of average productiveness and has many good qualities. It makes flour of fine quality.

(5) Red and White Mediterranean Wheat: These are both very good varieties, although their straw is sometimes too weak.

(6) Fullz Wheat: This variety is presently cultivated more extensively, perhaps, in the principal wheat-growing regions of America than any other. It is white wheat which possesses almost all desirable characteristics. The flour from this variety is of fine quality, and the quantity [p. 723] produced is large.

(7) Clinton Wheat: This is a variety of white wheat which has been developed quite recently. There seems to be a considerable diversity of opinion as to its value. All observers agree, however, that its straw is remarkably stiff, it is very hardy and much more productive than most other varieties, and that it has a very fine appearance. Some expert say, however, that it does not make as good nor as much flour as some other varieties. Others assert with equal positiveness that it makes flour of the very best quality. On account of its many good qualities, I consider this variety worthy of extensive trial.

(8) Egyptian or Mummy Wheat: This is a variety of wheat which in recent times has been introduced from Egypt. [p. 724] The ears of this variety are of a peculiar shape, being very much branched; on that account, the name "Fingered" wheat is sometimes applied. Because of this branching of the ears, this variety is very productive, but its quality is not equal to that of many other kinds; therefore, this variety is probably unworthy of extensive cultivation.

4) Cultivation

(1) The Best Soil for Wheat: Though wheat may be grown upon a great variety of soils, it is generally admitted that it thrives best upon a somewhat heavy clayey loam; the presence of a considerable amount of lime in the soil is also considered beneficial. On account of its liability to winter killing[6], the soil upon which wheat is to be raised must always be well drained. Soil of almost [p. 725] any kind, provided it is in a high state of fertility, will produce large crops of wheat, but the grain is not of as good a quality as that which may be produced upon clayey loam. Mucky soil is poorly suited for wheat, although if well drained it may produce a fair crop; however, that grain will be of very inferior quality.

(2) Preparation of Soil: The wheat plant is very fastidious in its habits of growth,

and unless the soil is well prepared the crop produced will be very small. The soil should be somewhat compact for growing wheat, and the surface made mellow for the reception of the seed. The subsoil should not be excessively hard, but of medium compactness. [p. 726] If the ground has been deeply pulverized about a year previous to the time when the crop is sown, and if it is then ploughed shallowly and thoroughly harrowed, it will if of the proper sort, be in a good condition for wheat. A summer fallow, with repeated plowings and harrowing, perfectly prepares a soil for wheat; in the majority of cases, however, it will not be profitable to allow the land to remain idle for one year.

Among the crops which in this latitude will be suitable to precede wheat, the potato may be regarded as one of the most important. Potatoes cultivated upon land where wheat is to follow should be of an early variety, and should be harvested at least as early as the first of September. After harvesting that crop, if the land is free from weeds, it can be well prepared [p. 727] for wheat by repeated harrowing; in most cases, it will probably be best to give the land a shallow plowing, after which it should be harrowed. Great care should be taken to free wheat lands from weeds; if any are present, they should be picked up and carried off as they are brought to the surface by harrowing.

The crop which in all parts of the world is regarded as the best to precede wheat is clover. That plant seems to take in nitrogen and make a large amount of plant food available for wheat. Clover may be allowed to occupy the land for either one or two years; if it is desirable to do so, one crop may be harvested or fed off, [p. 728] and the second crop ploughed in. Clover sod should be ploughed several weeks before the seed is sown, in order that it may be somewhat rotted by the time the young wheat plants send their roots down to it. Great pains should be taken in plowing to do the work nicely, so that all the sod may be covered to a uniform depth. Flat furrowing will accomplish this objective most perfectly. ⟨See p. 251⟩ After plowing, the land may be rolled to press the sod down compact; by and before sowing wheat, it should be very thoroughly harrowed.

(3)Fertilizer: The best manure (fertilizer) for wheat are those containing a considerable amount of nitrogen and phosphoric acid. Peruvian guano, ground bone, [p. 729] fish guano, and ashes are among the best organic fertilizers. Good barnyard manure is of course good always and everywhere. It is not usually profitable to use very large quantities of manure rich in nitrogen, since doing so would produce straw at the expense of grain. The soil for wheat must contain an abundance of all elements of plant food; if any of those which I have not mentioned are deficient, they also must be added. Manure for wheat should not be buried deep in the soil; it is best to mix it thoroughly with the first 3 or 4 inches of the surface soil. If the manure is kept near the surface, the roots of the wheat plant will mostly spread out near the surface, [p. 730] and the plant, in the case, will be less liable to be injured by frost heaving.

(4) The Proper Time for Sowing Wheat: This varies according to the climate of the country, the condition of the land, and the probability of injury from insects. In countries where the climate is severe and where little snow falls, wheat must be sown earlier than in countries where there is an abundance of snow. If sown early, it makes a considerable growth before the cold weather sets in, and will be better able to withstand its effects provided there is little snow. However, if there is much

snow, the wheat is likely to be smothered during the winter [p. 731] if its leaves are luxuriant. Upon soils which are likely to heave, wheat should be sown somewhat earlier than is necessary on other soils. There is one insect enemy of wheat, which was brought to America from Europe; that insect frequently does great damage to wheat crops, and is more likely to injure wheat sown early than that sown later. Early sown wheat tillers far more abundantly than that sown later, and on that account it is better to sow early. In this vicinity, September 12th to October 1st is the best time for sowing, but it is desirable to sow before the 20th of September. If the crop makes too luxuriant a growth before cold weather commences, a portion of the leaves may be cut off with a scythe, in order [p. 732] to reduce the danger of smothering.

(5) Techniques for Sowing Wheat: There are 3 ways in which wheat is sown, it is dibbled, drilled, and sown broadcast[7]. If the operation of dibbling can be economically and rapidly performed, it will be the best system, since in it one kernel is planted in one place, and the kernels are equally distributed over the ground. Dibbling, however, is impractical on a large scale because of the length of time it takes and the expense it involves. If a good machine for dibbling could be invented, it would be a valuable acquisition.

The drill system is, all things considered, the best. There are many machines which will perform [p. 733] the work rapidly and well, and it can even be economically performed by hand. The distance between drills may vary according to the condition of the land, but from 9 to 15 inches is the most common range. In rich soil, since the wheat tillers extensively, it is not necessary to sow as much seed as on poor land. The drill system is superior to the broadcast system because it makes it possible to cultivate the crop, but on soils which are free of weeds the broadcast system will give good results. Another reason for preference of the drill system is that it requires less seed per acre. If wheat is drilled, 3 to 4 pecks[8] per acre will be sufficient in ordinary cases. Sometimes, as little as one peck per acre is sown, and the resulting harvest is excellent. [p. 734] If sown broadcast, 6 to 8 pecks per acre are necessary.

Wheat is subject to several diseases caused by the growth of parasitic fungi, the spores of which often adhere to the grain in considerable quantities. On this account, it is a common practice to soak seed wheat for a short time in some preparation which will destroy those spores. For this purpose, a strong brine (salt water) may be used, but a solution of blue vitriol (SO_4Cu) is much more effective. The quantity of vitriol necessary is quite small; 1/2 pound dissolved in a few gallons of water will be sufficient for a bushel of seed. [p. 735] The best method for soaking seed wheat is to put it into a basket and immerse the basket into the liquid. It need remain just a short while, and then it may be taken out and a quantity of air-slaked lime mixed with the grain.

Tab. 1 Number of seed grains per unit area for different quantities of wheat

per square foot	per square yard	per acre	amount (in pecks)
4	36	174,240	1
8	72	348,480	2
12	108	522,720	3

(6) Selection of Seed: It is a principle of universal application that like produces

like, and the farmer, in selecting any kind of seed for planting, should keep this principle in mind; the better the seed, the better will be the crop, all other things being equal. A large, full, plump seed produces much more vigorous plants than small or shriveled seeds. It is, therefore, always profitable to exercise great care in the selection of seed.

Wheat to be used for autumn sowing [p. 736] should be grown the same year, and newly threshed grain is better than old. It is often found profitable to go through the wheat field before the crop is harvested, and select the largest and best ears of grain for seed. This selected seed should be stored so carefully that it will not be heated up or be affected by weather. Just previous to sowing, it should be threshed, and the large ones separated from the small, the large only being sown. This separation may be effected by the use of a winnow having a sieve with a mesh of the proper size[9], or by throwing the grain in a thin stream towards the wind. In the latter method, small and imperfect grains will be blown to one side and may be easily separated from the larger. When soaked in brine or vitriol solution, some of the imperfect kernels [p. 737] will rise to the surface; these should all be carefully removed.

(7) The Proper Depth to Cover Wheat: This is a matter about which there is much diversity of opinion, some advocating deep covering, and others shallow. The depth should be varied somewhat according to the nature of the soil, the lighter and drier the soil, the deeper it should be covered. In general, a depth of $1\frac{1}{2}$ to 2 inches will be best.

(8) Cultivation: There are some farmers, even at the present time, who hold the opinion that the cultivation of wheat does no good [p. 738] even if it does no harm, but the more intelligent ones generally advocate cultivation. This should be done early in spring and until the grain reaches such a height as to make it inconvenient. On a large scale, cultivation is done with the use of implements designed for the purpose which are drawn by horse power, but the work may be quite economically performed by hand. Harrows are now generally used for the cultivation of wheat in the earlier stages of its growth.

(9) Harvesting: The proper time for harvesting wheat is matter about which there is some difference of opinion. It should be varied somewhat according to the climate of the country in which the crop is raised. It is the general practice in most countries [p. 739] to harvest wheat when it is "in the dough"—when the grain is fully formed, and of about the consistency of very thick paste. At that time, most of the straw will have a yellow color, although a portion of it may still be somewhat green. Harvested at that time and properly cured, the grain makes more and better flour than if it is harvested later, but this method of harvesting can be followed only in those countries where the weather at the time of the wheat harvest is sufficiently dry that unripe grain will cure without injury.

After being cut, the wheat should at once be bound into sheaves and put in shocks. The shocks should contain about 12 or 13 sheaves, set upon end close together with 1 or 2 sheaves laid over the top of these in such a manner as to shed rain; [p. 740] or what is even better, the shocks may be covered with hay caps whenever there is danger of rain. After having laid in shocks for a length of time sufficient to allow the grain to become hard and the straw well cured, the grain may be made into stacks and carried into barns, or threshed at once. If it is not convenient to thresh the grain immediately, it should be put into carefully made stacks rather than stored in large masses in large piles, because it will be less liable to damage in stacks. The stacks of grain should be raised a little distance above the ground, so that air may have a chance to circulate beneath it. There should be a ventilator, which may be made of poles running through the center of the stack. A suitable foundation for a stack may be made of pieces of timber or rails. [p. 741] The bottom of the stack should be made small; as the stack become taller its size should be very gradually increased until about the middle, from which point it should grow smaller to the top (see above figure). A round stack is better than one of any other shape, since there is less surface exposed in proportion to the volume of the stack of any other shape. Care should be taken to lay the bundles of grain very carefully, so that the butts will be even and the center of the stack always a little higher than any other portion. The butts of the bundles should of course be placed outward. When completed, the stack may be thatched, if it is to remain [p. 742] for a considerable length of time. Wheat straw is excellent for this purpose.

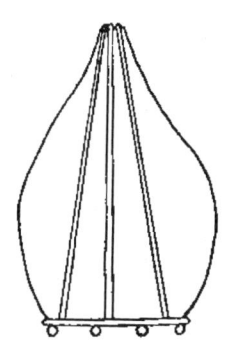

Threshing can now be most economically performed by means of machines for the purpose, of which there are many very excellent ones. Threshing by means of hand implements is laborious and expensive. Still, where labor is cheap and the quantity of grain not large, it will not be profitable to buy machines for threshing.

There are 3 principal ways of cutting grain by the use of sickles, cradles, and reaping machines[10]. Cutting by the use with sickles is the most expensive method of doing the work, but it can be done very excellently; therefore, in countries where labor is cheap, hand reaping may sometimes be advantageously followed. [p. 743] Cutting by the use with a cradle is also expensive, but the work can be more rapidly performed than with sickles, and therefore this method of harvesting may sometimes be best. A self-rake reaper is generally the most economical and best method of harvesting grain at the present time. The Buckeye and Wood reapers are two of the best machines. There are also self-binding reapers, which are coming into extensive use; wherever labor is expensive or scarce, it will be best to use such a machine.

[参考図]

5) Diseases

Some of the most notable diseases of wheat are Ergot, Chlorosis, Bunt, Smut, Rust, and Mildew. [p. 744]

(1) Ergot[11]: This is a disease caused by the parasitic fungus *Claviceps purpurea*. It only affects only a portion of the kernels of an ear of grain. Instead of a perfect kernel of wheat, the ovule develops a long spur-like projection, to which is given the name of "ergot". This disease is most prevalent in wheat grown in unfavorable locations, but it never affects wheat to as great an extent as it does rye. The amount of grain destroyed is not often sufficiently great to cause any serious

diminuition of the crop, but a very few ergotted grains—if ground with perfect grain and eaten either by man or beast- cause a terrible disease. Too much care, therefore, cannot be exercised in separating the affected grain from that which is [p. 745] to be used as food. Equal precaution should be taken not to sow wheat from ears in which any kernels have been affected by ergot. Care in selecting seed, and the soaking of seed as previously described, will ordinarily prevent this disease almost entirely.

(2) Chlorosis[12]: When affected by this disease, the growing grain turns yellow or white, and in extreme cases dies. This disease is caused by unfavorable conditions of the soil; the remedy is to avoid sowing under such unfavorable conditions.

(3) Bunt[13] or Smut Ball: This disease is caused by the parasitic fungus *Uredo caries* or *foetida*. With careless cultivation, this disease sometimes [p. 746] proves very destructive. The spores of this fungus probably adhere to the grain, so if the precaution of soaking it in vitriol or brine is neglected when the grain is sown, the spores begin to grow and the mycelium of the fungus soon penetrates the growing plant. Affected plants can be recognized by a fat, slightly crumpled appearance. These plants grow several inches taller than unaffected ones, but there is little left of the grain except the cuticle, the interior being filled with a black powder or paste which has a very disagreeable odor. As far as is known, a small quantity of this powder ingested into the stomach produces no very bad effects. Its odor, however, renders so a very small amount of it injures the appearance of flour. [p. 747] Too much care, therefore, cannot be taken in separating the diseased kernels from the good. In order to (as far as possible) prevent the spread of this disease, great care should be taken in selecting and soaking seed in some preparation that will destroy the spores in all cases where any grain is affected. Blue vitriol is perhaps the most effectual remedy, but solutions of Glauber's salt ($NaSO_4$) or ordinary salt ($NaCl$) will also be good. Quicklime in boiling water is also good.

(4) Mildew[14]: This disease is also caused by a parasitic fungus, *Uredo linearis*. It is most prevalent in light soils, and increases most rapidly in foggy or moist hot weather. It is more likely to affect heavy crops of grain than smaller ones. The fungus develops beneath [p. 748] the surface of the growing plant, and makes its way through the cuticle in the form of small, flat, deep brown sort.

(5) Rust[15]: This disease is also caused by a fungus, *Uredo rubigo*. Plants affected are covered with reddish spots, which are usually found to be most numerous upon the leaves (where they do comparatively little harm), but they are sometimes found on the glumes or other portions of the inflorescence, where they do more injury. This disease is favored by the same conditions which are likely to produce mildew, so the remedies must therefore be the same. In both cases, rank manure must be avoided.

(6) Smut[16]: This disease is also caused by a fungoid growth. It is called "smut" from the fact because a large number of spores in the form of a black powder [p. 749] develop upon the ears of grain. This disease commences before the spike bursts its sheath. In some seasons, smut is very abundant, but it is not generally much feared. The same remedies as those used to prevent bunt will be useful, but will prove less efficacious because the spores of smut are more widely distributed.

6) Insect Enemies

There are many different insects which prey upon the wheat crop. Almost all parts of the plant—blossoms, leaves, stems, roots, and even grains after they are harvested—are affected. There are several kinds of insects which prey upon harvested grain, among the most [p. 750] destructive of which is the weevil[17]. The most effective means of preventing the ravages of all insects that prey upon stored grain is kiln drying. The high temperature to which the grain is subjected in kiln drying is sufficient to destroy the life of all insects that may be present.

Among those insects which prey upon growing plants, the one most dreaded in America is the Hessian fly (*Cecidomyia destructor*)[18]. That fly lays its eggs upon the leaves of young plants in autumn. In a short time, the eggs hatch and the young larva makes its way into the interior of the stem, where it fastens itself near the bottom joint and lives by sucking the juices of the plant. If only one larva is present in a stalk of grain, [p. 751] little injury will result, but there are usually several larvae, and these will so lessen the vitality of the plant as to very much check its growth, or in extreme cases to kill it. The larvae remain in the stalk through the winter, and early the following season develop into perfect insects which lay another lot of eggs upon the same grain. The eggs soon hatch, the larvae make their way to the interior, and they live as previously described, most of them being at the bottom of the stem. When the grain is harvested, since the larvae are mostly near the bottom of the stem, they will be left in the stubble; therefore, burning the stubble after the crop is harvested is found to be one of the most effective means of preventing [p. 752] the increase of that insect pest. There are several parasitic insects[19] in America which prey upon the Hessian fly, and those are doubtless very useful. [p. 753]

7-2. Rye (Secale cereale)

Rye is a cereal grass distinguished from wheat by its narrow glumes and consistently twin florets having a membranous abortion between them; otherwise it is much like wheat in structure, although the straw is usually taller and tougher. Rye is at present found wild in the mountains of the Crimea at an elevation of 5,000 or 6,000 feet. There, however, the ears are no more than $1\frac{1}{2}$ to $2\frac{1}{2}$ inches in length.

In nutritional value, rye is generally considered inferior to wheat, but bread made from rye retains moisture longer and is more easily digestible. In color, rye flour is much darker than wheat flour. Rye bread has a peculiar taste agreeable to many, which taste is believe to lie in the outer portion of the grain. It is not, therefore, [p. 754] customary to bolt rye flour, only the very coarsest portions being sifted out. Americans and Englishmen much prefer wheat bread, and use that almost exclusively, but many of the continental nations of Europe rely almost entirely upon rye as a bread grain. Germans, in particular, eat a great amount of rye bread, and German workmen even think bread of rye is more substantial and better than wheat bread. A mixture of rye and wheat flours, in the proportion of '1:2' makes a very good flour. Rye is also sometimes used as feed for animals, but it is probably inferior for such uses to corn, barley, and oats. It may, however, be frequently used with economy as a portion of the grain allowance for animals.

As I have said, the straw of rye is longer and tougher than that of wheat. It is not, therefore, of much use as animal feed because it is not easily digested. [p. 755] In actual nutritional value, however, it is equal if not superior to wheat straw. The very qualities which make rye straw of little value as feed make it extremely valuable for numerous other purposes, such as thatching for the roofs of buildings, stuffing collars, filling mattresses, bedding animals, and making straw hats and numerous other articles manufactured from straw. Sometimes rye is cultivated exclusively for its straw, being cut somewhat early, carefully bleached, and manufactured into hats which are a very good imitation of "the Leghorn" in Italy.

As a fodder crop, either to be fed off the ground or for soiling[20], rye is of great value, particularly because of its early ripening, but its yield is large and this is also, of course, a valuable characteristic. As a green crop for plowing in, rye has many qualities to recommend it—its rapid, luxuriant growth even on comparatively poor soils, [p. 756] and the fact that it will make some growth where almost any other crop would fail as upon a very sandy soil. We therefore can see that rye, although not as highly esteemed as wheat, is a crop of no mean value, and should find a prominent place among the crops of every temperate climate region

1) Soil and Climate Adaptation

Rye thrives upon a rather light, sandy loam and will grow upon quite poor, sandy soils. It will do fairly well upon any well-drained loam, although heavy, clayey loam is to be avoided. Rye will succeed anywhere in the temperate zones, and is able to withstand a more considerable degree of cold than wheat. It is not quite so liable to winter killing as the latter crop, but wet soil and protracted heavy rains in winter will prove fatal to success. [p. 757]

2) Cultivation

(1) Time and Method of Sowing: There are only two distinct kinds of rye, namely, winter and spring rye. The former is sown in autumn and the latter in early spring. The difference between the two kinds of rye is not great, some agronomists maintaining that either may be readily converted into the other by changing the time of sowing. Winter rye grows more luxuriantly, usually yielding larger crops, and is therefore much the most extensively cultivated. The preparation of soil for a rye crop should be similar in all respects to the preparation for wheat. With rye as with wheat, there are two kinds of sowing commonly practiced, namely, broadcast and in drills. The latter is to be recommended in all cases where grain or well-grown straw is the objective, and the former when the rye is cultivated for pasturage [p. 758] or for soiling purposes. When grain is the objective, the drill system is preferable because it allows cultivation which favors growth. When pasturage or fodder is the objective, the broadcast system is best, because the thick growth possible in this system favors fineness of the stalk.

(2) Quantity of Seed: If cultivated for grain and sown in drills, about 5 pecks per acre is sufficient; if sown broadcast, 2 to 2½ bushels. If cultivated for fodder and pasturage, 3 to 4 bushels should be sown broadcast.

(3) Cultivation: Much of the rye raised throughout the world is left entirely without cultivation, but it is to be recommended since by it the weeds are kept down, tillering is favored, and the yield is increased. The methods of cultivation are the same as for wheat. [p. 759]

(4) Harvesting: When grain is the objective, the time of harvesting should be just

as it enters what is called "in the dough". At that time, most of the straw will have a decidedly yellow color, though portions of it may still be greenish. Grain harvested at that time makes more and better flour than that allowed to stand later. There is also much less liability to waste and what is perhaps equally important, the straw is of much better quality. Methods of harvesting may be precisely like those recommended for wheat.

Rye which is to be cured for fodder should be cut just as the grain is beginning to form. If it is to be used for feed, it is customary to begin to cut as soon as the heads commence [p. 760] to show themselves. When cut this early, if the season is wet, a second crop will grow from the stubble. You may continue to cut for green feed as long as the stems and leaves remain succulent. When only rye straw is the objective, the time of cutting is just at the period of blossoming. The straw is at this time made into small bundles, steeped awhile in hot water, and then carefully bleached, when it is of much value for hats and similar articles.

3) Rye as Pasturage

It is sometime customary to pasture, to a limited extent, even that rye which is cultivated for its grain; some alleging that this pasturage diminishes the crop very little, if at all. Except in cases where the growth in autumn or in early spring is extremely luxuriant, I do not think such pasturage is advisable. [p. 761] As I have stated, however, rye is frequently raised for pasturage only. In this case, it is best to put animals upon it quite early in spring, before it begins to send up tall stems, since those are not relished by any animal. It is best to confine sheep pastured upon rye to a limited space until they have eaten everything growing therein, giving them fresh space from day to day. If they are allowed access to a large field, they will trample underfoot great quantities of forage, which is consequently wasted.

4) Diseases

A fungoid growth, to which the name Spermaedia has been given, and which is commonly known as "ergot"[21] or "spurred rye", is the only formidable disease of rye. Those grains affected by that fungus send up a long spur-like growth [p. 762] which, if broken open, is found to be full of light-colored dust. Those abnormal growths are found to be most prevalent upon rye grown in wet, cold, and unfavorable situations, and are also most common in wet, cold seasons. Ergot is of considerable value as a medicine, and in Europe is often collected and sold at a high price. If taken into the animal system in large quantities, ergot causes a terrible disease, in the course of which portions of flesh and sometimes whole limbs rot and drop off. Ergot also exercises a strong influence upon the womb of females either animal or human, causing it to contract violently, and if the animal is pregnant, to expel the fetus. Abortions in cows are often attributed to the eating of grass or grain affected by ergot. There is no positive, certain remedy known. All that can be done is to carefully select the seed and to plant it under circumstances favorable [p. 763] to healthy growth. [p. 764]

7-3. *Oats (Avena sativa)*

The oat is an important member of the Grass family. It usually has three flowers in the each spikelet, which are mostly shorter than the glumes[22]. These flowers are perfect. The plant is soft and smooth, with a loose panicle of large spikelets, the palea investing the grain; one flower with a long, twisted awn on the back, while the others are awnless. It is not known for certain from what wild plant cultivated oats originated. At present, there are several species known, *Avena orientalis, A. nuda, A. brevis, A. sterilis, A. fatua*, and *Danthonia strigosa*. The *A. orientalis* has one-sided panicles, and several valuable varieties of oats belong to this species. *A. fatua* is now found wild in Great Britain, being in some places a very troublesome weed. This species is considered by many agronomists to have been [p. 765] the origin of the present commonly cultivated oats. The grain in *A. brevis* is very short. With *A. sterilis*, the quantity of grain is very small. The 3 species last mentioned are useful only for hay, seldom being cultivated for grain. *A. nuda* is a species of naked oats, but is of little value.

The oat is a grain cultivated in most parts of the world as feed for animals, but in some countries it is extensively used as human food. In no country is it so extensively used for human food than in Scotland. Many of the peasants in that country formerly lived almost entirely on oatmeal. Certainly, there are no people on the face of the earth physically superior to the Scotch, and their large size and strength is attributed by many to the fact of their having lived so much on oatmeal. It is stated by some doctors that a too exclusive diet of oatmeal [p. 766] causes a skin disease, but others contend that the cutaneous eruptions so common among some oat-eating peoples rather than to their food. It is as animal feed, however, that oats are most valuable. All over the world, oats are considered to be the best feed for horses. They are also excellent for young, growing livestock and for breeding animals, being much superior for these uses to corn. Oats develops muscular strength and gives life and spirit, while corn produces fat. As a fodder crop, oats are deserving of attention. My experience in Hokkaido has led me to form a very high opinion of oats as a fodder in this country. The yield per acre is large, the season when it is fit for harvesting is usually favorable, and the quality of well-cured fodder most excellent. The straw of oats, on account of its softness, is more readily eaten and [p. 767] digested by livestock than almost any other kind.

1) Varieties

With regard to color, oats can be divided into 3 classes, namely, White, Black, and Gray or Dun. There are two kinds of white oats, Early and Late.

(1) Common White Oat or Late Angus: One of the most productive and most extensively cultivated of all. The straw is tall and stiff, and the grain of good size. In the extreme north, however, on account of its lateness, this variety does not succeed as well as some others.

(2) Sandy Oat: A late variety of the white kind, extremely productive, and extensively cultivated in Great Britain. The kernel is smaller than in some other varieties. The straw is quite stiff, and hence this variety is less likely to lodge[23)] than many other sorts. [p. 768]

(3) Early White Oats: Among the early white oats is the Potato Oat, which is very highly prized in Scotland. This is early, very productive, and of the best quality. A measured bushel weighs 46 pounds, and a quart of grain makes 28 pecks of flour. This variety shells rather too easily, and must therefore be cut before it is fully ripe.

(4) Black Oats: Among black oats, the Black Tartarian seems to be the only

variety worthy of our attention. These oats are particularly heavy and make meal of the finest quality, but the color of the husk is objectionable; since portions of it must always be left in the meal, those black specks injure its appearance. This variety is excellent for horses or livestock of any kind.

(5) Oregon Oat: A white American variety which was cultivated upon [p. 769] the Sapporo Agricultural College Farm last year (1878) and seems to be of excellent quality.

2) Soil and Climate

The best soil for oats is a good, strong loam containing some clay, and the soil should be deep. In a dry climate, oats should grow very well upon a loam containing considerable organic matter, but in most climates they are likely to lodge if cultivated upon such soils. Climatic adaptation varies with the soil, but upon a strong, clayey loam success is most certain in a rather moist and cool climate.

3) Crops Best to Precede Oats

In Scotland, where the best oats in the world are produced, it is considered [p. 770] that oats follow grass better than any other crop. In some other parts of Great Britain, they follow hoed crops as leave the land in good condition, and free from weeds. This last consideration is of considerable importance.

4) Cultivation

(1) Preparation of the Soil: This is the same as with other grains. The soil should be well pulverized, and free from the roots or seeds of weeds.

(2) Fertilizers: Oats are extremely likely to lodge, so the application of strong nitrogenous manure is to be avoided. It is indeed better, in most cases, to make a liberal application of manure to the crop preceding the oats, and not to apply any directly to the oats. If any manure is to be applied, however, it should be [p. 771] well-rotted compost or mineral manure. The best method of application is to spread it broadcast and thoroughly mix it with the surface soil by repeated harrowing.

(3) Sowing: As with other grains, two methods of sowing are in common use, broadcast and in drills. The former is perhaps the most common, and upon soils free from weeds it answers the purpose well, but upon most soils of cultivation it is to be preferred. If sown broadcast, the quantity of seed oats necessary per acre is $2\frac{1}{2}$ to 3 bushels. If sown in drills, about 1/3 will be sufficient. In selecting seed, it is impossible to exercise too much care. Always take the heaviest, earliest oats you can find and those quite free from disease. [p. 772]

(4) Cultivation: It will be profitable to give sufficient cultivation to prevent the growth of weeds, and this work may be best done by the use of grain hoes drawn by horses.

(5) Harvesting: On account of the propensity to shell, which is very great with fully ripened oats, it is best to harvest them quite early. When about half of the straw has turned yellow, the grain will usually be fully formed and just beginning to harden. It is at this time that the crop should be harvested. A delay of only a few days often causes great loss. The methods of harvesting are about the same as with wheat, although sometimes the crop is mowed and not made into bundles. That method, however, is not to be recommended, since often only the head is chopped. [p. 773]

5) Diseases and Insect Enemies

Few crops cultivated by farmers are less subject to diseases and ravages of insects than oats. The only disease which ever does much injury is caused by a fungoid growth which affects the panicles; that disease is commonly called smut[24]. At about the time of blossoming, the panicle develops a large quantity of fine black smut-like powder. That is in reality the spores of the fungus. Upon peaty, poorly drained soils and in cold, wet seasons, smut is most common. The best remedies, then, are preventive. We should avoid sowing under unfavorable conditions. Soaking the seed for a short time, in a solution of common salt or blue vitriol, will probably prove useful in destroying the spores in all cases where we are obliged to sow grain from a crop which was affected by smut. [p. 774] The only insect which does serious damage is one called in England "wire worm" (Elateridae, Coleoptera)[25]. The perfect insect is a spring beetle, six-legged, reddish brown in color, and hard. This insect lives upon the fruits of oats. The only known remedy is to destroy the preceding grass crop some time before sowing the oats. [p. 775]

7-4. Barley (Hordeum[26])

Barley is a grain belonging to the genus Gramineae. That genus is distinguished by its one-flowered spikelets invariably arranged in threes, so placed that the 2 glumes belonging to each spikelet are more in front of the palae than at their side. The spikelets always stand in threes and back to back; thus, every ear of barley consists of 6 rows of spikelets. If only the middle spikelet of each set of 3 is perfect (the side spikelets being abortive), we have *H. distichum* (two-rowed barley), the species from which it is supposed all cultivated varieties may have originated. The glumes are usually narrow and even-awned, and the inferior palet is extended into a long awn.

1) Varieties

(1) *Hordeum distichum*: There are 2 varieties commonly cultivated, Common or Early English Barley and Chevalier. [p. 776] The former is good, early, adapted to a great variety of soils, and is most commonly cultivated. In the latter, the straw is stouter, and the grain is much liked by the manufactures (brewers) of ale, porter, and beer.

(2) *H. vulgare*: I will mention only 2 of the varieties, Common Bere and Victoria Bere. In this species, there are 2 single and 2 double rows of grain in each ear[27]. This variety may produce larger crops than the *distichum*, but the quality is not so good. The quantity of its alcohol is much lower. Both of these varieties must be regarded as inferior to most of the *distichum*, but Victoria Bere is rather more productive and better in quality than Common Barley.

(3) *H. hexastichum*: All 6 rows are equidistant in the ear. It is hardy and prolific, but in quality is inferior to the two-rowed barley. [p. 777]

(4) "Naked Barley"[28]: In all of the species of the barley, there are varieties which are nearly naked; some are extremely early and can be cultivated far to the north. These also make very good flour, but aside from those 2 qualities, they have no merit.

2) Cultivation

(1) Soil and Climate: A rather light, rich loam is best. An excess of either sand

or clay is unfavorable, but still a fair degree of success is attainable on any well-drained soil in fine tilth. The climate should be warm and dry. Heavy rains are extremely injurious, especially soon after sowing, or at the times of blossoming, ripening, or harvesting. [p. 778]

(2) Preparation of the Soil: For the barley crop, it is rather more and unusually important that the soil be finely pulverized, and it must be plowed and harrowed until it is brought into a very fine condition.

(3) Fertilizers: Barley is a crop which occupies the land but for only a short time: it is, therefore, highly important that any manure applied to this crop be in an available form; hence, well-rotted composts or soluble mineral manure are best. It is, however, more advisable to let barley follow some hoed crop such as turnips, beets, or potatoes, to which a liberal amount of manure has been applied, and not to apply any manure directly to the barley.

(4) Sowing: Barley should be sown [p. 779] as early in spring as the land is brought into suitable condition, since it needs the cool weather of early spring to extend its roots and to tiller. The relative merits of the drill and broadcast systems are the same as with other grains. The seed must, of course, be carefully selected; if sown broadcast, $2\frac{1}{2}$ to 4 bushels per acre will be necessary, and if in drills, half to 1/3 less. In all other respects, the cultivation of barley is the same as with other grains.

(5) Harvesting: Barley is least likely to shell; hence, harvesting may be deferred until it is quite ripe. Thoroughly ripe grain is much better for malting than poorly ripened grain, [p. 780] and since a large portion of barley is used by malt liquor manufactures, farmers usually find it more profitable to allow their grain to ripen thoroughly, but although the straw becomes very brittle. Harvesting methods are the same as for other grains. The threshing of barley is quite difficult, because it does not readily separate from the ears, and the long awns must be broken off. Therefore, the use of a threshing machine is almost absolutely essential for the economical performance of this work.

3) Diseases and Insect Enemies

Smut[29] is the only formidable disease, but even it seldom does serious damage. The preventive measures are similar to those already recommended. Among insect enemies, [p. 781] the wire worm[30] already spoken of, is one of the most destructive in Western countries. In Sapporo, there is an insect which bids fair to prove extremely formidable; it first made its appearance in the summer of 1879. It is a small white worm living in the soft inner cellular tissue of the leaves, sometimes one worm in a leaf, but often several. Thus, it seriously injures the crop. After completing its growth in the leaf, it undergoes its transformation. The means of destruction is not yet devised, because its habits are now yet unknown, but burning the stubble soon after harvesting the crop may prove efficacious. [p. 782]

7-5. Indian Corn (Zea mays)

Indian Corn is a member of the grass family with pithy stems varying in height from 3 to 20 feet in different varieties. The spikelets of the inflorescence[31] are arranged in spikes, staminate and pistilate being separate and in different spikes. The stem terminates in the clustered slender spikes of staminate flowers (the tassels), with 2 flowered spikelets; the pistilate flowers are in a dense and many-rowed spike borne upon a short auxilliary branch, 2 flowers within each pair of glumes but the lower one neutral, but the upper pistilate having an extremely long style (silk).

Indian corn is a native of America[32], and from that country it has been introduced into various others, and is at present distributed almost all over the world. It is an extremely nutritious grain, [p. 783] but is particularly rich in oil and carbohydrates. It is, therefore, better suited to produce fat or to sustain animals in the performance of severe labor than to build up the bony or muscular framework of their bodies. As food for humans, it is much used in some parts of the world, particularly in some portions of the southern states of the United States. The Negro slaves used to live almost entirely upon corn and pork, and even at present many Negroes in the South live very largely upon it. White men also use large quantities of corn; throughout the whole United States, its use in small quantities is very common. Corn is made into bread and various kinds of cakes; it is hulled and boiled; during summer, while it is green, much is either boiled or roasted; it is also parched or popped, and made into confections; in these several ways, corn is much used. [p. 784] It is as feed for cattle, however, that corn is most extensively used in the United States; thus, indirectly, it is a means of supplying the people there with a very large portion of their food. Although grain is better suited for fattening animals, great numbers of cattle and hogs raised in the western part of the United States are fed largely upon corn. Equal quantities of corn and oats make a fine food. Corn is also extensively used for the manufacture of whiskey. It is also made into starch, and recently sugar has been manufactured from it. That industry, in some parts of the West (Midwest) where corn is cheap, is found to be profitable.

The corn crop is one of the most valuable crops of the United States, and may well be called "the king". In 1875, the U.S. produced over 1,300,000,000 bushels, the average price of which was $0.42 per bushel, [p. 785] making a total value of $546,000,000; in 1879, the yield was considerably greater.

1) Climate and Soil Adaptation

Corn is almost tropical in its habits of growth, flourishing best at a temperature between 90 and 100°F. Different varieties vary much in the period of time required to complete their growth, but none of them can be successfully cultivated in any country where there are several weeks of very hot weather. Thus, although the average temperature in England is higher than in the northern United States, it is found that corn will not thrive there because the summers are too cool. The climate of Hokkaido, though not the most favorable, is such as to allow fair success with the earliest varieties of this crop. Corn requires considerable moisture, but this requisite must be accompanied by heat or failure will result. [p. 786] During the autumn months, when the crop is ripening, warm and dry weather is favorable. In this respect, Hokkaido is sadly deficient.

The soil should be a warm, rather light, and rich loam, though in one well drained and warm, success may be expected from a great variety of soils.

2) Varieties

For convenience of consideration, we may divide into 5 classes, namely, Dent, Flint, Rice, Pop, and Sweet[33].

(1) Dent Corns: These are large, late varieties. The kernels are considerable softer

than in the flint corns, and at the end of each kernel is a small hollow or dent which gives this class its name. [p. 787] Its varieties are so late as to make it impossible to cultivate them in Hokkaido. Throughout the west and south of the United States, dent corns are most extensively cultivated, the yield being greater than that of flint corn varieties. To the dent corn class belong all the large varieties, sometimes 20 feet high, but they are less nutritious than flint corn.

(2) Flint Corns: These must be cultivated in northern regions. Flint corn is characterized by a hard kernel of medium size; its hardness gave it its name. There are many varieties of flint corn, such as red, white, yellow, red-and-yellow, and white-and-yellow. There are also black or bluish black varieties, which are cultivated more as curiosities than for any other purpose. Among the red, white, and yellow varieties, which are the most common, there is a great variation in the number of rows [p. 788] upon an ear, the most common being 8 or 12, although in some cases 20 or more. All other things being equal, 8- or 12-rowed varieties are better in cold climates, since the cob is small and the corn dries and ripens more rapidly and perfectly.

(3) Varieties of Flint Corn: "King Philip" is an old variety formerly extensively cultivated in New England, and still cultivated quite largely. This is partly red and partly yellow, 8-rowed, medium in length and in earliness[34], and ranks highly in productiveness. These qualities taken together, it is very good, though surpassed by some other varieties. "Waushakum" is an 8-rowed variety of yellow corn named after the farm of the Sturtevant Brothers in South Framingham, Massachusetts, by whom it has been much improved. The ear is of medium length, the kernel large, cob small, quality productive, earliness medium, 8-rowed, and color yellow. [p. 789] It was cultivated on the College Farm and found to succeed here very well. Still, each year, there are many imperfectly ripened specimens and an earlier variety is therefore desirable. "Stebbins" is a variety of 12-rowed yellow corn raised on the College Farm, the seed of which was originally obtained from Mr. Stebbins of Massachusetts. The ears are long and large, the kernel deep, and cob not excessively large. Although mainly yellow, there are frequently a few white kernels. In earliness, it is about the same as the Waushakum, but in productiveness is far inferior. The name "Stebbins" may not be in general use. "Compton's Early" is an 8-rowed variety of yellow corn which has recently come into prominence. The ears are of medium size, and the kernel large. [p. 790] In productiveness, it is hardly up to the average. Its extreme earliness is the characteristic which most recommends it for cultivation. It is said to have perfected itself in 74 days from the time of planting. "Longfellow's Field Corn" is one which has been cultivated for many years in Massachusetts, and there it is very highly esteemed. It is 8-rowed and yellow, with remarkably long ears which are often as much as 15 inches. The size of the kernel is medium and the cob small, but this variety is somewhat later than the Waushakum; therefore, I do not think it will be well suited to this locality.

(4) Sweet Corn: The number of varieties of sweet corn is great, there being much difference in the size, and time of ripening, of different varieties. Sweet corn differs from other classes by having a sweeter taste, and when dry [p. 791] a wrinkled, instead of a smooth skin. I will only mention a few of the best varieties. "Marblehead Early" is one of the very earliest, comparatively small but very excellent.

"Crosby's Early" is a variety ranking second in earliness, considerably larger than the Marblehead Early, and of fine quality. "Egyptian Sweet" is a new variety of medium earliness, said to be very productive and of remarkably fine quality. "Evergreen" is one of the best late varieties, but this will not ripen without considerable extra care in Hokkaido. "Marblehead Mammoth" is a very large and late variety, but of good quality. It will probably not ripen here.

The last 2 varieties mentioned, although they may not ripen sufficiently for seed, can be cultivated for eating purposes, the seed being raised [p. 792] a little further south. With the varieties I have mentioned, you can have corn suitable for eating from sometime in August (and possibly in July) until the coming of severe frosts in autumn.

(5) Rice Corn: Some varieties of corn are known by the name of "rice corn", probably because the kernels bear a considerable resemblance in shape to those of rice. It is very hard in texture, like flint corn; there are several varieties, some being white, others red and less commonly yellow. Rice corn is not as productive as many other kinds, but it is very good for parching, and for this purposes it is cultivated. Most of its varieties are probably rather late for cultivation in Hokkaido.

(6) Pop Corn: There are several varieties of corn of the same shape and texture of kernel as flint corn, but much smaller. [p. 793] Those are cultivated exclusively for parching or popping. The best variety has rather small, 8-rowed ears and white grain.

3) Selection of Seed

Success in raising this crop largely depends upon the careful selection of the seed planted. Due to the large size of the ears, the methods of harvesting, and the fact that a comparatively small quantity of seed is necessary, this selection is easier than with many other farm crops. A perfect ear of corn is long (for the variety), of nearly uniform size from butt to tip, well covered with grain at both ends, with even, regular rows of large-sized grain. The top should be small, the kernel deep, and the cob such that it can be easily separated from the stalk. [p. 794] For this latitude, it is of the highest importance that the seed be selected from among the earliest and best-ripened perfect specimens. All other things being equal, seeds should be selected from those borne upon stalks producing 2 or more perfect specimens. Having selected the best ears, it is desirable to shell off and reject corn at each end, since these kernels are less perfect in shape and smaller than those from the middle portion. Seed corn must always be selected with the greatest possible care, and therefore should invariably be shelled[35] by hand. Machine shelling cracks the kernel.

A disease known as smut is oftentimes prevalent in cornfields; in order to destroy the spores of the fungoid growth which causes smut, it is advisable to soak seed corn for 10 or 12 hours in strong brine. [p. 795] After being taken out of the brine, the seed corn should be put in a small quantity of tar (only sufficient to give each kernel a very thin coating), and the whole very thoroughly stirred. Then, sufficient air-slaked lime should be mixed with it to prevent the kernels of corn from sticking together, and to make it convenient to handle. The objective of this coating of tar is to protect the corn from crows, and it is very effective.

4) Preparation of the Soil

Being an easy crop to cultivate, corn may well follow a crop which cannot be so easily cultivated; grass, therefore, is often followed and with the greatest advantage. Root crops, among which are potatoes, rank next. Both deep and shallow plowing have their advocates. [p. 796] It is the opinion of Dr. Sturtevant who has given the subject of corn cultivation much study that shallow plowing just before the seed is planted is best, but that the previous year it is good policy to plough deep. Corn sends some of its roots deep into the soil, but most if its spreading roots are near the surface; since corn needs heat, they should be encouraged to stay there by shallow plowing and the application of manure from the surface. In addition to ploughing, the land must be well harrowed to reduce it to fine tilth.

5) Cultivation

(1) Fertilizers: Good barnyard manure, especially that of the swine is well suited to the corn crop; the best method of applying it is to spread it broadcast after the land has been ploughed, and to thoroughly mix it with the surface soil by harrowing. [p. 797] If the manure is very coarse, however, it may be better to plough it in lightly. Corn is a crop which, to give good returns, must either be planted upon rich land or receive a liberal application of manure; 5 or 6 cords[36] per acre will be a fair amount. There are numerous commercial fertilizers that are well suited for the corn crop. Superphosphate of lime, used in small quantity on the hills or drills in connection with barnyard manure applied broadcast, gives excellent results. Herring guano is a good fertilizer for this crop, and its beneficial results are greatly enhanced by an application of wood ashes in connection with it. Both the guano and ashes may be spread broadcast and thoroughly harrowed in; 500 or 600 pounds of guano and 25 or 30 bushels of ashes are sufficient in most cases. Rapeseed oil cake, [p. 798] another fertilizer that may be obtained here, is also well suited to this crop.

(2) Planting: The American Indians had 2 very good rules as to the time for planting corn: "Plant corn when the leaves of oaks are as large as the ears of a mouse", and "Plant corn when the leaves of the walnut tree look like a crow's foot". Corn will germinate and grow at temperature varying from 48° to 115°F, but it grows most rapidly at a temperature from 90° to 95°. Around Sapporo, the time for planting is usually the very last of May or early in June.

There are 2 principal systems of cultivation, in drills and in hills, the former being undoubtedly superior to the latter. It is confidently stated by those [p. 799] who have given both systems a trial that drill cultivation will give a much larger yield than cultivation in hills. Corn may be planted quite rapidly and cheaply both by hand and machine labor, but the latter is of course superior to the former, especially in countries where the cost of labor is high. The distance between drills or hills must vary according to the variety of corn and the nature of the soil and climate. Large varieties, rich soils, and moist soils demand the maximum space; while small varieties on poor soils and in dry climates may be planted much closer. Around Sapporo, with the varieties of flint corn commonly cultivated, a distance of $3\frac{1}{2}$ to 4 feet between drills, and 6 to 8 inches between plants in the drill, will be suitable in the majority of cases. If planted in hills, [p. 800] a distance of 3 to $3\frac{1}{2}$ feet between drills will be best. Another very good way of planting is in rather long hills about 2 feet apart, in rows $3\frac{1}{2}$ to 4 feet apart. The proper depth for planting varies according to the principles well known to you all, but it is usually $1\frac{1}{2}$ to 2 inches

(3) Cultivation: The cultivation of corn must be commenced early, and be thorough up to the time when the grain becomes so large as to that it shades the ground completely and renders the use of implements difficult. Soon after the crop comes up, the ground may often with advantage be thoroughly harrowed with the "Thomas Smoothing Harrow". After that, when the crop reaches the height of 2 or 3 inches, the cultivator should be used; the use of that implement should thereafter [p. 801] be sufficiently frequent to keep down all weeds and keep the soil in a mellow condition.

Deep cultivation at certain stages of growth, for the sake of root-pruning, is recommended by some experts, but its advantages are not yet thoroughly attested. If that method can ever be advantageous, it is with a crop growing very vigorously just before the time of blossoming. Very little hand cultivation is necessary, but if weeds are numerous and large, and so close to the pants as to make their destruction by the cultivator impossible, they should be destroyed by the use of the hand hoe.

(4) Harvesting: The system of harvesting this crop is very different in various countries. Here, on account of the shortness of the growing season, it is best to cut corn down to the roots as soon as the grain is glazed, and then to put it in shocks of medium size, [p. 802] there to stand until it becomes ripe and dry, or as long as the season will allow. As soon as the corn gets dry and hard, it should be rapidly husked out in any manner which is most convenient, and stored in narrow, well-ventilated bins. Such a bin should not be more than $3\frac{1}{2}$ to 4 feet in width, but may be of any length and depth convenient. Both the sides and bottom should be made of strips of board wood not more than 3 to 4 inches in width and with a space about 3/4 inches wide between the strips. Such a bin cannot be too open, since there is always great trouble in this climate in preventing corn from becoming moldy. The corn bins should be exposed on both sides, if possible, to the open air.

After the corn is husked, the fodder should be made into small bundles, and these bundles should be stored either in a barn or in small stacks. The stacks should not be made large, or else the fodder will rot. [p. 803] If stored in barns, it must not be put in large masses. The best way of keeping corn is to stand the bundles on end upon an open floor below which the air can circulate, though upon a tight floor it will usually keep very well.

In view of the difficulty of properly curing corn fodder, it seems appropriate to note a French method of preserving it, known commonly as "Ensilage"[37]. That method consists of simply storing the fodder, which has previously been cut into fine pieces, compactly in large masses in a compartment made as airtight as possible. Considerable pressure is usually applied to make the mass compact, and sometimes a small quantity of cut straw is mixed with the fodder. Stored in that way, it undergoes a certain amount of fermentation and its tissues are considerably softened. It is said to be more digestible and nutritious for livestock. [p. 804] Its taste is slightly acidic, but most cattle eat it readily and all soon learn to like it exceeding well. It is stated, by those who have tried it, not to exert any injurious effect upon the animals' health.

The form of receptacle most convenient and reliable is an underground room, with bottom and walls made of bricks laid in cement. The room should, of course, be made in well-drained ground, and there must be some convenient arrangement for sealing it up nearly airtight. Fodder to be stored in this way should, when green, be cut with a machine into pieces not more than 3 to 4 inches long. I have only seen this method of preserving fodder when it is cut green; I have great doubt that it would be successful with the fodder from a ripened crop. [p. 805]

6) Fodder Corn

On many accounts, corn is one of the best crops that can be raised for soiling cattle. The yield of corn fodder is very great, and cattle are exceedingly fond of it. It is especially valuable for dairy cows, causing a very large flow of milk. Sweet corn is considered to furnish better fodder than ordinary corn, and so, a large variety is selected. If needed for soiling throughout a number of weeks, it is best to plant corn at different times, at intervals of about a week. That method will enable the farmer to use the fodder when it is at its best condition, which is at about the time of blossoming.

The general preparation of the soil for fodder corn should be the same as for field corn, except that more nitrogenous manure may be applied, since the aim is to cause a very vigorous growth. The seed should be planted rather thickly, in drills about 3 feet apart, [p. 806] but not so thickly as to prevent the free penetration of sunshine on all growing parts of the plants. Fodder grown in shade is much inferior in nutritive quality to that otherwise fodder grown in sunlight. [p. 807]

7-6. Buckwheat (Fagopyrum esculentum Moench)

Buckwheat belongs to the Polygonaceae, not withstanding its common name which is likely to mislead; it is widely separated from wheat botanically. Buckwheat is characterized by alternate entire leaves having stipules[38] in the form of scaleous or membranous sheaths at a strongly marked, usually tumid, joint of the stem. The flowers are perfect and white or nearly so, in corymbose panicles; stamens 8, with as many honey-bearing glands interposed; styles 3, the seed is an acutely triangular achene quite large in size. The plant is nearly smooth, leaves triangular, heart-shaped, inclining to halberd-shaped or arrowhead-shaped on a long petiole; the sheaths are half-cylindrical.

1) Varieties

There is another species of buckwheat known as *F. tartaricum*. [p. 808] This is like the above, but the flowers are smaller and tinged with yellow, while the grain is half as large and has less acute angles, which are very wavy. Buckwheat is usually 2 to 3 feet in height. Our cultivated varieties are believed to have come originally from Central Asia. There are other varieties cultivated in some parts of the world. One of them is notch-seeded buckwheat *F. femarginatum*, and another is perennial buckwheat *F. cymosum*; neither of these are of any particular value. *F. esculentum* is the kind commonly cultivated. The *tartaricum* is said to be hardier, but its seed is small. The cymosum produces many leaves, and might possibly be valuable as a green crop for plowing in. Buckwheat is, in various parts of the world, used as food by humans. In the northern part of Italy, the poorer classes live largely [p. 809] upon it, making it into a food they call *polenta*[39], of which they are very fond. Buckwheat is not well suited for bread, because it is deficient in gluten. As human food, this grain must be considered inferior to those we have spoken of. It is used as food for horses, cattle, hogs, and poultry, especially good for the last. The state of Rhode Island is remarkable for its fine turkeys, and these are fattened largely upon buckwheat. Another use is as a green crop for plowing under, and for this purpose it assumes the highest grade.

2) Soil Adaptation

Buckwheat thrives best upon rather light, sandy loams and will often succeed well where hardly any other crop could grow. Almost any light, well-drained soil might be expected to give good crops of buckwheat. [p. 810]

3) Cultivation

(1) Fertilizers: It is not customary, in many countries, to apply any manure directly to the buckwheat, the soil being sufficiently fertilized at the time of manuring the preceding crop. Buckwheat grows rapidly, and perfects its seed in a few weeks; on that account, the plant food, directly applied or not, must be in an available form. Well-rotted stable[40] manure or commercial fertilizers are good. The best method of application is to spread it upon the surface of the ploughed land and thoroughly mix it with the soil by repeated harrowing.

(2) Planting: Buckwheat grows rapidly in hot weather, but will not grow well in cool weather; hence, it must not be planted until quite late in the season. It may often succeed an early harvested crop [p. 811] such as rapeseed or very early peas. In this vicinity (Sapporo), July is the usual time for planting; the best method is doubtless in drills, although it is sown in light, sandy soils. The broadcast system is not as objectionable as with other crops, and also since it grows so rapidly as to outstrip any weed.

(3) Harvesting: Buckwheat does not send out all its blossoms at the same time; but as usually sown, it continues to blossom for several weeks, and consequently there is great irregularity in the degree of its maturity, so much so that we find perfectly ripened grain and blossoms upon the plants at the same time. Therefore, this crop should be harvested when it contains the greatest quantity of perfect grain; this is usually before the plant has finished blossoming. [p. 812] Buckwheat is very likely to shell out; hence, care in handling is essential. It is customary to handle it only when it is wet with dew or a slight rain. After being carried to the place where it is to be threshed, it is allowed to dry and then immediately threshed. If cultivated as a crop for plowing under, buckwheat should be sown broadcast and ploughed under when it is judged to contain the most plant food, which will be when it is well developed. [p. 813]

7-7. Millet (Setaria and Panicum)

1) Varieties

The different varieties of cultivated millet belong to Italian millet (*S. italica*), small foxtail millet (*S. germanica, S. italica* Beauv. var. *germanium* Trin), and common millet (*Panicum miliaceum* L.), all being members of the grass family. In the Italicum, the spikelets are in clusters of contracted spike panicle which is usually 6 to 9 inches long, and nodding when ripe. Each spikelet has a single perfect flower, the palea of which are coreaceous or cartilagenous in texture, while by the side of flower are 1 or 2 palea of a sterile (usually neutral) flower. The bristles upon the panicles are short and few in clusters: the palea of sterile flower is smooth. The other species of millet do not differ widely in botanical characteristics from the one now described. Asia is its native country. In the Western world, its only use is as fodder, for which it is very valuable. The Western varieties do not develop [p. 814] nearly as large seeds as those of Japan. For use as human food, it is inferior to cereal grains, and for that end its cultivation is not recommended. On account of the smallness of its seed, it has considerable value for poultry. Its merits are that it may be sown late, and thus may take the place of some crop that has failed; that it grows with great rapidity; and that it furnishes a large amount of fodder per acre which in quality, if well cured, is not much inferior to English hay.

2) Soil and Manure

Millet succeeds best upon rather well-drained, rich loams containing a large percentage of organic matter. Much of the soil in Hokkaido is very well suited to this crop. Millet is a crop which will not succeed unless the soil is rich, which is done by a liberal application of manure to the crop preceding it, because of [p. 815] its rapid growth which demands an available plant food. Any manure, if applied directly, must contain soluble plant food. It should be spread broadcast and thoroughly mixed with the pulverized soil.

3) Cultivation

About here in Sapporo, it is sown during June, the middle of the month being preferable. If cultivated for seed, it should be sown in drills about one foot apart; if for fodder, it must be sown broadcast and rather thickly, so that the growth will not be too coarse, but not too thick or else it will be shaded. If sown broadcast, the surface must be smoothly harrowed and the seed covered by brushing or rolling. A fodder crop needs no cultivation, but that raised for seed must be scrupulously weeded. [p. 816]

For fodder, millet should be harvested when in blossom, or even a little before. Harvested then and cured in the same way as grass is cured, it makes excellent hay, but if allowed to stand much later it becomes hard, wiry, and quite unfit for fodder. Indeed, some farmers who have used such hay have concluded that millet is poisonous, or at least highly injurious. If cut early and cured well, such will not be the case. Millet for seed should be allowed to stand until it is perfectly ripe, at which time the whole plant will have turned yellow or brown. The method of harvesting is not at all different from that pursued with the cereal grains. [p. 817]

7-8. Hungarian Grass

This is a plant closely related to millet, but it does not grow as high nor as large, nor does it develop as much seed. It is cultivated solely as a fodder crop, and is regarded by many as superior to millet for this use, being much finer and not so likely to become tough and wiry. The soil, manure, cultivation, *etc.* may be the same as for millet grown as a fodder crop, except that it may be sown somewhat thicker since the plant is smaller. [p. 818]

[*Lectures for the Fourth Semester*]
7-9. Potato (Solanum tuberosum)

The potato is an annual plant belonging to the natural family Solanaceae, of which it is one of the most important members. It has pinnate leaves[41] consisting of several ovate leaflets, with some minute ones intermixed. The flowers are blue or white with 5 petals: the stamens have anthers equal or mostly longer than the very short filaments: the anthers are not united, and their cells open by holes apices. Although it develops seed, it is not cultivated for those, but for its tubers. Potatoes are usually propagated from tubers, although propagation from seed is possible. Its native country is America[42], from where it was introduced into Europe, and then distributed all over the world. It is still found wild in some parts of tropical America, [p. 819] but the tubers of the wild plant are small and almost unfit to eat. The principal use of potatoes is as an article of human food, filling a very conspicuous park in the diet of many nations. It was early introduced into Ireland, where the soil well suited for it, and there it has long constituted almost the sole article of food among the poorer community. While the potato cannot be recommended as the sole article of diet, it is nevertheless certain that the Irish who live largely upon it are very strong and large. For the manufacture of starch, it is one of the most valuable crops; in many parts of the world, where potatoes grow well, this industry is extensively and profitably carried on. The potato contains a great deal of starch, and its separation into a comparatively pure state is quite easy. It is also used to a slight extent for the manufacture of whiskey. [p. 820]

1) Varieties

The number of varieties of potatoes is very great, and many of them are of almost equal excellence. They differ greatly in the length of time required to complete their growth; some are very early, and others very late. Here I will mention only a few of the leading American varieties.

(1) Early Rose: The variety deserving to be mentioned first, it was developed in America about 12 to 15 years ago. It is of pinkish color, oval to oblong in shape, with comparatively few and shallow eyes[43]. It is very early, of most excellent quality, and average in productiveness. All things considered, it is the very best. This is the variety cultivated on the College Farm.

(2) Early Ohio: A white, round variety. It is claimed that it is to be both earlier and more productive than the Early Rose, and in quality is good.

(3) Dunmore: A white potato, rather round, with few eyes. It is very productive, large in size, late, and good in quality. It is of comparatively recent origin; but is well spoken of. [p. 821]

(4) Burbank's Seedling: A white variety of medium earliness, it is said to be very prolific and of the best quality. Almost all the tubers are good-sized, very rarely hollow, and of fine appearance. It is extremely hardy, and is considered by many experts as one of the best varieties.

(5) Snowflake: Medium in earliness, superior to most others in quality, and worthy of cultivation.

(6) Bresee's No. 6 or Peerless: White, rather roundish, large and extremely productive; in quality, inferior to many other varieties, but on account of its great productiveness, it may be raised for starch or for feeding to livestock.

(7) Davis' Seedling: An older variety than any I have described. [p. 822] The skin

is red, the shape roundish, the eyes rather deep, late but very fine, and extremely productive. It is surpassed by some other varieties, but is still very good.

2) Soil Adaptation

The best soil for potatoes is a rather light loam, and a new soil is better than an old soil. Upon clayey soil, not only the crops produced are smaller but they are much inferior in quality to those grown upon lighter soils. The land around Sapporo is well suited for potatoes.

3) Cultivation

(1) Preparation of the Soil: Potatoes follow best after grass, and may often be most profitably cultivated upon land which has been used [p. 823] several years for pasturage or mowing. Its sprouts being strong, and its cultivation by horse power easy, a less careful preparation is admissible than with some other crops; still, the land should be deeply ploughed, and harrowed at least once. Plowing may be done either in the autumn or spring, whichever is most convenient, but the latter, when the grass has started a little, is preferable on account of the manure (fertilizer) it furnishes if plowed in.

(2) Fertilizing: Barnyard manure, if applied, may be undecomposed, and is best spread broadcast and harrowed in. Among the best special fertilizers for potatoes are ashes, plaster, superphosphate of lime, and potash salts. Guano, night-soil[44], and other azotized manures should be used sparingly, in connection with some material containing [p. 824] a great deal of potash; otherwise, the crop will be inferior and more likely to be affected by disease. Of the potash-salts, chloride must be rejected because it injures the quality. Most of these may be applied broadcast, although if the quantity is small, they may be put in hills or drills.

(3) Seed Selection and Preparation for Planting: We should select for seed potatoes those of medium size and of the shape considered to be the most perfect of that variety planted. Small potatoes, if used for 1 or 2 years, will not degenerate the crop, but if such are constantly selected, the size will surely diminish; excessively large ones are equally unsuitable. Selected a week to 10 days before planting, each potato should be cut into 2 or 3 pieces, [p. 825] the best section (as experience teaches us) being lengthwise. After being cut, they should not be piled in large masses, because heating up may take place. The quantity of seed potatoes necessary per acre varies according to the way it is prepared and the system of planting, but on the average it is 6 to 8 bushels.

(4) Planting: Potatoes should be planted very early in spring, since, all other things being equal, early planting will produce a superior crop to late planting. They are planted in 3 different ways, drills, hills, and ridges. Upon well-drained land which is easy to cultivate, the first method is best, being made 3 to $3\frac{1}{2}$ feet apart, or even a little less, if the seed is cut very fine. Upon new land, or where it is difficult to cultivate, I prefer the second method, since it allows [p. 826] the free passage of a cultivator. Upon poorly drained land, which, by the way, should be avoided if possible, the third system is the best. The distance between ridges should be 3 to $3\frac{1}{2}$ feet; their height should be 8 inches to 1 foot, and their breadth at the top about the same. The distance between the potatoes upon the ridges may be the same as in the drill system.

(5) Cultivation: If planted early, as they should be, potatoes require considerable time to send their sprouts to the surface; hence, weeds often get a good start before

they come up. If that is the case, the land may be harrowed without much injury to the crop. In this work, no attention is paid to the positions of the potatoes, because if planted sufficiently deep (3 or 4 inches), they will not be disturbed. As soon as the potato sprouts show above ground, a cultivator should be passed [p. 827] between the rows; unless the land is quite free from weeds, it should also be hand-hoed. From this time until the time of blossoming, every weed must be kept down; and during the last cultivation, it is good to pull a considerable amount of earth towards the plants, so as to prevent the tubers from growing above ground. Cultivation after blossoming is not advisable, since it is believed to lessen the crop.

Most of the cultivation work may be performed by horse power.

(6) Harvesting and Storage: Provided the intention is to keep them for any length of time, potatoes should not be harvested until they are perfectly ripe. This is indicated by the complete drying up of the tops, the hardness of the skin, and their endurance of ordinary handling; if they are harvested too early, [p. 828] the skin is tender and easily broken, and the risk of decay is very great. Machines have been invented for digging potatoes, but from their limited circulation I conclude that their use has been found profitable. If potatoes are planted in drills, a skillful plowman may, by running his plow first along the sides of the drills and then directly under them, throw most of them to the surface, where with a rake nearly all of them may be secured. The most convenient implement for hand picking is the "potato-hook"; another is the garden fork; an ordinary hoe may

[参考図]

Potato showing perfect flowers from Brooks, *Agriculture*

also be used. After being dug, the potatoes must be allowed to lie upon the ground until they become dry. On a pleasant day, those dug in the morning may be picked up in the afternoon. A long exposure to sunlight, or to rain or dew, is extremely injurious.

Potatoes should be stored in rather small masses in [p. 829] such a way that air can freely circulate through them. Bins 4 or 5 feet wide and about 3 feet deep, with slatted bottom and sides, will answer well. Straw bags are also excellent. Potatoes, though they can bear considerable cold, must not become frozen, or else they are entirely spoiled. The room they are stored in should be dry and cool, the nearer the freezing point (provided it is not reached), the better. In this latitude, a well-drained cellar is most excellent for winter storage; in case they cannot be stored in a cellar, pits dug in well-drained soil may be resorted to. Such a pit should not be more than $1\frac{1}{2}$ feet deep and 3 feet wide, but of any length desirable. At the bottom, it is good to cover the bare ground with a layer of dry straw; the potatoes may then be piled up about 3 feet high, and the pile covered with straw, [p. 830] over which should be placed sufficient earth to prevent freezing. Around Sapporo, one foot of earth will be sufficient. Another, safer way is to leave in the pit, at intervals of $3\frac{1}{2}$ feet cross-walls, so that there is less danger of heating up.

4) Production of New Varieties

This may be readily done by planting seeds. Potatoes growing from a certain seed are not like its parent, some are better, others are equivalent, *etc.* The chance of getting something better being quite good, many farmers have engaged in this business, hence a prodigious number of different varieties of potatoes. The seed may be planted either in autumn or spring (the latter season is better). Planted in well-prepared soil, in drills 15 to 18 inches apart, the tubers produced the first year will be small, none of them [p. 831] perhaps larger than a walnut. You cannot judge their quality at first, so they are carefully saved and planted the next year, like the ordinary potatoes. The second year, some of them are nearly full-size, and you may judge something of their value, but those must again be planted another year before you can arrive at a correct estimation of the exact value of your new varieties.

It often becomes desirable to propagate new varieties, which may be rapidly done as follows: the potatoes are planted in a bed made very warm and rich; as fast as they sprout, strong well-developed sprouts with roots attached are separated from the parent and carefully transplanted. The parent potato will then put forth other sprouts, which may again be separated [p. 832] and planted by themselves; it will continue to send out sprouts for some time, and thus it is possible to raise a large number.

5) Diseases and Insect Enemies

Potato rot[45] (sometimes called "potato murrain") check sp is the only serious disease of this crop. The first symptom is the appearance of little brown spots upon the leaves. Those rapidly increase in size and number, become confluent, and at last affect the stem also. Sometimes, in a very few days both the leaves and stems are entirely killed. At about the same time, the potatoes begin to be affected; and if the disease is extremely severe, the crop will be totally destroyed, and all the tubers become soft and of emit a very rank, offensive odor (farmers say they are "rottened"). The most generally accepted opinion [p. 833] is that the disease is caused by a fungus (*Botrytis infestans*). The fungus first gains a foothold upon the leaves, and then sends its myselium down through the stem and to the tubers, causing the decay. Cutting off the tops of the plants down to the ground, or even pulling them up at the very first sign of the disease, is believed to prevent the decay of the tubers in a measure. Another method proposed is to lay the tops flat upon the ground at the very first rumor of the disease and cover them almost entirely with earth. Still others advocate digging up the crop immediately, and that is perhaps the most certain way of saving it, but if that is postponed a little too long, the potatoes will decay after being harvested. Rotten potatoes should be immediately used; such potatoes are useful for starch, but if they cannot be disposed of in that way, [p. 834] nor immediately fed to livestock, they may be boiled and closely packed in boxes or barrels, and in this way may be used at any subsequent time as feed for swine. Though they becomes sour if the weather is warm that will not be injurious to the swine. It has been ascertained that poor drainage, nitrogenous manure (in large quantities), and warm, sultry, moist weather all tend to predispose the crop to become afflicted. Avoiding those conditions is the only preventative known at present.

The only insect which does any serious damage to potatoes is a beetle[46] (*Doryphora decemlineata*) which of late in recent years has proved very destructive in America. Its first abode was east of the Rocky Mountains, where it fed upon wild plants of the natural family Solanaceae. As soon as civilization reached the beetle's native habitat, it at once conceived a great liking for potatoes, feeding upon them [p. 835] wherever it was possible to do so. It first proved seriously injurious in Colorado, hence common name "Colorado Potato Bug". It made its way eastward, going a few hundred miles each year; thus, in Massachusetts we heard of its ravages and even foretold just when it might visit us; sure enough, in the fulfillment of the prophecy, it made its way to the potato fields in 1875. Even in the latitude of Massachusetts, it produces three broods annually. Both the perfect mature insect and the larvae feed upon the potato vines, but the larvae are by far the most destructive. The beetle's eggs are orange-colored and are laid in little patches on the underside of the leaves. The perfect mature insect is about half an inch in length; upon its wing covers are 10 lines, from which give it comes species name *decemlineata*. Among the most effective means of preventing the ravages of this insect are the picking off [p. 836] and destroying of both insects and eggs, and sprinkling the potato leaves with "Paris green" or arsenite of copper; the latter is the surest protection. It must be mixed with several parts of some fine, dry powder such as flour or plaster of Paris. That mixture is then sprinkled upon the plants from a muslin bag or a sieve. Attach the bag or sieve to the end of a long pole, and walk upon [along] the windward side of the row. Although it is a poison, it does no harm to the crop if proper precautions are taken. Allowing hens, turkeys, and ducks to roam over the potato field is also recommended. [p. 837]

7-10. Beans (Phaseolus)
("Adzuki"=*Phaseolus radiatus*, "Tsura Adsuki"=*P. radiatus* v. *pendula*, "Bundo"=*P. radiatus subtriloba*, "Sasagi"=*D. uniflorus*, "Sengoku-mame"= *Dolichos cultratus*, "Hata-sasagi"=*D. umbellatus*)

Some beans belong to the genus *Dolichos*, but those commonly cultivated belong to the genus *Phaseolus*. Some are climbers, while others retain a bushy form. Many of the climbers belong to the species *vulgaris*, others to *lunatus*, and a few others to *multiflorus*. Many of the bush beans belong to the species *nanus*. As food for humans, beans are exceedingly wholesome and nutritious. They are eaten both when green (together with pods) and when mature. As animal feed, too, beans are good, being especially rich in nitrogen. Horses and sheep eat them readily, the latter even uncooked. Hogs will eat beans if they are ground into meal. Fowls like them only if cooked. Beans cannot be used as a complete substitute for grain, but they may constitute [p. 951] quite a large portion of livestock feed.

1) Varieties
(1) Bush, Dwarf or Snap: The first 3 mentioned below varieties are well suited for garden cultivation, being very early and having tender pods when young; hence, they are valuable as "String Beans". "Early China" or "Red Eye" is an old, popular variety which is early. "Early Valentine" is the standard variety of early bean in the United States; it is extremely early, sometimes fit for use in 6 weeks; good quality. "Early York Six Weeks" is very early, productive, and of good quality; much cultivated in the U.S. "Refugee" or "Thousand to One" is a new variety which is remarkable for productiveness, whence its name. "White Pea

Bean" is long cultivated. It may be regarded as the standard field variety; small, round, of good quality, and very productive. [p. 952]

(2) Pole or Running Beans: The early varieties mentioned below may fill a place unoccupied by any Japanese variety, but the same cannot be said of the field beans. "Large White Lima" is very fine, but rather too late for northern climates. It would probably not ripen here in Sapporo. "Early Lima" is about 2 weeks earlier than the preceding variety, yet rather late (being called "early" only in comparison with the Large Lima); in quality, it is all that could be desired, being extremely rich and of fine flavor. It is rather doubtful whether it could ripen completely in the vicinity of Sapporo. "London Horticultural" is very productive, excellent either as a string or shell bean. It resembles a common large bean cultivated in Japan. I do not think that many of the American varieties of beans are superior to, or even equal to, the Japanese varieties. [p. 953]

2) Cultivation

(1) Soil Adaptation: For early varieties, a rather light soil is best; for the later field varieties, a stronger soil is preferable.

(2) Crops Best to Precede: Growth habit of the beans is such that they may be preceded by almost any crop, as far as the physical condition of the soil is concerned; it is better to precede this crop with one which is liberally manured. Foliaceous or root crops, such as cabbages or turnips, are among the best. Ordinary preparation is sufficient.

(3) Manuring: Undecomposed nitrogenous manures must not be used [p. 954] in large quantities. The nitrates and sulphates of soda and potash, phosphates, plaster, lime, ashes, and well-rotted barnyard manure or composts; small quantities of herring guano used in connection with ashes; night-soil well composted with muck or loam; and various other fertilizers of the same general character are all good. I am an advocate of broad application, though if the quantity is small, more direct benefit to the crop will result from application in close proximity to the seeds.

(4) Planting: Since beans are injured by the slightest frost, from May 20th to about June 10th afford the best opportunity for planting in this vicinity. The bush varieties are best planted in drills $1\frac{1}{2}$ to $2\frac{1}{2}$ feet apart. [p. 955] It is good to allow sufficient width for the use of a horse cultivator. It is possible to produce large crops by sowing seeds thinly in closer drills, but it is doubtful whether any increase in yield will cover the expense of the increased hand labor required.

In America, the running varieties of beans are planted around poles set firmly in the soil (their length is usually 7 to 8 feet above the ground). The poles are set about 3 to 4 feet apart each way, and about 6 to 8 beans are planted around each pole, though their number is subsequently thinned to 4 healthy plants. The beans are planted eyes downward because germination is more certain and rapid with that method, and the young plants are less liable to be deformed. With other varieties of beans, no particular care with regard to position is taken. The depth of covering for the seeds is, in all cases, about 1 inch. [p. 956]

(5) Cultivation: Never cultivate when the vines are the slightest bit wet from rain or dew, otherwise a disease known as "rust" may seriously lessen the crop. Do not draw earth toward the vines, since that is injurious.

(6) Harvesting: Beans to be eaten green are harvested just before use, or else they

lose much in quality. Pods should be picked as soon as they get large enough. In harvesting the ripe beans, sometimes the pods are picked, sometimes the vines are pulled out, and sometimes the vines are cut off near to the ground. Whichever be the method, the beans are allowed to become very nearly dry and ripe, otherwise many will be inferior and some quite worthless. [p. 957] On the other hand, if beans are allowed to become perfectly ripe, many will shake out, unless the handling is very careful. Cutting vines close to the ground is best, since that is less laborious and secures the vines for use, avoiding the dirt, little stones, *etc.*, mixed with the leaves when pulled. The cutting is best done with a "Kama". After being cut, the vines are laid upon the ground until partially dried; then they should be put into small stacks thatched with straw. Set 3 poles in the form of a tripod in the ground, and place some timbers and poles upon the ground around the tripod, in such a way as to make an open platform a few inches above the ground for the circulation of air. The beans should then be piled around 3 poles, pods inward and ends outward, exposed to the weather. The diameter of such a stack should not be more than 3 feet. [p. 958] Beans must not be threshed until they get perfectly dry and hard, or else many will be crushed or broken. When threshing them the common flail is often used. Machines are also made for this purpose.

3) Diseases and Insect Enemies

The only formidable disease is "rust"[47], the principal cause of which has already been spoken of. Other causes seem to be unfavorable soil and climatic conditions. Prevention is the best policy. Raise beans only upon well-drained and properly manured soil, and be careful in the manner of cultivation. A species of plant lice (*Aphis fabae*)[48] sucks the juice of the plant; when numerous, they seriously lessen the yield. Sulpher, carbonic acid may prove useful. Among other injurious insects are *Bruchus granarius*, *B. flavimanus*[49], and weevils (*Coleoptera*). The larvae of weevils prey upon the beans, [p. 959] sometimes in the field and sometimes after the harvest. Some larvae remain among the beans through the winter, and are planted with the seeds the following spring; hence, destroying them and planting unaffected seeds is the best preventive measure. [p. 960]

7-11. Peas (Pisum sativum)
(Nata-mame" (Overlook pea)=*Canavalia incurva*, "Rakkasho"=*Arachis hypogaea*)

The pea is an annual, smooth, and glaucous plant having very large, leafy stipules; commonly, they have 2 pairs of leaflets, branching tendrils, and peduncles bearing 2 or more large flowers. The corolla is white, purple, or partially colored. The pods are rather long in most varieties and fleshy. Peas are presently found growing wild along the seashore in many countries. Whether our cultivated peas originated from them or not is unknown. The early cultivation of this plant is lost in antiquity. It is raised for its edible seeds, which are most extensively eaten when green sometimes with the pods, too. In the dry state, peas are much used as food for humans in some countries. Peas are also extensively used as animal feed, being extremely nutritious. The vines, if properly cured, [p. 961] are also readily eaten by animals.

1) Varieties

(1) Very Dwarf: "American Wonder" is new and extremely productive; it is the

earliest among the wrinkled varieties. "Tom Thumb" is very early and extremely productive; the pods are well-filled; quality good; vines 10 inches high. "McLean's Little Gem" is another wrinkled variety; nearly as early as the Tom Thumb; quality first-rate; vines 12 inches high. "Commodore Nutt" is a remarkably dwarf variety of the sugar (or string) pea; its pods are also eaten; vines 10 inches high.
(2) Dwarf: Some varieties of this class will do well without bushing, except upon rich soils, where it will be profitable to do so. "Hancock" is very early and, all things considered, one of the best of the hard peas. "Carter's First Crop" is the earliest in this class; quality good; height of vines about 2½ feet. [p. 962] "Extra Early Dan O' Rourke" is one of the earliest among the standard market; very productive; vines about 2 feet high. "Dwarf Blue Imperial" is long cultivated, and of good reputation; quality good.
(3) Tall: All of the following must be bushed. "Dwarf Sugar" is a string pea with edible pods; of large size and good quality. "Champion of England" is an old favorite, richly flavored and very prolific; vines 4 to 5 feet high. "Popular" is an improved Champion of England; prolific, and of the best quality. "Black-eyed Marrowfat" is an old favorite having large pods; very prolific, though not of the best quality, and profitable for marketing; vines 3 to 4 feet high. [p. 963] "Large White Marrowfat" is an old standard variety, but very late.
2) Soil Adaptation
   For early varieties, a rather light, rich soil is best; for the later varieties, a heavier (though somewhat sandy) soil is preferable. It must be well drained in all cases.
3) Crops Best to Precede
   The preceding crop must be well manured, especially for the tall varieties of peas; it is better to depend upon the natural fertility of the soil, rather than to apply a large quantity of manure directly to the crop. A grain crop such as corn, oats, or barley is good. [p. 964]
4) Cultivation
(1) Manuring: There is a danger in a too liberal application of manure. The dwarf varieties, however, may be liberally fertilized with the ordinary manures of the farm. Well-decomposed manures are preferred, and those containing much nitrogen should be avoided. Broadcast application is best.
(2) Planting: The pea is extremely hardy, and seeds may be planted just as early in spring as the ground can be worked; early planting generally makes for a better result, because peas thrive best in cool weather. Planting in autumn (just before the coming of snow) may succeed, but not without some risk. If the snow should melt, and there be 2 or 3 more weeks of warm weather, the peas would start sufficiently [p. 965] to be injured by subsequent cold. In field culture, peas are sometimes sown broadcast, but a drill system is better since it provides a chance for cultivation. The distance between drills is, for the smallest varieties, not more than 2 feet; for the medium ones, about 3 feet; and for the tallest ones, 4 feet. The seed in drills for medium-size peas is planted at the rate of 2 seeds per "sun" (Japanese unit) of length, being somewhat spread out, and not all in a single row. The depth for soil covering is 2 to 3 inches
(3) Cultivation: Cultivation may be done mostly with a horse cultivator, only a little hand hoeing being necessary. It is usually best to cultivate and hoe twice

before bushing; hoeing should be done when the pea plants are 4 inches high. The branches of birch trees (particularly of the young white birch) [p. 966] make good bushes. The height of bushes varies, being a little less than the height reached by the vines. After bushing, one more cultivation is usually sufficient, and during that cultivation earth should be pulled up along the rows from both sides.

(4) Harvesting: Green peas are picked from the vines as soon as they reach the condition in which they are most liked. The usual time is when the pods are nearly filled, but before they begin to change color. No pods should be allowed to ripen upon vines from which it is desired to get a large supply of green peas, or the yield will be severely diminished. In field culture, where dry peas are desired, the whole crop is allowed to remain until most of the pods begin to dry [p. 967] and the vines turn yellow. Then, the vines are cut, cured with as little exposure to the weather as possible, stacked almost in the same way as beans (or stored carefully in barns or sheds), and threshed by machine or hand at any convenient time.

5) Diseases and Insect Enemies

At present, pea crops in Hokkaido seem to be quite free from both diseases and insects. There is one insect of the weevil[50] tribe in America which is very injurious. Its eggs are laid at or near the time of blossoming, and larvae feed upon the peas themselves. Many remain inside the peas until spring, when the mature insect comes out and is ready to lay its eggs. Soaking the seeds in hot water before sowing is recommended. [p. 968]

7-12. Turnip (Brassica campestris)

The cultivated turnip is considered to be a variety of this species, viz., by some it is called *B. napus*. Another is known as var. rutabaga. Var. turnip includes the so called "English Turnips"; var. rutabaga is also known as "Swedes" (Swedish turnips). It belongs to the natural order Cruciferae. Cultivated turnips and rutabagas are biennials; the lower leaves are pinnatifid and rough, with stiff hairs; the upper leaves are auricled at the base and smooth. This plant has large, bright yellow cruciferous flowers in racemes, at first short but lengthening in fruition. The flowers have tetra-dynamous stamens and a single 2-celled pistil; the fruit is a silique, the pod being beaked or pointed; the seeds are globose. [p. 856]

The turnip is cultivated for its esculent root, which is much enlarged, becoming a storehouse the first year for the material which is to be used in producing seed the second year. It is used to a considerable extent as an article of food for humans, but it is as feed for livestock that it is most valuable. For human consumption, English Turnips are best, and for livestock the rutabaga is much relished by cattle, horses, and sheep (especially by cattle and sheep, being almost essential in such a climate as this as winter feed for sheep). The reason for its extensive cultivation is its enormous yield; and yet, in nutritive value it compares well with almost any other root crop.

1) Varieties of Turnips

We may divide all kinds of turnips into 2 classes: English Turnips and Rutabagas or "Swedes"; [p. 857] the last two names being used interchangeably. (1) English Varieties: "Early Red Top" is somewhat flattened but globular; the skin upon the top growing above ground is red. It perfects itself in a comparatively short time, and may be secured very early in the season by early planting; or, it may be sown quite late, and will yet have ample time to complete its growth. Planted upon the College Farm as late as the first of August, excellent crops were secured. It is very excellent in quality, and is highly prized as a table variety. In productiveness, it is about average, but none of the English varieties is equal to Swedes in that respect. It is not a good keeper; hence, it must be used soon after harvesting. "Early White Top" is similar in nearly all respects to the preceding variety, except that the part growing out of the ground is white. [p. 858] "White Egg" is shape, color, and general appearance are excellent; flavor sweet and mild; a choice variety for table use; later than the Early Red Top, but superior in keeping quality. This is a comparatively new variety, and is very well spoken of. (2) "Swedes" (Swedish Turnips): "American Rutabaga" is said to be well suited both for table use and for feeding livstock; reputed to be a good keeper. "Laing's Improved Swede" is one of the earliest Swedes; it is excellent both for table and livestock use. It does not reach quite as large a size as some other varieties of Swedes, but upon the rich soil of the College Farm, in this moist climate, it grows large enough. Although recommended only for table use in America, here it succeeds better than any other variety. It is a good keeper. [p. 859] "Skirving's Purple Top Rutabaga" is a standard variety for field cultivation, raised almost exclusively for livestock feed. "Carter's Imperial Swede" is one of the best for field cultivation as livestock feed. Raised upon the College Farm, it proved inferior to Laing's Swede.

2) Soil and Climatic Adaptation

The best soil is a strong, rich, deep loam containing a considerable amount of clay. Light loams will not yield as much crops as soils of heavier classes. A soil containing much peaty matter is not well suited for turnips. Drainage is absolutely essential in any kind of soil. Turnips thrive best in a rather cool, moist climate; the climate of Hokkaido [p. 860] is well suited for their successful cultivation.

3) Cultivation

(1) Fertilizers: The turnip is a very rank feeder, and must be supplied with plenty of strong, rich manure. Strong barnyard manure may be profitably applied in large quantities, both broadcast and in the ridges upon which the seeds are to be planted. Herring guano, superphosphate of lime, Peruvian guano, fowl dung, rapeseed oil cake, and numerous other fertilizers may all be applied in large quantities with profit. When small, turnips are particularly vulnerable to the attacks of insects; hence, it is a good practice to apply, in close proximity to the plants, a small quantity of forcing manure which will cause them to first make a very rigorous growth. [p. 861] For that purpose, a little ammoniated superphosphate, Peruvian guano, or night-soil in liquid form may be applied with advantage just beneath the seed, but not in contact with it. Most of the manures may be used broadcast and harrowed in; it is also thought a good practice to put a little manure in the drills.

(2) Crops Best to Precede: Turnips, being a root crop, should follow some grain crop—such as wheat, oats, barley, or corn. For turnips which will be planted late, the land may be ploughed and harrowed a sufficient number of times to subdue all weeds which appear very often in grain fields. Grass or clover is also good, furnishing plant food in its decay. [p. 862]

(3) Preparation of the Soil: The seeds of turnips being small, it is essential that the soil be thoroughly pulverized before planting. Wherever circumstances will allow, it is advisable to first plough in autumn, and again 2 or 3 plowing or harrowing in spring, the last just before planting. It may also sometimes be necessary to go over the land with a brush, roller, or clod-crusher. Very thorough cultivation will certainly benefit.

(4) Planting Turnips: The rapidly maturing English varieties may be planted either very early or very late, according to whether you desire early turnips for market or table use, or late ones for the same purposes or as feed for livestock. For early use, the English varieties may be planted [p. 863] just as early in spring as the ground can be brought into suitable condition (normally, late April or the first of May). For late use, you may plant from mid-July to mid-August. Swedish turnips for winter use should be planted from mid-June to the first of July. If circumstances make it necessary, a fair crop may probably be secured by a little later planting. Turnips make their best growth in the cool months of September and October, and they should be planted so that the roots begin to rapidly increase in size early in September.

Turnips should always be planted in drills—for small English turnips, 15 to 18 inches apart; and for large rutabagas, 30 inches apart [p. 864] They are planted either upon level ground or in ridges, according to whether the soil is deep or well-drained, or not. Here in Hokkaido, where the climate is moist and the soil level, I prefer planting in ridges for the large varieties. The seeds should be covered 1/2 to 1 inch deep with good, fine soil. As the plants increase in size, they must be thinned out to 3 or 4 inches apart for the small varieties and 8 or 12 inches for the larger. More seeds than needed should be planted to secure an extra number of plants; those may be used to fill vacant spaces, or even to plant another field, since they bear transplantation exceedingly well. If there is great trouble from insects when the plants are small, it may even pay to sow the seeds quite thickly in seedbeds, where they can be easily protected and forced to make a rapid growth. Those seedlings should be transplanted to the field after each has formed about 4 secondary leaves. [p. 865]

(5) Cultivation: This is very similar to that of beets, except that it is not customary practice to pull earth toward the plant, since the turnip naturally grows largely above ground. The amount of cultivation needed for turnips is also not as great as that for beets, since they grow very rapidly and soon get ahead of all weeds. The English varieties are sometimes sown broadcast and left without cultivation, but that is a rather slovenly method of raising turnips, and is not recommended.

(6) Harvesting and Storage: Turnips are extremely hardy, and in this vicinity may doubtless be left in the ground until winter with very little risk of injury. In a place where there is little snow but much cold weather, it is not safe [p. 866] to leave them in the ground over the winter. Turnips are quite uninjured by severe frosts; indeed, a few sharp frosts are considered by some experts as beneficial. Harvesting, therefore, is delayed as late as possible. Here it is done just before the coming of snow, which is usually very late October or the first of November. A time should be selected when the weather is pleasant and the soil dry, and the work of pulling and cutting off the tops can be rapidly performed. The roots should lie exposed to the sun until external moisture has evaporated, when they

may be stored either in a cool, dry cellar, or in pits (as already described for potatoes and beets).

Turnip leaves are a good feed for cattle, and if your supply of fodder is not great, you may well substitute those leaves. They should be fed in rather small quantities, and before they have wilted. The tops also make a good manure, and may be spread evenly over ground [p. 867] and allowed to remain where they grew, thus serving to enrich the soil for the next crop.

4) Diseases and Insect Enemies

The most formidable disease which afflicts turnips is "fingers and toes"[51]. The affected turnip, instead of forming a single large, smooth root, forms a much-branched, very rough, and straggling root which is likely to be hollow, and in bad cases quite worthless. That disease sometimes ruins entire crops; its nature is not well understood, but there are certain preventive measures which are usually quite effective. It has been observed that the disease is common upon land where 2 or 3 successive years have been devoted to this crop, but that an interlude of at least 4 or 5 years between any two turnip crops is followed by very little injury. Applications of fertilizer containing a [p. 868] considerable amount of lime also seems to check it.

The number of insects that prey upon the crop is very great. When very young, aphids[52] are often extremely numerous, sucking the juices of the plants, much weakening them, or even destroying them in extreme cases. Aphids are partially held in check by beetles of the ladybug tribe, which feed largely upon them. Every precaution, therefore, is to be taken to encourage the propagation of those useful assistants. Manuring the plants, so as to make them grow vigorously, is a good preventive. As the plants increase in size, cut-worms, if they are short of its food, is the next object of attention. The cabbage worm also sometimes preys upon turnips, but not usually to such an extent as to be seriously injurious. There is yet another insect [p. 869] which in some years has wreaked great havoc on this crop, having in some cases almost entirely ruined it. At that stage of its life at which it preys upon turnips, it is a little white worm, probably a maggot[53]. Those worms live upon the inner substance of the roots; in a single root, the number is sometimes very great, in which case the turnip becomes worthless. There is not yet any good remedy known, but late planting is found to succeed better than early planting. [p. 870]

7-13. *Cabbage (Brassica oleracea)*

The Cabbage is one of the most extensively cultivated and most valuable plants of the Cruciferae. It is originally a seacoast plant found in Europe, having a thick, hard stem and large, pretty pale yellow flowers. The leaves are very glabrous and glaucous, the upper ones entire, clasping the stem, and not auricled at the base. The fruit is a beaked pod; the seeds are spherical. It is cultivated as a biennial. The thick, rounded, fleshy, and strongly veined leaves collect into a head the first year atop a short, stout stem.

Var. Broccoli is a state in which a stem divides into short fleshy bearing abortive flower buds. Var. Cauliflower has its nourishing matter mainly concentrated in short, imperfect flower-branches collected into a short head. [p. 871] Var. Kohlrabi has the nourishing matter collected in the stem, which forms a turnip-like enlargement aboveground beneath the cluster of leaves. Var. Kale is closer to the original wild species, with the leaves not forming a head.

Cabbage is mostly used as an article of food for human food, and is eaten raw and cooked in a variety of ways. It is used in certain salads, and is also pickled. It is sometimes raised as feed for livestock, being particularly relished by neat cattle. It causes a large flow of milk, and is consequently good for milch cows. Swine, sheep, and poultry also eat cabbage, but it is not so often raised for them; for the latter, it constitutes a valuable winter food to keep hens in good condition during the laying period.

1) Varieties of Cabbages

[参考図]

(1) Early York Varieties: An old standard sort of very good quality. [p. 872] "Early Jersey Wakefield" is a good standard market variety. Raised upon the rich soil and in moist climate of Sapporo, however, it is likely to crack unless used early. "Little Pixie" is a small, tender sweet cabbage with pointed heads; somewhat earlier than the Early York; not so well suited for market, but for home use it is one of the best. "Early Wyman" is a very large early variety good for market. "Early Ulm Savoy" is very early, of extremely fine quality, and unsurpassed for family use. "Henderson's Early Summer" is large, quite early, and very popular.

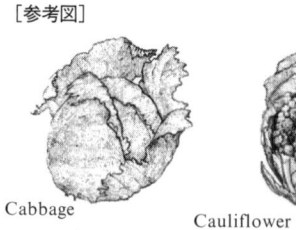

(2) Medium Varieties: "Fotter's Improved Early Brunswick" is the largest of a class of "drumheads"; very popular. Upon the College Farm, it has done extremely well. "Early Winning Stadt" [p. 873] is quality medium, but a very reliable header, even on unfavorable soils. "Newark Early Flat Dutch" is a standard market variety in New York.

(3) Late Varieties: "Improved American Savoy (Extra Curled)" is very remarkable for heading, and very fine both for family use and for market. "Marblehead Mammoth Drumhead" is the largest in size, but not the best. On account of its size and productiveness on suitable soils, it is often raised for market and for livestock feed. "Stone Mason Drumhead" is a standard variety in the Boston market, large, of fair quality, and very reliable in heading. "Improved American Savoy" is particularly sweet and tender; much esteemed for family use, and a very reliable header. "Red Dutch" is an old variety, [p. 874] used principally for making pickles (sauerkraut). "Red Drumhead" is larger than the preceding variety, hence more profitable; used for the same purposes.

2) Soil Adaptation

Cabbages need a very strong, rich, and deep soil; new land is preferable to old. Cabbages will, however, grow with a fair degree of success upon any land which will produce good crops of corn; but it is true that the stronger the soil, the better will it develop. Cabbages can endure considerable moisture, but very poorly drained soils should not be selected.

3) Cultivation

(1) Fertilizers: The cabbage is one of the rankest feeders, and requires a plentiful supply of available plant food; very strong, rich manures may be applied [p. 875] in large quantities. Strong barnyard manure, herring guano, night-soil, the dung of fowls, rapeseed oil cake, and rotten kelp are among the best. Ashes are not well

suited, unless applied in conjunction with strong nitrogenous manures. If the manures are well rotted and in a fine mechanical condition, a somewhat smaller quantity will answer.

The cabbage, as we have seen, was originally a seaside plant; thus, when cultivated away from the seashore, the application of some salt may prove beneficial. If the quantity of manure is small, put it directly beneath the plants, but the best way is to apply a large amount of manure broadcast and mix it, and a little forcing manure directly beneath the plants. Hog manure does not seem well suited for cabbages, often [p. 876] causing a root disease known by various names "club-foot", "fingers-and-thumbs", etc. Stable manure, upon which a limited number of swine has been allowed to run, does not produce the same effect. Salt is not, by any means, perfect manure by itself; from 10 to 15 bushels per acre are sufficient. From 5 to 12 cords of strong barnyard manure may be used, though the quantity varies according to the condition of the soil and the size of the variety cultivated.

(2) Crops Best to Precede: Cabbages should not follow cabbages, turnips, or any other member of the same family. At least 5 years a should intervene between successive cultivation of any of those crops on the same soil, or else the cabbages will be extremely likely to be affected by "club-foot". [p. 877] It is thought by some experts that with the liberal application of alkaline manures such as lime or ashes, the tendency toward that disease may be counteracted, and crops raised in successive years. Others, however, tried that in vain. It is stated by Mr. Peter Henderson that in some soils of Long Island which contain a large percentage of lime, rotation of crops for cabbages is not necessary. Except when absolutely necessary, it is better to resort to crop rotation. One of the best crops to precede cabbages is grass; among others, in the order of their adaptation, may be mentioned potatoes, beets, corn, cereal grains, and peas.

(3) Preparation of the Soil: It is best, in almost all cases but especially if the land is in grass, [p. 878] that it be ploughed the preceding autumn and again in spring. The plowing should be deep, since cabbages luxuriate in a deep soil. Sufficient harrowing, rolling, etc., must be done to make the soil fine. Manure, if well rotted, may be applied shortly previous to planting the crop; otherwise its application in the preceding autumn is preferable.

(4) Planting: Cabbages for early use are usually started in a finely prepared seedbed, which is sometimes covered with glass and made warm by the mixture of a liberal amount of decomposing matter such as leaf-mold or horse manure. For extremely early use, cabbages are started in hothouses. Cabbages can bear considerable frost and remain uninjured. [p. 879] Seeds for the production of plants for early use may therefore be planted at such a time that plants will reach the proper time of transplanting as early as the soil can be brought into suitable condition. In this vicinity, the proper time would be during the month of March, but if planted that early, the covering of the seedbed with glass will be necessary. In the seedbed, the seed may be sown quite thickly in rows 8 to 10 inches apart. The seedbed should then be frequently watered. All weeds should be kept down and the soil kept loose. When the plants have formed about 4 secondary leaves, they should be taken up and transplanted. If allowed to remain for a long time as close as they are sown in the seedbed, the stems of the plants become long and crooked,

and the whole plant weak and spindled, becoming unfit for the production of a good head. [p. 880]

The distance necessary between plants in the field [depends upon] the variety; for most of the earlier sorts, 2½ feet between rows, and 18 to 24 inches within the rows will be sufficient. In the field, the growing season varies from very early in spring until as late as the last of June, according to the time the crop desire to mature. In planting in the field, make hills of loose, fine earth at the proper distances apart; then plant in each hill, in a straight line (within a distance of 5 or 6 inches) about the same number of seeds. When the plants get large enough to need more room, those which are superfluous in one hill may be used to fill a vacant space in another, or (if not needed) thrown away. Some of the largest varieties of cabbage require a distance as great as 4 feet between, and 3 feet within the rows. The smaller the variety, the less the distance. [p. 881] A common distance is 3½ ft. between, and 2 to 2½ feet within the rows. The seed must be covered deeply; 1/2 or 3/4 inch is sufficient. In this vicinity, I prefer to start both early and late varieties in seedbeds due to the vulnerability of young plants to the ravages of insects. Seedlings can be more easily protected, and are less likely to be affected, in seedbeds.

(5) Cultivation: It is impossible to cultivate cabbages too much. They must receive very much cultivation. The soil must be stirred frequently and deeply, otherwise the plants do not form good heads. If keeping down weeds is not enough; in the early stages of the growth of cabbages, a horse cultivator can do most of the work. As the plants begin to head, [p. 882] it is often useful to run a light plow between the rows, throwing the earth towards the plants. Pronged hoes are also used.

(6) Harvesting and Storage: Early cabbages are harvested as they are needed. We cannot harvest and store them early in the season in such a way that they will keep well. As harvested for use during the summer, the heads are usually cut off with little or none of the stem and are either sold or used immediately. If needed for winter use, they keep best if pulled up by the roots. For this work, a dry day should be selected. As they are pulled, or just before pulling, as preferred, most of the loose outer leaves are broken off; those leaves are used either [p. 883] as feed for livestock, or are left upon the ground as manure, whichever seems best.

After being pulled, the cabbages may be stored in a cellar or in pits; in either case, they must not be piled up in large masses because, under such conditions, heating up and decay will be sure to follow. Whether in a cellar or a pit, the method of storage shown in the figure 〔図なし〕 will be best. First, a single layer of cabbages with heads downward should be placed as close as possible, either on the floor of a cellar or in a pit. Then another layer with roots downward should be placed upon the first. If the cabbages are stored in a cellar, no covering is needed, but if stored in a pit, a covering of earth sufficient to prevent severe freezing must be given. If there is any doubt as to the drainage of the soil around where the cabbages are stored, [p. 884] it is best to place the first layer upon the surface of the ground, and to pile the earth upon every 2 layers. The most perfect soil for this purpose is one containing a considerable amount of sand but not much organic matter. The width of a pit may be 3 to 4 feet, and the length as desired. This is the best way to preserve this vegetable. For a shorter time, larger masses may be stored together, but there is a considerable chance for heating and decay.

As cabbages are uninjured by quite severe frosts, the work of harvesting may be delayed up to the very beginning of winter. If the heads are once frozen and allowed to thaw gradually, they would be undamaged for eating purposes, but if once frozen through to the inner bud, they are worthless for seed purposes. The stalk, it is true, may send out some sprouts, [p. 885] but seeds raised from them are much inferior to those from the central bud. Another way of preserving cabbages is to put them in a barrel perforated with numerous holes.

4) Diseases and Insect Enemies

The cabbage is vulnerable to just one serious disease, "Fingers-and-Thumbs"[54]. The nature of that disease is not well understood, but it is altogether likely that the abnormal growths upon the roots (which characterizes it) are the result of either the presence of some microscopic insect or a parasitic fungus. Whichever it may be or even if it is neither makes practically little difference. Of its causes, and of the efficacy of using ashes and lime, I have already spoken. Drawing the earth about the stems encourages the emission of new roots higher up than they naturally grow; this course is recommended by some experts, [p. 886] it being stated that those new roots will be of material use.

The cabbage is subject to the ravages of numerous insect enemies, generally those that also prey upon turnips. In this vicinity, I have noticed upon cabbages the caterpillars of a butterfly of the genus *Pieries*[55], and cut-worms of the genus *Agrotis*[56]; aphids seldom prove seriously injurious. The best method of combating them is (as with turnips) to force the young plants to make rapid growth. In Sapporo, caterpillars of a butterfly known, in common parlance, as the "Cabbage butterfly"[57] (*Pieries*), has proved to be the most serious enemy of this crop. It is pale yellow in color, with some black markings upon the wings and abdomen.

In the warm days of early summer, those butterflies may often be seen in large numbers flying about cabbage plants. [p. 887] Here they pair and the female soon lays her eggs, usually only one, upon the lower sides of the leaves. Several eggs are often deposited upon the same plant, but they are not left in bunches. In a few weeks (usually 2 or 3 weeks) varying according to the temperature, the eggs hatch and the young caterpillars immediately begin feeding upon the leaves (but not the veins, unless hard pressed). The result is that the cabbage is very much checked in growth or, as is often the case, killed entirely. In about 3 or 4 weeks, the caterpillar reaches its full size, when it leaves the plant and crawls under some protecting leaf or into any convenient crevice and passes into the pupa state. Pupation occupies only a comparatively short time, and the butterfly comes out ready to deposit another crop of eggs. The number of broods in a year will vary with the climate, but in this latitude [p. 888] it is probably 3; those insects undoubtedly hibernate through the winter in the pupa state.

An application of lime, ashes, or plaster to cabbage plants while the dew is yet on may prove a slight protection, but is not perfectly reliable. Picking off the caterpillars by hand and destroying them, although somewhat expensive, is one of the best means of checking the ravages of that insect. Particular efforts should be taken to destroy the first brood, which is comparatively few in number, because many pupae died during the winter. The employment of boys and girls, with large nets, to catch the first butterflies of the season is also recommended. One female now killed makes a difference of many thousands of caterpillars at the end of the season. When about to pupate, the caterpillar crawls under some protecting material; [p. 889] it is, therefore, a good practice to place among the cabbage plants small pieces of boards or shingles, as may be convenient. Many of the caterpillars will avail themselves of the shelter thus afforded, and may subsequently be destroyed very easily.

5) Varieties of the Cabbage Family

(1) Cauliflower: This is quite extensively cultivated in some countries, being used for about the same purposes as the cabbage. Its use, however, is not so extensive as that of the latter, and its cultivation is seldom carried on upon a large scale. It is only as food for humans that it can be profitably raised, though the loose leaves and portions of the stems may be utilized as cattle feed. [p. 890] Varieties of Cauliflower are:

"Henderson Early Snow Ball" is very early and reliable for heading. "Mount Blanc" is new and very early, with magnificent heads well protected by the leaves; one of the best for heavy, somewhat wet soils. "Lenormand's Short-Stemmed Mammoth" is Dwarf variety; heads very large and fine; one of the best for general cultivation. "Improved Early Paris" is good, very extensively cultivated by market gardeners near Boston. Soil, climate adaptation, fertilizers, *etc.* are the same for cauliflower as for cabbage, except that this crop may be more liberally manured. It is useless to attempt to raise this crop without the application of sufficient manure to make the soil rich. [p. 891]

(2) Kohlrabi or "Turnip Cabbage": This is less extensively cultivated than most other members of the same family. When young and tender, it is used to a considerable extent as food for humans; the enlarged [portion of the] stem has somewhat the flavor of the turnip. When mature, it is too hard for humans to eat, but not so for cattle. It is not advisable, however, to cultivate it exclusively [as cattle feed, since] there are many other crops better for that purpose. Two of the best varieties of kohlrabi are Early White Vienna and Large Purple. Soil, fertilizers, *etc.* are the same as for cabbage. The proper planting distances are 2 feet apart in rows, and plants are thinned to 12 inches apart in row. [p. 892]

(3) Borecole or Kale: This plant is cultivated for its leaves, which are used as food and also for ornamental purposes. Its cultivation is not very extensive, being of much less importance than cabbage. Some of the best varieties are Green Curled, Tall Scotch, Dwarf Green Curled, and Carter's Garnishing. Soil, fertilizer, cultivation, *etc.* are the same as for kohlrabi. The proper planting method is in hills 2 to 3 feet apart.

(4) Broccoli: This is very much like cauliflower in form and use, though not so extensively raised, being regarded as inferior[58]. Some of the best varieties are: Walcheren White, Purple Cake, White Cake, Early Purple, and Knight's Protecting. [p. 893] Soil, cultivation, *etc.* are the same as for the preceding varieties.

(5) Brussels Sprouts: This is cultivated and used for human food in about the same way as cabbage. They are extremely hardy, and may be left in the open field throughout the winter. As cattle feed, its cultivation cannot be recommended, since it is inferior in productiveness to cabbage. Its good varieties are Scrymger's Giant Dwarf and Dwarf Improved. Soil, *etc.* are the same as for the preceding

varieties. The proper planting method is in hills 2 feet apart.
(6) Dalmeny Sprouts: The plant bearing this name is supposed to be a hybrid between the Drumhead Savoy Cabbage and Brussels Sprouts. [p. 894] Its use, soil adaptation, cultivation, *etc.* are the same as for the latter vegetable. [p. 895]

7-14. *Carrot (Daucus carota)*

In agriculture, the carrot is a biennial cultivated for its root. It belongs to the Umbelliferae. The calyx of the blossoms is adherent to the 2-celled ovary; petals, stamens, 5 styles, 2; the dry fruit splits into 2 seed-like portions; the flowers are white or cream-colored, in a regular compound umbel. Upon the seed are prickles in rows upon the ribs. The leaves are pinnately compound.

The uses of carrots are various, but it is as cattle food that it is most important, and widely raised. Carrots are very nutritious, and are much relished by horses, cattle, sheep, and hogs. In the popular estimation, it is considered the best root feed for horses, although some farmers assert that Swedish turnips are superior. Horses engaged in hard work must not be fed very largely upon carrots, [p. 896] but it is certain that an occasional small allowance of carrots exerts a very beneficial influence upon the general health of the horse. The yellow varieties may often be fed to milk cows, since the yellow color imparts a desirable appearance to dairy products.

As feed for sheep or hogs, carrots (though good) are undoubtedly surpassed in point of profit by some other vegetables. In nutritive value, carrots compare with hay of ordinary excellence at a ratio of 1:3—that is, 3 pounds of carrots are equal to 1 pound of hay; for oats, the ratio is 1:5. Aside from nutritive value, however, carrots are desirable for livestock feed, for the qualities of improving the health of livestock and aiding in their digestion of hay and grains. Carrots, then, must be considered to be worth more than 1/3 as much as hay. Now, while 4 tons of hay per acre is a very large crop, 30 or 40 tons per acre of carrots have been grown. [p. 897] The expense of raising carrots, it is true, is much greater, but not so great as to counter-balance their higher volume, and thus value.

1) Varieties of Carrots

Varieties are very numerous, varying in earliness, size, shape, and color.
(1) Danver's Carrot: Superior to any other in general cultivation; color good, shape fine, earliness medium, and extremely productive. Its shape is that universally considered the best; i.e. of large size, and quite uniform to the bottom, where it suddenly stemminates in a fine taproot extending further into the ground. This shape is typical; by it, we get the greatest bulk consistent with ease in harvesting; though these carrots are not so long as some other sorts, the yield may be equally great because of the large and uniform size near the surface of the ground. Upon the College Farm, [p. 898] it has proved most satisfactory.
(2) Early Very Short Scarlet: Earliest and smallest; color fine, quality good, but of no special value as a field crop. It is highly esteemed for forcing.
(3) Early Scarlet Horn: Early; used for forcing; excellent for table use; good quality and of very high color.
(4) Short Horn: The standard early variety, sweeter than the Long Orange and more solid. On account of its high color, it is well suited for coloring butter. It is not recommended for field cultivation as feed for cattle.

(5) Improved Long Orange: Richer, darker color than the Long Orange, of somewhat larger size near the crown, and shorter; very productive, and well suited for field cultivation as feed for stock.

(6) Long Orange: For many years, a standard field carrot in the United States; productive, of good color and quality; a good keeper. [p. 899] Now, it is perhaps surpassed by the Improved Long Orange and the Danver's. Both this and the Improved Long Orange are somewhat later than the Danver's.

(7) Large White Belgian: As its name indicates, it is large, white, and generally considered the most productive; in quality, slightly inferior to the Long Orange. On the College Farm, it has not proved more productive than the Danver's or Long Orange. It has a characteristic of growing partly above ground, and may be harvested easily.

2) Soil Adaptation

Carrots flourish best in a rather light loam; it is highly important that the soil be deep and well drained. Any loam which is not exceedingly heavy and compact may be made, with proper cultivation, to yield fair crops; clayey loams are to be avoided. [p. 900]

3) Preceding Crops

The seeds of carrots are very small, and the plant (when it first starts) is weak and grows quite slowly; hence, cultivation must be done by hand. For these reasons, it is of the highest importance that the preceding crop is one which will leave the land in a fine state of mechanical division, and that land is free from the seeds and roots of weeds. Anyone who plants carrots in a rough, weedy piece of land will later be sure, with an aching back, that he has done so. Coarse stubble must not be left upon a carrot field, since it greatly interferes with planting. Among the crops suitable to precede carrots are cabbages, peas, potatoes, and, though not equal to those, corn and all other grains. Having once brought a field into proper condition for the cultivation of carrots, it is good to continue to raise them for a number of years in succession upon the same land. [p. 901] In this respect, carrots differ from most other crops. To follow this system with success, the ground should be liberally fertilized each year.

4) Cultivation

(1) Fertilizer: Among the best fertilizers obtainable in this vicinity for this crop are barnyard manure, night-soil, fish guano, ashes, and rapeseed oil cake. Any of those may be liberally applied; if fish guano is used, ashes should be applied in connection with them. All fertilizers must be in a fine state of mechanical division; hence, if barnyard manure is used, it should be well rotted, or at least free from straw (or it should be applied the preceding autumn). All of those manures, upon light soils suitable for carrots, are best kept near the surface and applied broadcast. [p. 902]

(2) Preparation of the soil: Deep plowing and sub-soiling in autumn, harrowing and plowing in spring, then harrowing again twice or more, and after that going over the land with a clod-crusher, brush, or loader, are recommended. Plowing twice in the spring, with thorough harrowing and breaking up clods after each plowing, may be essential in some cases. While it is important that the soil be deep, too much of the cold, inert soil must not be brought up to the surface at once; the deepening process must be gradual. Raking the soil over with hand

rakes, after pulverizing as much as possible with the harrow, clod-crusher, *etc.* will be essential in some cases, especially if the seed is to be sown by a seed-sower. [p. 903]

(3) Planting: As a field crop, carrots are planted from about May 15th to June 10th. I prefer May for planting the later varieties, but the Danver's, Short Horn, and other early varieties may be planted in June (and some of the earliest, even later than the time specified, with certainty of success). Carrots which are not overripe keep better than those which are.

Carrots are always planted in drills, the distance between which varies with the variety. The most common distance is 10 to 15 inches (about 14 inches being best for varieties recommended for field culture). The work of planting may be done either by hand or with a seed-sower. Unless you have a careful workman [p. 904] to run the seed-sower, or can do the work yourself, hand planting is preferable. The proper depth is about 1/2 to 3/4 inch. The plants in the drills should stand 2 to 5 inches from each other (in most cases, about 3½ inches being suitable for field varieties). The amount of seeds sown, however, must be greater—the sprouts of carrots are weak; if planted too thinly, they are unable to grow, so it is desirable to plant rather thickly. The soil above the seeds should be made compact. When a seed-sower is used, that is done with a roller attached to it.

(4) Cultivation: Since carrots germinate slowly, and do not at first make rapid growth, cultivation must be commenced between rows about as soon as the plants can be distinctly seen. At that time, a scuffle hoe, wheel-hoe, or hand cultivator [p. 905] should be run between the rows, to cut up starting weeds and to stir the soil near the surface. As soon as the plants have sent out 1 or 2 secondary leaves, they should be weeded by hand and partially thinned. Complete thinning should be left until the next weeding, which must follow as soon as weeds have commenced to start much. Throughout the growing season, a hand cultivator, wheel-hoe, or scuffle hoe must be passed between the rows sufficiently often to prevent any growth of weeds. A two hand type weeder are usually sufficient. After the plants are thinned, cultivation is mostly done with a hoe (a triangular-shaped hoe is best).

(5) Harvesting and Storage: Since carrots harvested in cool weather keep better, they should be harvested as late as possible, but certainly [p. 906] before the coming of cold, severe weather freezes the ground. Unless the team is steady and the plowman skillful, the use of a plow in harvesting is likely to bruise and bury many roots. I prefer the use of a garden fork or a spade (the former is better). After the carrots are pulled, the tops should be cut close to the roots and the carrots exposed to the sun until exterior moisture is completely evaporated. They may then be stored in piles of moderate size in dry, cool, well-ventilated cellars. Keeping them in pits is possible, though attended with some risk. Storage pits for carrots should be smaller than pits used for other root crops. Carrots will not keep as long as turnips, beets, or potatoes; therefore, they should be used early. [p. 907]

5) Diseases and Insect Enemies

In Sapporo, carrots have thus far been quite free from both diseases and insects. Some of the earlier varieties have, upon one or two occasions, rotted considerably; the cause, I think, was the extreme wetness of the season, coupled with the fact that the seed was planted too early, resulting in the over-ripeness of the roots. Hence, do not plant too early. No insect but a cut-worm[59] has yet given any evidence of the catholicity of a taste for carrots. [p. 908]

7-15. Parsnip (Pastinaca sativa)

The parsnip is a vegetable belonging to Umbelliferae. It is cultivated as a biennial; its uses are about the same as carrots, but in Western countries it is better liked as an article of human food. It is very sweet and nutritious, and is relished by all animals. In some countries, it is extensively raised as feed for dairy cows, being especially valuable for such a purpose; it imparts a peculiar richness and fine flavor to milk. It is extremely hardy, and withstands the winters of this locality without the slightest injury. It may, therefore, be kept in the open field until spring, when it may be harvested and used after the supply of most other root crops is exhausted. Because of this, its ease of cultivation, and its productiveness, [p. 909] the parsnip is eminently worthy of a place upon every farm. Moreover, it has shown itself to be very well adapted to the climate and soil of this vicinity.

1) Varieties of Parsnips

"Hollow Crowned" is one of the best, especially for table use, and sufficiently productive to also make it valuable as feed for livestock. "Large Dutch" is not very different in size, shape, and quality from the above.

2) Soil and Cultivation

These are the same as with carrots; however, since the parsnip grows more rapidly and vigorously, less cultivation is necessary. Unless needed for use in autumn or early winter, parsnips [p. 910] should be left in the ground until spring. Here in Sapporo, they keep perfectly well and even improve in quality. They are seldom stored for any length of time, being harvested around the time they are needed for use. If they are to be stored, methods similar to those used with carrots are best.

3) Diseases and Insect Enemies

Parsnips are remarkably free from both diseases and insects. In Sapporo, they have never been affected by any disease. There is one insect, a beautiful caterpillar[60], which lives upon leaves, but it begins its ravages so late and increases in number so slowly as to never do any great injury. [p. 911]

7-16. Onion (Allium cepa, "Negi")

The Onion belongs to the Lineaceae. It is a perennial, but is mostly cultivated as a biennial. The plant is herbaceous, with parallel-veined, cylindrical hollow leaves. The flowers are in a single globular umbel, which is many-flowered and comes from the spathe of a one or two-leafed scallion. The flowers are regular and symmetrical, with a perianth of 6 parts; each flower has 6 stamens with 2-celled anthers and a 3-celled ovary; the style is persistent and slender, the stigma entire, and the flowers white (or bulblets in their place). The bulbs are large and depressed.

The onion is an important article of human food, being extremely wholesome. It is especially prized aboard ship or in the army, or indeed anywhere where men are obliged [p. 912] to live largely upon salted or preserved food. In such situations, onions prevent a disease caused under the conditions mentioned. For animals, onions have no great value; they are eaten by some, but are excelled by many other root crops; hence, it is not raised much for such purposes.

1) Varieties of Onions

(1) Danver's Onion: Originated in the town of the name in Massachusetts, it is large, thick, yellow, good-flavored, and productive; it is medium in earliness, keeps fairly well, and is good for garden cultivation.

(2) White Portugal: Medium in size and white, but flatter than the above; productiveness medium, quality particularly good, flavor delicate, but less strong than most others; late, with poor keeping quality. As raised upon the College Farm in 1879, it proved rather too late and not very productive. It is only good for private use, [p. 913] on account of its delicate flavor; it is not good for the market.

(3) Early Red Globe: As large as Danver's, but not as thick; red in color, good flavor; keeps well. As raised on the College Farm, it seems to be the most productive; being early, it has formed a large proportion of well-ripened bulbs. I consider it the best for cultivation in this vicinity as a market crop, though to some people its color is objectionable.

(4) Wethersfield Large Red: Very similar in general appearance and quality to the Red Globe; but a little flatter; under good cultivation in a suitable climate, it reaches a larger size. It is a later variety, and on the College Farm it has proved too late. Under favorable circumstances, it is extremely productive; though I consider it inferior to the Danver's and the Red Globe, here it proves quite profitable. [p. 914]

(5) Early Cracker (or Large Yellow): Similar in general appearance to the Danver's onion, but thinner, being one of the flattest and thinnest of any variety of onion; good quality, keeps well; very early, and as cultivated here, very reliable; in productiveness, inferior to the Red Globe, Danver's onion, and Wethersfield. Still, upon somewhat unfavorable soils and in cool climates, it is well worth cultivating.

(6) Yellow Dutch (or Strasburg): Very extensively used in America for propagation from sets[61]. For any other purposes, it is doubtless excelled by several other varieties I have mentioned. The little onions grown for sets keep remarkably well, and that is the chief recommendation of this variety for this purpose.

2) Soil Adaptation

The best soil for onions is a well drained, [p. 915] very rich, and rather light loam. Fair success can be expected in any soil suited for corn and potatoes.

3) Cultivation

(1) Crops Best to Precede Onions: It is of even greater importance with onions than with carrots that the preceding crop leave the land in fine condition. Carrots are perhaps the best crop to precede onions, followed by potatoes, turnips, and beets. Onions may, without disadvantage, be raised in successive years on the same piece of land, provided fertilizers are abundantly applied. Indeed, because of the fine condition of soil onions need, that is a common practice.

(2) Fertilizers: It is useless to attempt to raise onions in a poor soil. The land must be very rich and plentifully fertilized. [p. 916] Upon old, rich onion fields, 6 to 8 cords per acre of fine, well-rotted stable manure is sufficient; upon poorer soils, 8 or 10 [cords will be needed]. About the same quantity of well-rotted

compost (consisting of 1 part night-soil to 3 parts good muck or loam) is also good; near the seashore, the use of rotten kelp upon lighter soils is recommended. In this vicinity, herring guano used in connection with ashes may probably be used. About 1,000 pounds per acre of guano and 50 to 100 bushels of ashes are a good proportion.

Any of those fertilizers may be used if in a fine state of mechanical division and thoroughly incorporated with the surface soil. If barnyard manure is coarse, it may be applied in the autumn and lightly plowed in, where it will decompose before spring. In the case of well-rotted or fine manures, broadcast application and harrowing in spring is best. [p. 917]

(3) Preparation of the Soil: As with carrots, it is of the highest importance that the soil for onions be perfectly pulverized but it need not be deep, about 5 inches being sufficient; in all other respects, soil preparation is the same as for carrots.

(4) Planting: The Onion is hardy, and succeeds best if planted early in this vicinity from about May 1st to 15th. Success will also follow later planting, but I would always endeavor to plant at the time specified. Onion seeds are always sown in drills 12 to 15 inches apart; the seeds are planted by hand or with a seed-sower. In either case, the seeds should be covered about 1 inch deep. Onions can tolerate considerable crowding; therefore, bulbs growing above the ground, [p. 918] however much they may push each other, grow to a good shape and size. However, excessively dense planting is not advisable. It injures the onions in the ground (more than with most other plants) to pull up the superfluous plants growing near them. Plant no more than is necessary; if carefully sown, 3 to 4 pounds of seed per acre is sufficient in most cases, $3\frac{1}{2}$ pounds per acre is suitable.

(5) Cultivation: As soon as an eagle eye can see the onions in the rows, it is prudent to begin cultivation. First, run a scuffle or wheeled hoe between the rows. Then, when the plants get a little larger and you conclude that all of them have come up, the onion bed should be hand weeded. Throughout the season, all weeds (whether between or within the rows) must be kept down, by implement or by hand, respectively. [p. 919] Shallow cultivation is sufficient, indeed better.

(6) Harvesting and Storage: As soon as the majority of onion plants has fallen over and commenced to die, I find it advisable, in this vicinity, to roll the bed lightly with a hand roller. This breaks the tops still standing, starts the bulbs a little from the ground, and considerably hastens maturity. As soon as the majority of tops are dead, the bulbs should be pulled or raked out of the ground and left lying in thin windrows until the tops of good onions are quite dead and completely dried up. While the onions are lying there, occasional turning is advisable.

If rains are frequent, the onions should be covered with straw mats or similar material, but such provision is not necessary against occasional rains. [p. 920] If the onion field is inclined to be moist during the harvest period, the onions must be carried to a drier place. After they are well dried, they should be broken or cut close to the bulbs. All onions which have large, thick necks, and the tops of which are not yet dry, should be separated from the rest because they are not yet perfectly ripened and are likely to decay, and to cause decay even among the good. Such imperfect onions should be used immediately, or sold at a low price. Well-ripened onions should be stored, until the coming of extremely cold weather, in a cool, dry room above the ground, where they may be spread in layers about a foot thick, kept in bins which are well ventilated and not too wide, or packed in casks or straw bags. Of all these methods of storage, spreading in thin layers [p. 921] is the safest and best; but as the weather grows cold and there is danger of freezing, they may be packed in straw bags.

Onions can bear considerable exposure to cold, and indeed some farmers keep them by allowing them to freeze early in the winter and then covering them up so that they will not thaw before spring, when they are gradually thawed. That practice, however, I do not recommend. The onions should be kept in a room above ground as late as possible, and may be protected from frost by a covering of hay or straw. If they must finally be carried into a cellar, that should be done in cool, dry weather.

4) Raising Onions from Sets

This method is often practiced when it is desired to obtain [p. 922] very early onions, either for home use or for market. Onions from sets must be used quickly, since they keep poorly. The method of raising from sets is as follows: Early in spring, seeds raised at least 1 to 2 degree of latitude further south should be sown thickly in broad, shallow drills about 12 inches apart. 30 pounds of seed per acre is the proper amount. The soil should not be made rich, but it must be perfectly dry, and should be as warm as possible; therefore, a sandy soil is best in most cases. This should be prepared as with an ordinary onion field; but here in Sapporo it is even more important that the soil be free from weeds. If we plant seeds raised in this latitude or further north, the sets will not mature properly, even under the most favorable circumstances. Hence, many market gardeners near northern cities in the United States follow the same system I have recommended. [p. 923]

The first year, onions grown from seeds will make little growth—the size of good sets vary from that of a pea to that of a small nut. The sets should be ready for harvesting in August, but harvesting and dying should be done in the same manner as recommended for other onions. These little onions should then be spread in thin layers upon the floor of a cool, dry room. If they are of the White Portugal variety, the depth of the layer should be no more than 2 inches (if of the yellow varieties, 4 inches). Frequent stirring and turning over facilitates drying and tends to prevent sprouting. As winter approaches, the sets must be protected from cold as already described. As early the following spring as fields can be worked, the sets are planted in rich, finely pulverized soil about 2 inches apart in drills 10 inches apart. [p. 924] Growth soon commences, the bulbs enlarge, and when they reach the proper size, the crop is harvested and immediately used or marketed.

5) Potato Onions[62] or Multipliers

This is a hard-fleshed and thick variety of onion which never produces seeds. Its flavor is mild, and flesh very tender early in the season; later, it becomes hard and tough. It keeps very poorly, and is raised mostly for summer use. The method of propagation is as follows: Early in spring, in well-prepared soil, plant small and large bulbs in drills 10 to 12 inches apart (the distance apart in the drill varying according to the size, from 2 to 3 and up to 6 inches). The large bulbs divide and produce several smaller bulbs of irregular shape. The average number of small bulbs is 6 or 8 (though sometimes as many as 16). [p. 925] The irregular-shaped small bulbs are used for planting the next year. After planting, they increase in size until they become, under favorable conditions, 2 or 3 inches in diameter. A portion of these bulbs are used, but some must be kept for the production of small bulbs the following year; thus, each spring, you must plant both small and large bulbs again.

(1) Shallots (*Allium ascalonicm*): These differ from potato-onions in that they always multiply by division, never growing into large onions. They also sometimes, though seldom, produce seed-shoots, which potato-onions never do. Shallots exceed every other variety of the onion family as good keepers; with little care, they may be kept the year round, yet their flavor is mild and pleasant. [p. 926] Propagation is conducted in the same way as explained above.

(2) Top Onions (*Allium capa* ver. *viviparum*)[63]: These are propagated from little bulbs which grow in place of seeds. They are planted in early spring, in the same manner as potato-onions. They then increase in size, and are soon ready for use. In structure, top-onions are somewhat loose and coarse. Their flavor is mild, and they keep poorly.

(3) Garlic (*A. sativum*): These are propagated from bulbs which become clustered and are pointed. The leaves of this plant are lance-linear and keeled; flowers are few and purple in color, or bulblets[64] are present in their place; filaments all broad and 3-clefted. [p. 927] Garlic should be planted in extremely rich soil in onion fields. The bulbs should be planted about 6 inches apart in drills or ridges 14 inches apart. The general manner of cultivation is the same as for onions.

(4) Leeks (*Allium ampeloprasum*, "Nira")[65]: The garden leek has an elongated single bulb; its leaves are broadly linear, keeled or folded; the flowers in a head are white with rose-colored stripes. Two of the best varieties of leeks are Broad Scotch and Very Large Rouen. Select good onion soil, and plant the seeds in drills 6 to 8 inches deep and 18 inches apart. Cover the seeds to the same depth as onion seed. Later, thin down the plants to 9 inches apart. As their size increases, gradually pull earth around the plants until the drill is filled even with the surface of the ground. [p. 928] The leeks will be fit for use in October. The seeds need not be planted until quite late; hence, leeks may be grown as a second crop. [p. 929]

7-17. Squash (Cucurbita)

All squashes do not belong to one species of Cucurbitaceae. Plants of the genus *Cucurbita* are succulent herbs with tendrils; the leaves are alternate, and palmately lobed or veined; flowers are monopetalous, dioecious and usually yellow in color; fruits are pepo; seeds are large, usually flat and destitute of albumen, containing only 2 cotyldons. The Squash is a tropical or sub-tropical plant; but some varieties have now been developed which may be raised in quite northern latitudes. The best kinds are highly prized as human food, and the larger and coarser for animals. They are fed raw to cattle, especially dairy cows, with profitable results. When cooked, they are much relished by swine and are good for fattening such animals. [p. 930] Squash seeds act upon the kidneys, causing the secretion of an unusual quantity of urine, hence they must be carefully avoided.

1) Varieties of Squash

The varieties are very numerous, the variations in size, shape, and earliness being very great.

(1) White Early Bush: Small in size; vines assume a bushy form, never sending out runners to any great length; very productive, good in quality, and earliest among squash varieties in the U.S.; fit only for summer use.

(2) Summer Crookneck: Considerably elongated; neck usually crooked; vines bushy; of good quality, very early, and well adapted for summer use.

(3) Boston Marrow: The standard autumn squash [p. 931] with a rich orange color; very productive and well suited for autumn use, but not a good keeper.

(4) American Turban: Regarded as the best of all autumn squashes; flesh sweet, fine-grained, and dry.

(5) Marblehead Squash: A new variety which is highly esteemed; generally light blue in color; characterized by a very hard outer shell; flesh very sweet, dry, and fine flavored; keeps well. As cultivated on the College Farm, it has even excelled its high American reputation.

(6) Hubbard Squash: For many years, this variety has been very highly esteemed in the U.S.; very dark green or almost black; outer shell extremely hard; flesh dry, fine-grained, sweet, and of excellent flavor; keeps well; extremely productive. Upon the College Farm, it has succeeded well.

(7) Butman: This is a very recently developed variety said to have been produced [p. 932] by the cross between the Hubbard and a variety known in America as the "Yokohama Squash". It is very highly recommended, but a year's experimental cultivation upon the Farm was not very successful. It is perhaps too late for this locality.

(8) Large Winter Crookneck: This variety was formerly held in very high esteem, but is now surpassed in quality by several others; still, it is one of the best keepers and is worth cultivation.

(9) Yellow Mammoth: Squashes of this variety sometimes reach enormous size (some specimens weighed 300 pounds); quality inferior; cultivated only for livestock feed.

2) Soil Adaptation

Since the squash is a tropical or sub-tropical plant, we must select for it the warmest soil possible. A well-drained sandy loam sloping southward is best. [p. 933] The soil should be enriched by the application of manure to preceding crops. It has been found in the U.S. that sod land is very well suited for squash, the decaying grass and grass roots furnishing much available plant food. The squash grows rapidly, and soon gets ahead of all weeds, so that having the ground clean is not so important.

3) Cultivation

(1) Fertilizers: The manures for this crop should not contain much nitrogen. Well-rotted barnyard manure or composts are among the best. Wood ashes also give good results. Herring guano, rapeseed oil cake, the dung of fowls, or undecomposed human excrements should be sparingly used, if at all. Squash revels in heat; in all possible ways, therefore, [p. 934] we should encourage the extension of roots near the surface of the ground, since the soil is warmest at or near the surface. It is, therefore, folly to plough manure in deeply; the greater portion of it should be broadcast and lightly harrowed in. It is also a very common practice to put some well-rotted compost in the hills.

(2) Preparation of the Soil: This work is in no way peculiar, a rather shallow

plowing and ordinary pulverization being sufficient.

(3) Planting: The squash is killed by even a very slight frost. Nothing is gained by planting it early, since it requires a high temperature. We must wait until the soil becomes thoroughly warm [p. 935] and the danger of frost passes (in this locality, usually about the first half of June). Squash is usually planted in hills, the distance varying from 5 to 8 or 10 to 12 feet, according to the variety. A broad, shallow hole not more than 4 inches deep should be dug, and in the hole should be spread a shovelful of well-rotted compost. That is then covered with fine earth, so that the hole is not more than 1 inch deep. In that earth, about 6 seeds are sown and covered with fine earth, which should be compacted with a hoe.

(4) Cultivation: Squashes rapidly germinate, and cultivation should begin early. The field should be sufficiently cultivated to keep down all weeds, especially before the plants begin to run, after which provided cultivation has previously been thorough little work need be done. [p. 936]

(5) Harvesting and Storage: Summer and autumn varieties of squash are picked from the vines only as needed for use. Winter varieties must be harvested before the coming of even the slightest frost, since squash is very sensitive to frost. The method of harvesting is to pull or cut the squashes from the vines carefully, leaving the stems attached (if not, decay will soon begin at that point). They must be very carefully handled, because any bruise will cause early decay. They should be stored in rather small piles in open, airy sheds as late as they can be kept in such places without freezing; a little covering should be added at night, or when it is very cold.

When the weather gets colder, the squashes must be removed to a warmer, more enclosed [p. 937] but well-ventilated room; there they are most perfectly kept by the aid of artificial heating, just sufficient to prevent freezing. In such a room, they may be kept all winter. Few farmers, however, can afford to do that; therefore, it is customary to keep squashes in such a place only as late as they can be protected from frost by a moderate amount of covering, and afterwards to remove them to a dry, cool, well-ventilated room. Dryness and coolness are especially important. Squashes soon decay wherever it is the slightest degree warm and damp. They will never keep well in large masses.

4) Diseases and Insect Enemies

As cultivated in this vicinity, squash has shown itself thus far to be almost entirely free from diseases and insect pests. The most formidable insect affecting squashes in America is one commonly known [p. 938] as the "squash bug", a hyemeiptera (*Coreus tristis*)[66]. It has a peculiarly strong, disagreeable odor when disturbed. It is dark brown or dull black in color, and when fully grown measures about 1/3 inch long. Hymepterae females lay their eggs in patches upon the undersides of leaves. The young soon hatch, and when present in large numbers make great havoc among the vines, eating the leaves and, if hard-pressed, even the fruits. When about full grown, they leave the vines at night and hide in crevices. Many of those insects may be destroyed by placing chips, small pieces of board, or shingles among the vines and lifting these every morning, and destroying the bugs found under them. They are large and conspicuous, and hand-picking early in the growing season, if done persistently, will prevent serious injury to the plants. Another insect (Lepidoptera, *Aegeria cucurbitae*)[67] sometimes does serious injury to squash plants. The larvae bore into the stems [p. 939] near the ground. The best way to prevent their rapid increase is to destroy them by cutting off the affected portions of the plants.

7-18. *Pumpkin (Cucurbita pepo, "Tonasu")*[68]

The Pumpkin is botanically very similar to the squash, but is characterized by a smooth skin, and also by the fact that its flesh never hardens. Its uses are about the same as those of the squash, but it is never eaten after a simple boiling. It is made into pies for which it is well suited. In nutritive value, it is probably inferior to most kinds of squashes.

The principal varieties are: Large Field, Michigan, Mammoth, and Sugar. The 2 former varieties are good for livestock, and the 2 latter for humans.

The soil, cultivation, *etc.* for pumpkins are all the same as for squash, provided they are raised alone; however, that is very seldom done, since they are usually planted together with corn. One pumpkin seed may be planted [p. 941] in every 8 to 10 hills of corn. Thus cultivated, it costs very little, and it is believed not to lessen (to any considerable extent) the yield of corn, since each feeds upon different elements in the soil. [p. 942]

7-19. *Melon*
("Shiro-uri"[69] = *Cucumis conomon*, "Niga-uri"[70] = *Momordica charantia*, "Kamo-uri"[71] = *Lagenaria dasistemon*)

Melons are plants belonging to the Cucurbitaceae, and are planted quite extensively as an article of luxury wherever the soil and climate will allow. Some varieties are used in various preserves and pickles, but most are eaten uncooked. There are several species, among which we shall consider only the Muskmelon and Watermelon.

1) Muskmelons (*Cucumis melo*)[72]

This is a plant having leaves which are round, heart-, or kidney-shaped with simple tendrils; corollas 5; petals acute and almost separate; stamens separate; anthers have only one head; stigmas 3, blunt; fruits [p. 943] have a smooth rind and sweet flesh, the edible portion being the inner part of the pericarp; the thin and watery placentae are discarded with the seeds; the seeds are not margined. Its varieties are numerous:

"Green Citron" is round, with green flesh; sweet, melting, and good tasting. "Early Nutmeg" is round, with green flesh, fine, and very highly scented. "New White Japan" is apparently the common Japanese melon, which has recently been introduced into the United States. It is very well spoken of; its flesh is greenish white; sweet and fine. "Casaba" is Very large and long, with thick green flesh, delicate flavor, and fine quality.

2) Watermelon (*Citrullus vulgaris*)[73]

The Watermelon, [p. 944] though generically different from the musk-melon, requires about the same kind of soil and cultivation. The leaves are deeply 3-5 lobed, with the divisions lobed or sinuate pinnatifid; pale blueish; the edible portion consists of the thickened placentae, which is usually of a reddish color and in which the seeds are imbedded. The so-called "Citron" variety has firm flesh and is used only for preserves. The common watermelon is nice and wholesome when fully ripe.

"Excelsior" is a comparatively new variety of good size and fine quality, having for several years taken the first prize wherever exhibited; it is early, and has a thin rind. "Ferry's Peerless and Vick's Early": These varieties are new and highly spoken of. "Ice Cream" is white-seeded, very early, and of superior quality. "Mountain Sweet" is an old standard, very early and delicious; [p. 945] this is one of the best varieties for northern climates. "Black Spanish" is an old standard variety of good quality. "Citron" have various sorts, but are not fit to be eaten raw. [p. 946]

7-20. Cucumber (Cucumis sativus)

The Cucumber, in field cultivation, needs the same treatment as the Squash. Its leaves are more or less lobed (acutely); the middle lobe is most prominent, and is often pointed; the fruit is rough when young, smooth when mature. It is usually eaten when unripe, both raw and pickled.

1) Varieties of Cucumbers

"Improved White Spine" is an old standard variety. "Early Russian" is the earliest of any. "Improved Long Green Prickly" is often 18 or 20 inches long, of good quality. "West India Gherkin" is used for pickles only.

2) Cultivation

All of these cucurbitous crops need [p. 947] a rather light, warm soil. Thorough preparation of the soil is essential. Planting in the open field must not be done until the soil becomes warm and all danger of frost has passed. The manner of planting is the same as for the squash. The distance between hills is about 7 feet for watermelons and 4 to 6 feet for muskmelons or cucumbers. Crowding the plants produces many (but small-sized) fruits. Cultivation is done only to keep down weeds and to loosen the soil.

Melons and cucumbers are short-lived fruits, and are harvested as needed for use. The watermelon may, with care, be kept a few weeks; if a portion of the crop remains in the field so late that there is danger of frost, the ripe melons may be carefully picked and kept in small piles in a cool, dry room. All of these fruits are very sensitive to frost, and should be quickly disposed of. [p. 948] Cucumber vines will continue bearing fruits for a long time, provided the fruits are carefully picked before reaching maturity. If a few cucumbers are allowed to ripen on the vines, they will produce little more fruit; therefore, even though you do not need all that the vines yield, it is good to pick them all to make your vines productive for a long time.

Pruning the vines of cucurbitous plants is often recommended; provided the growth is luxuriant (as it will be if the soil is too heavy and the manures too rich in nitrogen), pinching back may do good, checking the growth of plant and inducing greater fruitfulness. The degree of pruning must depend upon the degree of luxuriant growth, which will be influenced by the weather as well as by conditions already alluded to.

Artificial fertilization, particularly of squash blossoms, is recommended. [p. 949] If any of these plants is to be raised under glass, as cucumbers frequently are, artificial fertilization is absolutely necessary. Cucumbers and melons are frequently started under glass, and afterward transplanted to the field, but they do not bear transplanting well, so unusual precautions are necessary. A very good method is to plant each seed upon a little square of inverted sod, then cover it with fine earth. The plant starts and sends its roots into the sod; at the proper time, the sod attached to the plant may be transplanted into a field without much disturbance to the roots. [p. 950]

7-21. Celery (Apium graveolens)

In its wild state, this umbelliferous plant is very strongly scented, and acrid if not poisonous; it is found along the seacoast of Europe. The "*var. Dulce*" is a variety rendered bland; the bases of the leaf-stalks are enlarged, succulent, and edible if branched, having been modified by cultivation. The leaves are pinnately divided into 3 to 7 coarse, wedge-shaped, cut or lobed leaflets or divisions; both the umbels and the fruits are small. "Turnip-rooted Celery" (*Var. Rapaceum*) is a variety with enlarged roots which are edible. Celery is now quite extensively cultivated in America and Europe, where the blanched leaf-stalks are eaten raw with a little salt, or in salads and soups. Almost everyone who tastes it becomes fond of it; celery is a wholesome vegetable. [p. 969]

1) Varieties of Celery

"Sandringham Dwarf White" is the smallest in size, but good. "White Solid" is the old standard variety. "Boston Market" is short, compact, solid, and of the finest quality. "Hood's Dwarf Red" is red in color; consequently it is ornamental; not of the best quality. "Seymore's Superb" is very large in size, and best suited for southern climates. A common name for a turnip-rooted variety is "celeriac". Celery is almost always raised as a garden crop, following some crop that may be harvested early in July, such as early beets, peas, radishes, cabbages, *etc.*

2) Soil and Fertilizers

Celery requires a very rich soil. Usually, no manure is directly applied. When applying well-rotted manure [p. 970] is mixed into the soil to a considerable depth. Many farmers apply large quantities of compost manure in trenches, cover them slightly with earth, and set the plants directly on top, but that method is not good.

3) Cultivation

(1) Planting: Seeds are always sown in beds consisting of rich, finely pulverized earth free from weeds. The seeds are small, and must be covered lightly. They may be planted in rows 5 to 6 inches apart, the seeds being dropped upon the surface and covered by patting the earth with a spade or the hand. If the weather is dry, the beds must be properly watered. They must be well weeded; when the plants are about 3 inches high, they are transplanted to another bed consisting of rich, finely pulverized soil, and set at a distance of about 4 inches apart each way. When first planted, watering and protection [p. 971] from the hot rays of the sun will be essential until they are rooted. Some farmers recommend shearing off the leaves almost to the ground once or twice before the final transplanting to the field, upon the assertion that this makes the plants more stocky. The season of planting in the seedbed will depend upon when the celery is desired, but it is usually from late April to the first of May. The soil of the field having been thoroughly prepared during the first half of July, those plants which have become well established in the second bed are transplanted to the field. A former practice was to plant in

trenches, the subsequent blanching being easier; presently, however, the plants are placed on level ground at a distance of 8 to 12 inches apart in rows 3 to 5 feet apart. The dwarf varieties are now generally cultivated. The usual precautions in transplanting must be observed. [p. 972]

(2) Cultivation and Blanching: Early cultivation is in no way peculiar. When the stem becomes tawny enough to fall over, they are made tough by a process technically called "handling", that is pulling a sufficient quantity of earth around the plants and pressing it against the stems in a compact bunch, and so keeping them in an upright position.

Later in the autumn, at a time varying as to whether the celery is wanted early or late, sufficient earth is thrown over against the plants to cover all except the very top of the stem. This causes the stem of most varieties to twin white; in other words, they are "blanched", and are now fit for use. Celery blanched in the field is fit for late use in autumn or early winter. For spring use, it is subjected only to the handling process. [p. 973] It is allowed to remain in the field until the coming of quite cold weather because celery can withstand 10 or 12° Fahrenheit of frost, provided it is not disturbed when frozen. When dry, it is then removed to narrow trenches as deep as the celery is tall dug in well-drained soil. The celery is placed in an erect position, roots downward, as compactly as possible, without bruising or breaking the stems.

As the weather becomes colder, straw, hay, or some similar material may be thrown over the tops of the trenches to prevent freezing. This covering must be done gradually, or else heating up and spoilage will result. The trenches are made 10 to 12 inches wide and of any length. Another way of blanching is to stand the plants in long, deep, narrow boxes, in the bottom of which a few inches of sand are placed. The size of [the boxes is such that they] may be kept in any cool cellar. [p. 974] The celery is put in them in a similar way as in trenches. [p. 975]

7-22. *Tomato (Lycopersicom esculentum)*

The Tomato is a solamaceous plant having many varieties, and originated in tropical America. It is annual, hairy, and rank-scented; its leaves are interruptedly pinnate, the larger leaflets cut or pimsatifid; the corolla is wheel-shaped, lobed, or divided into 5 or sometimes more parts; the tube is very short; the anthers connive around the style; flowers are yellowish by cultivation having their parts often increase in number; the esculent berry, usually red but sometimes yellow or white, becomes multi-celled. From tropical America, it has by now been transmitted to many countries. At first, it was not eaten; even at present, few people like its taste the first time, but most soon become very fond of it. It is now generally acknowledged [p. 976] to be an extremely wholesome vegetable, and in many countries, it now comes to the table in some form at all seasons of the year.

1) Varieties of Tomatoes

"Boston Market" is particularly smooth and of fine quality. "Paragon" is so named from the estimation in which it was held. "General Grant" is a new, good variety. "Trophy" is one of the most productive varieties.

2) Soil Adaptation

For early tomatoes, a somewhat sandy and not excessively rich soil should be selected. For late varieties, if a large yield is preferred, a strong rich loam is best;

it must be warm and well drained. Ordinary preparation and fertilizing with well-rotted manures is/ [in the original notebook, the text is interrupted here. [p. 977]

3) Cultivation

(1) Planting: Even under the most favorable conditions in this latitude, tomatoes seldom ripen all their fruit. Even the earliest variety ripens but little when the seeds are planted in the open field. For the best success possible, tomatoes must be started early in a hot-house, hotbed, or warm room in an ordinary house. The seeds are planted at such a time that the plants will reach the proper size for transplanting to the open field as soon as the soil becomes thoroughly warm, and all danger from frost has passed.

Here in Sapporo, tomato seeds may be sown in a hotbed around the middle of March. The soil should be made rich and thoroughly pulverized. Sow the seed about 1/2 to 3/4 inch deep and rather thinly in drills 5 inches apart. As soon as the plants are about 3 inches high, transplant them, giving each plant [p. 978] either a small plot or space 3 to 4 inches square in a well-drained soil, which should be well enriched. When the plants are 6 to 7 inches high, transplant them to the field. The proper distance between plants is 3 to 4 feet, and it is best to plant each upon a mound, so that the stems may spread apart as they lie upon the ground, thus giving sunlight full access to the fruit.

In garden cultivation, the vines are sometimes trained upon a trellis; Figure shows a very good one. It may be made by driving 3 stakes around the plant, to which are fastened 3 hoops. Another kind of simple trellis can be made of light strips of board fastened to stakes of about the same height as the plants, 3 cross-strips being sufficient. A cloudy or rainy day should be selected for transplanting; [p. 979] when that is impossible, the plants must be liberally watered and shaded from the sun for 5 or 6 days, until they become established. Cultivation is the same as with other plants.

(2) Harvesting: This is very simple, consisting only of picking the fruits as soon as they are ripe. Tomatoes are short-lived, and must be used soon after being picked. They will keep at best 2 or 3 weeks. At present, tomatoes in Sapporo are entirely free from diseases or insect enemies. [p. 980]

7-23. Asparagus (Asparagus officinalis)

Asparagus is a perennial plant belonging to the Liliaceae. The stem is erect and branching, about 5 feet high; the leaves[74] are needle-shaped. The asparagus is hardy; its natural habitat is the seashore, and it is found wild in many parts of Europe and Asia. The flowers are greenish; the drooping fruit, a red berry; the seeds small, black, and somewhat triangular. Asparagus is cultivated for its esculent spring shoots, which are cooked and eaten in many ways. It is undoubtedly very wholesome, and is fit for use earlier in spring than almost anything else.

1) Varieties of Asparagus[75]

"Oyster Bay (Conover's Colossal)" is very large and most profitably cultivated. [p. 981]

306 Chapter 7 Crop Cultivation

2) Soil and Fertilizer

The best soil is a deep, rich, and rather sandy loam; the sub-soil, especially, should be loose and porous. A soil with clayey sub-soil is ill-suited. Well-rotted manures should always be selected. It is also generally regarded as necessary to apply some common salt when the soil is prepared, at the rate of 2 to 3 pounds per square yard.

3) Cultivation

Asparagus is usually started in a seedbed consisting of thoroughly prepared, enriched soil; the planting time is very late in autumn or early in spring (the latter is probably better). Seeds are planted rather shallowly, in rows 1 foot apart, and the plants later thinned to 3 inches apart. In the seedbed, they should be well cultivated and allowed to remain for 1 year [p. 982] though some farmers prefer 2 years. In the spring of the first year, the young plants are removed to the permanent bed. Lay out the field in beds 5 feet wide, leaving paths of convenient width between them. In each bed are planted 3 rows of asparagus, as shown in the figure. The plants should be about 9 or 10 inches apart within the rows. The soil should be thoroughly pulverized to the depth of 2 feet, and well-rotted manure thoroughly incorporated with it. The crowns should be placed 3 to 4 inches below the surface. In transplanting, general rules must be observed. The work is done as early in spring as the ground can be worked. Sufficient cultivation is given to keep down all weeds.

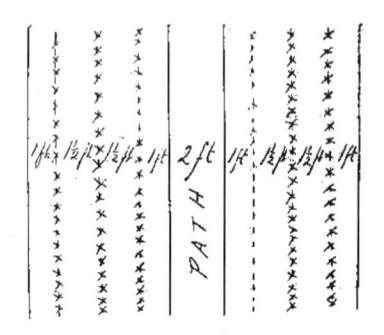

In the autumn, about 2 inches of well-rotted manure and [p. 983] a little salt are applied. Early the following spring, with a spade or pitchfork, the manure is mixed with the topmost 3 to 4 inches of the soil, care being taken not to injure the roots. The second spring, the soil is similarly treated. The third spring, a little asparagus may be cut for use, but the amount should be small, or else the plants will be unduly weakened. The treatment thereafter is the same from year to year; each year, the bed should yield a plentiful supply of tender shoots. It is customary to cut them all for several weeks, but the cutting must cease when the plants send up stems and leaves, otherwise the roots will die. Cutting commonly ceases about the middle of June or, at the latest, the first of July. Freshly cut asparagus is far better than that cut some time before eating. [p. 984]

7-24. Lettuce (Lactuca sativa)

Lettuce is an annual herb of the Compositeae. In the head, all the flowers are lingulate; the plant contains milky juice, and when full grown is 2 to 3 feet high; the leaves are alternate, and the flowers ovate, with imbrecated scales of unequal lengths. This plant is said to be of Asiatic origin, though it was introduced into America from Europe. Its broad, tender root-leaves are used for salads. The stem leaves are heart-shaped and clasping, but are not used as food for humans. Lettuce is generally acknowledged to be a wholesome vegetable, though it contains but little nourishment. It is rather a narcotic, and is good for nervous men.

1) Varieties of Lettuce

"Early Curled Simpson", "True Boston Curled", "Tennis Ball", "Boston Market" [p. 985]

2) Soil and Planting

The soil should be rich and rather moist. Unless lettuce is required very early, it is planted directly in a field or in a seedbed in open ground. Transplanting is thought by many farmers to increase the size of the head, and to lessen its tendency to send up seed-stalks early. In the majority of cases, however, lettuce is not transplanted. If sown in the field or garden where it is to stand, the proper distance between rows is 12 inches. If the plants are cultivated for heads, they should be thinned to 8 to 12 inches apart, but if only leaves and not heads are desired, less than half that distance is suitable.

The time for planting is very early spring; or, if a succession of lettuce is wanted, [p. 986] it may be planted at intervals of 2 or 3 weeks throughout the season. The seeds are small, so the depth of covering should be no more than 1/4 inch. Lettuce is cultivated as other crops; if early lettuce is desired, the seeds are sown in hotbeds, or in boxes placed in a warm, sunny room in a house. The plants should be transplanted to the open field as early as the ground can be worked. Lettuce is very hardy, and without difficulty may be had every month of the year. For winter use, it is only necessary to start them in the autumn, and to keep them in a cold-frame banked up so that the soil will not freeze, and protected at night or during cold weather by covering them with thick mats or blankets. [p. 987]

7-25. Eggplant (Solanum melongena, "Nasu")

The Eggplant is an annual herb which is more or less prickly. It is native to Africa, tropical America, and perhaps Asia; consequently, as raised in a temperate climate, it is rather tender. The leaves are ovate, rather downy, and obscurely sinuate; the stamens have anthers mostly equal in length to or longer than the very short filaments; they are usually not united. The eggplant is cultivated for its large, oblong or ovate, violet-colored or white esculent fruit, which is 2 to 6 inches long. It is prepared for eating in various ways, and is doubtless both wholesome and nutritious. Two good varieties are New York Improved and Black Pekin; the latter, I judge from its description to be similar to the Japanese variety. The ordinary American varieties cannot be [p. 988] successfully raised in Hokkaido because they require hotter weather.

The soil should be warm, rich, well drained, and thoroughly pulverized. The plants are usually started in seedbeds, being planted at about the same time and in the same manner as tomatoes, these two vegetables being quite closely related. The eggplant is less hardy than the tomato, and recovers slowly from any injury given it by cold weather. It must not, therefore, be transplanted to the field until both the soil and weather have become warm. The distance in the field is 2 to 3 feet apart. Cultivation after transplanting is by no means peculiar. The fruit is harvested for use as needed. Eggplant does not keep well. [p. 989]

7-26. Endive (Cichorium endivia)

Endive belongs to the Compositae, and was introduced into America from the East Indies[76]. Its leaves are smooth, sometimes slightly deeply serrated; the flowering stems are short and leafy. Endive is cultivated for its leaves, which are used for salads, particularly in autumn. Varieties of Endive are "London", "Green Curled", "Moss Curled".

The soil and its preparation are the same as for lettuce. Seeds are planted from early spring to midsummer, according [p. 990] to when it is needed for use. The method of cultivation is also the same as for lettuce. Endive is usually used only after blanching; that is done by gathering the leaves into a bunch and fastening them with a string, and also pulling a little earth around each plant. [p. 991]

7-27. The Culture of Vegetable Seeds

To secure good seeds is very important to a farmer in any country, but especially to one in Hokkaido, where good seeds cannot be gotten from good seed-growers. Most farmers seem to appreciate their importance, but they are also very likely to overlook this fact, and to incline to the opinion that wheat is wheat, peas are peas, squash is squash; thus, to get and to plant the cheapest seeds is the most expensive policy. I read in one seed catalogue of a gentleman who paid 12 dollars for a single ounce of a certain seed, simply because that seed was good, and because he knew the importance of securing good seed. A successful farmer, once asked "What are the essentials in the art of agriculture?", is said to have answered: [p. 1035] "(1) good seed; (2) good seed; (3) good seed". So much did he, and do all good farmers, esteem the importance of good seeds. Now, to secure good seeds it is necessary to know: How to raise good seeds, and How long seeds will remain good. Also, it is important to ensure the purity of varieties. For convenience of consideration, all farm crops are divided into 2 large categories, namely Annuals and Biennials, which can be further sub-divided.

In certain respects, the methods of agriculture necessary to secure good seeds differ widely between the 2 categories; however, it may be stated as applicable to both categories that good cultivation upon good soils is the first essential for the production of good seeds. That will be evident from the consideration of what a good seed is. A good seed is a large, well-formed, well-ripened seed. Such a seed is full of nourishment for the young plant; [p. 1036] consequently, a strong, vigorous plant may be expected from such a seed. From small, inferior, imperfectly developed, poorly filled seeds we cannot expect strong plants, because such seeds do not contain enough nourishment to produce strong, vigorous plants. It is only necessary to add that, with seeds, it would pay to take abundant care. With these introductory remarks, I will proceed to consider the subject at hand.

1) Annuals
(1) Leguminosae: The principal crops under this heading are peas and beans. Concerning their cultivation, I have little to say, but a few remarks as to the precautions necessary to keep varieties distinct may be useful. The blossoms of peas and beans are perfect, and due to the shape of the corolla, the wind probably has little effect in distribution of the pollen. [p. 1037] Looking from this starting point, it would seem that varieties of peas and beans might be planted quite near to each other without danger of mixing, but the fact is that the flowers of these vegetables are much visited by insects. It is found, therefore, in practice to be necessary, in order to prevent mixing, to plant varieties of these vegetables at considerable distances from each other.

There are very great differences in earliness among different varieties, and it is possible to plant extremely early and late varieties side by side without any danger of mixing, because the early variety will have finished blossoming before the late one begins. Peas and beans, though closely related, will not hybridize. Both peas and beans retain their vitality for a long time, but fresh seed is considered better than old.

(2) Cucurbitaceae: These embrace squashes, pumpkins, cucumbers, and melons. All of these vegetables are dioecious, and [p. 1038] from the depth of the corolla it would seem that fertilization by the agency of the wind must be well nigh impossible. Since these plants depend upon insects for the conveyance of pollen from male to female blossoms, we should therefore naturally expect what is found to be true in practice, namely, that varieties of any of these vegetables must be widely separated in order to secure their purity, because insects can fly long distances. It is, indeed, found extremely difficult to keep pure the different varieties of these plants; squashes, cucumbers, pumpkins, and melons. The distance necessary between varieties, in order to secure purity, depends greatly upon the nature of the intervening countryside. If a thick forest intervenes, the distance need not be great, but if the countryside is smooth and level, unobstructed by tall growing crops, the distance must be much greater. In practice, it is seldom possible to have a suitable distance for plants, but ensuring the comparative purity of any [p. 1039] variety of these vegetables is highly desirable. The squash and pumpkin are very closely related, and may probably hybridize. The cucumber and muskmelons are also closely related, and though it is not certain that they would hybridize in all cases, it is best not to cultivate them side by side.

The seeds of these vegetables should be taken out when they are perfectly ripe, thoroughly washed to free them from all adhering pulp, and then dried (either by exposure to sunlight or fire heat) to such an extent that there will be no danger of subsequent moulding. That condition is indicated by the manner in which a seed reacts to being broken. If thoroughly dried, it is brittle and will snap into pieces; if not thoroughly dried, it will bend. The seeds retain their vitality for several years, those of squashes and pumpkins considerably longer than those of cucumbers and melons. Pumpkin seeds which are 3 or 4 years old are quite esteemed as more able to produce fruitful vines than new seeds. [p. 1040]

(3) Solanaceae: These include potatoes, tomatoes, and eggplants. The seeds of potatoes are of little importance, being saved and planted only when it is desirable to develop new varieties. Tomatoes, however, as you know are always raised from seed; the same is true of eggplant. The flowers of these plants, as you remember, are perfect and probably are usually self-fertilized. There is, therefore, rather less danger of mixing than in many other cases. Varieties planted at the distance of a few hundred feet are unlikely to become mixed. The seeds must be taken from ripe fruit, and thoroughly washed and dried, but the amount of drying necessary is not as much in the case of squash seeds. Those mostly retain their vitality for at least 2 years, but fresh seeds are considered better than old.

There are other crops belonging to a group which I shall designate "Miscellaneous".

(4) Miscellaneous(Lettuce, Asparagus, Radish): [p. 1041] These vegetables are comparatively less important; not much can be said about their seeds, but the

separation by a few hundred feet is sufficient, in most cases, to ensure purity. Their seeds are small, and retain vitality only for a comparatively short time. Fresh seeds are better than old.

2) Biennials

This category includes many of the most important vegetables, such as onions, carrots, turnips, cabbages, beets, parsnips, cauliflower, kohlrabi, and celery. For the production of good seed from any of these plants, our care must extend though 2 years, the 1st year for the production of the vegetable itself, and the 2nd for the production of seeds; it is through the failure of proper care during the 1st year that many [farmers] make mistakes. They may give the most careful culture the 2nd year, knowing that however careful their management that year may be, it is impossible to get good seed unless the seed stock is good. [p. 1042]

It is highly important, then, in the first place to give the most careful culture the 1st year in order that we may have good-sized, well-formed, well-ripened vegetables from the seed produced. Having such seed, with good culture we can get good seed. We must bear in mind that "Like produces like". From the seeds of good vegetables, we can only expect good vegetables; from the seeds of poor vegetables, we can only expect poor vegetables. Having good seed, good soil, and abundant fertilizers, selecting the very best of the yield for seed stocks is very important. As we know, our crops have been brought to their present standard by man's careful selection and cultivation alone. We cannot expect to keep them in their present condition, or to improve them, because among them there is the tendency of degradation to their original state. [p. 1043]

3) Selection of Seed Stocks

In order to select a good seed stock, we must have clearly fixed in our mind what a good vegetable of the variety under consideration is as follows.

(1) Onions: The perfect onion (the only one therefore fit for the propagation of seed) is of good average size, fine regular form (with a small neck, and the top rather flat, or even concave, on the bottom as well as the top). It should also be one which ripened early, is firm in texture, and has the normal color and shape of the variety. With onions, there is a constant tendency to grow a large neck and top, but a small bulb. This tendency we can counter only by constant selection of those with smaller necks. [p. 1044] Avoid selecting any which are much convexed on either the upper or lower surface, because conspicuous convexity or the tendency to be pointed is an indication of a tendency towards the production of scallions, as onions with large necks and small bulbs are called.

We selected a large quantity of onions and stored them in a new cellar, but the cellar was wet and damp, and almost all of them decayed. The 2nd year, onions were again selected in large quantities, but very carefully dried, kept in a 2nd-story room until as late as was thought safe, and then stored in straw-bags in the cellar; that year, the result was better, and yet about 1/4∼1/3 of the onions decayed. The 3rd year (that is, last year), all precautions were directed to be taken, and were probably taken, but through a cause unknown more than 3/4 of the onions stored decayed. [p. 1045] That being my experience with the cellar thus far, I wish to follow another system this year. In this system, the place selected should be cool and dry, and the onions stored in small bulk. Large masses will certainly heat up and decay. The best place will be a room above ground just sufficiently warm to prevent freezing. If kept out of doors, onions should be stored in the following manner: a piece of well-drained, rather sandy soil should be selected, and the onions placed in a single layer, then very lightly covered with earth. Upon that earth, it would be well to place brush to the depth of about one foot, and upon the brush, place a small quantity of hay or straw. The purpose of the brush is to prevent snow from packing too closely upon the onions. The slight covering of earth, brush, and straw together with snow will be ample protection from the cold. [p. 1046] I cannot yet speak with confidence about this method, but I think that it will keep onions better than the method hitherto pursued.

(2) Carrots: All that has been said about onions applies equally to carrots. For seed purposes, a carrot of fair average size and typical form should be selected. A good seed carrot should have the following characteristics: be smooth, regular in form, of good color, without a large quantity of fibrous root, and have a rather small neck and top. It should also be firm in texture and perfectly solid, without any cracks or hollows. With most varieties of carrots, it is desirable to get the largest bulk within a short distance of the surface of the ground on account of greater ease in harvesting. A carrot which ripens too rapidly from top to bottom should never be selected. The best method of preserving carrots is as follows: after harvesting, cut the tops off just above the point where the leaf stalks unite. Then, insert the carrots at an angle of about 45° into dry, sandy soil, covering the whole root with earth but leaving the ends of the leaf stalks exposed. Then cover the ground in which the carrots are placed with hay and brush, as done with onions. Preserved in this way, carrots last 2 years on the College Farm, and there was absolutely no loss.

(3) Beets: From a good crop of beets, select only those of the most perfect form and appearance, and of average size. Avoid in this case, as in all others, those specimens having excessively large necks and tops. It may, however, be remarked that the constant selection of small necks and tops induces a decrease in the size of the vegetables. Beets are best preserved as is directed for carrots.

(4) Parsnips: These, being perfectly hardy, are seldom harvested before spring. When the best specimens (according to the principles pointed out with regard to carrots) are selected, they should be planted at once. Any delay seriously diminishes the vigor of the plants. [p. 1048]

(5) Turnips: From the best specimens, as directed in the case of beets, and preserved in the same manner as carrots.

(6) Cabbage: The heads of cabbage selected for seed should possess the following characteristics: fair average size, regular and symmetrical form, a flat solid head, short stem, and few loose leaves. The method of preserving cabbages is that already pointed out.

(7) Cauliflower: The best specimens are those with the largest, most compact heads and the fewest sprawling leaves. Cauliflower is preserved in the open air, being planted as it grows, but perhaps a little deeper, and protected by a covering [p. 1049] of loose material which will not be packed down too closely by the snow. Kohlrabi and celery are preserved in a similar manner.

4) Cultivation of Seeds

All the kinds of vegetables under consideration begin to grow rapidly early in spring; therefore, it is important to plant them just as early as the melting of snow will allow. There is perhaps little or no danger of them being injured by cold, and I dare to do so very soon after the snow has melted, provided the weather permits. It is highly injurious to plant sprouted shoots, because those shoots are slender, pale, irregular, and unhealthy; hence, subsequent growth is never as good as it would have been if the vegetables were planted before sprouting. [p. 1050] This being the case, it is wise to prepare good soil for seed purposes the preceding autumn. Good, rich loam should be selected and a liberal application of well-rotted manure be made (except perhaps in the case of beets, with which there is a tendency, in rich soil, to produce an excess of leaves with comparatively little seed). The soil for beets should not be so highly enriched, but pulverization should be thorough and deep.

(1) Onions: With onions, there is a great danger of mixing varieties because the flowers are much visited by insects. The greater the distance between varieties, the better. The minimum distance at which comparative immunity from mixing may be expected is about 1/3 of a mile, but here again that depends upon the nature of the intervening crops. Onions should be planted in trenches about 4 inches deep, and covered at first only slightly with earth; later, as the seed stock grows, the earth should gradually [be pulled] around them to assist in holding them in an upright position. The distance between trenches should be 3 to $3\frac{1}{2}$ feet, [p. 1051] and the distance between onions in the trenches 3 to 4 inches.

Cultivation throughout the growing season should be thorough. The proper time for harvesting is indicated by the gradual turning yellow of the seed-stalks. When the majority of seed-stalks are quite yellow, the seed umbels (together with 2 to 3 inches of the seed-stalks) should be cut, and then, thoroughly dried, either by hanging them in a dry, airy place or by spreading them in a thin layer and exposing them during pleasant weather to the sun. When the umbels are thoroughly dry, the seeds should be beaten out carefully, winnowed to free them from all chaff, and exposed in a thin layer to sunlight for 1 or 2 days, after which they should be carefully stored in a dry, cool place. Tightly closed paper bags, or boxes are best for the storage of seeds. If quite perfect seeds are required, it is best to immerse them in water, and to separate those which sink to the bottom and those which float upon the surface of the water. The heavier the seeds are, the better they are formed. [p. 1052]

(2) Carrots: The blossoms of carrots are perfect, but are considerably visited by insects; the danger of mixing is probably not as great as in the case of onions, but still the greater the distance in ordinary practice, the better. Carrots should be planted in rows at a distance of $3\frac{1}{2}$ to 4 feet apart and within $1\frac{1}{4}$ to 2 feet apart, the distance varying of course with different varieties. The best method of planting is with a strong, heavy bat (by preference, of iron, though wood may answer) to make holes in the ground of the full size and depth required to receive the carrots, which should be planted so that the crowns are just at the surface; the earth should then be made compact about each root. Cultivation is the same as for an ordinary crop.

The period of ripening of the seed is very different in different umbels, the terminal umbels of the principal stalks ripening first, with the others in succession from within, centrifugally outwards. The earliest seeds from the central umbels are considered the best. [p. 1053] The ripening of seeds is indicated by the brown color assumed by the umbels. It is necessary, therefore, to harvest seeds at several different times; the umbels should be cut and dried in the manner pointed out in the case of onions. The seeds are then separated, winnowed, dried, and stored in the same manner.

(3) Turnips: There is a great danger of mixing varieties of turnips, which should therefore be separated as widely as possible. It is also thought by some farmers to be best not to plant turnips for seeds in close proximity to cabbages; principally, however though fear of injury to the seed of latter, it being the idea that the tendency to head will be diminished through the possible hybridization of seed. Turnips should be planted in rows $3\frac{1}{2}$ to 4 feet apart and within the rows 2 to $2\frac{1}{2}$ feet apart; the planting depth should be the same as for carrots.

The seeds of turnips do not all ripen at one time, and in order to prevent waste [p. 1054] through shelling of the first ripened seeds, it is necessary to harvest at least 2 separate times. The time for harvesting is indicated by the yellow color assumed by the stems and seed pods. The method of harvesting is to cut off those portions in which the seeds are ripe, and dry them by exposure in open air. When thoroughly dry, the seeds are beaten out, winnowed, dried, and stored as in the case of onion seeds. Too much drying is injurious, but it is not as bad as too little drying.

(4) Cabbage: Here again, the danger of mixing between varieties is considerable, being about the same as in the case of turnips. The distances necessary between rows of cabbages and between cabbages in the rows are the same as for turnips. The cabbages should be planted a little deeper than they grow the 1st year, the earth being packed very compactly around the roots. The time of harvesting and method of treatment of the seeds are precisely the same as for turnips. [p. 1055]

(5) Beets: The danger of mixing in this case is not great, though considerable separation of varieties is advisable. The proper distance between rows is $3\frac{1}{2}$ to 4 feet, and within the rows 15 inches to $2\frac{1}{2}$ feet. The method of planting the roots is the same as for carrots, though beets are generally not buried quite so deep. The time of harvesting is indicated by the yellow or brown color assumed by the seed capsules. As soon as the majority has assumed that color, the seed-stalks are cut, dried by exposure in the open air, and the seeds separated by rubbing; then they are winnowed, cleaned, and dried in the ordinary manner.

(6) Parsnips: The danger of mixing is about the same as in the case of carrots. The proper distance between rows is $3\frac{1}{2}$ to 4 feet, and within the rows $2\frac{1}{2}$ feet. The ripening is precisely the same as with carrots. The method of harvesting and treatment of the seeds are the same as for carrots. [p. 1056]

5) Vitality of Seeds

(1) Onion: Onion seeds are considered to be best when only 1 year old, though they are heavy and well-developed. Well-ripened seed will mostly germinate the 2nd year, though germination will be somewhat slower. It is never safe to plant onion seeds which are more than 2 years old.

(2) Carrots: Seed is much better the 1st year; germination is very uncertain the 2nd year; seeds 2 years old are seldom advisable.

(3) Turnips and Cabbages: Seeds are good for several years, but fresh seed is considered best.

(4). Beets: Seeds of the 1st year are much better, but germination is at best rather uncertain on account of the thick, hard, dry capsules in which the individual seeds are enclosed. Seeds which are 2 years old should not be planted, except in case of absolute necessity; even then, there is a great danger that the crops will be very small. [p. 1057]

(5) Parsnips: The vitality of these seeds is the same as that of carrots. [p.1058]

7-28. Beet (Beta vulgaris)

The Beet is the most valuable member of the *Chenopodiaceae*. Its leaves are ovate, oblong, smooth, often wavy margins, and are sometimes purple-tinged. The flower-clusters[77] are spiked, sometimes almost globular. The flowers are clustered, perfect, very inconspicuous, and of a greenish color. Each flower has 3 bracts and a 5-cleft calyx, becoming indulated in fruit, enclosing the hard achene, the bases of the two conherent; stamens 5, style short, stigmas mostly 2. The plant is a biennial; its native habitat is the seashore, and it is found wild in Sicily and the Barbary States.

1) Varieties of Beets

Beets may be divided into [p. 838] Table Beets and Mangel-Wurtzels. The latter may be subdivided into 2 classes, namely, Sugar Beets and Stock Beets. Table Beets are considerably sized in most Western countries as an article of human food. It is boiled until it becomes soft, whereupon the outer skin readily separates from the inner part. Both leaves and roots are eaten only when they are young. As a source of sugar, beets have rapidly gained in importance. In France in 1870, 300,000 tons of beet sugar were produced; in Austria, 80,000 tons; in Russia, 100,000 tons; and in Germany, too, a vast quantity was produced. There is also one factory in successful operation in Maine, in the United States. Very large portions of the U.S. are well suited to the growing of beets rich in sugar. In temperate climates, where soil is of the right kind and labor is cheap, [p. 839] it is the best and the most economical plant for the production of sugar. As feed for livestock, mangel wurtzels are valuable. The yield per acre upon suitable soils is large, the expense of raising them is not great, the nutritional value is considerable, and the keeping quality is good. Hence, it is particularly suitable for late winter and early spring feed; 40, 50, and even 60 tons per acre have been raised.

(1) Table Beets: "Early Blood Turnip" is comparatively small, shaped like the English turnip, dark red, and extremely early. "Egyptian" is small, shaped like the preceding variety, red or white, and extremely early; together with the preceding, it is well suited for summer and autumn use, but not for winter keeping. "Long Smooth Dark Blood" is large, long, conical, dark red, late, and well suited for winter use. [p. 840]

(2) Sugar Beets: "Vilmorin's Improved French"; This is cultivated in Sapporo and this vicinity, it has proved the most excellent. It is considerably smaller than the White Imperial, but of better quality. In color, this variety, the White Imperial, and the Electoral are all white or nearly so, sometimes slightly tinged with pink. In the manufacture of sugar, the coloring matter in the beet is more or less troublesome; hence, all other things being equal, colored beets should be avoided.

(3) Stock Beets: "Yellow Globe" is quite similar to the following variety, but larger; superior to any other variety for growing in this vicinity; yield large; roots usually solid and firm in texture; quality good; keeps well.

(4) Red Globe: "Red Ovoid" is not much different in appearance from the above; [p. 841] general characteristics and value about the same. "Norbiton Giant Long Red" is very large and productive; quality not equal to the Yellow Globe, and therefore not so profitable.

2) Soil Adaptation

According to the purpose for which the beets are raised, the soil should vary: for table beets, a light, warm sandy soil, filled with manure and deeply pulverized; for sugar beets, a deep, rich and thoroughly well-drained loam, neither too sandy nor too clayey, and containing only a small quantity of organic matter; for stock beets, the kind of soil is not of such vital importance, since any strong, deep and well-drained loam will yield fair crops. Most of the soil in Sapporo [p. 842] and this vicinity seems to be very well suited for the production of both table and stock beets. Large quantities of sugar beets, too, may be raised, but in quality they are inferior. As the quantity of organic matter in the soil diminishes, it is not unlikely that beets of better quality may be raised. The climate of Hokkaido, being extremely moist during the late summer and autumn months, induces a very rapid, luxuriant growth and produces beets large in size but poor in sugar. Thorough drainage may partially counteract the injurious effect of excessive rainfall, but in my opinion, the latter will always be quite an obstacle to the successful cultivation of good sugar beets [p. 843].

3) Crops Best to Precede

The preceding crop should be one which requires and admits of very thorough cultivation and abundant manuring. It is highly important that the land for beets be free from weeds, deeply pulverized, and full of essential plant food in an available form. Wherever possible, grain crops are desirable; for this vicinity, corn is very good.

4) Cultivation

(1) Fertilizing: For either table or stock beets, manure may be liberally applied in the spring just previous to planting. However, for sugar beets, a large application of manure, particularly animal excrement or herring guano at that time is not advisable [p. 844] because it causes the growth of beets which are poor in sugar and contain nitrogenous compounds that are very troublesome for the manufacture of sugar. Manuring should be done very heavily the preceding year; it will then become decomposed, and will not injure the quality of the beets. If manure is to be applied to sugar beets in spring, it should be mineral fertilizers, ashes, sulfate of potash, nitrate of potash, magnesium sulfate and superphosphate. If animal excrements are to be used, they should be in the form of well-rotted composts.

In raising table or stock beets, the objective is not the roots. Excessively rich in sugar, but simply a rapid luxuriant growth, hence the manure applied may contain much nitrogen. Good barnyard manure, fish guano, Peruvian guano, the dung of fowls, night-soil *etc.* may be [p. 845] used advantageously in large quantities, but remember that their excessive use would produce many hollow beets, and also impair their keeping quality. It is well known that salt exerts a very favorable influence upon the growth of mangel wurtzels, especially at some distance from the seashore. A dressing of 400 to 500 pounds of salt per acre together with other manure will be found profitable. Due to its cheapness and

equal effectiveness, coarse, dirty salt will answer the purpose. Salt may also perhaps increase the quantity of sugar beets, but doubtless not of sugar. All manure should be applied broadcast and thoroughly mixed with the surface soil.
(2) Preparation of the Soil: It is highly important that the soil be deep; hence, plowing should be done to the preceding crop [p. 846] and again just before the beets are planted. The sprouts of beets are small and weak, and cannot make their way above the surface unless the soil is finely pulverized; hence, great efforts in harrowing, brushing, or rolling must be taken.
(3) Planting: Beets should be planted very early in spring (in this vicinity, the first half of May), although fair success might be expected from seeds sown in drills, either level or on ridges; the former method is better in all deep, well-drained soils. For smaller varieties of table beets, 12 to 15 inches between drills is sufficient; for sugar beets, 18 inches to 2 or $2\frac{1}{2}$ feet; and for mangel wurtzels of larger varieties, 2 to $2\frac{1}{2}$ feet. Table beets should be thinned to 5 to 6 inches in the drills. [p. 847] Sugar beets and mangel wurtzels should be thinned to 18 inches to one foot. It is customary to sow beet seeds in beds, and afterwards to transplant to the field, the objective being to escape the ravages of insects which prey upon the young plants. On account of the considerable labor and expense required, that method cannot be pursued on a large scale. There are various machines for planting beet seeds. The most common type is run by man-power and sows a single row at a time. There is a large German machine, drawn by 2 horses, which sows several rows at a time, and about 20 acres in a single day. Machine planting is not so reliable as that done by hand with trustworthy workmen; where the cost of labor is cheap, I prefer the latter. A good average depth of planting [p. 848] is about $1\frac{1}{2}$ inches. The soil above the seeds should be made compact, in order that moisture may be brought into constant contact with it.
(4) Cultivation: Beet seed is rather slow in germination, and as soon as young plants make their appearance, it will be necessary to begin cultivation. The first thing to do will be to run a scuffle- or wheel-hoe between the drills, cutting the weeds as close to the young plants as can be done without disturbing or covering them. Then must follow a careful hand weeding. When the plants become a little larger, if the space between drills is sufficient, a horse cultivator may be used. By the time of the second cultivation, the plants will have become large enough to need thinning out; [p. 849] that may be done largely with the hoe, although some hand labor will be necessary. The vacant spaces may be filled with plants from places which were thickly sown. Transplanting is best done on rainy or cloudy days, or in the afternoon; a portion of the leaves must be stripped off beforehand.

Subsequent to thinning, all necessary cultivation may be done with the cultivator and the hoe (mostly with the latter). In cultivating sugar beets, if they tend to grow up out of the ground, it is advisable to cover them with earth, since that will improve their quality.
(5) Harvesting and Storage: Early table beets are mostly harvested as needed for use during summer and autumn; those not needed then, and those for winter use, [p. 850] are allowed to remain in the field until there is danger of severe frost. Mangel wurtzels for livestock feed are harvested at the same time. Sugar beets are harvested as early as that time, but if needed in the refinery, as soon as they are ripe which is indicated by the death of their outer leaves. Late harvested beet keeps

much sugar better than early harvested beet. The method of harvesting must differ with the variety, but most can be pulled out without much injury or difficulty. Those beets which are difficult to pull out may be started by a garden fork.

After being pulled, the beets are laid in a row with the tops even, and then the tops may be rapidly cut off using a large knife. They should not be cut below the point where the leaf-stalks unite, if the crown of the beet is cut, decay soon ensues. [p. 851] Before being stored, beets should be allowed to lie in the sun until the external moisture is evaporated, and then immediately stored. They may be kept in somewhat larger masses than potatoes without risk of decay, and may be kept in quite large piles in dry, well-ventilated cellars, or in pits in well-drained soil. The safest kind of pit is one constructed as described under the heading "Potatoes". Although beets can bear considerable rough handling, bruised places are likely to decay, so a reasonable amount of care must be exercised. A box full of small holes, placed inclined through a mass of beets, will serve a good ventilator, provided one end communicates with the air above ground. [p. 852]

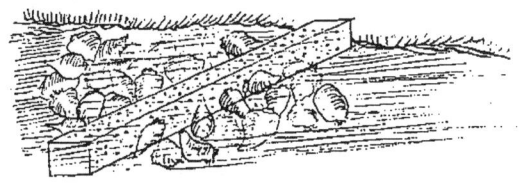

5) Insect Enemies

In most countries, beets are quite free from insect enemies, but in Sapporo there is one whose ravages are so great as to make it very difficult to raise this crop. That insect is known in America as the "cut-worm" and here as "*nekiri-mushi*"[78]. The mature insect is a moth which flies and doubtless lays its eggs at night; the larvae also have nocturnal habits. Feeding in the evening and very early in the morning and perhaps sometimes on very cloudy days, it lays its eggs in the ground, probably in July or August. The eggs doubtless mostly hatch sometime during the autumn, and the young larvae feed upon the leaves of different plants, often upon beets. The insects are, however, so small—and the beets so large—at this time that not much injury is done. [p. 853] Upon the approach of cold weather, the larvae go into the ground a sufficient distance to be beyond the reach of frost. They next usually make their appearance sometime in June, when they are about half an inch (or a little more) in length. It is at that time that the insect does great damage; every night, it crawls along the surface of the ground, cutting off the young and tender beet plants just at or a little beneath the surface. One insect can thus kill many plants. I know of no effective remedy which can be economically applied on a large scale. Hand picking can only be done on a small scale. Strips of stiff paper about 1½ to 2 inches wide may be bent around each plant; since the worm travels upon or very near the surface, these strips will probably protect most of the plants from injury. [p. 854] It is my opinion that an application of "Paris green" around the stem of plant might be effective, but those would need renewing after each severe rain, and might possibly adhere to the roots in such a quantity as to kill the animals fed upon them. The worm is of a dark brown color (very similar to the color of the soil), which is a great advantage to them in hiding. The mature insect is a very light brown moth—the shade differing among individuals —with darker markings upon the wings. [p. 855]

7-29. Sorghum and Imphee (Sorghum vulgare var. *saccharatum* and var. *imphee)*

Sorghum and Imphee are well-known plants belonging to the Gramineae family. According to Gray, they are specifically the same, being merely varieties. According to some other agronomists, the Imphee is called *Healcus caccharatus*. The var. *Saccharatum* is often known simply as "Sorghum"; it is of Chinese origin, and hence is sometimes known as "Chinese Sugar Cane". Imphee is of African origin; hence, it is called "African Sugar Cane". The true tropical sugar cane is generically different, being known as *Saccharum officinarum*; this cannot be raised in Hokkaido.

1) Varieties

Some varieties of Sorghum and Imphee can be cultivated here in Japan. They are well suited for the refining of fine syrups and sugar [p. 992] when grown under favorable conditions. At present, a large percentage of the sugar used throughout the world is made from tropical sugar cane and sugar beets. The growing of sorghum (of the kind which can be raised in a temperate climate) has received great attention in the United States; the amount of sugar made from sorghum is rapidly increasing year by year.

The number of varieties of Chinese and African sugar cane is immense, but only a few of them can be raised in Japan. Early Amber Sugar Cane, one of the newest varieties, was developed in the state of Minnesota, and the name "Minnesota" is often prefixed to it. It is not a true sugar cane, but is supposed to have resulted from a cross between a Chinese variety and an African Imphee. It is the earliest of any variety, and in America has shown itself [p. 993] to be extremely well suited to the refining, without expensive equipment, of fine-quality syrup and also —under favorable conditions—for the production of dry sugar. The name "amber" is indicative of the color of its syrup.

Some of the unfavorable conditions which attended the failure of sorghum experiments on the College Farm were: (1) late planting, (2) an unusually wet and cold autumn, (3) failure to get the building and machinery ready at the proper time, and (4) the inexperience of the workmen. The climate of Hokkaido, I do not consider as very favorable for the production of any sugar plant. However, by the most judicious treatment of the soil, and with the utilization of what little sunshine we have to the utmost degree, the plants will suffer injury from excessive moisture. [p. 994]

2) Soil Adaptation

Sorghum grows vigorously upon any soil that will produce good corn, but it will not upon all such soils have the maximum sugar content. The most suitable soil is a deep, rich loam which is rather sandy, does not contain much organic matter, and is well drained. Mucky, heavy, clayey soils are to be avoided. Alluvial soil is also good. The soil of Hokkaido, though it fulfills many of the needed conditions, is generally very flat, hence poorly drained, and contains too much organic matter. However, those characteristics can be done away with by a few years of cultivation and artificial drainage.

3) Cultivation

(1) Fertilizers: The fertilizers for sorghum and imphee should be similar to those applied to other sugar plants. Their application in large quantities [p. 995] to the preceding crop is highly recommended. Mineral fertilizers, such as bone meal and

ashes, may be directly applied. Sulphates of potash, phosphate of lime, nitrate of soda, or potash in small quantities are extremely useful; but they cannot be cheaply obtained here. Well-rotted barnyard manure may produce fair results. In all cases, the fertilizers are applied broadcast.

(2) Soil Preparation: Autumn plowing is advisable, since it exposes the soil to the action of frost. Barnyard manure, if used, must be applied in the autumn; then, early in spring the ground must be ploughed and harrowed.

(3) Planting: Sorghum seeds are small. Sorghum resembles corn and other winter grains. [p. 996] After germination, it is for some time apparently very tiny, but during that time it is rapidly extending its roots to a considerable distance, as well as storing material for future growth. When it begins to grow again, it shoots up with wonderful rapidly. Now, for the first period, rather cool weather is favorable. If the weather is hot, the sorghum shoots up too quickly and makes an inferior growth. In habits of growth, sorghum is quite different from corn. Sorghum, therefore, is planted early. From May 5th to 20th will perhaps be the best time for planting. Since the soil in this vicinity is mostly very flat, and the climate is rather cool, I recommend planting upon ridges $3\frac{1}{2}$ feet apart, 6 inches high, and 8 to 10 inches broad. Upon each ridge, a sharp furrow not more than 1/2 to 3/4 inch deep is made. [p. 997] In those furrows are planted the seeds, rather thinly, say 1 to 2 inches apart.

(4) Cultivation: In the early stages of its growth, sorghum should be very thoroughly cultivated, because if once checked by weeds it will never fully recover. That is done in about the same way as with corn, early, before the plants get large enough to crowd each other in the rows. All of the suckers which start, especially upon rich soil, are best broken off, or else the main stems will not be as well developed; the juice contained in the suckers, being of inferior quality by not having time enough to ripen, is not of sufficient value to make good the loss in the main stems. Cultivation should continue until the cane begins to shoot up rapidly, [p. 998] and if it has been thorough up to this time, very few weeds will grow thereafter.

(5) Harvesting: Examination of the juice of the cane at different stages of its growth has shown that it contains the largest percentage of cane sugar when the seeds are perfectly ripe. We should, therefore, harvest as nearly as possible at that time. The cane may be cut with a corn-cutter (use *kama* in Japan), and immediately thereafter all the leaves and 2 to 3 feet of the top of the cane are removed (those parts being valueless for the production of syrup). The leaves make good fodder, if well cured. That portion of the seeds not needed for sowing is valuable for livestock feed, being estimated to be worth about half as much as corn.

As soon after cutting as practicable, the juice should be extracted and rapidly evaporated to the desired density. [p. 999] The juice of freshly cut cane, provided it is ripe, may be boiled without the addition of anything to clarify it. If the juice, however, is inferior in quantity, either because of imperfect development or long keeping of the cane, it may to a certain extent be clarified by the addition of a small quantity of freshly slaked lime. If for any reason it is impossible to process the cane at once after harvesting, it may be piled up in small heaps in an open, airy shed having a tolerably free circulation of air. There it should be kept, at as cool and equitable a temperature as possible, without freezing. [p. 1000]

7-30. Broom Corn (Sorghum vulgare, "Morokoshi")

Broom Corn is a plant the same as the kinds of sorghum cultivated for sugar, yet it is very different from them in the character of its juice, and consequently in its uses. It is quite valueless for the production of sugar, but is raised for its long stalks that bear the blossoms, those being best suited for the manufacture of brooms than any other known material.

The seeds of broom corn may be of slight use as feed, but it is for the manufacture of brooms that it is most cultivated. As far as I know, raising it in Hokkaido has not yet been tried; there are many kinds being successfully cultivated in Massachusetts, and those can probably be raised here. [p. 1001] The soil, fertilizers *etc.* are all the same as for Indian corn. The rows should be 3 to $3\frac{1}{2}$ feet apart, and the plants thinned to about 8 inches apart in the rows.

Harvesting is done at different times according to the kind of brush desired. The best brooms are made from brush harvested while it is still rather green, since then it is pliable and tough. If ripe seed is desired, the brush is left standing later. Usually only the brush is cut, leaving the remainder of the plant in the field, since it is not of much value for any purpose except fertilizer. [p. 1002]

7-31. Hemp (Cannabis sativa)

Hemp belongs to the Urticaceae, and is a tall, coarse, rough-furred plant having watery juice and a tough, fibrous bark. It is for fiber that it is cultivated. The flowers are greenish and dioecious; the stamens are 5, with drooping fertile flowers in irregular, spiked clusters; the leaves are 5 to 7 in number, and lancelate; the leaflets are irregularly serrated. This plant was introduced into America from the Old World, but whether it was originally native to both Europe and Asia is unknown.

1) Soil Adaptation

The best soil for hemp is a deep, rich, alluvial soil, though any deep, rich soil containing a considerable amount of organic matter will produce good hemp. Soils which have been in grass for a long time, such as old pastures or mowing fields, [p. 1003] and in this country, newly broken up soils will be found well suited for it. The best climate is rather moist and cool. Hence, we can see that Hokkaido is a "chosen land" for the production of hemp.

2) Cultivation

(1) Fertilizers: Hemp grows with great rapidity, and so we should apply readily available plant food. Well-rotted barnyard manure, night-soil, and herring guano are all good. If coarse manures must be used, they should be incorporated with the surface soil the preceding autumn. If the manures are well-rotted and fine, broadcast manuring in spring and mixing with the surface soil by harrowing will be best. The quantity must be liberal, because the fibers of slowly grown, imperfectly developed hemp is short and inferior in quality. [p. 1004]

(2) Planting and Harvesting: The soil having been thoroughly pulverized by ordinary methods of tillage, the seeds are thickly sown as soon as the weather becomes fairly warm, and all danger of severe frosts has passed. In this vicinity, sometime during the first half of May is usually best. Broadcast sowing and planting in closely arranged drills are both practiced. Upon clean soils, the former method is even preferable, because the stems are less likely to branch, and

branching is injurious to the quality of the fibers. Seeds must be sown thickly so that the plants do not branch. About 3 bushels of seeds per acre is usual in the broadcast system. If sown broadcast, the seeds are covered by a thorough harrowing, and if in drills, rather lightly. The drills should be no more than 4 to 5 inches apart. [p. 1005]

Cultivation is almost unnecessary, beyond a little weeding when the plants are small, since hemp grows very rapidly and soon gets ahead of weeds. Where both seeds and fibers are desired, the male plants are pulled soon after they shed their pollen, and the females as soon as the seed is fully formed. Here in Hokkaido, hemp ripens very little. The time for harvesting the fibers varies somewhat, according to the use to which the fibers are to be put, but for most purposes it is at about the time of blossoming. At that stage of growth, the plants are usually pulled and made into small bundles, which are allowed to dry for a few days. Then they are usually immediately retted, [p. 1006] though they are sometimes kept through the winter and retted the following spring.

(3) Retting: One of the earliest methods of retting hemp, and one which is still commonly used, is putting it into water, where it is left until the fibers readily separate from the inner woody portion of the stem. The length of time necessary varies with the temperature, but is often about 10 days. When you think that the plants have been in the water nearly long enough, you should frequently examine them and see if the fibers separate readily. As soon as this is the case, the hemp is taken from the water and allowed to stand in piles for about one day to drain. It is then taken to a grassy area where the grass is no more than 3 to 4 inches high. There it is spread and carefully turned from day to day until it is thoroughly bleached [p. 1007] and dry, when it may be stored in stacks, sheds, or barns and the fibers separated whenever convenient. The length of time occupied in bleaching varies much with the weather, but it is usually about 10 days. [p. 1008]

7-32. *Flax (Linum usitatissimum)*

Flax belongs to the small family of Linaceae. It is an annual, with simple leaves which are usually sessile and entire. The flowers open for only one day, and in the sunshine are regular and symmetrical. The cultivated species of flax is valued for its fibers, from which fine cloth is made; it was introduced into the United States from the Old World. Ireland and Flanders produce the best linen.

Flax thrives best upon a rather light but rich loam. It may be raised with fair success upon a great variety of soils, but all heavy soils or those containing much clay are ill suited. The climatic conditions are the same as for hemp; hence, Hokkaido is favorable for its cultivation. [p. 1009]

The fertilizers are the same as for hemp. The soil having been brought into a state of fine tilth, the seeds are planted rather late in the season—here, in late May or early June. Broadcast planting is preferred. About 3 bushels of seeds is the usual quantity per acre, because the plants must be thick, otherwise the fibers are deficient in both quantity and quality. The seeds must be covered very lightly; a rake or brush is used.

Harvesting usually takes place when the seeds are fully developed, at which time the lower leaves and the lower part of the stems will have turned yellow. [p. 1010] The flax should then be thoroughly dried, after which (at a convenient season) it is subjected to a process known as "rippling", in which the seeds are separated out. The seeds of flax, aside from their value for planting, are very important in the production of oil, and the refuse for animal feed. After the seeds are separated, the flax is retted and bleached in much the same way that hemp is treated. The greatest care must be taken that the stems do not come in contact with the bottoms or sides of bricks laid in cement. The water used should be clean, pure, and free from minerals in solution because the presence of mineral impurities often causes staining of the fibers. The water in which flax has been retted is of much value as a fertilizer. [p. 1011]

7-33. *Hops (Humulus lupulus)*

The Hop is a plant having perennial roots and annual stems. The stems are almost prickly near the bottom; the leaves are heart-shaped and conspicuously 3-7 lobed; the plant is dioecious. Its aromatic bitterness resides in the yellow resinous grains which appear upon the drooping calyx, akenes, *etc*. In the wild state, hops are extensively distributed, being indigenous to America, Europe, and Asia. The plant has long been cultivated in Europe and North America, being extensively used in the brewing of malt liquor, beer, porter, *etc*. Hops are also used to a slight extent in medicine, and to make yeast, and sometimes paper. The varieties are quite numerous: White Vines, Godlings, Grapes of several sorts, and Jones are all good. [p. 1012]

1) Cultivation

Hops require a deep, very rich, and well-drained soil, though they will flourish fairly well upon almost any land which will produce corn, provided the soil is properly prepared and enriched. The soil should be ploughed and sub-soiled in autumn. Apply a liberal dressing of fertilizer, and mix it with the soil (unless it is already rich). Then, early in spring, plough again to pulverize the soil, and apply a dressing of fertilizer again if it is still too poor. In preparing the soil, well-rotted composts of animal manures are best; those should be mixed with [p. 1013] the soil to a considerable depth. Subsequently, it will be necessary to make yearly applications of fertilizer. Animal manures are good, but if their supply is insufficient, herring guano, ashes, rapeseed oil cake, *etc*. may be used. If a deficiency of lime in the soil is suspected, it is better to apply some, at the rate of 200 bushels per acre. Cultivation of hops must be of the most thorough kind. No weeds should be allowed to grow, and the soil must be kept loose by often repeated cultivation. Very early in spring each year, the earth should be pulled away from the plants so that the crowns and upper roots are exposed. All of the suckers should then be cut off clean, along with the last year's vines and 2 to 3 inches of the roots. The hole should be left open for 2 or 3 days until the wounded surfaces have dried, after which the earth is put back.

2) Propagation of Hops

Hops are propagated by separating the roots—that is, by taking the natural underground layers. The layers are separated from the parent plant in spring, and the preferable method is to plant them temporarily in well-prepared soil; they are then transplanted to a field the following autumn, or early the next spring. Very often, the slips are planted at once in the field, but it is uncertain that all would start; hence, the merit of the first method. [p. 1014] Three to 5 of the slips usually form a single hill (3 are sufficient if all are strong and vigorous). The hills are usually 7 feet apart each way. A deep hole is dug, and the young plants are set so that the crowns are about 6 inches below the surface. [p. 1015]

3) Training Hops

There are 2 systems of training hops, on poles, and on poles and strings; the latter is usually cheaper, since fewer and much shorter poles will suffice. If poles alone are used, they must be 14 to 20 feet long; about 3 are used in each hill. If poles and strings are used, only 1 pole on each hill, and that only 10 to 12 feet long, is sufficient. As the hops' growth progresses, the vines must from time to time be tied to the stakes; if they are much inclined to branch, they should be cut off. Since hops are dioecious, proper proportions of male and female plants must be sown. Usually, one male plant per 200 females (or 6 males per acre) should be evenly distributed. The male plants should, in any case, be trained upon tall poles because [p. 1016] the distribution of pollen under such circumstances is more perfect.

4) Harvesting Hops

The best time to harvest hops is when they become yellow, but it is difficult, where large quantities are raised, to harvest them all at just that time. It is customary to begin a little earlier, and the picking is not usually completed before some have turned brown. Unless heavy rains are frequent, the quantity of lupuline will not be seriously diminished, even though the hops become brown. The hops are rapidly picked from the vines, and should be at once quickly dried by artificial heat. The essentials for successful hop-drying are: that a constant current of hot air be made to pass through them; that there be abundant provision for the escape of moisture. [p. 1018] As soon as the hops are picked, the vines should be cut down to the ground; they are then utilized for bedding livestock, or for manure, being very rich in plant food. The poles should then be taken up, and either piled compactly or set on end (bottom upward) in large bunches, there to remain through the winter. [p. 1018]

7-34. Rape or Colza (Brassica napus, B. campestris, B. oleracea)
("Abura-na" = *Brassica scinensis*)

Rape belongs to the Crucifereae, and is much like the turnip and cabbage in its botanical characteristics. It is extensively cultivated in many parts of the world, mostly for its seeds, which yield a large quantity of valuable oil, the refuse from the processing of which is an excellent fertilizer. Rape is also sometimes raised as a fodder crop, being pastured on by sheep. The practice of pasturing sheep by confining them within a definite area is most common in England, where that is found profitable.

Rape thrives well upon that class of soils suited to turnips, though it is not absolutely necessary [p. 1019] that the soil be so rich as to be suited for turnips. The preceding crop should be one that may be harvested early. Early potatoes or peas would answer well. Rape may also follow a grain crop, such as wheat or rye, but those do not leave the land in as good a condition as the crops first mentioned. Any ordinary farmyard manures may be expected to give good results. Manures should be applied broadcast in moderate quantities, then mixed with the surface soil. [p. 1020]

The soil having been prepared as for turnips, the seeds are planted from the last of July to about the first of September, in drills 14 inches apart. If rapeseeds are desired, the plants should stand rather thinly in drills, say one about every 2 inches. If intended for pasturage, broadcast sowing and a more liberal allowance of seeds are recommended. If sown in drills, about 2 quarts of seeds per acre is sufficient. If the soil is in good condition, this crop needs little cultivation. Some weeding, and stirring the soil between the drills once or twice may be useful in the autumn. In spring, the crop shoots up with such rapidity that it gets ahead of all weeds, so cultivation is unnecessary. [p. 1021]

Harvesting takes place as soon as the majority of pods are fully developed and commencing to turn yellow. The seeds shell very easily; hence, with late harvesting a great loss is incurred. The plants should be cut close to the ground, carefully handled until dry, and immediately threshed usually in the field, or—if inconvenient to do so there—stored in stacks or in barns until a convenient time. The straw should be utilized for bedding, or in the yards, for cattle or swine. [p. 1022]

解　題

1. 食用作物関係の講義とその後の継承発展

1) 食用作物分野に関する講義の特徴

　第7章のCrop Cultivationは作物各論で，食用作物ばかりでなく野菜(17作物)のほか，テンサイなどの工芸作物(7作物)も含まれている。食用作物はコムギ，ライムギ，エンバク，オオムギ，トウモロコシ，ソバ，ミレット(アワ，キビ)，ハンガリアングラス(ごく簡単な記述のみ)，バレイショなど9作物で，その記述は禾穀類が中心となっている。マメ類はアズキ，インゲン，ササゲなどが'Beans'として総括して述べられており，記載順序からも野菜に含められている。食用作物の講義は，当時のアメリカの主要作物であるコムギ，トウモロコシ，バレイショについて多くの頁が割かれており，他の作物については比較的簡単な記述に止めている。記述の内容は，主要作物については，起源や分類(学名や分類は現在と異なるものもある)，欧米での利用状況，形態，生育特性，品種，土壌および気候適応性，栽培法，病虫害などの順に述べられており，全体的に栽培法に関する記述が多いほかは，現在の教科書(食用作物)の原型をうかがわせる。しかし，日本の主要作物であるイネやダイズについての記載がなく，作物ごとに取り上げられている品種もすべて欧米のもので，講義の内容は欧米の農業を対象にして述べられている感はまぬがれない。

　特徴的なのは，北海道への提言が含まれていることであろう。例えば，エンバクは，「馬の餌として最良で，北海道に住んだ経験から，北海道では飼料作物として非常に価値があると確信した」と述べている。事実，エンバクは，わが国では第2次大戦前は軍馬の飼料として軍事上の重要作物となり，戦後，栽培は減少したものの1940-44(昭和15-19)年では全国で約13万haが栽培され，そのうち90％近くが北海道で栽培されている。また，現在，北

海道の主要畑作物となっているテンサイについても「札幌やその近辺の土壌は生産に非常に適している」と指摘しているほか，トウモロコシについても「気候は必ずしも適しているとは言えないが，極早生品種を用いれば成功しよう」と指摘し，今振り返ればきわめて的確な提言を行なっている．

2) イネに関する講義と開拓初期の稲作

前述したように，講義ノートにはわが国の主要作物であるイネについてまったく触れられていないが，当時の学生はどのように感じたのであろうか．ブルックスの最初の学生で，その後ブルックスの後を継いだ南鷹次郎(札幌農学校2期生，1877(明治10)年9月入学，第2代総長)の伝記(南鷹次郎先生伝記編纂委員会発行)に，このことについて面白いエピソードが語られているので，そのいくつかを抜粋して当時の事情を紹介しよう．

まず，南自身が「教師は，漢学，地質学の外は総て外国人であったから，農学というても自然妙な教育をうけたものだ．例えば，作物でも日本には何等の関係のない外国のものなどは教わっても，日本に最も必要な米作などのことは少しも聞かされたことがなかった」(今昔雑話，『文武会報』第55号)と述べており，当時の学生の気持がうかがえる．また，1889(明治22)年卒業の藤田経信の回憶によれば「当時米を北海道に作るのは天理に背くと考えられていた時代で，ブルックス教師も米作は北海道に適せずと断じて居った．或る討論会が開かれ，自分は不幸にして米作が北海道に適するという側に廻された．その頃，師範学校に稲の試作地があったが，その成績〔当時の単収は，現在の1/3以下の150 kg/10 a程度であったと思われる〕を見て居ったので討論には全く自信がなく，「稲藁ができる」とか「東北地方では出来る」など，奇怪なことを論じて演壇を降りた」とあり，当時の北海道における稲作事情がうかがえる．さらに，「卒業生が内地に就職してから，よく言うて来ることは，母校は稲を作ることを教えないので駒場出〔東大の前身〕の者に圧倒される．米国式だけでは困るから，札幌農学校でも稲を作ることだけは教えてくれ」というのであった．そこで自分達も随分とやかましく学校に要望したので，ついに日本農学という講座ができた．この講座を担当したのが南先生であった」とあり，稲に関する講義が始まった経緯が述べられている．ちなみに，アメリカは現在ダイズの世界主要産出国であるが，最初の栽培の記録は1804年で，1896(明治29年)に初めて農務省が試験に着手し，実際に普及したのは1920(大正9)年以降である．ブルックス自身も日本の品種をアメリカに導入している．

3) ブルックスと南鷹次郎

南は1881(明治14)年7月に卒業し，2年間東京で獣医学，農学を研究した後，1883(明治16)年に母校の助教となり，当時の農園(農場)長ブルックス(第2代，初代はクラーク)の下で農園副監督として農場実習の助手を務めた．実践の場として重要視されていた農場における実習は，当時は週2回行なわれており，南鷹次郎伝に同期生の言として「農場実習の担任者は農学教師ブルックス氏なりしかば，氏は米国風の思想をいだき，労働は相当の報酬を与うべしと主張し，学生毎週両度づつ手業の課〔農場実習〕を受くれば，農園のために幾分かの労働をなすこと故，1時間5銭ずつの割合を以て，月末の勘定をすることを定めり」とあり，学生はお金をもらうことに抵抗があったようであるが，「然るに1度壱円以上の金を掌に握

り,「これぞ吾が額にかきたる汗の貸金なり」,「これだけあれば5度や10度蛇足〔当時南1条にあった菓子屋〕に通うも苦しからじ」と心に悟開きたる後は,誰一人として手業料の貴賤を論ずる者なく,数月ならざるに課業外にも手業を願い出る輩さえ顕れたり」とあり,手業料は「一大富源」となり,官費学生であった当時の学生の姿が垣間見える。

ブルックスは,1888(明治21)年10月に帰国し(在職12年),その後をブリガムが招請されて継いだが,南も翌年に教授に昇進した。次いで南は,ブリガムの帰国(1893(明治26)年)後,作物学,園芸学,土壌学,肥料学,農具学,土地改良学,畜産学などの講義と農場責任者(当時は農事部長)の職をすべて引き継いだ。南は,卒業当初は獣医学,畜産学を目指したようであるが,農学全般の講義を担当するについては,「ブルックス先生は学校の教授以外に北海道の農業経営の研究などを熱心になされ,その結果は北海道の農業開発には大いなる貢献をしておられる。とにかく,自分などが農学全般にわたってこれ程までに興味を持つように教育せられたのは,全く同先生のお陰であると思っている。その御恩は深く感謝して忘れることができない」(前掲今昔雑話)と語っている。

4) その後の展開——講座の成立

実質的にブルックスの後を引き継いだ南は,長年にわたって農学全般の講義を担当したが,その後,卒業生のうちから優秀な新進の学徒が輩出し,いずれも海外留学を終えた橋本左五郎が畜産学を,時任一彦が農具,土壌,土地改良学などを,星野勇三が園芸学を担当するようになった。さらに後には種子学,育種学を明峰正夫が,工芸作物学を東海林力蔵が担当するようになった。

札幌農学校は,1907(明治40)年に東北帝国大学農科大学に昇格し,大学官制により講座制が敷かれ,農学,畜産学,農芸化学,林学の4学科12講座体制となった。農学科は1913(大正2)年に生産分野,農業経済分野,農業生物分野の3分野に分けられ,選択科目を設けて教育の専門化が図られたが,1918(大正7)年に北海道帝国大学が発足すると,生産分野だけを農学科と呼ぶことになり,農学科は,農学第一講座(担任:南鷹次郎),農芸物理学講座(時任一彦),動物学昆虫学養蚕学第三講座(須田金之助),園芸学講座(星野勇三),農学第三講座(明峰正夫),農学第四講座(東海林力蔵)の6講座体制となった。この体制はしばらく続くが,1936(昭和11)年に園芸第二講座(花卉園芸学)が増設され,翌年には動物学,昆虫学,養蚕学の第三講座が農業生物学科へ所属換えとなり,さらに1949(昭和24)年に農業物理学科(後の農業工学科)が新設されると農業物理学講座は同学科へ所属換えとなった。この結果,農学科は,農作物を対象とした食用作物学(農学第一講座),工芸作物学(農学第四講座),作物育種学(農学第三講座),果樹・蔬菜園芸学(園芸学講座),花卉・造園学(園芸第二講座)の5講座構成となり,この体制は1992(平成4)年に学部改革が行なわれるまで継続した。

5) 開拓当初の作物栽培状況と北海道農業の発展

(1) 稲 作:北海道の稲作は,1685(貞享2)年に吉田吉右衛門が文月村(現在の大野町)に試作したのが始まりとされるが,気象的悪条件を克服するため,府県の在来種のなかから早熟性と寒さに強い耐冷性品種を農民みずからが探し出すことから始められている。イネの試作研究は1876(明治9)年に官園で開始されているが,開拓使に招請されたケプロンらの外国人

顧問団は「稲作は，技術的，経済的に北海道には適せず」とし，輪作，有畜，機械化という輪作式畑作農法を主張したため，開拓使は稲作にきわめて消極的であった。しかし，道央での稲作導入試作に成功したことから，1893(明治26)年に上白石村，真駒内(ともに現在の札幌市内)，亀田村(現在の函館市)に水田試作農場が設けられ，積極的な稲作試験研究が開始された。北海道の稲作は基本的には，早生，耐冷性品種の育成によって拡大している。1873(明治6年)，中山久蔵が島松村(現在の北広島市島松，記念碑や住宅と復元田が公開されている)で品種「赤毛」の試作に成功し，1886(明治19)年には稲作は石狩平野まで広がり，作付面積は約1800 ha であった。1894(明治27)年になると，「赤毛」のなかから出穂期が1週間ほど早い変異株が発見され，芒(のぎ)がなかったことから「坊主」と名づけられた。この品種の出現により稲作は上川盆地まで拡大し，稲作面積は1902(明治35)年には1.6万 ha に達した。1923(大正12)年には初めての交配品種「走坊主」が育成された。この品種の出穂期は，「坊主」よりさらに1週間早かったため，十勝や上川地方でも栽培が可能となり，稲作面積は1937(昭和12)年には18万 ha に増加した。さらに，同年には当時としては世界で最も出穂期の早い極早生品種「農林11号」が育成され，稲作はほぼ北海道全域で可能となった。昭和40年代に入って，北海道は全国一，二を争う米の生産地にまで発展し，1969(昭和44)年には作付面積が26.6万 ha にも達したが，食生活の欧米化に伴う消費の減少により畑作物への転換政策がとられ，近年の稲作面積は約14万 ha である。

(2) 畑　作：北海道は，近年では国内総生産量のうち，コムギで約50％，バレイショで73％，ダイズで21％，アズキで85％，インゲンで92％を生産しているほか，砂糖の原料となるテンサイはすべて北海道で栽培されており，北海道は畑作王国と言われている。

　ムギ類は，元来，北海道のような冷涼な気候を好み，コムギやオオムギは安政年間に根室国野付郡で栽培された記録があるが，広く栽培されるようになったのは開拓使によって七重，札幌，根室に官園が設けられ，欧米の種苗，機械技術を導入し，その普及に努力したことによる。開拓当初はその適応性から農家の自給作物，また輪作の基幹作物としての地位を占めてきたが，コムギの在来種は稈が短く，収量性や品質も劣っていたことから，明治時代はもっぱら導入品種のなかから選ばれた品種が栽培されていた。本格的な交雑育種による品種改良事業が開始されたのは1919(大正8)年で，収量性，病害抵抗性，耐倒伏性(短・強稈)，品質などの改良が進められ，多くの品種が育成されたほか，近年ではグレンドリルやコンバインなどによる大規模，省力化栽培が進められ，コムギの作付面積は次第に増加した。

　ダイズは1857(安政4)年に日高で，1871(明治4)年に十勝で試作された記録がある。開拓当初は主に道南で栽培され，明治中期までは七重官園と札幌農学校が試作の中心であった。北海道の気候風土に適し，所要労力が少なく大面積，粗放栽培に耐えるほか，土壌を選ばず少肥で済むことから北海道が主産地となった。アズキは，北海道に入植した人たちが故郷から持参した種子を植えつけ，そのうち北海道に適応したものから栽培が始まったと考えられている。作付面積は，1886(明治19)年では1700 ha にすぎなかったが，1900(明治33)年には空知，上川を中心に約3万 ha が栽培されている。インゲンが栽培され始めたのは明治初期で，明治末から大正初期にかけて欧米に輸出されるようになると急増して，1915(大正4)

年には道央，道南を中心に3.5万haが栽培された。

　開拓使は，バレイショを適作物として奨励したばかりでなく，開拓当初から海外より多数の品種を導入し，七重官園や札幌官園で試作を行ない，種いもの配布を始めている。1880(明治13)年には札幌農学校でもイギリスやフランスから品種を導入，試作しており，1901(明治34)年に北海道農事試験場が設立されるまでは，開拓使，北海道庁，札幌農学校が中心となって品種の導入や普及指導にあたった。導入品種のうち，「男爵薯」や「メークイン」は今日でも栽培されているが，「男爵薯」は函館の川田男爵がイギリスより多数の品種を導入し，七重の農場で試作した結果，選ばれたものである。1916(大正5)年に，農事試験場において系統分離，純系淘汰により優秀な個体や系統の選抜が開始され，1918(大正7)年にはわが国で初めて人工交配が行なわれたが，最近の育成品種には遺伝子源として近縁野生種が積極的に利用されている。

　テンサイは開拓当初から官園で試作が開始され，1880(明治13)年には伊達に，1890(明治23)年には札幌に製糖工場が設立されたが，原料確保が困難で明治30年前後には操業を中止している。しかし，北海道庁はテンサイを重視し，1920(大正9)年に増産施策を展開し，1935(昭和10)年を中心にその栽培が普及した。現在ではテンサイはすべて北海道で栽培され，処理能力の大きい製糖工場で砂糖が作られ，国内生産量のうち約75％はテンサイから生産されている。

　このように，札幌農学校は，開拓初期は単に教育機関としての役割ばかりでなく，北海道農業の発展に大きな役割を果たしている。特に，農園(農場)では，各種作物・品種の導入試験や栽培適応試験が行なわれ，その普及に大きく貢献した。事実，南が農場で行なった試験研究の数は144件にも及び，その対象となった農作物は，食用作物はむろんのこと，果樹，野菜，工芸作物，飼料作物の多岐にわたり，南は北海道農会長を務め，その普及にも努力している。

[中世古公男]

2. マメ類の学名について

　7-10, 11はマメ類について述べている。世界的に見れば，食用(野菜)のマメ類の代表がbeans(インゲンマメ)とpeas(エンドウ)であることは講義当時も現在も変わらない。ただし，beansは*Phaseolus*(インゲンマメ属)だけを指すのではなく，*Faba*(ソラマメ属，現在は*Vicia*が主流)や*Dolichos*(フジマメ属，現在は*Lablab*)などを含めて使われていた。しかし，種を特定するためにつけられる学名(属名＋種小名)も120年の経過とともにかなり変化している。かつて*Dolichos*には，フジマメ，アズキ，クズ，クズイモ，ササゲ，ダイズなど多数の種が含まれていたが，これらは現在それぞれ別属として扱われ，*Dolichos*はコウシュンフジマメ属に限定して使われることが多い。

　*Phaseolus*は「小舟」の意味で，種子を含む莢(さや)の外観を表現した属名と思われる。その「曖昧さ(莢はマメ科植物の一般的特徴)ゆえに」というわけではないだろうが，現在では*Vigna*(ササゲ属)に属する作物でも，最初はほとんどが*Phaseolus*に入れられていた。この

農学講義録にも，冒頭に Beans の仲間として追記したと思われる作物名と学名がいくつか見られる(括弧内)。細かなことをあら探しするようだが，そこには綴りの誤りを含め，当時の分類上の混乱によるもの，ブルックスによる引用違い，新渡戸の転記違い，と思われるものがある。以下に気づいたことを列記してみよう。ただし，[ ] は現在の学名を指す。

(1) Adzuki＝*Phaseolus radiatus*：アズキ，小豆。*radiatus* は本来リョクトウ(緑豆)につけられた種小名であるが，別名の *Ph. mungo*(ケツルアズキ)と同様に，見たこともないヨーロッパの学者が外国の文献を頼りに誤って分類したものと見られる。[*Vigna angularis*]

(2) Tsura-adzuki＝*Ph. radiatus* v. *pendula*：ツルアズキ，蔓小豆。Tsura- ではなくて Tsuru- の間違いだろう。アズキの1変種になっているが，形態的にはかなり異なり，同じく分類上のミスと考えられる。[*Vigna umbellata*]

(3) Bundo＝*Ph. radiatus subtriloba*：ブンドウ，文豆(緑豆の別名)。*subtriloba* は変種名(v. が抜けた)。異名として通じるが，本来 *radiatus* がリョクトウで，当時は上述のようにアズキやツルアズキがその変種を指していたはず。[*Vigna radiata*]

(4) Sasagi＝*Dolichos uniflorus*：ササゲ，豇豆。地方によってはササギと呼ぶ。この学名はインドに多いホースグラム(horsegram, *Macrotyloma uniflorum*)の異名である。ササゲはアフリカ(エチオピア)原産とされており，これも分類上の混乱と見られる。[*Vigna unguiculata*]

(5) Sengoku-mame＝*Dolichos cultratus*：センゴクマメ，千石豆(藤豆の別名)。フジマメの異名として確かに存在するが，*Dolichos* を使う時は，*Dolichos lablab* を用いる方が多い。[*Lablab purupreus*]

(6) Hata-sasagi＝*Dolichos umbellatus*：ハタササゲ，畑豇豆。この学名が何の異名か残念ながら分からない。*Vigna umbellata*(ツルアズキ)と混同されたのかもしれないが，形態はかなり異なる。現在はササゲの亜種とされている。[*Vigna unguiculata* ssp. *cylindrica*]

同様に，Peas(エンドウ)の項の冒頭にも追記があり，Nata-mame(Overlook pea)＝*Canavalia incurva* と Rakkasho＝*Arachis hypogaea* と読める。後者はラッカセイ(落花生)である。*Canavalia* はナタマメ属であるが，*incurva*(内曲した，の意)のついた種を筆者は知らない。ナタマメ(鉈豆)は普通 *C. gladiata* であり，overlook pea (bean)と称する作物は *C. ensiformis*(タチナタマメ，立鉈豆)に相当する。当時は実用上両者を区別していなかったのかもしれない。ラッカセイ(peanut)とともに pea がつくのでエンドウの項に入れたのであろうか。bean も pea も総称的に使われることが多い。

ところで，これらの和名(ローマ字)＝学名が，新渡戸が自ら調べて追記したものか，あるいはブルックスが持参した書物を参考に補講した内容なのか，どちらであろう。和書にしろ洋書にしろ，当時，名前を対比させた専門書があったということか。そんなことを想像すると，混乱した内容もさほど気にならなくなり，新渡戸が新知識を得ようと懸命に記録したであろうこの講義録に，あらためて親近と敬意を覚える。

「農学講義」作物各論でブルックスは約30の作物を紹介し，品種，土壌，前作物，栽培法(施肥，播種，中耕，収穫など)あるいは病虫害について述べている。マメ類の項においても，

品種や肥料の内容は別として，播種法や収穫・調整法等は基本的に現在と変わらないことに，最初は少しショックを受けた。しかし，考えてみればたかだか120年，作物栽培法が大きく変わるはずがないと言えるかもしれない。

［由田宏一］

3. 果樹および野菜について

　この講義のなかで取り上げている果樹および野菜などの園芸作物についてみると，第6章にある果樹12種類と，7-12～27で述べる野菜25種類である。この講義は総合農学的に講述されており，果樹および野菜は，他の一般作物のグループに含めて講述されている。またイチゴについては，野菜のなかではなく，果樹のなかで述べるなど日本の場合と異なっている（アメリカでは，現在でも試験研究ならびに講義などでは果樹のなかで取り扱われている）。

　また取り上げている果樹・野菜の栽培，収穫物の調製，輸送，貯蔵などの技術についてみると，例えば，栽培過程における施肥については，時代的背景もあるが，堆厩肥，鶏糞，海草，魚かす，腐葉土，草木灰，石灰などのほか，人糞尿（下肥 night soil）などを用いると述べており，化学肥料については，わずかに，硝酸ナトリウム，硝酸カリウム，硫酸ナトリウム，硫酸カリウムについて触れているにすぎない。別の見方をすれば，このことは，現在，関心が高まっている有機栽培あるいは持続可能型農業に通ずるところがある。また，環境調節についてみると，温度の人為的調節については述べているが，水や光のそれについては触れておらず，現在とは大きく異なっていることが分かる。さらに収穫物の調製，貯蔵，加工，輸送などについても，わずかに言及しているにすぎないが，北海道が寒冷地であることを意識して，相対的に貯蔵方法に力点をおいた記述が見られるのは注目に値する。

　技術的な面についてみると，果樹の栽培において，矮性台木を用いて樹体を矮化させると，比較的若齢で結実し，病害虫防除にも効果があり，収穫作業も行ないやすいことが強調されているのは，現在，各種の果樹で矮化栽倍が進みつつあり，矮性台木の開発が行なわれていることと結びつけて考えると興味深いものがある。さらに寒冷地の果樹栽培では重要な意味を持つ凍害，霜害について述べており，南斜面では，昼中は直射日光が当たって樹体温度が上がり，夜間は北斜面と同様に冷却されるなど，昼夜の気温較差が大きく，樹体内組織の温度が激変し，組織・細胞が凍結・融解を繰り返すことになるため，凍害の発生が多いが，これに対し北斜面では温度変化が比較的少ないため，凍害が少ないことを指摘している。このことは，凍害発生機構についての重要な例証を栽培面から提起しているもので，現在の植物の凍害発生に関する理論的見解および実験結果とも一致しており，注目されるところである。

　一方，果樹，野菜の病害については，マサチューセッツ地方で認められるものについて触れており，北海道においても同じ病害が発生することを予想して講述されている部分があり，札幌農学校の「農黌園」（農場，当時の college farm の訳）における具体的な症例をあげるまでには，さらに期間を要したものと推測される。

　この講義は，わが国の当時の時代的背景の下に，また，アメリカと日本との条件の違いが存在するという状況下で行なわれたものであるが，北海道における果樹・野菜の本格的な栽

培技術の定着と食生活の向上につながることを意図して行なわれたものであることが強く印象づけられ，興味深いものがある。

　札幌農学校は，実学を重んじており，「農学」と「農学実習」を担当し，校園監督でもあったブルックスは，『札幌農黌年報』中の「農学科報文」にもあるように，学生とともに北海道内各地を訪れて，これにより，実態を調査するとともに，学生の見聞を広め意欲を向上させることに意を注いでいる。さらに，これにより，北海道の実情とマサチューセッツ州のそれとの違いを把握し，後者の技術を講義体系のなかに位置づけることに努力したものと思われる。講義録のなかにもたびたび挿入されているそれぞれの果樹・野菜については，マサチューセッツ州における栽培法などをもとに講述されている。一方では，可能な限り農黌園にもその作物を栽植し，札幌地方における栽培適性について比較を試み，これを講義のなかに取り入れたが，マサチューセッツと札幌とでは，同一の果樹または野菜の生育が異なっている部分がかなりあったことが，講義録の記述からうかがい知ることができる。特に果樹では，結実年齢に達するのに長い年限を必要とし，両地の生育を比較検討するのに苦労のあったことが述べられている。

　札幌農学校では当時は全寮制となっており，寄宿舎に隣接して「復習講堂」があったので，受講後のノートをここで整理したものと思われる。このノート中の記載によれば，onion (*Allium cepa*)を"negi"とし，leek(*Allium porrum*)を"nira"としているが，これらの食い違いは，前述の一連の状況の下で生じたものと思われる。一方，メロン(*Cucumis melo*)のなかに，スイカ(*Citrullus lanatus*)を含めているなど，やや奇異に思われる部分もあり，現在のアメリカおよび日本の体系のいずれとも異なっているところも見られるが，着任早々における講義の緊急性を知ることができる。このように講義当初には混乱もあったが，ブルックスは，講義とは別に地元農家を集めた技術指導にも精力を注ぎ，札幌村丘珠(現在の札幌市東区)にタマネギ栽培を広め，国産タマネギの過半数を生産する北海道タマネギの発祥地を育成したし，ブルックスの馬丁を務めた吉田某を同村に入植させて帰国したなど，今も同地区の農家に語り継がれていることを特記せねばならない。

　次に，当時行なわれた果樹や野菜についての講義に関連する諸事情を推察すると，次のようなことがあげられる。

(1)　わが国で栽培されている果樹や野菜の起源はほとんど日本以外の世界各地にあり，原生地から，それぞれの伝播経路を経て導入されたものである。当時の実際の栽培という側面から見た場合，日常的に栽培され，すでによく知られていた作物と新たに導入された作物，または日本国内では栽培が一般化していないが北海道開拓使の試験場(官園)で導入し試作されていたものが多数あった。わが国においては，1868年以前(江戸時代)と，それ以後の明治時代，大正時代，昭和20年(第2次大戦終結)以前または以後など，時代によって食されてきた果樹および野菜の種類は異なっているが，講義のなかで述べられている果樹・野菜のうちで，当時すでに北海道内で栽培され，食生活に取り入れられていたものとしては，果樹ではリンゴ，モモ，オウトウ，スモモ，ブドウなどがあり，野菜ではカブ，ニンジン，スイカ，キュウリ，カボチャ，ナス，エンドウなどがある。一方，当時，日本国内で栽培され食され

ていたものとしては，果樹ではカンキツ類，ニホンナシ，ウメ，アンズ，クリ，クルミ，カキなどがあり，野菜ではハクサイ，ダイコン，ネギ，トウガラシ，ゴボウ，シュンギクなどがあった。また，フキ，ウドのような山菜類もあったが，これらについて講義のなかで取り上げられていないのは，当時の状況から推測して止むを得なかったことと思われる。

(2) 当時の受講生の出身地や生活体験が異なっており，果樹・野菜の栽培および利用に関する実際の見聞または食体験の違いから，受け止め方や実感に差異があったものと想像される。例えば，それ以前に導入され，通常の食生活のなかにすでに取り入れられており，見聞経験・食体験のあったものよりも，講義中あるいは校園で初めて見聞するものの方がはるかに多かったと推察される。

(3) 一方，ブルックスは，野菜の種子をいつまでもアメリカから取り寄せるのではなく，北海道において早期に種子生産・採種を行なうことの必要性を痛感し，この講義のなかでは相対的にかなり多くの講述を行なっている。おそらく，自ら種子を生産する態勢を整え，その種子を用いることが，北海道における一般栽培の普及ならびに試験・研究の推進にとって最も重要であると，強く感じていたと推察される。

(4) 本講義では，野菜育苗時における環境保護（温度・光などの人為的調節）および本圃での栽培におけるマルチング（藁などで地表面を被覆すること）が重要であることを述べているが，これらについても，その後現在に至るまで，資材の進歩によって大きく変化し，現在では施設園芸(protected horticulture——この講義のなかではまだ用いられていない)が発展している。

(5) 野菜栽培における栄養供給に関連し，その後新たに開発され発展した技術は，水耕栽培，礫耕栽培などをはじめとする養液栽培(nutriculture, solution culture, soilless culture, hydroponics などの語が用いられる)の発達がある。これは 1930 年代にカリフォルニアにおいて始まっているが，わが国では，第 2 次大戦後の 1947(昭和 22)年にアメリカ駐留軍の提唱によって始められており，したがって講義のなかでは述べられていない。

(6) 札幌農学校では当時，しばらくの間は洋食が奨励されており，これは果樹や野菜の利用，特に当時としては珍しい「西洋野菜」と言われるものの利用につながることであるが，講義のなかで，Ripe fruit is one of the most healthy articles of food, and it is much cheaper as well as more agreeable for the farmer to raise fruit than to pay doctor's bills と語っているのは，現代生活において重要視されている果樹・野菜の利用と健康との関連と併せて考えると興味深いものがある。

(7) 本講義中では寒冷地帯で栽培される果樹・野菜について述べており，暖地で栽培されるものについては言及していないが，このことはマサチューセッツ州と北海道との気候の類似性に基づき，北海道における速やかな知識の普及と技術の定着を重要視していたこととも関連があると推察される。

いずれにしても，果樹・野菜に関するこの講義は，欧米では新たな多種類の果樹や野菜が栽培され食用に供されており，それらに関する技術が進んでいることを受講生に鮮明に印象づけ，当時の学生たちの目を外に向けさせ，意識を高め，知識を広めることに大いに寄与し，

きわめて意義深いものであったと思われる。

[原田　隆]

4. 工芸作物について

　7-28以降に述べる工芸作物の多くは，食用作物とは異なって商品作物あるいは貿易作物として，主に大航海時代以降，歴史の舞台の正面で大きな役割を演じてきた。すなわち，綿，ゴム，香料，茶，ケシ，タバコ，キニーネなどがある。

　ブルックスの講義で取り上げているのは，テンサイ，モロコシ，ホウキモロコシ，タイマ，アマ，ホップ，セイヨウアブラナである。

　砂糖原料のテンサイは，ナポレオンのヨーロッパ大陸封鎖(1806年)により，甘蔗糖の輸入が止まって砂糖価格が暴騰したため，ヨーロッパ諸国で国家保護策によって栽培が始まり，急速に広がっている。わが国では1870(明治3)年東京開墾局で試作するが，うまく栽培できなかった。そこで北海道開拓使が取り組み，アマやビール用2条オオムギとともに試験栽培を始めた。1877(明治10)年春にクラークは，札幌郊外島松で 'Boys, Be Ambitious !' と叫んで学生たちと別れた後，開拓に実績をあげていた伊達に立ち寄り，北海道の畑作における輪作の重要性を説いてテンサイ栽培を進言した。それが1880(明治13)年の官営伊達製糖所(後に民間に払い下げられて紋鼈製糖㈱)の開設につながり，テンサイ糖の生産を始めたが，原料不足や製糖技術の未熟さなどのため，1896(明治29)年に工場閉鎖して事業は失敗した。ブルックスの講義では，ダンとともに糖分分析を行なった結果を用い，北海道の気象や札幌近郊の土壌はテンサイの栽培に向いているものの，晩夏と初秋の多湿によって収量に有利だが，糖分の蓄積に不利であるとしており，事業の失敗を裏づけているようにも見える。しかし，第1次大戦によってヨーロッパの砂糖が暴騰するのを見たわが国は，甘味原料としての砂糖を自給することが不可欠と考え，1919(大正8)年にテンサイ糖の再興に着手させた。その後，いろいろの困難を経験するが，製糖と栽培技術の高度化とテンサイを寒冷地輪作作物に位置づけるなどによって今日に至った。日本のテンサイは，北大農学部OB等の活躍によって世界唯一のペーパーポット育苗移植栽培法を採用し，わが国の工芸作物でいちばん大きな栽培面積の約7万haで栽培されている。日本の砂糖は，沖縄の甘蔗糖と合わせて86万トン生産して自給率が約3割となるが，このうちテンサイ糖が道内8工場で精製して67万トンの砂糖を生産している。また，精製時の副産物21万トン余のビートパルプは，多汁質飼料の飼料ビートに代わって乳牛飼料に使われている。

　アマは，プロシャ人ガルトネルが函館七重で栽培していたのを開拓使顧問アンチセルが視察し，製麻事業化を提言したのが始まりとされ，その後の紆余曲折を経て，1887(明治20)年に北海道庁が官立の北海道製麻会社を創設して，北海道の重要な畑作物に位置づけた。特に第1次大戦時の軍需物資として重要性が認識され，札幌農学校南鷹次郎教授(主に工芸作物学講座の開祖となる東海林力蔵が担当)の活躍などによって適正品種を選定し，栽培法が確立されて本格化した。その後も栽培が奨励されて面積は年々増加し，製麻会社も複数となり，繊維用ばかりか種子から油も取った。第2次大戦後も機械化などの支援が続けられたが，

1965(昭和40)年頃から合成繊維が登場して市場が狭められ，ついに撤退するに至った。タイマ(アサ)もアマと同様であるが，雌雄異株と明記されている。現在問題となっている麻酔剤，幻覚剤としての記述はない。

　ソルガム(モロコシ)類は砂糖モロコシとホウキモロコシを取り上げている。

　ホップは，明治維新前の横浜駐在外国人向けビール用を無視すれば，1876(明治9)年に官営札幌麦酒製造所(後に民間に払い下げられて現在のサッポロビール㈱)の開設に合わせて北海道で試作されたのが始めである。そのため，ブルックスは栽培され始めたホップの指導に関与した可能性もあり，講義で雌雄異株と明記している。1915(大正4)年に長野で本格的に栽培が始まり，北海道にも栽培が広がってきた。今日ではビールの消費の増大によりホップの需要は増加しているが，輸入が多くなって国内の栽培は少ない。

　ナタネ(アブラナ)は主として油用で，飼料にもしていた。わが国には $B.\ campestris$ は9世紀頃に導入され，灯火用として利用され，石油の輸入まで続いた。$B.\ napus$ は明治の初めにヨーロッパから導入された。収量，含油率とも $B.\ napus$ が優れている。講義ではこの区別がはっきりとしない。

　このほか，いろいろな重要な工芸作物があるが，茶，タバコ，綿，ハッカ，除虫菊などは触れられていない。特にハッカは，当時，日本の主要な貿易品として話題になっていたが，本来，日本の栽培品であり，ブルックスもよく知らなかった可能性がある。

[中嶋　博]

5. 病虫害の講義およびその後の継承・発展
1) ブルックスの病虫害講義

　本章の作物各論において，各作物の最後に病虫害の項が付けられている。これが作物の病気(害)あるいは虫害が札幌農学校で講義された最初であると考えられる。

　コムギでは病害6種，虫害2種，ライムギでは病害1種，エンバク，オオムギでも，それぞれ病害1種，虫害1種があげられている。ライムギでは麦角病(ergot)があげられ，その人畜に対する毒性が説明されている。

　トウモロコシでは病害1種，虫害2種があげられている。ソバ，アワ，ハンガリアングラスでは病虫害の説明はなされていない。

　ジャガイモでは，最も恐ろしい病気として，疫病が病原菌 $Botrytis\ infestans$ の学名であげられている。この病原菌は現在では $Phytophthora\ infestans$ として知られている。この病気は，ジャガイモの茎葉ばかりでなく，塊茎をも腐敗させる病気で，1840年代にヨーロッパで大流行した。特にアイルランドでは惨状をもたらし，100万人が餓死し，150万人がアメリカなど海外に移民し，人口の3分の1を失う大飢饉となった。ブルックスの講義の時点(1870年代)では，まだ生々しい記憶であったと思われる。この病気の原因，防除対策に関する研究が植物病理学の確立に寄与したことで知られている。虫害は1種があげられている。

　テンサイでは病気の説明はない。虫害1種があげられている。

　カブの病気として，原因不明の Fingers and Toes があげられている。その説明を見ると，

今日では根こぶ病 club root と呼ばれている病害ではないかと考えられる。このことはその防除に石灰が有効であるという説明からもうかがわれる。虫害 2 種がある。キャベツの重要な病害として Fingers and Thumbs があげられているが，カブの Fingers and Toes と同じ病気と考えられる。病名もたぶん同じであったと考えられる。Toes あるいは Thumbs となっているのはブルックスあるいは学生の思い違いか，間違いではなかろうか。虫害では 2 種があげられている。

ニンジン，パースニップでは病気の説明がないが，これらは病害に侵されにくいものとされている。それぞれ虫害 1 種がある。タマネギ，飼料カボチャ，カボチャ，メロン，キュウリなどにも病気の説明はない。飼料カボチャで虫害 2 種がある。

インゲンマメでは病害 1 種，虫害 3 種があげられているが，エンドウでは虫害 1 種で，北海道には病気がないと述べている。セロリ，トマト，アスパラガス，レタス，ナス，チコリ，モロコシ，ホウキモロコシ，アサ，アマ，ホップ，セイヨウアブラナ等では病虫害の説明はない。

当時の知識からして当然ではあるが，病虫害の説明には精粗があり，また防除法まで述べているものもあれば，ないものもある。防除法としてあげられているものは，いわゆる耕種的防除法(適地適作，健全種子の使用など)が主であり，なかには薬剤の使用も述べられているものもある。大変興味を引くことは，コムギの害虫ユスリカに寄生する昆虫があり，これが害虫の生物的防除に役立つと述べていることである。このような寄生昆虫による虫害防除は後に大発展する。

この講義が行なわれた 1800 年代後半は，菌類が病気の原因になることが証明されて，植物病理学(plant pathology, phytopathology)が確立される時期である。当時の病気に関する知識は当然限られたものであったが，そのなかで病原菌，害虫の学名まであげて，学生に講義していることには感銘を受ける。ブルックスは新世界アメリカから来日した教師であり，地理的，気候的，歴史的に類似性を持つ新天地北海道にあこがれてきて，農学を学ぼうとする若い学徒の意気に感ずるところがあったのではないだろうか。この点でアメリカから多くの教師を招請した日本側の意図・判断は正しかったと思われる。

植物寄生病の病原として知られている細菌，ウイルス，ウイロイド，ファイトプラズマ，および線虫は，まだ知られておらず，ここに説明されているのは菌類による病気のみである。細菌による病気が報告されたのは，1884(明治 17)年が最初であるとされる。ウイルス病の研究が本格的になるのはもっと後であり，線虫病についても同様である。

講義で説明された病害虫は，ブルックスの母国アメリカで知られていたものが中心のようであり，北海道で実際に発生していたものは少ないと考えられる。北海道あるいは日本で発生する病害虫については後の研究に待たねばならない。

2) 北大における植物病理学の継承・発展

以下には植物病理学についてのみ，その後の北大における発展について述べる。ブルックスは，その後農学のほかに植物病理学の講義をも担当した。日本の植物学，植物病理学の開祖とされる宮部金吾は，1877(明治 10)年に札幌農学校の 2 期生として入学して，ブルック

スの農学は聴いているが，この植物病理学は聴いていないと思われる。

　ブルックスのアメリカ帰国後，1889(明治22)年ハーヴァード大学に留学していた宮部金吾が帰国して教授となり，植物学と植物病理学の講義を開始した。ここに札幌農学校ばかりでなく，日本の近代植物学と植物病理学が始まったと言える。このうち植物学の系譜は，工藤祐舜・舘脇操へ引き継がれ，北方植物の分類・分布・生態の研究が，附属植物園を基盤としてなされたが，この研究室はついに講座の形態をとることなく，応用植物学教室としていわゆる不完全講座のままで終わった。

　植物病理学についてみると，明治期は病原論の時代と言われ，植物の病気の原因が何であるかを知ろうとすることに情熱が燃やされた。北海道においては，開拓が進められ，北海道に適しているとされた作物が普及するにしたがって，それに寄生する病害が猖獗をきわめるようになった。北大における研究はこれらを対象にその原因究明と防除対策に向けられ，以後の実学の伝統を確立した。初期のアマ，テンサイ，ホップ，リンゴなどの病害の研究は，宮部を中心にその門下生によって精力的に行なわれた。宮部自身による植物病理学に関する著書はないが，その弟子，出田新の『日本植物病理学』の校閲を行なっている。宮部は，ブルックスが説明しているジャガイモの疫病について，日本での初発生がちょうど1900(明治33)年に北海道洞爺湖畔であったことを紹介し，海外の本病に関する研究を紹介している。この病気は1980年代に入ってから新しい系統の病原菌が蔓延し，世界的な大問題となっているが，日本も例外ではない。

　当時の北海道における植物病理学研究者の活躍の舞台は，札幌博物学会会報，北大農学部紀要，札幌農林学会会報などであり，1916(大正5)年に日本植物病理学会が設立され，翌年学会報が創刊されると次第にそれに移っていった。しかし，農学部紀要は依然として重要な研究発表の場を提供していた。北大では，その発展により研究室の充実も図られ，植物病理学関係では，植物学第一講座が宮部金吾担任，植物学第二講座が伊藤誠哉担任となった。これにより北大は，わが国の国立大学で唯一，植物病理学2講座を持つ大学となったのである。

　植物病理学の系譜は伊藤誠哉，栃内吉彦へ引き継がれた。大正期に入ると，病原菌の同定ばかりでなく，その生理学的研究が進められ，松本巍，逸見武雄，栃内がその先駆者となった。松本は後に，台北帝国大学(現台湾大学)教授となり，多くの世界的研究者を育てた。また，逸見は京都大学教授となり，京大の植物病理学研究室の基礎を築いた。ここではいちいちあげることはできないが，多くの卒業生が各大学・農学校の教授になり，日本の植物病理学，菌学の基礎を築いた。

　植物の病因となる主要な微生物は菌類である。菌類は生物五界説の一界を構成する大きな生物群であるが，それに関する知識が向上するに伴い，純菌学的な研究が，菌類の特定の科・属などを対象に行なわれるようになった。それらの研究の中心は伊藤であり，その集大成は『日本菌類誌』として出版され続けた。さらに，伊藤の逝去後は，その弟子大谷吉雄によって継承された。

　昭和に入り，イネのいもち病に関する研究が伊藤・栃内両教授を中心に展開された。いもち病は今日でもイネの最も重要な病気であり，適正な防除対策がとられているものの，最近

でも年々10万トンから60万トンの被害がある。この研究に参加した坂本正幸は，後に東北大学教授となり，引き続きいもち病の研究を進めた。北海道農業試験場とともに行なわれたいもち病の研究は，総合防除の試験研究としてあまねく知られているものであり，画期的な成果をあげた。栃内はいもち病の研究とともにアマの立枯病などいわゆる土壌病害の研究も行ない，これを継承した研究室は，宇井格生を中心に土壌病害の研究室として，内外に知られるようになった。栃内には，『植物病理学通論』があり，名著として知られている。

ウイルスが一部の病因となることが明らかになるや，いち早く福士貞吉はこれに取り組み，北大での植物ウイルスの研究が開始された。研究室では菌学とともにウイルスの研究が行なわれ，やがてウイルス中心にシフトしていった。福士には『植物バイラス』の著書があり，これはわが国ウイルス病に関する最初の著書となった。福士の後継者，村山大記は北海道主要作物のジャガイモのウイルス病を研究し，北海道農業に多大の貢献をした。

植物の病気には，その防除が非常に難しい難防除病害と称されるものがあり，その多くが土壌病害とウイルス病害である。北大の2研究室においては，今日もこの両者を対象として研究を行ない，農学と農業に貢献しようとしている。

明治初期に始まったわが国の植物病理学は，現在までに約8000種類の植物の病気を記載した。ブルックスのあげた作物についてみると，その病気の種類は，コムギ53，ライムギ17，エンバク24，オオムギ56，トウモロコシ36，ソバ15，アワ23，ジャガイモ45，テンサイ34，カブ24，キャベツ26，ニンジン24，タマネギ35，カボチャ24，メロン38，キュウリ39，インゲンマメ44，エンドウ30，セロリ16，トマト53，アスパラガス14，レタス23，ナス41，チコリ1，モロコシ22，アサ13，アマ18，ホップ11である。

[生越　明]

注
1) コムギ属(*Triticum*)には20以上の種が含まれるが，現在世界で最も広く栽培されているのは，6倍体の普通系コムギ(*T. aestivum*)である。普通系コムギは野生の一粒系コムギ(*T. boeoticum*)とクサビコムギ(*Aegilops speltoides* Tausch)が自然交雑し，倍数化によって二粒系コムギが生まれ，二粒系のエンマーコムギ(*T. dicoccum*)と畑雑草のタルホコムギ(*Aegllops Squarrosa* L.)が浸透交雑を繰り返し，その間に起こった倍数化によって成立したものと考えられている。
2) 現在では同一種と見なされ，このようには分類されない。
3) 催芽種子や幼植物体の時期に一定の期間低温にさらすと，冬コムギが春コムギに転化する。これを春化(バーナリゼーション)と言う。
4) わが国ではこのようには分類されない。
5) tillering：分けつする，分けつが出る，tiller；分けつ(学術用語)
6) winter killing：冬枯れ
7) dibbled：穴を掘って播く穴播き，drilled：ドリルdrillを使って播くこと，ドリル播き，条播，broadcast：散播
8) peck：穀物などの計量単位，＝1/4 bushel，約9 L
9) Winnowは唐箕(とうみ)であるが，ここでは西洋唐箕を言う。唐箕は，中国古代前漢時代(河南省博物館)の世界的な発明品で，東に伝わって日本にきた形態は今日も変わらず，鼓胴で覆った手回しの翼車(送風装置)の前方に漏斗から選別する穀物を少量ずつ落とし，風によって1-3番口(順に精粒，再選別材料，ゴミ)に区分する。一方，シルクロードを通って西欧に伝わった西洋唐箕は，本文のように金網を使った篩(ふるい，sieve)を併用し，風による比重選別に加えて網の目による形態選別もする形式である。これに続く本文は，ゴミの混ざった穀物を箕に入れて人が高々と捧げ，少量ずつ落として

338　Chapter 7　Crop Cultivation

　　　　風で選別する方法であり，唐箕がない時に最も容易な方法である。
10)　この直前の threshing machine は Thresher(スレッシャ)と言い，穂付藁(わら)を投げ込み，脱穀，選別，精選を行なってきれいな穀粒と藁を分離して排出する農機具であり，現代のコンバインから刈取部を除いたのと同じ形式である。なお，水稲は難脱粒性でムギ類のように容易でないため，日本の脱穀機は藁の根本を支えて穂先だけ脱穀する。この後の数行に登場する農機具名は，sickle：日本の鎌と同様な形式の西洋小鎌(シックル)。cradle：刃渡り 1 m 内外で柄長 1.5 m ほどの西洋大鎌 Scythe(サイス)に麦稈を集めるレーキをつけた集稈装置付き西洋大鎌(クレードル)であり，半円形に振り回すと刈り取られた麦稈が鎌の上に集まって小束相当の堆積ができる。Self-rake Reaper: reaper(リーパ)はムギ類を刈り，それを直後のプラットホーム上に倒し，レーキで間欠的に掻き出して大束相当の堆積を作って進む畜力牽引作業機。当初は人力掻き出し式であったが，1860 年代に巧妙な機構で機械掻き出し式の改良機が出たため，講義時点では self-rake をつけて呼んでいた。文中の Buckeye と Wood は，当時の代表的メーカ 2 社の社名。Self-binding reaper は，その後 binder(バインダ)と呼ばれるが，reaper のプラットホーム上に倒れたムギ類を結束装置部に送って大束に結束して進む収穫機で，講義時はまだ開発中で試験利用の状況であった。なお，ここで価格の高低を言っているのは，人件費を入れた作業料金である。
11)　麦角病
12)　病名不明
13)　なまぐさ黒穂病
14)　黄化萎縮病
15)　さび病
16)　裸黒穂病
17)　ゾウムシ
18)　ヘシアンバエ
19)　寄生性コバチの一種(*Platygaster hiemalis*)
20)　grind：粉に引く(過去形 ground，なお mill は製粉する)，soiling：青刈り(青草を飼料とする)
21)　11) に同じ
22)　この前後の用語は，spikelet：小穂(しょうすい)，glume：頴(えい)・護頴(ごえい)，panicle：穂，palea：内頴，awn：芒(のぎ)，awnless：無芒，いずれも学術用語
23)　lodge：倒伏する
24)　黒穂病
25)　ハリガネムシ
26)　Barley(*Hordeum*)には *Hordeum vulgare* L.；6 条種，*Hordeum distichum* L.：2 条種がある。数行後の two-rowed barley：2 条オオムギ
27)　ear：穂，雌穂(トウモロコシ)
28)　Naked Barley：裸麦
29)　24) に同じ
30)　25) に同じ
31)　inflorescence：花序，spike：穂，穂状花序，staminate(-flower)：雄蕊の(雄花)，pistilate(-flower)：雌蕊の(雌花)，axillary：腋性の，style：花柱，以上学術用語，silk：絹糸(けんし，トウモロコシの穂の先端から出る花粉管)
32)　Indian corn is a native of America：トウモロコシの成立については多くの研究があるが，いまだ解決されていない。発祥地は中南米と考えられている。
33)　トウモロコシは頴果の形状と性質によりデントコーン Dent corn，フリントコーン Flint corn (硬粒種)，スイートコーン Sweet corn (甘味種)，ソフトコーン Soft corn (軟粒種)，ワキシーコーン Waxy corn (蠟質種または糯種)，ポップコーン Pop corn (爆裂種)，スターチ・スイートコーン Starchy-sweet corn (軟甘種)，ポッドコーン Pod corn (有稃種)の 8 つのグループに分けられる。Rice corn はポップコーンと想定される。
34)　earliness：早晩性
35)　shell：穂から脱粒する，husk：穂の皮をはぐ，i.e.: corn sheller, corn husker
36)　cord：体積の単位 3.6246 m^3。8 行後の rapeseed oil：ナタネの搾りかす
37)　ensilage：サイレージにする，サイロに詰める。生牧草をサイロなどに詰めて乳酸発酵をさせる貯蔵法。18 世紀にヨーロッパで土中貯蔵法が発明されてサイロ詰に発展し，アメリカに 1880 年頃，日本に 1900 年頃に伝わった。第 8 章解題参照。

解題 339

38) stipule：托葉，sheath：葉鞘，corymbose panicle：散房花序，achene：痩果(そうか)，petiole：葉柄，いずれも学術用語
39) polenta：ポレンタ(イタリア料理の一種，トウモロコシがゆの類)
40) stable：馬小屋，stable manure：厩肥
41) pinnate leaves：pinnate compound leaf：羽状複葉，leaflet：小葉，anther：葯(やく)，filament：花糸，いずれも学術用語
42) Its native country is America：バレイショの原産地は南米アンデス山系の山岳地帯である。
43) eye：目(塊茎の窪んでいるところ)
44) night-soil：下肥(しもごえ)，人糞尿
45) 疫病
46) コロラドハムシ
47) 15)に同じ
48) マメクロアブラムシ
49) マメゾウムシ
50) ゾウムシの一種
51) 根こぶ病と思われるが，正確なところ不明
52) アブラムシ
53) ウジ(ハエの幼虫)
54) カブの Fingers and Toes と同じものかと思われるが，正確なところ不明。
55) シロチョウ属
56) ヤガ属
57) モンシロチョウ
58) ブロッコリーは，現在，消費量の多い人気のある野菜となっている。
59) ヤガの幼虫
60) キアゲハの幼虫であろう。
61) set は「子球」「幼苗」の意味があり，特にタマネギ栽培では移植する幼苗をオニオンセット(onion set)と呼んでいる。寒冷地では生育期間が短いため，本講義のように直播でなく，あらかじめ育てておいた種球(幼苗)を移植する。ブルックスが丘珠に紹介した前述のタマネギは，農黌園の栽培記録から Yellow Danver's onion と Wethersfield Large Red と想定されるが，それらは直播栽培で札幌近郊に広まった。しかし，1960年代にタマネギ移植機が開発されてから，今日では100％移植栽培に変わっている。なお，裏作で栽培する府県のタマネギでは，大阪泉南地域で Yellow Danver's onion を導入したのが始まりとされる。
62) ポテトオニオンは分球性で，ロシアなど寒冷地で栽培される。
63) トップオニオン(*Allium cepa* var. *viviparum*)は球が比較的大きく，花序に小球(頭球)が着き，これによって繁殖する。
64) 子球
65) リーキ(*Allium ampeloprasum*)とニラ(*Allium tuberosum*)を同じとしているが，異なる種で間違いであり，当時の誤認かもしれない。また，リーキの学名は本講義で用いられている *Allium porrum* もあるが，現在では前述のものが用いられている。
66) ヘリカメムシ
67) スカシバガ科の一種
68) ニホンカボチャ(*Cucurbita moschata*)の別名としてトウナスが用いられることがあるが，本講義の学名は，これとは別のペポカボチャであって Tounasu と同じでない。
69) シロウリは日本と中国で栽培され，甘味はなく，漬物にする。現在では学名 *Cucumis melo* var. *Conomon* が用いられる。
70) ニガウリはツルレイシとも言われるが，沖縄県でゴーヤ(苦いウリの意)と呼び，つい最近，全国に広まりつつある。
71) カモウリはトウガンの別名であるが，京都の地名「加茂」に由来するとの説がある。現在では学名 *Benincasa hispida* が用いられる。
72) マスクメロン muskmelon の musk はジャコウシカの分泌物「麝香(じゃこう)」を意味しているが，このメロンは芳香があり，果皮にネット(網目)ができる「網メロン(net melon)」である。ネットは，表皮細胞が硬化した後の果実肥大によって生じた亀裂に癒傷組織としてのコルクが形成されたものである。

73) 現在ではスイカの学名として *Citrullus lanatus* が用いられる。
74) 植物形態学で言う本来のアスパラガスの葉は，若茎または生長した主茎の節部に着生している三角形の鱗片状のものである。ここで leaf(葉)と呼んでいるのは cladophyll(日本語で擬葉)と言われるもので，形態学上は茎の末端に相当し，針状となっている。
75) 当時はきわめて品種数が少なかったが，現在では多くの品種が育成されている。アスパラガスは多年生であり，形質が多様で，新品種の育成に長期間を要するためと思われる。
76) East Indies＝East India，東インドは，インド，インドネシア，マライ諸島を含むアジア南東部地域を指す。
77) flower-cluster：花房(かぼう)，bract：苞葉(ほうよう)，calyx：がく(萼)，へた(蔕)，style：花柱，stigma：柱頭，以上学術用語，biennial：2年生
78) ネキリムシ(ヨトウガ類の幼虫)

参考文献
〈食用作物系〉
1. 田口啓作・村山大記監修(1977)，馬鈴薯——Potatoes in Japan，グリーンダイセン普及会
2. 北海道農業試験場編(1967)，北海道農業技術研究史，同試験場
3. 北海道農業試験場編(1982)，北海道農業技術研究史：1966～1980，同試験場
4. 三分一敬監修，土屋武彦・佐々木宏純(1998)，北海道における作物育種，北海道協同組合通信社
5. 北海道小麦今昔物語編集委員会(2000)，北海道小麦今昔物語，ホクレン農産事業本部
6. 北海道大学(1980)，北大百年史〈部局史〉
7. 嵐 嘉一(1975)，近世稲作技術史——その立地生態的解析，農山漁村文化協会
8. 南鷹次郎先生伝記編纂委員会(1958)，南鷹次郎(南鷹次郎先生誕生百年記念)
9. 農業技術大系(作物編，基本編・イネ基礎編)(1975)，農山漁村文化協会
10. 日本作物学会50年の歩み刊行部会編(1977)，日本作物学会50年の歩み——創立50周年記念，同会創立50周年記念事業委員会
11. 戸苅義次・菅 六郎(1957)，食用作物，養賢堂

〈果樹および野菜系〉
1. 青葉 高(1991)，野菜の日本史，八坂書房
2. 小林 章(1996)，果樹園芸の世界史——果物の世界と野菜の世界，養賢堂
3. 杉山直儀(1998)，江戸時代の野菜栽培と利用，養賢堂
4. 板木利隆(2001)，ぜひ知っておきたい昔の野菜今の野菜，幸書房
5. 北海道果樹百年史編集委員会編(1973)，北海道の果樹百年史，北海道果樹百年史事業会
6. 佐藤公一他編著(1984)，果樹園芸大事典，養賢堂
7. 野菜園芸大事典編集委員会(1985)，野菜園芸大事典，養賢堂
8. 塚本洋太郎監修(1994)，園芸植物大事典 I・II，小学館

〈工芸作物系〉
1. 佐藤 庚他(1983)，工芸作物学，文永堂出版
2. 津田周彌(1988)，工芸作物——その光と翳，私家版
3. 西村周一編(1960)，実用工芸作物ハンドブック，朝倉書店
4. 細川定治(1980)，甜菜，養賢堂
5. 森 義忠(1995)，ホップ——ホップの基礎科学と育種，栽培について，私家版
6. 砂田 明編著(1986)，薄荷物語——北の華，北見観光協会

〈病理学系〉
1. 栃内吉彦(1956)，植物病理学通論，誠文堂新光社
2. Carson, Rachel (1962), *Silent Spring*, Houton Miffin Co., N.Y.
3. 成田武四(1981)，北海道農作物病害総覧，北海道防疫協会
4. 北海道農業フロンティア研究会編(1991)，土は求めている，北海道大学図書刊行会
5. Schumann, Gail L. (1991), *Plant Diseases; Their Biology and Social Impact*, American Phytopathological Society, St. Paul
6. 脇本 哲編(1994)，総説植物病理学，養賢堂
7. 日本植物防疫協会編(1996)，植物の病気——研究余話，㈳日本植物防疫協会
8. Agrios, George N. (1997), *Plant Pathology* (4th ed.), Academic Press, San Diego, USA

[*Lectures for the fifth Semester*]
Chapter 8   Stock-Farming

8-1. Importance of Stock-Farming
  From the earliest historic times, it has been a practice among many nations to rear and care for domestic animals of many sorts for the safe of the food, work, and many other useful products afforded by them. In Japan, for numerous reasons fully known to you and me, the keeping of domestic animals has been much less than in most other agricultural countries. At present, the hold of Buddhism upon the people seems to be weakening, and the importance of raising domestic animals seems to be more and more appreciated from year to year. Indeed, that importance is extremely great cattle supply people with food, work, many articles of clothing, and numerous other useful products. Almost every part of an ox is at present turned to some useful purpose. The keeping of cattle also enables man to utilize lands which are otherwise of very little value for agricultural purposes.
  At present, in many areas of southern Japan, we find only the plain lands (susceptible of irrigation) fully occupied, only a short distance from a city such as Tokyo, and further from the city, there are large acres of land unoccupied. That land, although not available for rice culture either because of natural features which render irrigation impossible, or because of the distance from cities or towns from which supplies of manure can be obtained might doubtless profitably be occupied for the wise keeping of domestic animals more common. Due of the most important consideration in favor of livestock husbandry is the fact that rightly followed, its waste products in the shape of animal excrement are sufficient to maintain the fertility of a farm. The following statistics will give you some idea of the very great importance of stock farming in a country, the United States, in which it is very extensively pursued. [p. 1061]

Statistics[1], January, 1878

Kind of Livestock	Number of head	Value
Oxen & other cattle	19,223,300	$329,541,703
Sheep	35,740,500	$80,603,062
Hogs	32,262,500	$160,838,535
Horses	10,329,700	$600,813,681
Mules	1,637,500	$104,329,939
Milch-Cows	11,300,100	$298,499,866
Total		$11,774,626,786
Average value for each man is about $40 yearly		

8-2. Climate and Soil Best for Stock Farming
(1) The best climate is one which is mild and equitable, allowing pasturage the whole year. The latter consideration is important in the economical raising of stock. Cattle thrive equally well in countries considerably colder, where they must fed under cover during several months of every year. To feed in that way, however, is much more expensive than to feed in the pasture the whole year[2]. Nevertheless, a temperate climate possesses numerous advantages over a subtropical or tropical

one. Chief among these is the fact that the temperate regions are the home of perennial grasses, most such grasses fail to grow well in tropical regions. Hence, although the vicinity of Tokyo and regions further south possess an advantage in the length of time during which pasturage is possible, it is nevertheless true that, in this vicinity on account of the much better growth of perennial grasses there is advantage if not quite counterbalances those of more southerly regions. Rainfall is also extremely important for stock-farming. For the best growth of grasses, it is essential that rainfall be considerable in amount and evenly distributed throughout the growing season. In cooler climates, too, cattle are perhaps rather less subject to disease than in warm, and this also gives them an advantage. For all these various reasons, we find that it is more extensively carried on in temperate climates.

(2) The soil best for stock-farming is, of course, that which is best for the production of grass upon which the success of this industry mostly depends. The soil, then, should be well drained, firm, somewhat retentive of moisture, and contain a considerable admixture of clay, such as I have already described when speaking of pastures. It is of the highest importance, too, that in a region selected for stock-farming there will be plenty of pure running water. It is not necessary, as I have pointed out, that the region be plain. Hilly or mountainous country is often even superior to plain land, on account of its being unsuited for other kinds of farming, it is usually selected, but it should not be too steep or too abrupt. [p. 1064]

8-3. *Locations Best for Stock-Farming*

While a location near a market as in any other business is of course a great advantage, it is never true that stock-farming more than most other kinds of businesses affords the chance for profit even in regions quite remote from markets[3]. A little thought will show the principal reasons for that. Animal products are much more valuable, pound for pound, than other farm products, while the cost of transport is not necessarily very much greater for a given weight. It is true that fresh meat cannot be transported without extraordinary precaution, which may overbalance the advantage of its high value. Yet meat may be easily preserved in various ways and then taken to a distant market, at an expense not much greater than the expense of transporting a like weight of any other agricultural product. It is stated that the prosperity of several great American cities such as Chicago and Cincinnati, has been principally due to the discovery of methods of putting 400 pounds of meat into 100-pound packages.

Another advantage of stock-farming over grain-raising is that the animals can be driven long distances where there is no other means of transport. Another instance is that the greater part of the beef, pork, cheese, *etc.*, exported from America to Europe does not derive its origin from along the coastline, but from inland farms. In the year 1877, the exports of the United States to European countries had a total value of $140,564,066; since that time, the trade has very much increased. Still another instance of profit is that the greater portion of canned beef, fruit, cheese, butter, pork, *etc.* in Japan almost all take their origins from very distant countries such as America, England, France, *etc.* Another important importance is that almost all the beef, pork, *etc.*, consumed in the eastern part of America is brought there from western states. [p. 1066]

8-4. Hokkaido as Adapted for Stock-Farming

From what I have said concerning the soil and climate for stock-farming, you have been able to ascertain the adaptability of stock-farming to this country. Yet, in some respects, I think that you have been led into error. Experience in raising cattle, sheep, and swine in Hokkaido has abundantly shown that there is no natural obstacle to the successful raising of any of these animals, as far as climate is concerned. All under proper management are perfectly healthy. We may confidently say that there is no disease to be regarded necessarily to climatic conditions. The climate, however, poses one considerable obstacle to profitable stock-farming, namely, the length of the winters. This is an obstacle only in the sense that it increases expenses, and is an obstacle not regarded as by any means insuperable.

The soil, although not naturally producing the grasses best for stock-farming, may easily be made to produce good grasses; this has been abundantly shown by experiment. The natural grasses, though as already pointed out not of a very high degree of excellence, are not be regarded as worthless. Those are, in many cases, quite well suited to the needs of horses and neat cattle, although not well suited for sheep. The principal objection to the continuous pasturage of wild land, even for horses and cattle. is that a large area is required to feed a given number of animals. That must be regarded, as I have stated. It is an obstacle which is met with to a great extent in many countries where stock-farming is carried on. In conversation recently, with a man sometimes engaged in Colorado in the U.S., I was told that a common estimate in that country was that it required 30-40 acres of land to pasture a single animal. Notwithstanding this fact, stock-raising is profitable in Colorado. Now in Hokkaido, the number of acres necessary can in no case be so large. For instance, at Nii-Kap[4], I have been told that the amount of land used as pasturage for about 20,000 acres, while number of horses now kept there is 1,400, making, as you can see, about one horse to 15 acres.

There are, however, other obstacles to the successful prosecution of stock-farming with profit than those posed by climate and natural features of the soil. Among those may be mentioned, as of considerable importance, the savages of bears and deer in certain places[5]. They are particularly dreaded in sheep-husbandry, and considerably in the case of horses, but much less in the case of cattle, which are largely able to defend themselves by means of horns. The almost entire extermination of these beasts is quite possible. Still another is the extremely high cost of transport to other markets, at present doubtless due to the monopoly held by one steamship company. That monopoly, however, may soon be broken, in which case transportation of agricultural products to southern Japan and also to a portion of China will become cheap. Now, weighing all the advantages and disadvantages, Hokkaido is not naturally well adapted for stock-farming, but it is sufficient to supply the home consumption and a portion of China. [p. 1069]

8-5. Neat Cattle

Neat cattle[6] belong to the subkingdom Vertebrate, class Mammalia, order Artiodactyla, Division Ruminantia, family Cavicorniae, genus Bos; the common domesticated animals in this country, and in Europe and America belong to the species *Taurus*. In Asia and other Asiatic countries, the species *Indicus* has been domesticated, while in Africa the species *Bubalus*. The first 2 species were doubtless natives of Asia, the former of North Asia and the latter of South Asia. The latter is characterized by humps just behind the shoulders, and is commonly known as the Zebu. The species *Taurus* is humpless, though in its wild state it differed considerably from the highly improved breeds of the present day. The species *Bubalus* is the Buffalo, which differs greatly from either of the others. In America is found another species running wild in vast numbers, namely, the American bison or *Bos Americanus*; this species has been domesticated, but is not kept to any extent. All of these species, though differing much from each other, interbreed freely, and the offspring is fertile. All of these animals, in the wild state, are gregarious going into flocks or herds, and all are characterized by having two horns with a bony core inside of the bone and bifid hoofs. Their dental formula is as follows:

$$i\frac{0-0}{3-3}, \ c\frac{0-0}{1-1}, \ m\frac{6-6}{6-6}$$

The stomach is complex, and is divided into 4 distinct cavities, namely, the paunch or rumen, reticulum, omasum or manyplies, and abomasums or rennet. The paunch is by far the largest. Its office is to receive and store food, which first is but little chewed, and mix it with the water which it contains. The reticulum has an inner appearance like honeycomb, and is much smaller than the paunch; in it, a certain part of the liquid portion of food is absorbed. The manifold or manyplies is so named because it consists of many folds; in it takes place further digestion of soluble saccharine and fluid portions of food before it passes into the fourth stomach. The rennet or fourth stomach is much larger than the second and third, ranking next in size to the paunch. This is the true digestive stomach.

A rough classification of neat cattle is based upon their horns; according to that classification, there are long horns, middle horns, short horns, and hornless. The best breeds belong to the middle-horned class. [p. 1073]

8-5-1. *Devon*

1) Description

The Devon is one of the oldest of English breeds, having long been bred to a high degree of perfection in Devonshire, England. It is generally considered that the Devon has been brought to its present degree of excellence without admixture of foreign blood, and that these cattle have been built up by careful selection and management from the old native livestock of the country. As might be expected from

[参考図]

Devon Cow (from Brooks, *Agriculture*)

such a system of breeding, therefore, we find the Devons to have their characteristics very firmly fixed; and with a high degree of certainty, their characteristics are transmitted to their progeny. Indeed, so firmly fixed are certain characteristics of the Devon that some have almost been inclined to call the Devons a race, rather than simply a breed.

2) Points of Excellence[7] in a North Devon Cow

The head should be small, lean, and bony; the forehead wide and flat or from fullness of the frontal bone over the eyes somewhat directly; the face straight, muzzle fine, nostrils open, lips thin and rather flat. Upon the scale of 100, the perfection of parts of the head is entitled to 4 points; nose of a light delicate orange, 4 points. The eyes should be bright, prominent, and clean, but mild and gentle in expression, as indicative of that spirited but tractable disposition so necessary to cattle that must bear the yoke. A beautiful orange-colored ring should surround each eye, 4 points. The ears thin, with a rich orange color within, of medium size, with a quick and ready movement expressive of attention, 2 points. The horns light tapering, of a waxy color toward the extremities, and gayly as well as symmetrically situated on the head. The occipital bone narrow, thus bringing the base of horns nearer together, 2 points. The neck of medium length, somewhat light in substance, very clean, and well set upon the shoulders.

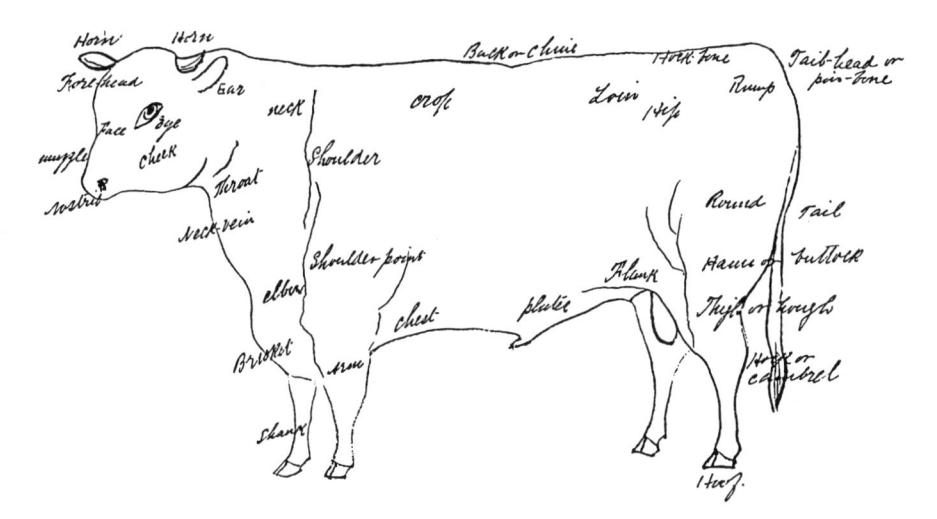

The chest is deep and round, carrying its fullness well back of the elbow. This is afforded by the aid of springy ribs and abundant internal room for the action of the heart and lungs, and also without any extreme width forehead and between the points of the shoulders which might interfere with the action of the animal, 14 points. The brisket, since it adds nothing to the internal capacity of the chest, must not overload the breast but be sufficiently developed to guarantee a feeding property attended with a full proportion of fatty secretion, 4 points. In this breed, the shoulder is a very beautiful and important point, and should to a certain extent approximate in form to that of a horse. It should take a more sloping position than is found in most other breeds, with its points projecting and angular, and the scapula more curved. This blending with and forming a fine wither rising a little above the level of the back line, 4 points.

The crops full and even, forming a fine line with a somewhat rising shoulder and level back, without either drop or hollow, 3 points. Back, loin, and hip broad and wide, running on a level with the setting on of the tail, 9 points. The rumps lying broad and apart, high and well covered with flesh, 5 points; the pelvis wide, 2 points;. the twist full and broad,3 points. The quarters long and thoroughly filled up between hip bones and rump, with good muscular development down the thigh to the hocks, 6 points. The flank moderately deep, full, and mellow in proportion to condition, this portion has a very good propensity to fatten or not, 3 points.

The legs not too short, standing as straight and square behind as may be compatible with activity; the bone quite small below the hocks and knee; the successors large and clean; the forearm well developed, 5 points. The carcass round and straight, its posterior ribs almost circular, extending well back and springing nearly horizontally from the vertebrae, giving in fact much greater capacity than it would at first appear, 2 points.

The udder should be such as will afford the best promise or capacity, and product, 1 point. The tail level with the back, long, very slender, and finishing with a tassel of white hair. The hair should be short and thick; if showing on its surface a fine curl or ripple, it looks richer in color and is supposed to indicate a hardier and most thrifty animal, 1 point.

Color in its shades and degrees is more or less governed by fashion, but in Devon it is always red. Formerly, rich blood red was the favorite, but now a somewhat lighter shade is preferred. In all cases, the colors grow lighter around the muzzle. Dark color about the head, especially if accompanied by a dark nose, are indicative of purity of blood.

Carriage, with Devons, given their excellence for the yoke and other destinates besides the butcher's block, it is important that the animal's carriage should indicate as much; to obtain this, something of the heavy, squarely moulded frame of the mere beef animal must be relinquished for a lighter and more active frame, 3 points.

Quality on this thriftiness, the feeding properties, and the value of the animal depend and upon the touch of this quality rest in a good measure the grazier's and the butcher's judgement. If the touch be good, some deficiency of form may be excused, but if it is hard and stiff nothing can compensate for such an unpromising feature. In raising the skin from the body between the fingers, it should have a soft, flexible, and substantial feel; when beneath the outspread hand, it should more readily with and under it as though resting on a soft elastic cellular substance which however becomes firmer as the animal "ripens". A thin, papery skin is objectionable, especially in a cold climate, 15 points.

3) Points of Devon Bull

The points desirable in the female are generally so in the male too, but must of course be attended with that masculine character which is inseparable from a strong, vigorous constitution. Even a certain degree of coarseness is admirable, but it must be so exclusively masculine in character as to never be found in females of his gets. The head of a bull may be shorter, and the frontal bone broader, than those of cows; the occipital bone may be flat so that it may receive and sustain the horn, and if it is somewhat heavy at the base, that may be excused provided that its upward form, quality, and color be right. Feather is the looseness of the skin attached to and depending from the under jaw to be other than a feature of the sex, provided it is not extended beyond the bone, but leaves the gullet and throat free from the dewlap. The upper portion of the neck should be free, full, and muscular since it is an indication of strength and constitution. The skin should be strong, the bones of the loins long and broad, and the whole muscular system wide and thoroughly developed even whole.

4) The Devon as a Dairy Animal

As a dairy animal, the Devon cow is excelled by many other breeds. The quantity of milk is medium, or perhaps a trifle above medium. In quality, the milk takes a very high rank, being extremely rich and of good color. The Devon must therefore be regarded as, upon the whole, a little above average as a dairy animal; still, if the production of milk or other dairy products is the sole objective, it is seldom advisable to select cows of this breed.

5) The Devon as a Working Animal

As a working animal, the Devon is superior to all others in the following respects: activity, docility, ease of matching, hardiness, ability to subsist and thrive upon moderate feed, and quantity of work performance. In size, the Devon is excelled by numerous other breeds, which may therefore under certain circumstances be superior to Devons for drawing extremely heavy loads; yet for all general work, especially on the farm or road, the Devon will undoubtedly yield a greater return in labor for the feed given him than an ox of any other breed. In rapidity of walking, the Devon approaches the horse; hence, it is very much employed for plowing, harrowing, *etc*.

6) The Devon as a Beef Animal

In several respects, the Devon excels all other breeds as a beef animal. One of the most important is quality, Devon beef being superior to beef of almost any other sort. One of its great points of superiority is the mixture of fat with lean meat, the fat of the Devon, instead of being laid mostly on the outside of the carcass, is found between the fibers of muscular tissue. This gives its beef the much-desired marbled appearance. In rapidity of growth and early maturation, the Devon is excelled by some other breeds also in size. The Devon seldom exceeds, at the age of 3 or 4 years, the dressed weight of 1,000 pounds, while that of some other breeds will considerably exceed it. The Devon, however, is an extremely hardy animal and will thrive tolerably well under circumstances in which the Shorthorn would be very poorly. The animal, being of medium size and active, is well adapted to getting his living upon somewhat hilly or scanty pastures. Owing to its superior quality, Devon beef usually commands from 1 to 2 cents per pound more than beef of most other breeds.

In arriving at a just estimate of the Devon, we must take into account the following points: as a dairy animal, the Devon is about average; as a working animal all things considered, superior to any other; as a beef animal, excelling in quality though deficient in quantity; hardy and able to subsist under somewhat unfavorable circumstances; its characteristics, in a very remarkable degree, to animals of other breeds with which it is crossed. Taking all these points into consideration, it is evident that we must regard Devon as a very superior breed of cattle. According to Professor Brooks's estimate, the ordinary Japanese bull weighs 1,000 pounds alive the minimum weight is 800 pounds; when dressed, 50 to 55% is lost. The cows weigh an average of 500 pounds less. The Ayrshire bull on the College Farm weighs 1,647 pounds, and cows 900 to 1,000 pounds. The Shorthorn bull weighs 1,995 pounds, and cows of this breed 1,600 to 1,700 pounds. [p. 1083]

8-5-2. *Hereford*

1) Description

This is a very old breed, also of English origin, but within the last century the principal improvements in this breed have been made. It is claimed by some that these cattle originated in Wales about the year 900 A.D., but that cannot be clearly established. 100 years ago, cattle of this breed were dark red, almost brown, in color, with mottled faces. Now they are red, with the shades running into light or yellowish red, with white faces, throats, bellies, and sometimes backs. Occasionally, a roan (mixture of two colors) of red and white mixed, and more rarely an almost clean white with red ears is found. Give a Devon, a quarter more size, somewhat more proportionate horns and bones, somewhat shorter legs, a longer body, and a little coarser in every except color.

2) Points in a Hereford Cow

The points in a Hereford cow are: head moderately small, with a good width of forehead, tapering to the muzzle, the cheekbone rather deep but clean in the jaw, 3 points. Nose light in color, and the head free from fleshiness, 2 points. Eyes full, mild, and cheerful in expression, 2 points. Ears of medium size, 1 point. Horns tight and tapering, long and spread with an outward and upward turn, giving a gay and lofty expression to the whole head, 2 points. Neck should be of medium length, full in its junction with the shoulders, spreading well over the shoulder points, and tapering finely to the head, 2 points. Chest broad, round, and deep, its floor running well back of the elbows which, with a springing for rib, gives great interior capacity to this all-important portion of the body, 14 points.

Brisket when in flesh largely developed, descending low between the legs and deep, covering the anterior portion of the breast bone but never interfering with the action of the animal when in working condition, 14 points. Shoulder lying smoothly and closely in toward the top, spreading toward the points, with the blade sloping somewhat back and running pretty well up into the withers, which by rising a trifle above the level line of the back, gives the ox a very upstanding and beautiful fore-end; the whole shoulder well-clothed with muscle, 3 points. The crops filling up evenly behind the shoulders and smoothly blending in with the muscles of the back, 3 points. Back, loin, and hips should be broad, wide, and level, 8 points.

Rumps should lie nearly or quite level with the back, and their covering should be abundant, loose, and freely moving under the hand ,this showing great aptitude to fatten, 4 points. Pelvis roomy, indicated by wide hips and the space between the rumps, which should stand well apart; giving a general breadth to the posterior portions of the animal, 3 points.

Twist broad and full, extending well down each side of the thigh, with corresponding width; a broad twist is a good indication of a butcher's animal, 5 points. Hind-quarters large and thoroughly developed in their upper, more valuable portions as beef ; thigh gradually tapering to the hock, but muscular, 6 points.

Carcass round throughout, full and capacious, with the underline of the belly level or nearly so, 3 points. Flank full and wide, 3 points. Legs straight and upright, firmly placed to support the superincumbent weight; a strong back sinew, but by no means a large, coarse cannon-bone, 3 points. Plates; of belly strong, this preserving a nearly straight underline, 2 points.

Udder broad, fully extending forward and well up behind; teats of good size, squarely placed, with a slightly oblique pointing out; veins large and swelling, 3 points. Tail large, and full at its point of attachment, but fine in its cord, 2 points.

Hair thick, close, and furry, accompanied by a long growth and the disposition to curl moderately is more esteemed, but that which has a harsh, wiry feel is objectionable, 3 points. Color red or rich brown, often very dark, with a white face, are now the colors most fancied, though there are also gray and cream-colored Herefords, 1 point. Carriage prompt, resolute, and cheerful, and in the ox gay and lively, 3 points.

Quality: Exactly the same points as in the case of North Devon cow should be looked for, 15 points. Points of the Hereford Bull are the same as the Devon Bull.

3) The Hereford as a Dairy Animal

The quality of milk yielded by Hereford cows is not large, and the quality is not extraordinary good. For dairy purposes, these cows are not worthy of a moment's consideration

4) The Hereford as a Working Animal

As working animals, Herefords take a high rank, standing next to the Devons, all things being considered, and in some respects exceeding the Devons. Their size is larger, and their strength is consequently somewhat greater, while there is a good degree of activity. Herefords, too, are hardy, and capable of performing a large quantity of work. In color and general characteristics, not quite as uniform as the Devons, yet have a high degree of uniformity, which makes matching easy. Altogether an excellent working animal.

5) The Hereford as a Beef Animal

As beef animals, Herefords take a very high position, in general estimation, standing next at least to Shorthorns for general beef purposes, and in the opinion of many exceeding them. The size, it is true, is not as great, but the body is well formed and compact, with fine bones and a large proportion of meat in the choicest part. Herefords, too, are rapid in growth, come early to maturity, are good feeders, and afford meat of very fine quality. Further, owing to their somewhat smaller size and greater activity, Herefords are better able to subsist in pastures of only medium or rather below medium richness. Exact experiments to determine whether Herefords or Shorthorns yield the larger proportional return for the feed consumed are lacking, some claiming such to be the case for one breed, and others for the other. Whatever may be the truth regarding this point, we are certainly warranted in asserting that the Herefords as a beef animal are well worth the attention of breeders of such stock. [p. 1089]

8-5-3. West Highland Cattle or Kyloes

1) Description

This is a breed of rather small cattle which originated in the Highlands of Scotland and some adjacent islands. Those regions are cold and bleak, swept by strong winds. and are all hilly or mountainous, affording rather scant pasturage. The climate is, moreover, very rainy. As might be expected of a breed of cattle originating in such a region, the Kyloes are rather small. The following is a description abridged from a Scotch writer: the Highland bull should be black, his head not large, his ears thin, and muzzle fine, rather turned up. He should be broad in the face, the eyes prominent and the artless and calm and placed; the horns should taper finely to the point ascending, widely set on at the roots and of a waxy color. The neck should be fine, particularly where it joins the head; the brisket wide, and projecting well before the legs. The shoulders broad at the top, the crops full, the girth behind the shoulder deep, the back straight, wide, and flat; the ribs broad, the space between them and the hips small, the belly not sinking low in the middle, the thigh tapering to the back point, the bones in proportion to size than in the breeds of Southern districts. The tail should set on level with the back, the legs short and straight, the whole carcass covered with a thick coat of long hair, and plenty of hair also about the face and horns and that hair not curly. The Kylo has been called a "miniature Shorthorn", and it is true that in form and in the development of its choicest parts, it approaches in perfection the Shorthorn.

2) The Kylo for the Dairy

Cows of this breed yield a fair quantity of good milk for animals of this size, but their size being small, the quantity of milk is small in comparison with that yielded by large breeds. It cannot, therefore, prove profitable to adopt Kyloes for dairy purposes in any country capable of supporting a larger breed of cattle.

3) The Kylo as a Working Animal

Oxen of this breed are much too small for working purpose. We may discuss this topic without further consideration.

4) The Kylo as a Beef Animal

It is for principally for beef that these animals are bred, and for this purpose they have certain points of high excellence. The quality of their beef is the very highest excellence, considered by many as even superior to the Devons. The proportion of the choice part is large. The animals are extremely hardy, active, and able to subsist under circumstances in which Shorthorns should well nigh starve. It is, therefore, as beef animals in a cold, bleak, hilly or mountainous region that this small, active, hardy breed of cattle have their chief excellence. [p. 1091]

8-5-4. Ayrshire Cattle

1) Description

The Ayrshire is a breed of Scottish origin, but it is not very certainly known what the parent stock of this breed was. It is claimed by some that Ayrshire are from the cattle found wild at an early period in Great Britain. Those wild cattle are now found only in the parks of one or two British nobles where they are carefully preserved. Those wild cattle as now found are nearly

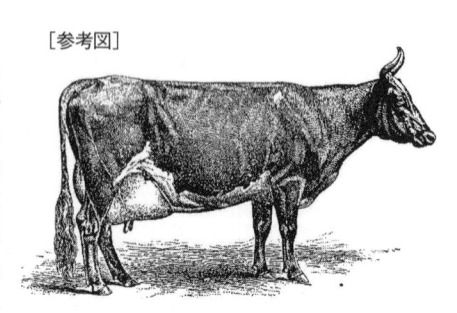

[参考図]

Ayrshire Cow (from Brooks, *Agriculture*)

pure white, only the ears being red. It is a more general opinion that Ayrshire breeds have been built up by cross-breeding, and it is probable that Shorthorns were at one time crossed with Ayrshires. Whatever their origin, it is now true that the Ayrshire is a breed with fixed characteristics which are transmitted with a high degree of certainty from generation to generation.

2) Points in the Ayrshire Cow

The head, as in other breeds, small; the face, long and narrow; the muzzle and nose variable in color, 4 points. Eyes placid, and strikingly large, 2 points. Ears of full size and of an orange color within, 4 points. Horns small, tapering with an outward and upward turn, and set on wide apart, the face somewhat dishing, 2 points. Neck of medium length, clean in the throat, very light throughout, and tapering to the head, 4 points. Shoulders lying simply to the body, thin at their tops and small at their points, not long in the blades, and not loaded with muscle, 6 points. Chest must retain sufficient width and roundness to ensure constitution; the lightness of the forequarter and the "wedge" shape of the animal from the hind- quarter forward arise more from a small, flat and thin shoulder than from any narrowness of the chest, 2 points.

The crops easily blend with so thin a shoulder, and prevent any hollowness behind it; the brisket not overload the fore and, but light, 4 points. The back should be straight, the loins wide, and the hips rather high and well spread, 8 points. The pelvis roomy, causing a good breath at what is termed the "thick" or "round" bone between the points of the rumps, 4 points. Quarters long, tolerably muscular, and full in their upper portion, but molding into the thighs below, which should have a degree of flatness, thus affording more space for a full udder; the flank well let down, but not heavy, 6 points.

Ribs springing out very round and full, affording space for a large udder, which by Ayrshire breeders is considered very essential to secure milking property, the whole carcass thus acquiring increased volume towards its posterior portion, 8 points. Rump nearly level with the back, projecting but little, 4 points. Tail, in its cord, of full length, light in its hair and set somewhat further into the back than would be admissible in some other breeds, 1 point. Legs delicate and fine in the bone, inclined to be short and well knit together at the joints, 3 points.

Udder: in this breed, it is of special importance since Ayrshires have been bred almost exclusively with reference to their milking properties. The great feature of the udder should be its capacity without being fleshy. It should be curved squarely and broadly forward, and show itself largely behind. As it rises upward, it should not too immediately with the muscle of the thighs, but continue to preserve its own peculiar texture of skin: thin, delicate, and ample in its face. The teats should stand wide apart and be lengthy, but not large and coarse, 12 points.

Hair soft and thick; in the phraseology of the country "wooly", 4 points. Color varies from a dark red to rich browns, a liver color or mahogany, running into almost black; very much broken color, and spotting at the edges on a white ground are the favorite colors at the present time. Light yellow is a color sometimes found on very good cows, but pale colors are objected to from an impression that such belong on animals of weaker constitution, 1 point.

Carriage should be light, active, and even gay; this latter appearance is much promoted by an upward turn of the horn, 1 point. Quality: handling will show the skin to be of medium thickness only, working freely under the hand and evincing a readiness in the animal to take on flesh when a drain on the constitution is no longer made by the milk pail, 6 points. Points of the Ayrshire bull are the same as the Devon bull.

3) The Ayrshire as a Dairy Animal

As already pointed out, Ayrshires have long been bred exclusively for dairy purposes. This being the case, we find, as we might have expected a very high degree of excellence in the Ayrshire cow. The yield of milk is extremely large. The following well authenticated yields are recorded: one cow in one day gave 74 pounds of milk, and in one month 1,902 pounds; another cow gave 84 pounds in one day, and in one month 2,168 pounds. It is estimated that an Ayrshire cow in Scotland should yield 6,258 quarts per year. In America, the following yields have been obtained per year: one cow 7,728 pounds, and another 8,159 quarts (one quart of milk weighs a little more than 2 pounds, or about $2\frac{1}{5}$ pounds). The average of a herd belonging to a well-known Ayrshire breeder, Mr. Miles, was 2,707 quarts per year. The average yield *per annum* for another herd belonging to Mr. Sturtevant in the year 1772 was 6,047 pounds. The average yearly production of each of the 34 best cows belonging to Mr. Sturtevant was 6,620 pounds. There are doubtless individual cows of other breeds which have produced larger yields of milk than are credited to any Ayrshire. There is one breed also, the Dutch, in which the average milk yield is probably greater than with Ayrshires, but the Dutch cows are very much larger than Ayrshire and consequently consume much more feed.

It is true that exact experiments to determine which among all breeds of cattle give the best return in milk for feed consumed are yet waiting. It is equally true, however, that there is a large preponderance of opinion in favor of the Ayrshire. It is conceded by most who have had good opportunities of judging that the Ayrshire cow is the most economical milk-producing animal among all our numerous breeds. In the best specimen, she is a more milk-making machine, and there are many individuals which in a single month produce much more than their weight in milk, even with only ordinary feed. One Ayrshire cow on the College Farm weighs 900 pounds and produces more than 40 pounds of milk a day[8].

(1) Quality of Milk: The milk of Ayrshire is especially suited to the manufacture of cheese, the butter globules being of comparatively small and uniform size, and not separating too rapidly from the milk. This property enables the cheese-maker to produce, with comparatively little trouble, a cheese in which the fat is well and intimately mixed with all parts of the cheese. The butter globules, being small, do not soon rise to the top; hence, after being allowed to, sometimes only a comparatively small quantity of cream is found to have risen upon the milk of Ayrshire. Little cream, especially when compared with the milk of such as the Jersey, gives to opinions that Ayrshire is poor in butterfat; that, however, is far from being the case. Careful investigation shows that Ayrshire milk contains fully the average amount of fat. The percentage of butter globules in the milk of Ayrshires is about 5%. There are numerous well-authenticated instances of yields of 14 to 18 pounds of butter per week from a single Ayrshire cow. They cannot then be regarded as poor animals for butter production; yet it is doubtless true that there is one breed, namely, the Jersey which is superior to the Ayrshire for butter production. That very property of Ayrshire milk which makes it well suited for cheese production makes it well suited also as milk for market purposes. The average cheese production of an Ayrshire cow is said to vary from 300 to 400 pounds annually.

4) The Ayrshire as a Working Animal

As work animals, Ayrshires are seldom or never used in America. Their size is somewhat too small, but they do have a good degree of activity and a reasonable degree of strength; hence, though not especially adapted to working, Ayrshires will probably serve well as working animals.

5) The Ayrshire as a Beef Animal

When giving milk, the Ayrshire cow throws all her energy into milk production and usually poor. Ayrshires, however, are extremely hearty and voracious feeders. Cows, when not on milk, and bullocks (castrated animals) readily take on flesh, and their beef is of good quality. However, the form of the Ayrshire is not best suited for beef, the proportion of the parts for beef being less than in numerous other breeds. For this reason and on account of its small size, the Ayrshire is not worthy of consideration for beef. [p. 1100]

8-5-5. *Jersey Cattle*

1) Description

In the English Channel off the coast of France are 3 islands: Jersey, Alderney, and Guernsey, belonging to Great Britain. Upon each of those islands has been developed a distinct breed of cattle, each differing considerably from the others, but all possessing to a large extent the same characteristics. Between the Jersey and Alderney there is really little difference, and those names are sometimes used interchangeably, though the cattle of Jersey are probably somewhat superior to those of Alderney, and have been most largely imported into America. Guernsey cattle are larger than Jerseys, and are regarded by some as equal to, if not superior to, that breed. Guernseys, however, have not been imported into America, though they are doubtless excellent specimens; beyond their descriptions, I shall not speak of them. The term "Channel Island cattle" is sometimes indiscriminately applied to cattle from any of the 3 islands which I have mentioned.

2) Points in the Jersey Cow

Purity of breed on both parents' sides reputed for producing rich yellow butter, 4 points. Head should be fine and tapering; eyes full and lively; face lean and of a smoky color; muzzle fine and encircled with white; horns polished, a little crumpled, tipped with black; ears small and of an orange color within, 8 points.

Back straight from the withers to the setting of the tail; chest deep, and nearly on a line with the belly, 4 points. Hide thin and movable, but not too loose, well covered with soft hair of good

[参考図]

Jersey Cow (from Brooks, *Agriculture*)

color, 2 points. Barrel hooped and deep, well ribbed, having but little space between the ribs and hips; tail fine, having 2 inches below the hock, 4 points.

Forelegs straight and fine, thighs full and long, close together when viewed from behind; hind legs short, bones rather fine, back small, hind legs not too crossed in walking, 2 points.

Udder full well up behind, teats large and squarely placed, wide apart, with veins large and swelling. Growth, 1 point; general appearance, 2 points. The total points for a Jersey cow are 31. Points of bull are the same as for the Devon Bull.

The Jersey is a well-established breed, having long been bred without admixture of foreign blood. As might be expected, we find in Jerseys the power of transmitting with great certainty all their characteristics to their progeny. The Channel Islands have a warm, equitable climate; that fact, taken together with the general appearance of the Jersey cow, has led to the quite general formation of opinion that Jerseys are rather weak, not hardy, and unable to withstand a cold climate. Experience in America, however, has abundantly shown that Jerseys are well able to bear all the exposure and hardship that any animal should be expected to bear. Their constitution is good, and there seems to be no unusual tendency to disease.

3) The Jersey as a Dairy Animal

For the manufacture of butter in the dairy, Jersey cows are undoubtedly superior to cows of any other breed. The quantity of their milk is not great, but the butter globules are of large and uniform size, and quickly separate from the milk. The percentage of butter is about average, and the color and quality of butter from the milk of Jersey cows is superior to those of cows of any other breed. The color of the butter is a deep, rich yellow, very pleasing to the eyes of all butter caterers. It is also claimed that the flavor and keeping quality of Jersey butter are superior to those of any other kind. The total quantity of butter that may be made from the milk of a Jersey cow is extraordinarily large; instances of 3 pounds per day from the milk of a single cow are by no means uncommon, while a yield of 2 pounds per day is very frequently seen. There is a well-authenticated record of a yield of more than 700 pounds of butter in one year from a single cow (Jersey Belle), and another, perhaps equally well-authenticated yield of about 800 pounds in a single year (Eurotas).

4) The Jersey for Work

Of little importance might be fair, but the fact is that in America almost all Jersey bulls are used for breeding; consequently, there has been no opportunity in that country to test the working qualities of Jerseys. Their small size and poor form indicate, however, that not much in the line of work can be expected.

5) The Jersey for Beef

For beef, Jerseys are even poorer than Ayrshires, being generally a little matter, but having a form giving an even smaller proportion of meat in the choice parts. [p. 1105]

8-5-6. *Dutch or Holstein Cattle*

This cattle is of Dutch origin, having been bred for very many years to a high degree of perfection in Holland, and perhaps in adjacent regions. Now, Holland is a low country with a rather moist soil, but with extremely rich and abundant pastures. The chief objective for which the cattle have been bred is for the purpose of producing milk which is manufactured into cheese, and also to a considerable extent into butter. As naturally be expected in a breed originating in a cow rich country, we find among the Dutch cattle of very large size; though perhaps generally not actually heavier than Shorthorns, Dutch cattle certainly have the appearance larger than that famous breed. That is because the legs are proportionally somewhat larger, and general form somewhat less compact in the Dutch. Although their form is less compact, it is by no means what might be called "loose". The head and horns are proportionately small; the neck is light, the chest good, the barrel tolerably round and well-ribbed up, though in this respect, it is inferior to the Shorthorn. The legs are not larger than is necessary to support so heavy an animal. The horns usually turn up and out. The color is black and white, the most fashionable of all being that which gives the appearance of black thrown over a white animal. It is, however, true that the black on the back and sides is very frequently broken by white spots. It is stated that some cattle have red hairs instead of black, but that is not common, at least not in America.

[参考図]

Holstein Cow (from Brooks, *Agriculture*)

1) The Dutch for Dairy

In total quantity of milk produced, there is perhaps no cow equal to the Dutch. Her large size enables her to eat and digest a large quantity of food, and hence make a very large quantity of milk. A Dutch cow owned by Mr. Chenery of Massachusetts yielded 4,018 pounds of milk from May 26th to July 27th, an average of 76 pounds 5 ounces (equivalent to $35\frac{1}{8}$ quarts) per day. In 6 days, 17 pounds 14 ounces of butter were made from the milk of that cow. The quality of the Dutch milk is good, but for butter-making inferior to Jersey milk. For cheese-making or for the production of milk for sale, the Dutch cow is the most excellent animal. Though its absolute quantity of milk may be greater than that of the Ayrshire, it is at least doubtful whether the cost of the production of a given quantity of milk is not less with the Ayrshire than with the Dutch. The Dutch cow, being much larger than the Ayrshire, demands a richer pasture and requires a greater amount of feed for the simple maintenance of life and repair of the wastes of the system.

2) The Dutch for Working

As a working animal, the Holstein, on account of its large size, may be expected to prove quite useful, but its form is not that best suited for work. The fact is that, thus far, this cattle has not been sufficiently numerous in the United States to be used as working animals. Consequently, no very positive statements can be made concerning them.

3) The Dutch for Beef

Their large size and rapid growth are points in favor of Dutch cattle, but as already pointed out, their form is not quite equal to that of the Shorthorn, the proportion of the choicest parts being smaller. In quality, the beef is about average. Their large size makes these animals unfit for any except the richest pastures. [p. 1108]

8-5-7. Hornless Cattle: Galloways

Galloways are a hornless breed of cattle of Scotch origin, having been brought to their present high degree of perfection by long, careful selection, breeding, and feeding. In size, the Galloway is rather above average, though not equal to the

Shorthorn. In general, the Galloway's form is excellent, with the proportion of choice parts being large; bones are fine, body is thick and compact; its color is usually black or very dark red. For the dairy, cows of this breed are of little value; for work, these cattle are of about average quality. It is for beef that the Galloways are principally valued. At the age of 3 or 4 years, they often reach a dressed weight of 800 pounds, and at the age of 5 years, the weight of 1,400 pounds. These weights, though less than those attained by Shorthorns, are greater than those of rich Devons. In quality, Galloway beef is scarcely to be excelled. The animals are extremely hardy, tolerably active, but of a very quiet, peaceful disposition, being very little inclined to fight or struggle with each other. They are good feeders, making an excellent return of flesh per feed consumed. Taking all things into consideration, it must be admitted that in the Galloway we have a beef animal of a very high degree of excellence. These cattle have not been much imported into America, but gradually tending to be imported, on account of the great advantage resulting from the destitude of horns. [p. 1109]

8-5-8. *Longhorn Cattle*

This is a breed of English origin, rather locally but not widely distributed, and never, as far as I know, imported into America. It is a breed having a red color, rather above-average size, and an excellent form, very similar to the Galloway. Indeed, these 2 breeds are similar in all characteristics, except color and horns. For dairy purposes, the Longhorn ranks high, though excelled for special purposes by the Ayrshire and Jersey. The Longhorn is a good working animal, but its extremely long, clumsy horns are a great inconvenience, there being the constant liability of struggling against obstacles of any sort, and even of injury to the drover. It is for beef that the Longhorn is principally valued, but it is exceeded in size by numerous other breeds and quality by some other breeds. These cattle, moreover, because they have the objectionable horns, will probably never prove profitable to import into any other country. [p. 1110]

8-5-9. *Shorthorn or Durham*

1) Description

This is a breed originating in certain counties of the northeast of England, principally Durham, Northhumberland, and York. Those counties were for a long time in the possession of Danes, whose ancestors brought with them cattle from the continent southern Denmark, Holstein and other provinces. Those cattle, except for a somewhat greater coarseness, were essentially like the Shorthorns of today. In color and size, they were the same. In the middle of the last century, those cattle were essentially kept along the banks of the River Tees, hence another name not infrequently applied to them: "Tees Water Cattle". Some of those early cattle reached very great size, in 1,740, an ox dressed 2,324 pounds and a cow 1,540 pounds.

Shorthorn cattle, however, had no great reputation before the time of 2 brothers named Charles and Robert Colling. By those men, the Shorthorns were much improved, and their coarseness was bred out. They now have a fine form, compact, and in almost all respects were made much better. Those gentlemen, for the sake of their cattle, fed one ox until at the age of 5 years, he weighed 3,024 pounds. When he was taken around the county for exhibition, people flocked in great numbers to see him, and the fame of Shorthorn cattle and their breeders the

Colling brothers became widespread. That ox was at one time valued at $10,000. At the age of 11 years, having suffered an accident in the consequence a loss of considerable flesh, he was slaughtered, at which time his weight was 2,620 pounds. The Colling brothers also produced a heifer which gave a dressed weight of 2,300 pounds. The Colling brothers attributed much improvement which they were able to make in the Shorthorn to a bull named "Hubback", from which they continued to breed for many years; even at the present time, it is the desire of Shorthorn breeders as be able to trace the blood of their stock back to Hubback.

2) Points in a Shorthorn Cow and Bull

Pedigree should show unbroken descent, on both sides, from known animals derived from English herds, as found in the English or American Herd Book; without this, an animal cannot compete in this class.

The head small, lean, and bony, tapering to the muzzle, 3 points. Face somewhat long, with the fleshy portion of the nose of a light, delicate color, 2 points. Eyes prominent, bright, and clear, "prominent" from an accumulation of a dipose substance in the socket, indicating a tendency to lay on fat; "bright" as evidence of a good disposition; and "clear" as a guarantee of good health, 2 points. The horns should be light in substance, waxy in color, and symmetrically set in the head; the ears should be large, with considerable action, 1 point. The neck should be short rather than long, tapering to the head, clean in the throat, and full at its base, their covering and filling out the points to the shoulders, 2 points.

The chest should be broad from point to point of the shoulders, deep from the anterior dorsal vertebra to the floor of the sternum, round and full just back of the elbows or, in other words, thick through the heart, 14 points. The brisket should be deep and projecting, indicating a disposition to lay on fat, 5 points. Where weight is an objective, as in the Shorthorn, the shoulders should be somewhat upright and of a good width at the points, with the blade bone just sufficiently curved to blend its upper portion smoothly with the crop, 4 points. Crops must be full and level with the shoulders and back, 8 points.

The back, loin, and hips should be broad and wide, forming a straight and even line from the neck to the setting on of the tail, with the hips round and well covered, 8 points. The rump should be laid up high, with plenty of flesh on the extremities, 5 points. The pelvis should be large, mediated by the width of the hips (as already mentioned) and the breadth of the twist, 2 points. The twist should be so well filled out in its "seam" as to form a wide, even plain between the thigh, 3 points. The quarters should be long, straight, and well developed downward, 5 points. The carcass should be round, with the ribs nearly circular and extending well back, 4 points.

The flanks should be deep, wide, and full in proportion to their condition, 3 points. Legs short, straight, and standing square with the body, 2 points. The plates of the belly should be strong, and thus nearly straight, 3 points. The udder should be pliable and thin in its texture, reaching well forward, roomy behind, teats wide apart and of convenient size, 3 points. The tail should be flat and broad at its root, but fine in its cord, placed high up and on a level with the rump, 2 points.

The coat should be thick, short, and massy, with longer hair in winter; fine, soft, and glossy in summer, 2 points. Carriage: if an animal shows style and beauty, its walk should be square, the step quick, and the head up, 2 points. Quality: the same may be said as with the Devon cow.

Points in the Shorthorn bull are the same as for the Devon bull.

3) The Shorthorn as a Dairy Animal

There has been much very earnest discussion among Shorthorn breeders as to the value of the Shorthorn cow for the dairy. Some claim that there is no breed superior to the Shorthorn for dairy purposes, while others claim that since the Shorthorn is preeminently a beef breed, it is impossible that it should at the same time be a superior milking breed. The fact is that originally most Shorthorn cows were deep milkers; even at the present time, there is one family of Shorthorns known as the "Princess" family, most of which are good milkers. There are also occasionally very superior milking animals in other families. Notwithstanding these facts, however, it is most certainly true that the great majority of Shorthorn cows are not good milkers. That is to be expected, since for many years Shorthorn breeders have been aiming principally at the perfection of form for beef, and it would seem to be most certainly true that the form best suited for beef is not well suited for milk. Milking breeds have thin shoulders and thin thighs, while beef breeds have thick shoulders, a square form, and thick, full quarters and thighs.

4) The Shorthorn as a Working Animal

Full-blooded Shorthorns are usually far too valuable to be castrated for work, but high grades of Shorthorn are very frequently used for work. From the well-known excellence of these animals, which differ in no great degree from Shorthorns, we are entitled to conclude that the pure Shorthorn would also be a good working animal. Those cattle have great size, weight, and strength. In activity, they are superior to the Devon, but for certain kinds of work they are even superior to that famous working animal.

5) The Shorthorn as a Beef Animal

In the Shorthorn, we have the nearest approach toward perfection for beef that we have been able to secure in any breed. The Shorthorn is superior to most if not all other breeds in the following points: rapidity of growth, early maturity, size, perfection of form, large proportion of the best parts, small percentage of offal, return in growth of flesh or fat for feed consumed; and the quietness of disposition so essential in the fattening animal. In quality, Shorthorn beef ranks high, but in one respect, namely, the mixture of fat with beef, it is inferior to others. In the Shorthorn, there is a tendency to the disposition of a great amount of fat upon the outside of the carcass just beneath the skin. The Shorthorn is sufficiently strong and hardy to withstand all reasonable exposure to any degree of cold or hardship in a climate such as this, but being a large animal it must for perfect development receive an abundance of rich feed.

There are several other different breeds in many different countries, but almost all of the breeds mentioned. Yet among them are some worthy to consider, such as the breed of Brittany in France. Although that breed is very small in size, and in beef- and milk-producing capacity, they can grow on rather scanty pasture and poor feed; hence, to some degree, it will prove profitable if kept by poor Japanese farmers.

Japanese Cow: Crossed with a foreign breed such as the Shorthorn, the Japanese cow produces a very excellent progeny which are greatly different from her in many respects. [p. 1118]

8-6. *Cattle Breeding*

The breeding of cattle must be regarded as one of the most interesting branches of agriculture. It demands great intelligence, constant care, and attention, but these united with natural law and an aptitude for business will enable the cattle breeder to produce wonderful results. We have only to look at wild cattle or at least unimproved specimens such as Texan or Japanese cattle and compare them with the square, compact, beautiful breed of Shorthorns, or to compare the small miserable udders of unimproved breeds with the large, finely formed udder of the Ayrshire, the Dutch, or the Jersey to see the improvements which have been attained by the skillful breeder of cattle. As I have stated, all breeds were doubtless derived from 1 or at most 2 common parentages or parent stocks. Cattle in the hands of a skillful breeder are as plastic as clay in the hands of a sculptor. With any breed, at the starting point, a skillful breeder may over the course of time produce a breed of any desired character. It would be a work requiring many years of most skillful management and most liberal feeding to change, for example, Japanese cattle into Shorthorns, yet certainly that might be done with good breeding; combined with the proper surroundings and feeding, cattle may be moulded precisely as the breeder wills. Such then being the possibility within the reach of every skillful breeder, you will see that cattle breeding must be an occupation of great interest. This is a subject about which very much may be said, but with a limited time at my command, I can only hope to cover a few of the leading principles. In commencing to breed cattle, the first point is the selection of breeding stock. The form, size, and general characteristics of animals best suited for breeding vary, of course, with the breed, but there are certain qualifications which should be possessed by animals of any breed. Among these are the following:

(1) Sound health and freedom from constitutional hereditary, or local disease, blemish or inferiorities of any kind;
(2) As much perfection of form as may be obtained in the breed, bearing in mind the chief use for which the animals are intended;
(3) Strong and marked characteristics of their breed in all points;
(4) Purity of blood established by pedigree;
(5) A good disposition, being gentle, docile, and free from vices of all sorts.

Cattle of all breeds should possess certain characteristics of form, among which are the following:
(1) a fine head, small and lean;
(2) a broad chest, full and deep, affording space for large vigorous lungs;
(3) length, breadth, and roundness of body, roomy and full from shoulder to hip, with low flanks, this giving room for the viscera and, in the female, also for the expansion of the fetus;
(4) a straight back, broad hips, and good length of loin;
(5) a fineness of bone and general smoothness of the carcass.

8-6-1. *The Proper Method of Increasing the Size of a Breed of Cattle*

This topic becomes of special interest in connection with the breeding of cattle from Japanese stock as the starting point. Japanese cattle, all will admit, are too small. What, then, is the best method of increasing their size? Theoretically, it is to select large females of some other breed to cross with the Japanese bulls. There is always more or less danger of trouble in the bringing forth of a calf which is the progeny of a large bull and a small cow, because the size of the calf will be greater than the cow is naturally fitted to bear. Practically, however, this method is open to the objection of being far too costly and slow for ordinary adoption. It is much cheaper to purchase a few large bulls and use them upon many small cows than to pursue the opposite course.

Suppose, then, that we desire, for example, to cross-breed Japanese stock with Shorthorns. It will be necessary, in that case, to use Shorthorn bulls upon Japanese cows. What kind of Shorthorn bulls is it best to select? Certainly not those of the very largest size, because in that case the danger of difficulty in parturition will be unnecessarily great. It will be wiser to select a Shorthorn of only average or perhaps a little smaller than average size. Animals of very compact form are preferable to those which are larger and coarser. Upon crossbred females born of the first cow, it will be possible to use Shorthorns of the very largest size, because those females will be much larger than their mothers, and better able to bring forth in safety calves of large size.

8-6-2. The Proper Age for Beginning Breeding

The age at which an animal should be allowed to begin breeding differs according to the purpose for which livestock is kept, and is also different for males and females of any breed[9]. A bull which has always been well fed and cared for may be allowed to serve a few females at the age of 1 year, but the number should be very limited, not more than 4 or 5 at most; the only objective which justifies using so young a bull for breeding is securing stock-getting quality. At the age of 1.5 to 2 years, a bull always supposing he is well fed may be allowed to serve double the number of females that he can serve at 1 year. From 2 to 3 years, he may be allowed to serve about 4 times as many as at 1 year; that is, about 20 to 25 cows.

After the age of 3, a well-fed bull may be allowed to serve yearly, according to his vigor, 30 to 40 females. Very vigorous bulls, after reaching maturity, which is usually at the age of 5 years, may annually serve about 50 cows. It is undoubtedly true that a mature bull, if not allowed to serve so many cows as to diminish his vigor, will beget excellent calves; in practice, it is probably true that most of the calves bred are begotten by bulls under 5 years of age. Bulls should remain virile and vigorous up to the age of 12 to 14 years, but since they are usually kept largely in confinement, most of them become so heavy and clumsy, and lose their vigor to so great an extent, as to make their use of doubtful advantage. An old bull usually becomes very heavy in the fore-quarters and neck; consequently, to support his weight greatly taxes the strength of even the strongest cows, and is impossible for small, weak cows.

A female for milk may be allowed to produce her first calf at the age of 2 to 2.5 years, provided she has always been well fed and has made a good growth. She will not, of course, have reached her full size at this early age, but the milking propensity is thought to be much increased by rather early breeding. If, after the production of her first calf, a heifer is considered to be much under-sized, a much longer than usual time may be allowed to elapse before she is served by a bull again. That will afford her the opportunity to grow. In a breed of cattle kept for beef or for milk, females should be allowed to become somewhat more mature before bearing the first calf. More size and fattening propensity are desirable; hence, to breed from small, immature females is not advisable. With cattle kept for those purposes, then, females should be allowed to bring forth the first calf at the age of 2.5 to 3 years. Ayrshire, Jersey, or Holstein may therefore generally produce the first calf at the age of 2 to 2.5 years, while Shorthorn, Devon, and Hereford at the age of 2.5 to 3 years.

8-6-3. Breeding in and in

Breeding in and in undoubtedly tends to intensify any points, either good or bad possessed by the family. If there is a constitutional tendency toward disease, such a tendency will be increased by breeding together animals of a close degree of consanguinity; hence, under such circumstances in and in breeding would be extremely unwise, no matter what the tendency toward disease may be, and no matter what bad point may be constitutionally inherent in the family. Such a tendency or bad point will manifest itself in the progeny of closely related animals to a higher degree than in the parents.

On the other hand, we find in certain animals points of a very high degree of excellence, which it is extremely desirable to perpetuate and to fix in the breed; in no way can these ends be so safely, surely, and perfectly attained as by in and in breeding. It is through such breeding that most of our most highly improved breeds of domestic animals have been brought to their present degree of perfection. To judicious in and in breeding the Shorthorn owes to a large degree his perfection of form, early maturity, and tendency to fat and flesh production. To the same kind of breeding, though probably to a lesser degree, the Ayrshire cow owes its milk-giving propensity; and the Jersey cow's excellence of milk for butter, the Devon for work. Hereford for beef. There is one bad consequence, however, which is likely to follow in any case from long, continuous in breeding, namely, the diminution of sexual vigor in both males and females.

There is a tendency to sterility in males and barrenness in females. Thus, we see in the Shorthorn which perhaps has been more in bred than any other breed of cattle, many individuals of both sexes with comparatively little sexual vigor; not a few of the cows prove barren, some never breeding at all, while others produce very few calves. That fault in the Shorthorn may not, however, be altogether due to in and in breeding; it is perhaps partially due to the great propensity to fatten, which is characteristic of the breed. We must, therefore, conclude that while in and in breeding under judicious management may be productive of excellent results, it should always be done with the greatest caution. In unskillful hands, it is as likely to be productive of bad results as of good. Wherever animals of equal excellence, unrelated to each other, are obtainable it is better and probably safer to breed from them than to breed from individuals of a close degree of consanguinity.

8-6-4. Management of Breeding Cows

The first point to be considered is the condition into which the female should be brought before she receives the service of the male; in this regard, it may be confidently stated that to secure the best offspring, it is necessary that the cow be brought into a thoroughly good condition, vigorous, healthy, and in average flesh not fat. Fatness, indeed, is an obstacle to conception; excessively fatty females frequently, though receiving the service of the male, fail to conceive. By liberal feeding and general good care, the skillful breeder therefore first brings all breeding animal into the condition just described. The cow, when not pregnant, usually comes into heat about once every 3 weeks, though there are individual differences; only when she is in heat will she receive the male. Heat is indicated in various ways. If the cow is running at liberty with other cattle, she will be constantly attempting to mount them and will allow any of them to mount her; if she is confined alone, she will indicate heat by general mulishness and perhaps lowing (crying). Another indication easily recognized by the experienced man is the peculiar appearance of the external genital organ the vulva, there is always to be noticed a slight discharge of a peculiar watery whitish fluid.

When a cow is in heat, one complete service by a bull is sufficient, if both are vigorous and in good condition to secure impregnation. Before the service, the cow should be allowed to see the bull fairly and fully. It is generally best to confine the cow in a rather small but smooth, hard yard, and to have an attendant lead in the bull, and then lead him out again after the service is completed. If it is important for any reason to utilize the sexual capacity of the bull to the utmost extent, one service to each cow is all that is allowed; however, if the number of cows to be served is not great, it is very common to allow the bull to serve each cow 2 or 3 times, for greater certainty. During the time of heat, it is best that a cow be confined by herself, because if allowed to run freely with other cattle, she will permit them to ride her, which not infrequently causes considerable injury.

It is especially important, in the breeding of pure stock of any kind, that the cows be carefully secluded while in heat from animals of a markedly different appearance. Since the cow at that time is peculiarly susceptible to various influences, not infrequently the progeny of a male and female of a given breed has been of an appearance altogether different from either parent, a marked resemblance being evident between it and some animal which the cow had seen before, or just after conception.

(1) Influence of Preceding Impregnation upon Subsequent One; This may be clearly shown from several instances. If we cross the cow of a certain breed with a certain bull, say a Shorthorn with red hair, then a certain progeny resembling that bull will be the result. Now, if we later wish to attempt to obtain progeny of another kind, say, a Galloway with black hair, we should cross that cow with a Galloway bull. Here we will not obtain progeny of a black color, but more or less will be brought forth. Subsequent crossing will produce progeny more resembling the first bull than the second above mentioned.

(2) Influence upon the Sex of Offspring; It has been stated by some men that if in the first hour of heat a cow manifests, she is allowed to receive a male, she will produce a female, and if she is served during the last hour of heat, she will bring forth a male; however, these contentions are altogether unproven.

8-6-5. *Duration of Pregnancy*

The shortest period for a cow is 220 days, the longest 320 days, and the average about 280 days. In a French agricultural school, out of 1,062, 15 periods were shorter than 241 days, 52 were 241 to 270 days, 119 were 271 to 280 days, 230 were 281 to 290 days, 70 were 290 to 300 days, and 32 were longer than 301 days.

Altogether, 544 periods were 271 to 300 days, the average being 283 days. American Medical Science reported the results of numerous observations, the shortest period was 213 days and the longest 336 days; the average for female calves was 282 days, and for males was 288 days. In 764 observations made in England with Shorthorns, no live calf was produced before the 220th day or after the 313th day and all borne before the 242nd day died during the attempts to rear them; the average was 284 days. The majority of calves born after the 290 day were males. From these numerous observations, it then seems to be evident:
(1) that a calf born much under 240 days is not likely to live, and neither is one which is born after 310 days; and
(2) that the average for male calves is several days longer than for females.

Farmers ordinarily reckon the period of gestation as being 9 months, and you would see very near the average time. The farmer usually desires to have his cows all produce calves at the same definite time and except business of farm to produce and sell milk equally at all seasons. To breed the greatest possible number of calves from the stock, it is generally best to have the calves born rather early in spring, because at that season the cows feeding upon fresh grass will yield the maximum milk, while the tender, succulent grass also affords the best feed for the young calves when they are weaned. This being the case, June, July, and the early part of August will be the season when most cows should be served .

8-6-6. *Treatment of Cows during Pregnancy*

It being, of course, desirable that cow shall bring forth a large, well-nourished fetus, while at the same time she must usually continue, during the first 7 or 8 months of pregnancy, to yield a large quantity of milk the larger, the better, you will perceive that there must be a very extraordinary craft upon her system. She not only must support her own body, but nourish the fetus; and so, large quantities of nitrogen, fats, and phosphates in her milk. This being the case, it is evident that a cow under such conditions should receive the most liberal food and careful management, otherwise she will become greatly reduced in flesh, strength, and vigor, and her period of usefulness will be much shortened. Since the nourishment fetus is a great tax upon the system of the mother cow, especially during the last weeks of the period of gestation, it is usually best to cease the milking of the cow at least one month before the time for the birth of her young. This allows her the opportunity for recuperation, and her yield of milk after the birth of the calf will probably be more than sufficiently greater to compensate for the short period of idleness.

With cows of great milk-giving tendency, there is commonly some difficulty in "drying them off" as it is called, because it is the nature of such cows to convert most of their feed into milk. When it becomes desirable, therefore, to cease milking such a cow, it is best to slightly diminish the supply of feed for a short time, and especially to avoid those ingredients of feed most calculated to promote milk secretion. Then, for a time, it is necessary to milk the cow at gradually lengthening intervals; cows being milked twice a day must, for a time, be milked only once per day, and then once every 2 days for a shorter time, after which it is usually safe to cease milking altogether. However, the cow's udder must be watched for a time, and if it seems to become very full, it may be necessary once or twice to draw from it at least a portion of the milk it contains.

8-6-7. Care of Cows Immediately before and at Parturition

For several days preceding the expected parturition, the cow should receive careful attention. If she is in good condition, the amount of feed should be diminished somewhat and, especially if she is naturally a great milker, the feed which promote milk secretion should be avoided. The feed should be such as to keep the cow's bowels well open, for costiveness at this time may cause serious trouble.

An experienced manager of cattle can determine from the appearance of the cow almost the precise time at which she will birth her young. The approach of parturition is indicated by several signs which it is rather difficult to describe. Of a few of them, I may be able to give you a fair idea; usually of the first importance, and probably most noticeable of all, is the change in the udder as parturition approaches, it gradually increases in size, and usually just prior to the birth of the calf the udder and teats will be observed to become somewhat tense and firm, being fully distended with milk. Another sign much relied upon is the formation of a hollow just forward of the rump bone, near the roots of the tail. The vulva too changes in appearance, becoming apparently larger, fuller, and fresh in appearance. Another sign is in the belly, caused by the change in position of the fetus, but that is very difficult to describe.

It is best that a few days at least prior to parturition, the cow be accommodated in a roomy, clean, light and airy box stall; one which is 12 feet square is better than a smaller one, although a stall no greater than 9 feet square will answer the purpose tolerably well. If the season is such that the cow cannot run at pasture, she should be placed in such a box stall and kept there most of the time, although a few hours each day in an open, dry and sunny yard unless the weather is extremely severe will be beneficial. If the season is summer, the cow may be allowed to run in the pasture most of the time; there is no need to separate her from other animals of her own kind, because if she gives birth to a calf in the pasture, she will select a suitable secluded spot, and other horned cattle will not molest the mother or her young.

As the time for the birth of the calf approaches, a careful breeder maintains a close, though not obtrusive, watch. Cows at this time like seclusion, and it is therefore not good to obtrude oneself too much upon the notice of the animal. Still, watch must be kept in order that in cases of difficulty in the delivery of the young, the breeder may be ready to render her any assistance which may be necessary. The proper presentation of the calf is as follows: fore feet and legs, with the head lying between them; next, the back, next to the back of the mother; and lastly, the hind legs, tail, and feet. If a calf is thus presented, in the great majority of cases there will be no difficulty in the birth, but occasionally a calf is presented in some other manner, sometimes the forelegs without the head, the latter being turned back, sometimes the head without the legs, sometimes one leg missing, sometimes the hind legs, and at other times the rump and tail. Indeed, calves are presented in almost all conceivable positions, those which I have mentioned being common.

If a faulty presentation is observed before the calf has been so far crowded backward into the pelvic orifice as to make such treatment impossible, it is best to carefully introduce a well-oiled hand and arm, and crowd or push back the fetus and turn it into the proper position. First, one foreleg may be reached and brought forward; to it should be attached a strong, but rather small, rope. It may then be allowed to slip back, when the other leg may be reached and secured in the same way. Lastly, the head may be brought into the proper position, and then all may be gently drawn up at every effort of the mother toward expulsion of the fetus that is, with each contraction of the womb. It is necessary to use much gentleness and care, otherwise the delicate tissues of the walls of the vagina and uterus may be so injured as to cause serious trouble from inflammation or hemorrhage, or even the death of the mother.

It is useless, in most cases, to use force except during the mother's contractions, while it is not best to interfere too quickly, it being far better to give nature the full opportunity to work at delivering the calf. Still, the careful manager will, in all cases other than unusually protracted labor, early satisfy himself that the presentation is normal, because if it is not normal, the sooner an effort is made to turn the calf into such a position that it may be delivered, the better. On the other hand, if the presentation is normal, in the great majority of cases no assistance will be necessary. Sometimes, however, due to the small size or malformation of the mother, or the extraordinarily large size or malformation of the fetus, delivery is attended with considerable difficulty. In such cases, a little judicious assistance given before the mother has completely exhausted herself in her efforts to give birth to her calf will often save the lives of both mother and calf. In such cases, with both normal presentation and faulty presentation, it is soon discovered to be impossible to save the lives of both; in such cases, it becomes the question which it is best to save. This is a question, the answer to which must depend altogether upon the circumstances; sometimes it is obvious that the mother can be saved but the calf cannot be, while in other cases the opposite is true.

The calf being safely born, the mother usually rises to her feet almost immediately and begins to lap it vigorously with her tongue, over all parts of its body, cleaning it, freeing it from all adhering tissues, drying it, and warming it. The rough tongue of the mother serves these purposes better than artificial means, but in case the mother due to weakness consequent to protracted labor, or from any cause, fails to perform this service for her young, it is advisable to wash the calf quickly in tipped water and to rub it briskly until dry with flannel cloth. In ordinary cases, the calf will soon rise to its feet, probably if everything is all right within 30 minutes, and sucks milk from its mother. It should in all cases be allowed to do that, but if the calf from weakness is unable to rise, or from stupidity fails to take milk from its mother, it should be assisted.

For the calf, there is nothing which can take the place of that first milk from its own mother. The first milk is quite different in composition from ordinary milk, and has received the name "Colostrums"[10]. It is a thick yellowish fluid rich in albumen and salt, but containing little casein; fat and sugar are also in relatively smaller proportions than in ordinary milk. The nutritive ratio, that is, the ratio between the quantity of digestible protein and carbohydrates, is narrower, and its digestibility apparently greater. The colostrums seems to be particularly suited to the needs of the young calf, apparently being necessary also to assist in the evacuation of certain fetal matter contained in its stomach and intestines.

Within one week, usually, the cow's milk becomes like ordinary milk; that change takes place earlier in cows yielding much milk than in those yielding little. After the young calf has taken his fill, all the milk remaining in the udder should be carefully drawn, this being extremely important with cows yielding a large quantity of milk. If this precaution is not attended to, inflammation and caking of the udder will almost certainly set in, and may perhaps cause the loss of a portion, or even the whole of the udder, and in extreme cases even the life of the cow. After the udder has been milked dry, it is usually best to give the cow some warm drink, such as bran stirred in warm water. Some even allow the cow to drink milk which has been drawn from her udder, but that practice is hardly to be recommended.

Both cow and calf should then be left to themselves, and if all things are as they should be, the cow will soon lie down and the afterbirth, the placenta, will be expelled from her womb. That should take place within a few hours, but from various causes it is not infrequently retained longer. Most cows seem to be possessed by an abnormal appetite after giving birth, and if allowed to do so will eat the afterbirth. Some have claimed that this is due to an instinct of the animal, and that the eating of the afterbirth is probably beneficial, producing some medicinal effect. That theory is not, however, well supported; although no injurious results follow eating of the afterbirth, it is the practice of most careful breeders to watch for its expulsion and to immediately remove it. If the afterbirth is obstinately retained, as it frequently is especially if the calf is born somewhat prematurely, there are various means which may used for removing it. If it is allowed to remain, putrefaction will set in after a time; rotting away, the placenta will gradually come out piecemeal. That is very likely to prove injurious to the cow, the putrefying mass causes irritation, inflammation, more or less general fever, and consequently usually the loss of flesh and diminution in the yield of milk.

If the afterbirth is retained, it is the practice of some to introduce a well-oiled hand and arm, and to carefully to break its connection with the walls of the uterus, and thus remove it. Care must be taken not to work violently, or too rapidly, otherwise severe hemorrhage may follow. If this treatment is resorted to, it must be done early, before the walls of the vagina and uterus fully contract and, if possible, before putrefaction commences. If the skin of any part of the hand or arm is broken, blood poisoning is likely to follow, through absorption of putrefied matter into the circulation. In the case of retention of the afterbirth, at least a portion of it will protrude; it is the practice of some to attach a small weight to the protruding portion, the constant tugging of the weight being of considerable assistance in the expulsion of the afterbirth.

As already stated, the first milk of the cow differs greatly from ordinary milk. It therefore becomes a question, what disposition can we make of it? Such milk is certainly not productive of imperious results upon people drinking it, yet to most people it is not agreeable; hence, it is not the practice of most to make use of such milk. It is commonly fed to swine for the first 48 to 72 hours; after that, though still not exactly like ordinary milk, it so nearly approaches ordinary milk that it is commonly used. For several days subsequent to parturition, the greatest care must be taken not to expose the cow to cold or wet, for inflammation of the udder, milk fever, and perhaps death might follow such exposure.

For the first few days, all drink given to the cow should be blood warm; do not be in too great in hurry to feed it, because for 6 to 8 hours after calving, the cow will need no food. For several days, the feed should be light, rather small in quantity, and of a kind calculated to promote the greatest yield of milk. Such precautions are especially necessary with all cows which yield a large amount of milk, or are in good flesh. On the other hand, poor bony animals, or cows which naturally yield little milk, may with advantage receive the full amount of feed almost from the time of parturition. As soon as any tendency toward inflammation and caking of the udder is perceived to have passed, the cow may receive full rations.

It is customary to allow the cow and calf to remain together for 3 or 4 days, the calf in such case, of course, sucking milk at will. It is necessary, however, that the manager of the cow draw all the milk remaining in the cow's udder at least every night and morning; in case there is any tendency toward inflammation, the milk should be drawn at noon also. One practice to be generally recommended, in all cases where there is a tendency toward caking of the udder, is to draw a large portion of the milk before the calf is allowed to suck; the hungry calf will then, in an effort to satisfy its appetite, remove very completely all that remains, and his vigorous sucking of milk from the udder and bunting it with his nose will tend to reduce the caking. If there is a tendency toward inflammation, frequent bathing of the udder with cold water, followed by vigorous rubbing until dry, often gives beneficial results; considerable manipulation of the udder is also recommended. [p. 1145]

8-7. Rearing Calves

There are many different ways of rearing calves. A very simple way, provided milk is not very valuable and also that the greatest possible growth of the young calves is realized, is to allow the young calves to run at liberty with the mother. However, that is the most uneconomical method, and the reasons why it is not followed are:

(1) The calf takes milk irregularly, and not always the whole amount in the udder, so the milk yielded by the cow is less.

(2) A calf brought up in this manner will, even when grown, often prove troublesome by sucking the udders of cows with which it is allowed to run, and does not soon forget the habit of taking milk from the udder of its mother.

(3) If the calf is allowed to stay with the cow and to suck her udder, she will not come into heat as often as if she were separated from her young, with the consequence that the number of calves that can be bred from her is necessarily fewer.

(4) Milk is usually valuable, and is desired for other purposes.

(5) The longer a calf is allowed to remain with its mother, the more fond she becomes of it; when the young calf must finally be weaned, the cow often refuses to give milk, and consequently the amount of milk she gives is much lessened.

A cow permitted to remain with her calf is not likely to come into heat within about 3 months after parturition, but she is separated from her calf she will come into heat within 2 weeks. In view of the disadvantages enumerated above, it is not customary to rear calves with cows. The calves are usually separated from their mothers when 2 or 3 days old, but are sometimes allowed to take milk from their mothers twice a day, morning and night; any remaining milk should be drawn by hand. Since it is very difficult to draw out all the milk remaining completely by hand, it is best to first allow the calf to suck milk from the cow's udder, then draw out milk from the cow's udder by hand, and then to allow the calf to drink completely what is left for him.

If a calf is extremely weak, milk may be given 3 times a day; the milk should be always fresh and warm. How do we teach a calf to drink milk? Fresh, warm milk drawn from the calf's mother should be given. Place it in a thin, shallow tin vessel, and introduce one finger into the calf's mouth. Then, with the other hand, bring the vessel to its mouth and remove the finger. The interval between morning and night should be equal, usually, 6 o'clock in the morning and 6 o'clock in the evening.

The following table gives the quantity of fresh milk necessary to nurse healthy, rapid growth in a young calf, together with the constituents of such milk.

By "nutritive ratio" is meant the ratio which exists between the nitrogenous, and the fatty and carbonaceous substances. This ratio must be adhered to for good, thrifty growth. The figures given in the preceding table must be regarded as being amply sufficient to promote very rapid growth; there is nothing which can be substituted for such a ratio which

Consumption per Day, and per 100 pounds Live Weight

Fresh milk	6.20 pounds
Total Dry Matter...	1.93 pounds
Protein	0.49 pounds
Fat	0.47 pounds
Carbohydrate (milk sugar)	0.84 pounds
Ash	0.13 pounds
Nutritive Ratio...	1:4
Average Gain per day	1.85 pounds

promotes healthier or better growth. Yet, as I have stated in many cases, to economize the amount of milk given by the cow, the experience of many person shows conclusively that it is perfectly possible to rear a young calf, keeping in view a thrifty growth, with very much less milk and even without milk at all, though the latter is seldom advisable. This principle should always be kept in mind; any substitute for milk should approach it as nearly as possible in composition, should be an easily digestible liquid if possible, and should be warm[11].

The nutritive ratio is the ratio between digestible protein and digestible carbohydrates. There are hydrocarbons. This must be reduced to an equivalent amount of carbohydrates by multiplying by 2.5; thus proteins are 49,

$$47 \times 2.5 = 117 + 84 = 201; \quad 49:201 = 1:4.$$

Hints for substitution; Sweet English hay steeped in hot water, and the resulting infusion with tea with the addition perhaps of some milk, and meal such as cottonseed meal or corn meal, is given warm to the young calf. Among all substitutes, cottonseed cake is best; it contains oil and, frequently mixed with skimmed milk, it is given to calves. Next to cottonseed meal, the best substitute or partial substitute is corn meal especially rich in fat. There are various substances which, combined so as to approximate cow's milk, can be fed to calves, but in ordinary practice it is best to feed calves with milk alone, and that is economical. A calf which we intend to make a beef animal may be fed abundantly with many substances, but a calf of a milking breed must be fed with caution.

If fresh milk is particularly valuable for the manufacture of butter or cheese, a calf may with ease be raised on very little of it. It is the practice of many to set new milk for 12 to 24 hours, and at the end of that time to remove the cream which has risen to the surface; then, after having warmed the skimmed milk, to feed it to calves. In that case, as already stated, it is good, as a substitute for the fat which has thus been removed, to mix with it some other fatty substance, such as cottonseed meal or corn meal. However, if you intend to substitute these foods for milk, you must take greater care than when you give a calf its natural food, in order to make a thrifty, healthy growth. The vessel of milk should always be kept scrupulously clean; it is best to wash it after every feeding, because old milk adheres to the vessel, and fermentation is very likely to ensue, which will cause great injury to the calf.

Since neat cattle are generally not kept for work, exercise must be regarded as being of less importance than in the case of working stock; yet, it is impossible for a calf to remain perfectly thrifty without exercise; therefore, it will be best to allow the calf to run in a small, smooth yard for a limited time per day. However, it is dangerous to let a calf which has reached the age of a few weeks out into the open air with full liberty, since at first they are foolish, quite insensible of danger, and they are apt to run at random, striking every object. That being the case, it is best to make the calf gradually accustomed to take exercise in the open air from birth, but never allowing liberty for a long time.

As a calf becomes older, we must change its feed, from liquid into solid feed. The change should not be made too suddenly, for the paunch of the calf is not yet fully developed at this time, and it cannot digest solid food as easily as mature cattle can. The change must be made gradually, after the calf has reached the age of 10 to 12 weeks, by gradually lessening the supply of liquid feed and at the same time increasing the proportion of solid feed. At first, the nutritive ratio should be kept close to that of milk, but be gradually widened; when at last the calf is deprived of milk altogether, the nutritive ratio should still be quite close to that of milk, say 1:5 or 1:6. The feed should be abundant in amount, in order to keep the animal in thrifty, rapid growth. In summer, calves may best be kept in a pasture affording plenty of rich, sweet grass, but if the most rapid growth is desired, some grain should be added to the calf's diet. [p. 1151]

8-8. Care of Neat Cattle
8-8-1. Stabling, Bedding and Carding
1) Stabling

In a climate such as this, it is necessary in winter to keep cattle under shelter of some sort. It becomes important, then, to consider what kind of shelter is necessary and most economical. As you very well know, it is essential to the health of the animal body that it be maintained at a certain temperature. Whatever the temperature of the surrounding air may be, the temperature of the body is always about the same, and always higher in ordinary cases than the temperature of the air. Higher temperature is maintained by the oxidation in the body of certain carbonaceous ingredients of the food; the lower the temperature of the surrounding air, the greater must be the consumption of such carbonaceous ingredients, in order to maintain the body at a proper temperature. It will at once be perceived from this that, to keep animal in a place which is extremely cold must cause the consumption of feed to maintain body heat, which at a higher temperature might serve some other purpose, for example, the formation of fat.

It therefore follows that to keep cattle reasonably warm in a comfortable stable is usually wiser than to keep them warm at the expense of the consumption of a greater amount of feed. While warmth in the stable is, therefore, important, there are certain other points which are equally important. The construction of the stable must be such as to afford light and ventilation, and it should be dry. There are some things which are essential to the health of animals, and must be secured whatever the method of construction, which may be as various as the circumstances of individuals are various. Sometimes, each animal is kept in a box stable large enough to move freely, but that is too expensive. The keeping of several animals in one large room is prone to many objections, the principal of which are:
(1) that the horns of the animal strike one another
(2) that the animal are liable to quarrel over food, the strongest always taking the best

With young calves whose horns are not yet fully developed, the above is a very good method. Cattle are, however, generally best kept by tying them with ropes, chains, yokes, or with stanchions. Tie each animal by itself in a space the dimensions of which vary according to the size of the animal. Common widths are $3\frac{1}{2}$ to 4 feet, and lengths from 4 to 7 feet. Of all methods of tying, I think a rope is best, on account of the fact that in case of emergency, such as fire, they are able to escape without the assistance of man. However, with yoke, chain, or stanchion, they are liable to danger if assistance is given.

2) Bedding

It is quite important, in caring for cattle, to furnish them with a soft, comfortable bed. This is important in the first place for the absorption of liquid excrement, but is of greater importance to secure the comfort and warmth of the animals. Materials which may be used for bedding are various, but among those easily available here may be mentioned the straw of grains, such as oats, wheat, barley, rye, *etc.*, the coarse wild grass of swamps and low ground, the dry leaves of trees, fine, dry muck or peat, sand, loam, and sometimes sawdust. Among these various materials, straw and coarse hay will perhaps be most commonly used; both are excellent so far as the comfort of the animals is concerned, and as absorbents. Straw is perhaps preferable, since it is less likely to contain the seeds of weeds which may later cause problems when manure is applied to fields. Both straw and hay contain a considerable amount of nitrogen, potash, and phosphoric acid, which will add to the value of the manure.

The dry leaves of trees, which may sometimes be collected in sufficient quantity for bedding, are better in all respects than straw or hay. They make a comfortable bed, and are about equally serviceable as absorbents, while they contain a much higher proportion of elements valuable as plant food. Straw and hay may be improved by cutting them into fine pieces, which make them very convenient in subsequent handling, but it is usually too expensive to cut them. Fine, dry muck is excellent as an absorbent, but cattle bedded in it are likely to become extremely dirty, the muck being moistened with urine, and adhering to the legs and body of the animals; consequently, its use, except in small quantities in the gutter behind the animal, is not recommended.

Fine, dry sand and leaves which do not contain a large percentage of organic matter are excellent for bedding; such materials perfectly absorb liquid excrements, and moreover are excellent to secure cleanliness, freedom from vermin such as lice, and general health. Sawdust is an excellent absorbent, and makes a sufficiently soft and comfortable bed; it is also one of the cleanest substances that can be used for bedding. Cattle abundantly bedded with sawdust will probably keep cleaner than if bedded with any other material. Sawdust, however, does not contain any appreciable amount of substances valuable as plant food; it consists, as you know, almost entirely of cellulose, but is very convenient of handing in the subsequent manuring.

3) Carding

"Carding" is the term applied to the operation of combing the hair of cattle, which is usually performed by the use of a little implement known as a "card". A brush, also, is sometimes though comparatively seldom used. It is the general opinion that a thorough carding at least once a day is productive of excellent results. I myself have heard an intelligent cattle-breeder state that he considered a good carding once a day equivalent to the feeding of one quart of corn meal. That statement is probably somewhat extravagant; indeed, it must be stated on the other side of the question that in one case where an experiment was conducted to determine the effect of carding, it seemed injurious that experiment was done at the Maine State Agricultural College. At any rate, it is certain that carding very much improves the appearance of cattle, and it is altogether probable also, from the evident pleasure the animals take in the operation, and their consequent greater comfort and satisfaction, that it is beneficial.

4) Exercise

Neat cattle being generally not especially valued for great strength or fine muscular development, exercise may be regarded as being less important than with horses, where muscular development and strength are everything. Notwithstanding this, however, a certain amount of exercise is evidently healthy, and should therefore in all cases be allowed; 1 or 2 hours per day, in an open yard which is dry and sunny, will be sufficient in most cases.

8-8-2. *Special Care of Breeding Bulls*

As already pointed out, it is necessary, in order to control breeding, to confine bulls, in most cases, altogether apart from other animals. This necessitates certain differences in management. In the first place, it is usually necessary in order to be able to control a bull, to put a ring through his nose. The best rings are made of brass or copper; such a ring should be put into the bull's nose when he is 1 to $1\frac{1}{2}$ year old. To inset the ring, a hole is made through the thin part of the partition between the nostrils. The bull's head having first been securely fastened, so as to prevent as far as possible all movement, the hole may be made either with a small, sharp knife or a hollow punch or chisel of the same thickness as the ring. If a knife is used, a simple slit of sufficient length to allow introduction of the ring is made. If a hollow punch is used, a firm, round block of wood is placed against the partition upon one side, and the punch against the opposite side in the proper place for the hole, whereupon a sharp blow upon the punch with a hammer will cut out a hole of the proper size. Until the wounded tissues have healed around the ring, the bull should not be led or tied by means of a rope attached to the ring.

It is not ordinarily possible to allow bulls to take exercise in an open yard; this being the case, it is best to give a bull a considerable amount of walking exercise every day. If it is convenient, he may be moderately worked to advantage. Moderate work only serves to increase his vigor and usefulness. Even under the best management, highly bred and well-fed bulls frequently become cross and treacherous, so much so that it is dangerous to care for them. In order to, as far as possible, avoid the danger of being injured by a bull, it is always best to lead him by means of a strong staff about 6 to 7 feet long. To one end of this staff should be fastened 6 to 8 inches of chain, for convenience in attaching it to the ring. In spite of the best precautions, as already pointed out, bulls sometimes become cross, but it is nevertheless true that such crossness may be caused by some neglect in management.

Anything like playing with a young bull, such as tickling his tail, hair, or ears, or tickling any part of his body, is almost certain to provoke crossness on the part of the bull. Anyone who approaches a bull and shows nervousness or fear is very likely to be injured. Anyone taking care of bulls should be perfectly calm, steady, and deliberate in his movements, and kind, firm, and fearless in demeanor. Such a person will seldom be injured by a bull. It is regarded by some, from experience, that even the work which the bull undertakes, causes crossness and makes him treacherous. Hence, a contrivance has been made in which a bull is attached to a movable rotating hub upon a vertical post; connected to that, he can walk round and round, making circles about the vertical post. It is only necessary that he be attended by a man or boy to drive him if he stops. Indeed, moderate exercise is important for a bull, to make him useful for a long time. The work should be moderate, at regular intervals, and at the same time he should be fed with regular amounts of food at regular intervals. That being done, he or she will be productive either in flesh or milk. [p. 1160]

8-9. Sheep
8-9-1. Description

The sheep is an animal belonging to the same family as the ox, the order Artiodaetyla, *Ruminantia Carvicorniae*. Its dental formula is the same as ox. Males in the wild state always have horns, and females also usually do. The horns have a central core which rises as an exotic from the frontal bone, and each is entirely coated with horn. The horns of sheep are usually beautifully twisted. In most domesticated breeds, the females are hornless, and in many breeds males are also hornless. There are 4 principal wild species known at present, namely, *Ovis Argali, O. Ammotragus,* and *O. Mouflon*[12]. The 1st of these species as found in Asia, in the Himalaya Mountains; the 2nd, the bearded Argali, is found in Egypt; the 3rd is found in the Rocky Mountains in the United States and known as the Rocky Mountain Sheep; and the 4th is found principally in the Caucasus Mountains.

The domesticated sheep, *O. Aries*, differs considerably from any of the wild species. Those differences are principally in the same areas as already pointed out in the case of the ox, namely, the comparatively small size of bones, the complete loss of bones, or great lessrounders in the size of horns, and further in the case of sheep, a great change in the character of the covering of the body. In most wild species, there is a considerable amount of hair mixed with wool[13]. In most highly improved domesticated breeds, the hair is now confined almost altogether to the nose, ears, and lower part of the legs.

From what I have said concerning the native habitat of wild species, you must observe that sheep are naturally adapted to living at high elevations, in mountains and comparatively dry regions. As might be expected, therefore, it is found that sheep always thrive best upon well drained, dry soil and in climates which are not excessively moist; the temperature does not make so much difference. Sheep are hardy, and able because of their woolly covering to withstand great exposure to cold. Also, however, due to the nature of their covering, they are very poorly adapted to withstand expose long continued rained.

8-9-2. Hokkaido as a Place for Sheep Husbandry

As far as soil is concerned, Hokkaido seems to be very well suited for sheep. Here there are good valleys and hills provided with fertility, making it quite possible to furnish plenty of rich grass (the wild grass is not considered good enough to use; grasses must be introduced from foreign countries). Since winters are long and the climate is generally moist, it requires great expenditure of money to protect sheep from those. Another consideration is that here wild beasts, such as wolves, bears, and dogs are serious obstacles to sheep husbandry. Instances are frequent in which 20% or even 50% of sheep were killed, even by a pair of dogs, within a few minutes. To protect sheep against that, we must build high, strong fences to prevent the incursion of beasts. Notwithstanding this, sheep kept at pasture have in Sapporo sufficiently proven that they can thrive very well, strong, and vigorously; moreover, the increase of lambs is estimated at about 100% yearly.

Is sheep husbandry profitable, then? Two objectives must be considered with regard to this question, wool and mutton.

(1) Wool, as we know, is very little liable to be wasted when made into cloth; no portion of wool may be regarded as useless when some remains after it has been used. Wool, moreover, may be pressed into a very small volume occupying a very small space; hence, its transport is very cheap, and also it can be treated with little care. From these facts, it is plain that we would be obliged to import wool from a foreign country such as Australia.

(2) Mutton; which meat, beef, pork, or mutton, can be cheaply and profitably raised here? Of course pork, then beef, and lastly sheep. The reasons are quite obvious.

8-9-3. Breeds of Sheep

(1) Merino Sheep: This breed is of Spanish origin, and many years ago there were to be found in Spain numerous flocks of these sheep of a higher degree of excellence than in any other part of the world, for wool production. At that time, the Spanish nation was prominent among the powerful nations of the earth, but since that time the relative power and intelligence of the Spanish nation has, as you know, greatly diminished. Merino sheep were, at an early

[参考図]

Merino Sheep (from Brooks, *Agriculture*)

time, imported into various other countries, France, England, America, and Australia. In France, the Merinos were considerably modified and increased in size, but on the whole not much improved. In England, Merinos never have gained much, the climate and soil of that country apparently being not well suited for them.

In certain parts of the U.S., notably in Vermont, the Merinos were much improved, the size of the body was increased, the length of the wool and the quantity thereof was much increased, while in fineness there was not much loss. From Vermont, Merinos have been sent in large numbers for breeding purposes to almost all parts of the U.S., and to Australia and Southern Africa. Today, in many parts of the U.S., as well as in Australia, much better Merinos can be found than in Spain. As evidence of this, I may state the following: at the World's Fair held in Germany in 1837, Merinos from Vermont took the highest premiums; also, at the World's Fair in Philadelphia in 1876, Australian wool was judged to be of the highest excellence.

In size, the Merino even in America, where as I have told you, their size has been much increased, they are still small, the live weights seldom exceed 100 pounds. Their legs are comparatively short, and the body short, rather round, and compact. In color, the Merinos appear to be very dark, almost black, this notwithstanding the fact that their wool is in reality white. The dark appearance is due to the presence of a great quantity of a substance known as "yolk". That substance, though not naturally black, collects a great quantity of dust and dirt; all of that hardening at the surface of the wool gives it the dark appearance. The face of the Merino is white, but there is much more wool around the head and face than in any other breed. In the best Merinos, the wool covers all except a small part of the nose; sometimes, indeed, it is so abundant upon the cheeks, forehead, and face as to interfere with the sight of the animal. This, of course, is not desirable, as there should be a clear space around the eyes sufficient to allow clear vision. The legs of Merinos, too, are much more completely covered with wool than any other breed. In the best Merinos, the wool extends as far, or almost as far, as the hoofs, while in other breeds there is seldom any wool below the knees or hocks, and usually it terminates some distance above them.

The abundance of wool around the head and legs is esteemed as an important point in Merinos, indicating a wool-bearing tendency not, of course, because of any value of the wool on the head or legs. Such wool is always short and inferior in quality. The skin of the Merino should have a deep rosy color and should be rather thin. Formerly, the skin, especially of the neck and forepart of the body of males, was very loose, lying in great folds or wrinkles extending over almost the entire body, and that was esteemed a good point. At present, the tendency with most breeders is against excessive wrinkling, but still in most Merinos that tendency is more or less evident. Thus, among Merinos in Vermont, there were animals named "Old Wrinkly", "Little Wrinkly", *etc*. If a sheep has a great quantity of wrinkles, the wool is of uneven quality, and that is one of the principal reasons for reaction against wrinkling.

The Merino is preeminently a fine-wooled sheep, and the fleece should combine the greatest possible length and thickness with the requisite fineness. American Merinos produce a wool perhaps a trifle coarser than the original Spanish

Merinos, but the length of the wool and the quantity produced per animal is much greater. Yet the length of the wool of Merinos never approaches that of some other breeds, seldom being more than 3 inches long. However, on account of its great thickness and evenness over the entire body, the weight of the Merino fleece is very great. Rams frequently yield more than 20 pounds of wool, and ewes 12 to 15 pounds. Those yields are, however, far above the average. The greater weight of the ram's fleece is in part due to the greater size of the animal, but in large part also to the far greater quantity of oil or "yolk" which it contains.

The wool of Merinos should always be fine, the finer it is, the more valuable the wool. It should also be, as nearly as possible, of even quality over all parts of the body. The wool on the belly and legs of other breeds is almost invariably somewhat coarse and hairlike, but on Merinos its fine woolly character should be well maintained. Each individual fiber of the wool should be perfectly round and true, that is, of precisely equal diameter at all points. Inequality of the diameter of fibers greatly lessens the value of wool for the production of the finest fabrics. Wool is considered "round" if it is strong and perfect throughout its entire length.

Regarding the "yolk" in the Merino fleece, I must give some explanation. The "yolk" is a substance which, in appearance and physical characteristics, resembles oil, so much so that it is often spoken of as "oil". If you thrust your hand into a Merino's wool, when it is withdrawn it will appear to be oily. Careful examination of the yolk, however, shows that it does not consist altogether of oil, but is somewhat soap-like in character. It is secreted by the skin, and its function is to lubricate the fibers of the wool, thus preventing undue friction between them; also, perhaps, it enables the fleece to repel water, and thus to keep the animal dry. The yolk is usually yellowish in appearance.

Merino sheep grow slowly and do not reach maturity until 2 or 3 years of age. By this, I do not mean perfect maturity. The Merino is strong, hardy, and better able than most other breeds of sheep to withstand considerable exposure, the animal being small, lightweight, and very active. These sheep can subsist in rougher and more mountainous district and upon scantier pastures than most other breeds. They are kept exclusively for wool; though their mutton is of good quality, it is small in amount, and the animal grows slowly. Consequently, for mutton the Merino is of no importance. Merino sheep are generally spoken of as "shy breeders". The meaning of this expression is that they are uncertain breeders; the percentage of increase from a given number of breeding ewes is commonly less than with most other breeds. The Merino ewe produces one lamb at a time, but twins are not common. The Merino ewe is usually able to support only one lamb.

(2) Leicester Sheep: This is a breed of English origin classed among the middle wooled sheep of large size and fine form. Among sheep, it is what the Shorthorn is among cattle, in size, early maturity, and aptitude for fattening. It is not, however, very hardy and is generally considered to require more careful management than other breeds. Its average weight at 22 months is said to be about 100 pounds, but mature animals have occasionally reached 300 pounds. Its fleece is rather long, though not the longest, and of medium fineness. The average weight of the fleece is 7 to 8 pounds. The Leicester sheep is said to be much improved by crossing with another breed, mainly, the Cotswold. In quality, the mutton of Leicesters is only medium; the proportion of fat is commonly large, and is as with

the Shorthorns among cattle usually deposited in large quantities on the outside of the carcass. Wherever high feeding is desirable, or in pastures of fair quality, the Leicester may be well chosen, but for rough treatment or in ordinary pastures, this breed is not desirable.

(3) Cotswold Sheep: This is an English breed of the so-called "long-wooled" class, with large size and fine form. Its average weight at 14 months is 100 to 120 pounds. Its face and legs are white. Both males and females are hornless. The fleece is rather light in color, and weighs 7 to 8 pounds. The wool is 6 to 8 inches in length, and is rather coarse, being valuable for manufacturing coarser fabrics, such as carpets. This breed is tolerably hardy, but on account of its large size requires rich pasture and abundant feeding. The ewes are good breeders, not infrequently producing 2 lambs at a birth.

(4) Lincolnshire Sheep: This is an English breed of long-wooled sheep, of very large size. Individuals at the age of 3 years have been known to weigh as much as 381 pounds; that, however, is far above the average. In general form, this breed of sheep much resembles the Cotswold. The hair on the nose, face, and legs is white, the wool is white. Both males and females are hornless. The fleece is long and heavy, weighing 8 to 15 pounds on the average, up to as much as 20 pounds in extraordinary cases. The length of the fibers is frequently from 10 to 15 inches. The Lincolnshire sheep are valued largely for their mutton, but also to a considerable extent for their unusually large wool, which is extremely useful for a certain kind of fabric. These sheep are tolerably hardy and prolific.

(5) Southdown Sheep: This is also a breed of English origin, of medium size, with very round and compact bodies, which are active and hardy. The hair on the face, nose, and legs is usually dark in color, sometimes almost black. The wool is usually white, but occasionally a black or brown animal is born. Both males and females are hornless. The Southdowns are among sheep much what the Devons are among cattle, especially as regards their proportional size, quality of meat, and activity. The fleece of Southdowns is of medium length, fineness, and weight from 5 to 7 pounds being the average. The mutton of the Southdown is esteemed to be superior to that of any other breed of sheep; the mixture of the fat with the lean, and the flavor, being regarded as usually fine. Southdowns are probably the most prolific among all breeds of sheep. Among well-kept mature ewes, 2 lambs at a birth is almost unexceptional, and occasionally 3 lambs at a birth are brought forth. As many as 3 lambs is not, however, desirable because the ewe has only 2 teats and never furnishes milk sufficient for 3 lambs. Southdowns are hardy and well able on account of their medium size, activity, and vigor to subsist on rather slight pasture. Ordinarily, the Southdown weighs about 100 pounds, but some weigh 150 pounds or more.

8-9-4. Breeding of Sheep
1) Proper Age of Breeding Sheep

This differs according to the breed. The Merino does not reach sufficient maturity for breeding before the age of about 2 years. The ewes are, therefore, usually first served by a ram when they are about 19 months old. Rams may first be allowed to serve a few ewes at the same age or size, but it is not best to use Merino rams much until they are $2\frac{1}{2}$ years old.

Southdowns are often allowed to bring down when 1 year old. The rams of this breeds, too, are occasionally used before they are 1 year of age, but much use is not desirable until the 2nd year, when the rams will be in the breeding season 19 to 20 months old. If Southdowns are always well kept, fed, and grown, there is no objection to permit service in the 1st year, but if they are being given ordinary feed, it is probably best to wait until the following years. Most of the large breeds, namely the Lincolnshire, Leicester, and Cotswold, will breed the 1st year, but in most cases it is best to wait until the 2nd year. Both rams and ewes are usually in the most perfect vigor from about 3 years of age to 6 or 7. Seldom it is profitable to keep breeding sheep beyond 10 years of age. By that age, the animal's teeth have usually deteriorated, and they are unable to perfectly masticate the feed. It is always best to fatten sheep before their teeth become poor, because fattening them afterwards is well nigh impossible. A vigorous ram, if allowed to run with the ewes, may probably serve 40 to 50 in a season, provided he is well fed. However, those numbers are probably too great for the best of the ram, 20 to 30 ewes for 1 ram are sufficient in most cases. If, on the other hand, the ram is kept apart from the ewes and allowed to serve each only once, the optimum number of ewes becomes considerably greater. Undoubtedly, a vigorous ram may beget in a single season 100 or more strong lambs.

2) Selection of Breeding Stock

Upon this subject it is necessary to add little to what I have said about cattle; the same conditions are to be sought in the case of sheep. It is especially desirable for the latter to select the ram with utmost care, since one male is capable of begetting such a large number of lambs.

3) Management of Rams in Breeding

It is a very common practice among sheepbreeders to allow rams to run with ewes, but more intelligent and careful breeders wish more perfectly to control breeding, or perhaps more perfectly to utilize the vigor of sexually valuable rams; hence, they keep the ram in confinement and bring to him those ewes which are to be served, allowing only a single service to each. The ewes are invariably served at such a season that the young will be brought forth the following spring. It is true that by high feeding they may be made to come into heat earlier than usual. October and November are the months in which the ewes are generally put to the rams. Heat is not displayed by ewes as it is by cows, provided rams are kept apart from them. It is thus necessary to resort to some method of ascertaining when the ewes come into heat. Formerly, a common method of ascertaining heat was to cover a ram's belly with a leather apron so that he cannot serve the ewes, and to allow him to run with the females. Then, any ewe in heat will permit the ram to mount her. By that means, those ewes which come into heat were identified and separated from the flock. Another method was to select those ewes which it was intended to have served by a particular ram, separate them from the flock and keep them in a yard, and then to allow the ram into the yard.

4) Treatment of Rams and Ewes during Impregnation

If large, strong, and vigorous lambs are desired, both rams and ewes must be first brought into fine, healthy condition. A poor ewe is likely to produce weak, poor young, and will seldom produce more than one lamb. Usually, single services are sufficient to ensure impregnation, but since that is not always the case, it is the practice of careful breeders to admit a ram into a flock of ewes already

378   Chapter 8   Stock-Farming

served; under such circumstances, those ewes which have failed to conceive the first time will be served again. This is to secure a large percentage of increase.
5) Period of Gestation

The average period of gestation for sheep is 152 days, but there may be a variation of 1 or 2 weeks either way. Males are generally retained or carried longer than females. "Five months" is the common expression.
6) Food and Treatment of Ewes during Pregnancy

During pregnancy, the ewe must be especially well fed if large, strong lambs are desired. It is best to feed so as to bring the ewe into good condition. The giving of milk to nurse her lambs is always a great tax upon her system. The ewe almost always becomes poor, converting much of her previously formed flesh into milk. In order, therefore, that the ewe may not become excessively poor and weak, it is best that she be made quite fat before the birth of a lamb. It must be remembered, however, that excessive fatness is likely to cause a difficult parturition.
7) Treatment of Ewes at Parturition

It is unnecessary that ewes be separated from each other at the time of parturition. One ewe will never offer the slightest violence which would be likely to prove injurious to the lambs of another. Breeding ewes are, therefore, usually kept in dry, spacious pens or yards; in this climate, of course, under shelter. For one small breeding Merino ewe, a space of 10 to 11 square feet is deemed sufficient; for sheep of the largest size, twice as much space is required. During the lambing season, the shepherd must exercise care and watchfulness if he would prevent loss. In the delivery of their lambs, ewes seldom meet with difficulty, though, as with cows, there are occasional false presentations. The proper presentation is the same as described for calves. In case of a false presentation, the shepherd may assist in the delivery of the lamb in a manner similar to that recommended for a cow.

After the birth of the lamb, the ewe soon rises to her feet and vigorously laps the lamb with her tongue; the lamb will soon struggle to his feet and find the udder, and sucking the milk from it will become warm and strengthened, so that no further trouble is anticipated. In cases, however, where the ewe gives birth to 2 lambs, it frequently happens that the ewe will owen the firstborn such lamb strays away from its mother and rubs itself against other ewes, bringing the peculiar smell by which its mother is able to recognize it. The mother, after delivering the second lamb, soon rises to her feet and cares for that young, but would not acknowledge the legitimacy of the firstborn when it is brought to her. The shepherd is much troubled by that, because the firstborn lamb is the strongest and best of the two. The refusal to recognize their own young is more often met with among young ewes than among old ewes.

Now, the question here arises whether the shepherd can do something to make the ewe acknowledge the forgotten lamb? Yes, much can possibly be done in that regard. The following are some of the methods by which that objective might be attained. Sometimes all that is necessary is to bring the firstborn to her while the young strayed away from its mother, care having been taken not to allow it to rub against other ewes, and also not to handle it much, thus retaining the peculiar smell; under such circumstances, she will at once acknowledge her lamb. But this result is not always obtainable; if not, then the breeder must resort to some other means.

In obstinate cases, the mother is taken with her twin lambs into a small (so that she may always be in contact with her young) and somewhat dark (so that she may not see clearly) pen separated from other sheep. In such cases, the shepherd must see to it that the lambs are suckled several times (6 to 8 times) a day. During the night, too, the lambs should not go without food; they must be given milk at about 10 or 11 o'clock at night and again early in the morning, in the cold season, at 3 or 4 o'clock in the morning. Under such treatment, the mother will recognize her lamb in a day or two, or in extreme cases 4 or 5 weeks later. Sometimes less extreme measures are successful. The mother is placed in the general pen and forced to acknowledge the lamb by disciplinary measures, by whipping her cruelly, for example. At first, the mother recognizes her young by smell, and afterward by voice; the lamb, however, apparently seems unable to recognize its mother at first.

Sometimes it may happen that a mother loses her young, while another ewe in the flock may have 2 or 3 lambs. Under such circumstances, it is extremely desirable that the former "adopt" one of the latter's. The "adoption" may be effected in this way: remove the lost lamb as fast as possible, put the lamb to be adopted which has been wellsmeared with the liquid and placental matter of the ewe and fetal matter of the dead lamb or which has been coated with the stripped coat of the mother adopting. In many cases, the surrogate mother will adopt the disguised lamb; in such cases, it is advisable to separate the mother from the flock. Sometimes, in a large flock, it happens that a ewe which gave birth to a live lamb dies, while the lamb of another ewe died. In such cases, the kind of "adoption" described above is absolutely necessary. A lamb which has become stiff and cold and to all appearances dead may be revived by putting it in an oven, stove, or putting it in a warm bath, and afterwards rubbing him vigorously and thoroughly with a flannel cloth, then placing it in a warm room heated by a stove. At such a time, giving the lamb a little whiskey or brandy is sometimes productive of good results. In this work, do not be too soon discouraged, for often you will be able to revive the lamb.

Notwithstanding the methods above referred to by which lambs are parceled out, as it were, among different ewes, there are cases in which a lamb must be reared by hand; this is not any harder work, it being said, than to rear the young of any other kind of animal. In rearing, cow's milk which come from recently calved cows is the best, may best be substituted. The lamb should be brought to drink, or it may be allowed to take the milk from a leather nipple, in a manner similar to suckling; thus it may be reared with comparative ease. With ewes, it is much less likely than with cows that the afterbirth is retained. Some very careful breeders remove the afterbirth from the pen, but the practice of most breeders is to leave it there, since it is small in volume and is soon carried out with the material used for bedding. Ewes are also much less liable to trouble with the udder than are cows; this being the case, precautions regarding the lightness and quality of feed which are urgent with cows are unnecessary. On the other hand, the feed given to sheep must be the richest in quality, well calculated to produce milk, and most abundant in quantity. It is absolutely necessary that roots also be supplied (no roots are better than turnips) during that time. In this latitude, in the spring (March and early April), turnips are almost indispensable. Some grains, such as corn, oats, or barley, together with early cut English hay are also necessary.

8-9-5. Weaning Lambs

Lambs are usually weaned at the age of 4 or 5 months. The season for that is consequently summer and early autumn, when the sheep and lambs are out in pasture. The pasture should afford an abundance of fine sweet grass. Then, when the sheep are led to a new pasture and the lambs left in the old, they seem to feel the loss of their mothers much less keenly, provided they are allowed to remain in a familiar place. The new pasture to which the ewes are removed should be at such a distance from the one where lambs are left that it will be impossible for lambs and sheep to see each other; it is even better if the distance between them is so great that they cannot hear each other's bleating. It is advisable to leave a few old wethers with the lambs for company, and in order to facilitate the driving or calling of the flock. It is best that the grass of the pasture to which older sheep are removed be rather short and scanty, because if it is rich the sheep are likely to produce a large amount of milk. Since that milk will not be drawn by the lambs, caking of the udder, *etc.* will result. Even if put in scanty pasture, it is extremely necessary to be watchful, in order to know the condition of the sheep; if you observe any ewe's udder full of milk, it should be immediately be drawn off. The ewes should be kept in scanty pasture for a short time after weaning, but then you must bring them into good condition before their next service by the rams.

8-9-6. Fall Feed of Lambs

It is of the highest importance that young lambs are abundantly fed during the autumn months. The change from grass to hay, even under the most favorable circumstances, is a severe one for a young lamb, and is often a severe cause of retarded growth. In order to avoid that, we must bring them into fine condition. This can only be done by giving them rich pasture, and some extra food such as grains. The feeding to lambs of small quantities of grains such as oats, corn, *etc.* during the autumn months is usually productive of very good results.

8-9-7. Washing Sheep

It was formerly a practice, well nigh universal, to wash sheep before shearing, the objective being to free the wool from adhering filth, dirt, dust, and a portion of the yolk. It is impossible to so perfectly wash the wool of a live sheep so as to require no subsequent washing to make it suitable for manufacturing. That washing was done because it partially removed the filth, and because washed wool commanded a higher price than unwashed wool. At present, washing sheep is not universal for several reasons, the principal among which are:

(1) If sheep are washed, we must necessarily delay shearing to wait for the production of yolk.

(2) The sheep thus sheared later do not make such a vigorous growth of wool as sheep sheared early.

(3) There is a great tendency toward the loss of wool and the discomfort of sheep.

(4) Washing is not practiced without a considerable risk to sheep if they catch cold.

(5) It is hard work for a shepherd to stand all day long in a cold stream.

However, if for any reason the sheep must be washed, take them into a small stream and allow them to stand until the dirt and filth become soft, then wash the wool as cleanly as possible. Sometimes the washing is done under the fall of a mill-dam, or sometimes below an artificial waterfall. It is absolutely necessary whenever washing sheep to have a pen with fine, smooth grass in which to put the washed sheep. Without such a pen, washing will only be a waste of time. Washing is usually performed 2 weeks before shearing.

8-9-8. *Shearing Sheep*

All sheep are sheared, almost universally, once a year; however, in some cases sheep are sheared once every 2 years, but with the latter case, there is much loss of fleece and much discomfort to the sheep. The season for shearing varies with different individuals; in most cases, however, shearing is delayed until the weather becomes somewhat warm, in this latitude, usually about the first part of June. There are many who advocate much earlier shearing. Some of the best breeders shear their sheep in March, before they are turned out to pasture. Those men claim that sheep sheared at this time, and kept in good warm barns or sheds, fare better than if sheared later and allowed to suffer inclement outdoor weather. I think that their claim is well founded. Newly sheared sheep exposed to long periods of rain, even during the month of June, suffer greatly from the cold; not infrequently, death follows.

It is claimed by many that the wool is greater in amount and better in quality in sheep which are sheared early, provided that good shelter is available; in such cases, early shearing is greatly recommended. There are 2 principal methods of shearing: (1) putting the sheep upon a table, and (2) having the sheep lie down on the floor. The place used for shearing should be clean and spacious. A skillful shearer can shear 20 sheep per day, and if the sheep are of small size such as Merinos, many more. After shearing, it is essential to keep the wool without tearing; it should look just as though the skin has all been removed; the fleece should be folded into a small sheet and pressed by a hydraulic press or an apparatus specially constructed for that purpose. In shearing a pregnant ewe, great care must be exercised in handling her. It is not advisable to shear such ewes; the time for shearing them should be arranged so that it is done subsequent to parturition.

8-9-9. *General Care of Sheep*

(1) Shelter: It is essential that the place where sheep are kept is dry and if possible sunny. It is not necessary that an enclosed warm shelter be provided. A shed facing south and provided with a spacious yard is very good for sheep. It will be convenient if the shed is so constructed that at the times necessary (such as the lambing season) it can be closed. For mature, well-covered sheep, it is not only unnecessary but injurious to enclose them, because sheep suffer much more from bad air than from exposure.

(2) Feeding: In feeding sheep, the same degree of regularity and liberality is advisable as recommended neat cattle. It is estimated that ordinary sheep need about 30% of their live weight in good hay, or its equivalent, per day. Ewes with lambs will, of course, feed somewhat rich food.

(3) Water: Sheep need a comparatively small quantity of water, but that small quantity of water they need with great regularity.

(4) Bedding: Pens and yards should always be liberally supplied with straw or

382 Chapter 8 Stock-Farming

hay for bedding. Sand, muck, and sawdust are not good for bedding sheep, since those materials make the wool filthy. [p. 1189]

8-10. Cattle Feeding
8-10-1. Composition of the Animal Body

The fluids circulating in the blood and lymph vessels constitute at most 7 to 9%, and in old and very fat animals, 4 to 6% of their total live weight. Other fluids, as for example the various digestive fluids, *etc.,* though produced in considerable quantities, can hardly be taken into account as being included in the total live weight because they are constantly changing in amount. The findings of numerous investigations of various farm animals, oxen, sheep, horses, and swine, provide the following average composition:

The remaining 27% includes the blood, hide, hair, and entrails, and the contents of the stomach and intestines. The volume and weight of the contents of the stomach vary greatly according to the nature of the food; for example:

Bones	9%
Flesh and Tendons	40%
Mechanically separable fat	24%
Other	27%

(1) in a sheep fed chiefly on straw, the contents of the stomach made up 23.2% of its live weight

(2) in another sheep fed on hay and a small amount of beans, 15.9% of live weight

(3) in another sheep fed on clover, potatoes, hay, peas, and corn. 9.04% of live weight

The preceding figures are according to the German investigator, Walff.

In an ox fed on straw, the contents of the stomach constituted 16.6% of live weight; in another ox fed on fattening fodder, 9.4% of live weight, according to Groven.

1) Non-nitrogenous Constituents of the Animal Body

Among the most abundant of these is water, which in a new-born animal constitutes 80 to 85% of its live weight; in a mature, but not fat animal, 50 to 60%; in a fat animal, less than 50%; and in one very fat sheep, 35.2%, according to Messrs Lank and Gilbert. It was at one time the general opinion that the amount of water in the animal body decreased in fattening, but although the amount is less, it is necessary to come to that conclusion, because the increase of other constituents, principally fat, would cause a decrease in the percentage of water, though the absolute amount remains the same. The following table shows the average composition of the live weights of fattened mature animals:

Composition of Fattened Livestock [14]

	Ash %	Protein %	Fat %	Total dry matter %	Water %
98 Oxen	1.47	7.69	66.2	75.4	24.6
348 Sheep	*2.34	7.13	70.4	79.9	20.1
80 Hogs	+0.06	6.44	71.5	78.0	22.0
Average	1.10	7.26	67.8	76.2	23.8
*probably too high on account of dirt in the wool.   +probably too low

You see, therefore, that in fattening there is an absolute increase in the amount of water, fat that its percentage is relatively small. The dry components of the animal body are composed of 2 classes of matter, organic and inorganic. The organic may be divided into 2 sub-classes, nitrogenous and non-nitrogenous.

Among the non-nitrogenous components of the body, fat is the most important. In appearance and taste, fat differs much in different animals, and even in the different parts of the same animal, but its composition is almost always the same. The average of a large number of analyses of beef, pork, and mutton fats gives the following figures: Carbon 76.5%, Hydrogen 12%, and Oxygen 11.5%. Analyses of other animal fats as those of dogs, cats, *etc* show them to also have nearly the same composition. The quantity of fat which may be laid up in the animal body is very large in fattened neat cattle and swine, it is often 25-40% of the live weight, or fully 2 to 3 times as much as all nitrogenous substances put together. In lean animals, on the other hand, nitrogenous substances predominate. Other nitrogenous organic substances in the animal body are very small in amount, though often of importance for the functioning of the organs and fluids in which they are found. Among the substances in this class may be mentioned lactic acid found in the gastric juice and some other parts of the body, sugar in the blood, and a few others of secondary importance.

2) Nitrogenous Organic Substances[15]

These may be divided into 3 classes, albuminoids, gelatigenous substances, and horny matter. The albuminoids are the most important, for upon them depends life and from them all other things are formed. Among the most important albuminoids may be mentioned the protoplasm, which is very similar in composition to vegetable protoplasm, and is equally important for the vital functions in animals.

(1) Albuminoids: These are found in all organs and fluids of the healthy body except the urine in many different forms. All, however, are much alike in many respects, and resemble in their many properties, albumin, white of an egg. Like it, they are amorphous; they exist in at least 2 modifications, soluble and insoluble. In both forms, they are destitute of taste and smell. The soluble may be converted into insoluble in many ways, as by heat, by acidic action, *etc*. For our present purpose, it will be sufficient to notice only 3 forms of albuminoids, albumen represented by the white of an egg, fibrin (in lean meat), and casein (the basis of cheese). Albumen predominates in all the animal fluids, for example, in the chyle, the serum of blood, and the juice in muscles and nerves. Albumen coagulates when heated above a certain point, pure albumen at 165°F, and the dilute solution at a higher temperature. Fibrin constitutes a large portion of the so-called clod of blood. Animal flesh contains as its chief constituent a form known as "flesh fibrin". Casein is found principally in milk, and is not normally a constituent of the general animal body. A large number of analyses show the albuminoids to have approximately the following composition: Carbon 52 to 54%, Hydrogen 7%, Nitrogen 15 to 17%, Oxygen 21 to 24%, and Sulfur 1 to 1.5%. Generally, the amount of nitrogen is considered to be 16%, so the amount of albuminoids in a substance is calculated by multiplying the amount of nitrogen by 6.25.

(2) Gelatigenous Substances: These constitute nearly as large a portion of body weight as the albuminoids. The nitrogenous organic substance of bones and cartilage, a large part of the mass of tendons, ligaments, connective tissues, and skin are composed of gelatigenous substances. The composition of these sub-

stances is very similar to that of albuminoids, but generally there is somewhat less carbon, 50-51%, while the substance of bones, tendons, and skin is richer in nitrogen, 18%. Sulfur is generally either present in a small quantity or completely absent.

(3) Horny Matter: This is found in the epidermis or scarf skin, hair, wool, horns, nails, hoofs, claws, feathers, *etc*. The composition is very uniform, namely, Carbon 50-51%, Hydrogen about 7%, Nitrogen 16-17%, Oxygen 20-22%, and Sulfur 3-5%. They differ from albuminoids and gelatigenous substances mainly in that they contain more sulphur. The factor 6.25 sufficiently accounts for the estimation of all nitrogenous substances, such as you see. Their composition varies but little, in a large number of cases showing a close agreement between the amount of nitrogenous substances as calculated and as determined by actual analysis.

3) Inorganic Matter

The total quantity of inorganic matter in the animal body differs considerably according to the condition of the animal, and in different classes of animals, but a large number of analyses show it to be, on the average, as follows;

In lean animals, we find the maximum amount of inorganic matter, and in fat animals the minimum percentage. Phosphoric acid and lime are present in about

Neat Cattle	4	5% of live weight
Sheep	2.8	3.5%
Swine	1.8	3.0%

equal quantities, and together make up about 4/5 of the total amount of ash. In the other 1/5 are found: potassium oxide, sodium oxide, magnesium oxide, chlorine, sulfuric acid, carbonic acid, and a minute quantity of silica. The ash and mineral matter in the animal body, though small in quantities, are yet absolutely necessary. Numerous experiments have shown that animals deprived as much as possible of mineral food, but furnished abundantly with organic nutrients, become first sleepy, then weak, and finally die.

In practice, in the feeding of mature animals, a deficiency of mineral matter is never to be feared; such matter is present to a large excess in common feeds. Young cattle and milch cows need more mineral matter, and feed rich in phosphate of lime must be given. Among mineral matter, common salt to a certain extent occupies an exceptional position. It has certain physiological functions. It facilitates the passage of the albuminoids in feed from the digestive canal into the blood; also, to a certain extent, it increases the energy of all total processes. Salt is especially necessary for stall fed animals abundantly supplied with foods rich in potash, because potash causes the excretion of salt through the urine.

4) Bones

8-10-2. Composition of Fodders

The animal body, though complex in its structure, may be considered to be made up of nitrogenous and non-nitrogenous organic substances, and mineral matter. Since these substances are constantly being destroyed in the body in the performance of vital functions, it is necessary that the animal receives from without a supply of substances identical or similar to those destroyed, and which can be assimilated by the tissues and fluids of the body to replace those lost, and enable vital processes to continue.

Any single chemical compound, such as albumin, fat, starch, sugar, *etc.,* which is capable of aiding to replace loss is called a "nutrient". Nitrogenous and non-nitrogenous substances are always found in the animal body in nearly the same proportion; moreover, each being destroyed in the body at a fixed rate within certain limits, it is evident that feed should contain the various nitrogenous and non-nitrogenous nutrients in certain proportions.

A fodder usually contains several or all groups of nutrients, but may not contain them in the proper proportion. For example, oil cake contains albuminoids in too great a proportion, and does not contain carbohydrates in a great proportion; hence, oil cake or roots, if capable of sustaining life, do so only at a great loss. By combining several on sided feeds, we may prefer a mixture containing several materials in the proper proportions to sustain an animal economically. Such a mixture we may designate a "ration" or "complete feed". Nutrients may be divided into 3 groups, corresponding to the 3 groups of substances in the animal body, namely, nitrogenous, non-nitrogenous, and mineral substances.

1) Nitrogenous Nutrients

Protein is the name usually given to designate vegetable albuminoids. Vegetable albuminoids may be divided into 3 classes, namely, albumen, casein, and fibrin. Vegetable albumen is found principally in the young growing parts of plants; in older parts, it is converted into other forms of protein. Vegetable casein may be separated in the following manner: make wheat flour into dough, then knead it in a stream of water; the starch is washed away, and the remnant consists of a mixture of poor albuminoids known as "crude gluten". If this remnant gluten is treated with dilute alcohol (60-80%), 3 of the albuminoids dissolve; the one remaining is gluten-casein, mixed with slight impurities. The same or similar substance is found in rye, barley, and perhaps buck- wheat, as well as in oil seeds. Oats contain an albuminoid which is more like legumin the main albuminoid in beans, peas, *etc*. Another albuminoid, known as conglutin, is found in almonds and corn. Both legumin and conglutin are much like gluten and casein, and all are very much like animal casein.

Vegetable fibrin is found in wheat, and is there known as gluten fibrin; it is also found in barley and corn. Its composition is much like that of animal fibrin. The 2 other albuminoids of wheat are mucidin and gliadin. The former is also found in rye and barley, and the latter also in oats. In composition, these albuminoids are not very different from fibrin. The composition of all the albuminoids mentioned differs considerably, but all share a strong resemblance, both in composition and properties. Whether all of them are equally valuable as nutrients is not yet known. Probably not, but the differences are doubtless not great, and given the lack of definite knowledge about them, all are generally considered to be equivalent. The albuminoids constitute the most important solid part of the animal body, the muscles, tendons, nerves, and in fact all its working "machinery" are composed of them. As far as is presently known, the animal body does not have the ability to create any of them.

The herbivores are directly, and the carnivores indirectly, dependent upon plants for albuminoids. Since vegetable albuminoids are, by reason of their great similarity, readily convertible into animal albuminoids, they are indispensable parts of any food. They are most indispensable indeed, for it has been determined by experiments that while they may to a certain extent take the place of non-nitrogenous substances, none can replace albuminoids. Therefore, in determining the value of any food, the proportion of albuminoids constitutes the most valuable factor. All other things being equal, fodders containing them in the highest proportions are the most valuable, since albuminoids are the most expensive ingredients to produce. Certain other albuminoids are found in plants, chief among these are: (1) nitrates, nitrites, and ammonial salts; (2) peptones; (3) alkaloids; and (4) ammines and amides. Most of these are found only in very small quantities, and their value as nutrients is not yet well understood.

1) Non-nitrogenous Nutrients

(1) Carbohydrates: The chief substances contained in this group of non-nitrogenous nutrients are cellulose, starch, dextrin, cane, grape, milk and fruit sugars, and gums. Those substances, especially cellulose and starch, form by far the greater part of the dry matter of the plant, and are distributed throughout all parts of it. All of those substances not only contain carbon, but also the elements of water in the proportions to form water, hence their name "carbohydrates". This similarity in composition, as well as the ready transformation of one into another, both artificially and in plants, show that they are chemically closely related to each other.

Cellulose: When pure, this is an odorless, tasteless, solid substance. It is distinguished from other substances of the group by its slight solubility in dilute acids, alcohol, water, or any other ordinary solvents; hence, the customary method of obtaining it in a tolerably pure state is to have various solvents act upon vegetable matter until all other substances are removed. The composition of pure cellulose is exactly the same as that of starch: 44.44% carbon, 6.17% hydrogen, and 49.39% oxygen. Cellulose is seldom found in a pure state except in the young, tender parts of plants. It is usually more or less impregnated with substances to which the collective name "lignin" has been given. The composition of lignin is: 55% carbon, 5.8% hydrogen, and 38.9% oxygen. The principal difference between cellulose and lignin is that the former contains less carbon. Cellulose was long thought to be entirely indigestible, but in 1854 it was proved that ruminants were capable of digesting large quantities of this substance[16]. Since that time, numerous experiments have shown that this ability is not confined to ruminants, but that all our domesticated herbivorous animals can digest much cellulose. The proportion digestible varies with the kind and quantity of the food, and the species of animal to which it is fed. Of the cellulose in ordinary coarse fodder, 30 to 70% is digestible by ruminants, while the cellulose in cereal grains is much less digestible. In general, the younger and more tender a feedstuff, the greater the amount of cellulose which is digested; in old woody plants, in which much lignin is found, its digestibility is considerably less. Lignin itself appears to be entirely undigestible.

The amount of cellulose in a feedstuff is determined by, in succession, boiling its fine powder with dilute acid, dilute alkali, and then washing with alcohol and ether. The solvents remove other constituents and leave impure cellulose. The residue, after the deduction of a small quantity of ash and albuminoids which it still contains, is called "crude fibre"; it is by no means pure cellulose, but is mixed with a quantity of lignin. Its composition varies somewhat; thus, for example:

crude fibre obtained from hay and straw contains 45-46% carbon; that from clover, hay and straw of legumes, such as peas and beans, 48-49%. The latter contains the most lignin.

Starch: Next to water and cellulose, starch is the most abundant substance of the vegetable world, being found in all our common plants. Pure starch is an odorless, tasteless white powder which consists of 2 substances, namely starch cellulose and granulose[17]. The composition of dry starch is the same as that of cellulose. In its common air dry state, starch contains 12 to 20 parts of water. Starch is rapidly dissolved and converted into sugar in the mouth by saliva, and also in the intestines by the pancreatic juice. It is among the most important of the non-nitrogenous nutrients, on account of its abundance, and the ease and completeness with which it is digested.

Dextrin: This is seldom formed in plants, but is often formed in the process of cooking substances rich in starch, as for example in baking bread. It is entirely digestible.

Sugars: There are 4 principal kinds of sugar, namely cane, grape, milk, and fruit sugar[18]. The compositions of the sugars are as follows:

All of the sugars are readily soluble in water, and are easily digestible. They are important nutrients, being formed in large quantities from other carbohydrates; in ordinary fodders, however, they are found only in small quantities.

Cane Sugar	C 42.11%, H 6.40%,
Milk Sugar	O 51.46%
Grape Sugar	C 40%, H 6.67%,
Fruit Sugar	O 53.33%

Gums: These are found in large quantities in grains. They appear to be digestible by domesticated animals, but their value as nutrients is little known; they are probably about equivalent to starch.

Pectine Substances: These are of uncertain composition; they are found in ripe fruits, and together with sugar constitute a large part of the non-nitrogenous organic substances in common root crops. Uncooked fruits and roots contain a substance known as "pectose". Boiling with water, or exposure to heat converts pectose into pectine, which is soft and soluble in water. All of these substances are digestible, and probably play about the same role in nutrition as do the carbohydrates.

(2) Fats: The fats found in plants have essentially the same composition as those found in the animal body, on the average, 76.5% carbon, 12% hydrogen, and 11.5% oxygen. They differ principally from carbohydrates in that they contain much more carbon and a small amount of oxygen; consequently, they require much more time for complete combustion, but give out about $2\frac{1}{2}$ times as much heat as do carbohydrates. Fat is found in small quantities in almost all plants, in root crops, 0.1-0.2%; in hay and straw, 1-3%; and in cereal grains, 1.5-3% except for oats, 6%; and corn, 4-9%. Fat is especially abundant in the seeds of certain plants, such as hemp, flax, and cotton, the amount in such seeds varying from 10-40%. In oil cake, rapeseed cake, cottonseed cake, *etc.*, the amount of oil remaining after processing varies according to the perfectness of its extraction, but is in the range of 8-12%. In ordinary fodders, fat has a rather subordinate part, but is an important aid in rapid fattening, though it is not the only source of fat in the animal body. Furthermore, fat aids in the digestion and resorption of albuminoids.

388 Chapter 8 Stock-Farming

(3) Inorganic Nutrients: These are found in the ash of plants. In ordinary cases, any ration (feed) which contains a sufficient quantity of organic nutrients will contain an abundance of inorganic nutrients. Therefore, the consideration of these is of no importance, except in the case of common salt.

8-10-3. *Chemical Analyses of Fodders*[19]

Albuminoids: Crude protein=Nitrogen×6.25

Cellulose: Already pointed out.

Fat: This is determined by dissolving it out of dry fodder with dry sulfuric ether; the operation is continued until nothing further is extracted. Next, evaporate the ether, dry the remnant at 100°C, and weigh it. The remnant thus obtained from grains will be tolerably pure, but that obtained from coarse green fodder, hay, straw, cornstover, *etc.* will contain, along with real fat, numerous waxy and tar-like substances, as well as chlorophyll.

Ash: This is obtained by burning at a low temperature as low as possible, to avoid the volatilization of alkaline chlorides, especially potassium chloride. Pieces of coal and carbonic acid remain in the ash thus deducted.

Nitrogen Free Extract: All that remains of dry matter after deducting the crude protein, crude fiber, crude fat, and ash called "nitrogen free extract". In all grains and roots, it consists chiefly of sugars, starch, substances of the pectine group, and sometimes vegetable, which is similar in composition to starch and exerts nearly the same nutritive value. However, in coarse green fodders there are found, in addition, varying quantities of gum-like substances and lignin, the latter is indigestible but has the same proportionate composition as starch. Therefore, the nonnitrogenous nutrients of fodders, with the exception of fats, are generally considered as carbohydrates.

8-10-4. *Nutritive Ratio*[20]

The ratio of digestible protein to digestible nonnitrogenous substances constitutes what is known as the "nutritive ratio".

Table of Digestion Coefficients

Ex. Average Meadow Hay	
Composition: water	14.3%
ash	6.2
protein	9.7
crude fibre	26.3
N free extract	41.0
fat	2.5
Digestibility obtained by averaging all available experiments;	
protein	56%
crude fibre	57
N free extract	63
fat	48

Digestible protein   =9.7×0.56=5.4 percent
Digestible crude fibre   =26.3×0.57=15
Digestible N free extract =41×0.63=25.8
Digestible fat   =2.5×0.48=1.2
   fat(1.2)×2.5=3; 3+15+25.8=43.8 5.4:43.8=1:8.1
or Digestible protein:Digestible carbohydrates[21] 5.4:43.8=1:8.1

The percentage of nutritive substances in the nutrients of several fodders, which is digestible is known from the results of a large number of digestive experiments; those results are found in the above table: [p. 1212]

8-11. *Digestion in Animals*

The feed is first masticated in the mouth and mixed with saliva, which converts starch into sugar. It is then received by the rumen, and morsel after morsel is taken until viscus is comparatively full. It is not the feed just ingested that undergoes this process, but feed which was swallowed 12 or 16 hours before. The partially digested feed is turned and shifted about in the stomach by muscular action, and becomes well mixed with the fluid secreted by its internal surface. The feed first enters the superior compartment, from which it passes to the inferior, and then again enters the former compartment, where rumination takes place. Before rumination takes place, it is evident that the feed must rise to the upper part of the viscus and enter the esophageal canal. What is its direction? The liquid portion passes on into the course of the canal. It is only the hard, indigestible feed that undergoes the process of rumination. The feed, on leaving the fourth stomach, enters the first of the small intestines, the duodenum from whence the chyle is principally absorbed. The "chyme", as the feed is termed, having first received the secretions of the liver and the pancreas.

8-11-1. Details of Digestion

(1) Mastication: The feed is first masticated in the mouth and mixed with saliva, which converts starch into sugar. The ground-up mass passes into the stomach through the gullet.

(2) Rumination: The feed, unless it is of a pulpy character, passes after a first imperfect mastication wholly into the first and second stomachs as shown by arrows B and D in Figure[22]. If it is of a pulpy character, it proceeds after this mastication in part into all the four stomachs. When the coarser feed which had passed into the first and second stomachs has become sufficiently dissolved by the aid of the saliva that is continually being swallowed in the intervals between eating and ruminating, it is thrown by the contraction of the stomachs, assisted by the action of the abdominal muscles and diaphragm, into the demi-canal. The canal then contracts and, moulding the pulp that has passed into it to the shape of its narrow and shortened form, converts it into pellets. The canal next throws this pellet into the gullet, by converted action of which the pellet is masticated to the reach for to rumination, that is to say, for a second mastication and insalivation.

(3) Second Digestion: In the second digestion, the ruminated aliment passes by way of the demi-canal into the third stomach represented by arrow A, but a portion passes at the same time into the first and second stomachs. That portion of the ruminated aliment that arrives at the third stomach is, after being acted upon there, passed into the fourth or proper stomach, where true digestion takes place. The portion of the ruminated aliment which goes into the first and second stomachs, being the new aliment, may possibly be partially carried to the mouth again. It appears that when the mass of feed is bulky, dry, and resistant to digestion, it is forced into the continuation of the esophagus; by its bulk and force the feed opens the lips of the canal, so that it is driven straight into the first and second stomachs. However, when the mass of feed after the second mastication is soft, pulpy, and yielding, it does not offer sufficient resistance to open the lips of the canal, so it is propelled onward to the third stomach, which communicates with the fourth stomach. Hence, when the ruminant drinks, the water mostly passes directly into the third and fourth stomachs.

(4) Gastric Juice: This is a clear, colorless liquid having a peculiar odor and a

slightly saline, acidic taste. Its essential constituents are muriatic acid and pepsin; the chlorine in the former is derived from common salt, and so it is that salt is necessary in animal feeds. The composition of pepsin is not yet known. Aided by the free acid, pepsin is concerned with digestion of the albuminoid portions of the feed, and exerts hardly any influence over its starchy and saccharine constituents. Hence, the feed is finally converted into a soft, pulpy mass called chyme. In this last conversion, it should be stated, the action of the stomach offers much assistance. Of this change sugar converted from carbohydrates by saliva, and the soluble peptone are resolved in the stomach, but the larger portion passes to the intestine through the pylorus.

8-11-2. *The Intestines*

(1) Intestinal Digestion: The intestines are very long, in horses, 10 to 12 times the length of the body; in oxen, 20 times; and in goats, 26 times. The more compacted the nature of the fodder the animal feeds upon, the less is the length of its intestines; thus, in carnivores, they are only 4 to 6 times the length of the body. The chief digestive fluids of the intestines are the bile and pancreatic juice. The principal action of the bile is upon fat, emulsifying a portion of its and forming a soapy salt with the soda from another portion of it. In addition, the bile hinders the decay of easily decomposable albuminoids, and has an alkaline character, neutralizing acids such as HCl. The pancreatic juice converts any starch which may have escaped conversions into sugar, and also acts upon albuminoids as pepsin. The pancreatic juice, too, acts upon fats, which it first emulsifies and more slowly decomposes into their constituents, glycerin and fatty acids. We see, therefore, that the solid constituents of the feed are changed into soluble forms in the body by water and digestive fluids, albuminoids by the gastric and pancreatic juices; starch, *etc.* by the saliva and pancreatic juice; and fats by the bile and pancreatic juice. The fat producible from albuminoids must never be lost sight of in estimating the result of feeding a certain amount of fodder.

Carbohydrates are represented in the body by sugar, into which all the others are converted if not otherwise decomposed. Fodders for herbivores contain carbohydrates in large quantity. A large ox may resorb 16 to 18 pounds of sugar daily, and yet never more than a very small quantity of sugar is found in its blood. The liver takes the sugar from the blood and converts it into reserve material, the glycogen, which is insoluble, is gradually reconverted into sugar; in other words, it is a reserve of carbohydrates. Glycogen is formed not only from carbohydrates, but also from protein; in the livers of dogs which have been fed purely on meat, glycogen is found. In what part, and by what, cellulose is digested is not yet known; the latest view is that cellulose, through fermentation, is broken up into marsh gas, CO_2 and various soluble products.

By the action of several fluids, and also by the action of the stomach muscles, chyme is produced. The ease of digestion is influenced by the condition of the feed, hard foods being less digestible than soft foods. Moreover, certain portions which are digestible are enclosed in the envelopes which prevent access to digestive fluids. The proportions of albuminoids, carbohydrates, and fats exercise an important influence on the digestibility of these groups. A moderate amount of fat assists digestion, but too great an amount hinders it. Furthermore, in feeds poor in albuminoids but rich in carbohydrates or starch, the latter may escape digestion; the loss is thus a total loss, because carbohydrates in animal manure are of no value. Any albuminoids that have escaped digestion are not completely lost, since they are of value in manure. In a proper proportional ration, the proportion of really digestible feed which escapes digestion is small.

(2) Resorption: This is the function of the living protoplasm of the epithelial cells of the intestines. Formerly, osmosis was called upon to explain this phenomenon, but that is now considered to be inadequate for the purpose. Mostly, only soluble substances are resorbed, but fat is an exception. Fat is reduced to an extremely fine state of emulsion. The whole is subjected to the constant wave-like, or persistent muscular, action of the intestines which forces their contents to their respective destinations. The nutritive portion or chyle is taken up by the absorbent vessels and conveyed to the blood, while the nutritious portion, together with some of the digestive fluids, the bile, *etc*, as well as some worn out portions of the walls of the intestines, is excreted as feces from the system. In addition to the substances named, the solid excrement of herbivores usually contains all the phosphoric acid from the metamorphosis of the tissues. In carnivores, that is found in the urine. [p. 1221]

8-12. Decomposition of Nutrients in the Body[23]

The albuminoids in the feed and body tissues are thought to be split up by numerous intermediate into urea and fat. In herbivores, there is also found a quantity of hippuric acid, the amount varying according to the kind of fodders and the species of animal, but the quantity is often very small. The urea is taken up by the blood, separated from it by the kidneys, and excreted in the urine. The nitrogen in 100 parts of water free protein may be separated from it in the form of 33 parts of urea. The remainder of the protein, 66.5 parts after taking up, and with 12.3 parts of water, contains the elements for the formation of 51.4 fat and 21.4 CO_2. Fat, whether derived from the albuminoids is contained as such in the feed is according to the circumstances, it is either deposited in the body, enters into the milk, or is oxidized completely during respiration, yielding CO_2 and H_2O. [p. 1222]

8-13. Excretion in Animals

Regarding the necessity of excretion of the fecal matter from the digestive system, I need say nothing; it is a subject familiar to all of you. The urine contains water, urea, and other nitrogenous substances coming from the decomposition of protein, and hippuric acid in herbivores or uric acid in carnivores.

1) Excretion of Nitrogen

It was formerly a large question whether nitrogen was excreted from the lungs and skin, but now it has been established that it is excreted from the body in excrements alone. The following experiment, done with dogs, shows that very conclusively.

392   Chapter 8   Stock-Farming

Experiments on Dogs

Duration of Experiment (in days)	Kind of Food	Nitrogen		Difference	
		In food g	In excrement g	g	%
49		2499.0	2525.6	+26.6	1.0
12	Meat	612.0	611.9	−0.1	0.0
23		1173.0	1176.9	+3.9	0.3
20	Mixed	340.0	335.2	−4.8	1.4
58	Food	986.0	982.8	−3.2	0.3

Experiments with numerous individuals of various species of animals showed the same results. The following experiment was done with a pigeon: A pigeon was fed for 14 days on peas containing 10 times as much nitrogen as the entire amount of nitrogen contained in its body. Over the duration of the experiment, the loss was found to be 2.0%, which was fully accounted for by a gain in weight of 73 grams. We are, however, more interested in cattle than dogs and pigeons; hence, the following experiment is introduced to prove the same fact in oxen;

Experiment upon Oxen

Number of Animal	Weight of Animal lbs	Nitrogen per day		Differences	
		In Food g	In Excrements g	g	%
1	1,403	135.5	135	−0.5	0.4
2	1,529	160.0	156.5	−3.5	2.2

Experiments upon milch cows, goats, and sheep have been numerous and all show the same results. In those experiments, the animals were all mature animal, there is no gain or loss. The fodders in those experiments were just sufficient to maintain the animals in their normal condition. These results, then, afford proof, though indirect, of the fact that nitrogen cannot be exhaled from the lungs and skin. We are not also in default of direct means of experimentation, we have the respiration apparatus. Some, after direct experiments, say that they have detected NH_3, but most other experiment doubt their statements. [p. 1224]

8-14. *Determinations of the Gain and Loss of Flesh*

The difference between the nitrogen in feed and that in the visible excrements, multiplied by the factor 6.25 will give the amount of albuminoids retained in the body, or lost. Excretion of Carbon: a slight quantity in the urine, but most of it through the lungs and skin. Excretion of Hydrogen: a little in the urea, but most of it in the water of the urine, through the skin in the form of perspiration, and water vapor in the breath.

1) Method of Investigation

While pure nutrients are theoretically entirely digestible, in practice some points being enclosed in is impregnated by indigestible matter escape digestion. Therefore, one of the most important subjects is to find out what proportion of any nutrients is digested. We may know the amount of nutrient contained in the feed; that amount, minus the amount of nutrients contained in the dung, gives the amount of nutrients digested. The difference between the total amount of dry matter in the feed and that in the dung gives the amount digested. We cannot make an analysis in each case, but must base our calculations upon the analysis of some samples (both feed and excrement) selected with much care that they would give for averages. There is one important source of error, that is, some portion of the worn-out intestinal matter, as well as digestive fluid, bile, pancreatic juice, *etc.* may be voided along with the actual excreta; hence, it follows that the amount digested is slightly greater than the amount found by the method just described, but the degree of such error is very small. Also, that extra matter may be separated by means of ether and alcohol, it being soluble in these. Thus, excreted matter derived from the feed and the pancreatic juice, bile, *etc.* is very small in ordinary cases, though when fodder very poor in protein ingested, the amount excreted may even exceed that in the body, because the decomposition of the system proceeds whether the feed given is rich or poor.

Digestibility of Fat: Results afforded by the ether process are somewhat inaccurate; since the products of the bile are soluble in ether, they are always too small in quantity. At present, however, there is no way of separating these products, that is, of finding the exact digestibility of fat.

2) Determination of the Nutritive Effects of a Ration[24]

(1) Production of Flesh: If a digestion experiment is conducted as described, the urine of the animal can be carefully collected and measured, and the quantity of elements in it be determined. We can also estimate the gain or loss of flesh. From the determination of the digestibility of the feed, we know how much nitrogen has entered the body, while the urinary nitrogen is the amount which has left it. The difference between the two is, of course, the gain or loss in the body. As previously indicated, $N \times 6.25 =$ dry protein. Now, fresh flesh free from fat has the following average composition: Water 75.9%, Ash 1.3%, Dry Matter 22.8%, and nitrogen 3.4%. Now, $3.4\% \times 29.4 =$ nearly 100%; hence, $N \times 29.4 =$ gain or loss of flesh, while the amount of urinary nitrogen will show how much protein or flesh has been decomposed in the vital processes. An example will make this clearer: Take 100 pounds; if 60 pounds are digested, and 50 pounds are found in the urine, there is 10 pounds gained: $10 \times 29.4 = 294$ pounds gain in flesh.

On the other hand, if 70 pounds are found in the urine, and 60 pounds was digested, there is 10 pounds lost: $10 \times 29.4 = 294$ pounds loss in flesh.

(2) Production of Fat: Since the production of flesh is measured by ascertaining the difference between the receipts and expenditures of nitrogen, so the production of fat is estimated be the gain or loss of carbon; however, as we have seen, most carbon escapes in the forms of breath and sweat. Hence, gaseous products have to be dealt with, and for such measurements a special apparatus is necessary; the one most commonly used is Petenkofer's Respiration Apparatus, which consists of a room of ample size for the admission the animal to be examined, provided with contrivances for admitting measured quantities of air, and of measuring the escaping air, *etc.*, in fact of all the contrivances necessary to secure the objectives of conducting such an experiment.

In this experiment, some aliquat part of air admitted and escaping is taken and analyzed the amount of water in H_2SO_4, CO_2 in baryta, water, whence the total amount is deducted. By the use of this apparatus, in connection with analysis and the weighing of feed and drink, urine and dung, we can identify and measure all substances which enter or come out of the body, and thus can know the exact effect of any ration. The amount of carbon in the digested portion of the feed, minus the amount of CO_2 emitted by the lungs and skin, and in the urea, *etc.* excreted by the kidneys, gives the gain or loss of carbon in the body; however, that gain or loss may be in one of 2 forms, fat or albuminoids.

If the amount of nitrogen in the feed and excrements has been entirely in the form of fat, the other non-nitrogenous concentrated constituents in the body are so small in amount that they may be ignored. Now, 100 parts of fat contain 76.5 parts of carbon; in other words, 1% C=1.3 of fat (76.5×1.3=nearly 100). The calculation of the gain or loss of fat is essentially the same, if there has been a gain or loss of albuminoids, except that the amount of carbon in the latter must be deducted from, or added to, that found in the experiment before multiplying by the factor 1.3. This will all be clearer by an example:

Experiment at the Wende Experimental Station on Sheep

The animals received 1,216 grams of hay per head per day, and the necessary amount of water. In the fodder and excrements, the following amounts of carbon and nitrogen were measured:

The amount of water can be measured by means of the Respiration Apparatus, by finding the difference between the amounts of water which enter and come out of the body.

In Fodder	Carbon	Nitrogen
Hay	460.1 grms	18.1 grms
Water	0.1	0.0
	460.2	18.1
In Excrements		
Dung	202.5	8.45
Urine	23.2	7.65
Expired air	213.8	0.00
	439.5	16.10
Retained in Body	20.7	2.00

N, 2 × 6.25 = 12.50 protein
N, 2 × 29.4 = 58.8 grams gain in flesh
Albuminoids are known to contain on the average 53% carbon; hence,
Protein: 12.5 × 0.53 = 6.6 grams C,
C 20.7 − 6.6 C = 14.1 C retained in the body (fat)
C 14.1 × 1.3 = 18.33 grams fat gained

The live weight of animals is not an accurate measure of the nutritive effects of a ration, since it is subject to great fluctuations on account of the varied amounts of feed and drink in the stomach and intestines. The following will show that:

The live weight, however, taken when the animal is in exactly the same circumstances and condition, *etc.* may be tolerably accurate.

(3) Formation of Flesh: The laws governing the formation of flesh have been most studied in the carni-

Experiment on an Ox

Weight		Weight	
May 23rd	1,298 pounds	June 3rd	1,271.1 pounds
24th	1,242.4	4th	1,210.7
30th	1,269.8	12nd	1,294.2
31st	1,288.0		

vores; however varied the food, the substances resorbed are the same, fat, protein, sugar, *etc.,* and salts. All of the organs of the Mammalia, at least, are very similar; hence, decomposition must follow the same course in all of them. [p. 1232]

8-15. Protein Consumption

In every living animal, a certain amount of albuminoids are destroyed, with that amount appearing in the urine. The amount may vary in different animals, and under different conditions in the same animal, but it cannot sink below a certain point without serious disorder to the condition of the animal. This necessary and constant process is called "protein consumption". From a suitably sufficient food, an animal takes more than it expends, and this surplus deposition of protein becomes a part of the body. Whatever increases the deposition and lessens the consumption is a gain, but protein consumption is not to be considered as a waste. The richer the food, the greater an improperly proportioned ration, extremely great, to make it as little as possible is best. There are 2 classes of protein, organized protein and circulatory protein. Under organized protein is included the protein found in muscles, bones, *etc.*; Circulating protein is that found in living fluid, its amount varying according to the kind of food, it is not necessary for the blood, it is only temporary.

1) Protein Consumption during Hunger

An examination of the preceding table shows that the excretions on the last day of feeding and the first day of fasting were very different, and that the amount of urea declined rapidly at first, then more slowly until about the 6th day, whence it continued nearly the same, its amount being represented by the excretion of about 12 grams of urea. A large number of experiments gave the same results. It is evident from this and many other similar results that protein is of 2 forms, "stable protein" which decomposes slowly at the rate of about 12 grams per day, and "easily decomposable protein", the amount of which depends upon the food, and which rapidly disappears when food is withheld.

Experiment on Fasting Dog (weight about 35 kilograms = 77 pounds)

Number of Experiments	14	15	14	15	16
Previous food per day	Meat 2,500 g	Meat 1,800 g Fat 250 g	Meat 1,500 g	Meat 1,500 g	
	urea excreted (grams)				
Last day of feeding	130.0	110.8	110.8	24.7	—
1st	37.5	29.7	26.5	15.6	—
2nd	24.9	23.3	18.2	18.6	15.6
3rd	19.1	16.7	17.5	15.7	14.9
4th	17.3	14.8	14.9	14.9	13.2
5th	12.3	12.3	14.2	14.8	12.7
6th	13.3	12.8	13.0	12.8	13.0
7th	12.5	12.0	12.1	12.9	—
8th	10.1	—	12.9	12.1	—

The "stable" is organized protein and the "easily decomposable" is the circulating protein. The first form is present in the body in much larger proportion, while the second form in poorly fed animals is very small in amount, in the herbivores 1%, and in well-fed carnivores it may rise to 5%. But whether present in larger or smaller quantity, the greater part (70–80%) of the circulatory protein is carried over the course of 24 hours; with the organized protein, not more than 8–10% over the same length of time.

2) Feeding with Protein Alone

The consumption of protein in the body is largely determined by the supply of protein in the food. This is evident from the following experiment made on the same dog as the preceding one. Several other experiments proved the same:

The Same Dog as the Preceding

Meat eaten per day	0	300	500	900	1,200	1,500	2,000	2,500	2,600
Urea excreted	12	32	40	68	88	106	144	173	181
Corresponding to flesh	165	442	552	938	1,214	1,463	1,987	2,387	2,498

An equilibrium between the amount of protein in the food and the protein consumed is usually soon established. Any gain or loss at first does not continue for long. Within 2 to 4 days, the composition of protein in the body becomes equal to that supplied in the food, and no further gain or loss takes place. The truth of this statement is shown by the following experiment:

Experiment on Dog

Food g	Composition of Flesh				
	Previous Food g	Day before g	First day g	2nd day g	3rd day g
2,500 meat	1,800 meat	1,800	2,153	2,480	2,532
2,000 meat	2,500 meat	2,500	2,229	1,970	

Protein consumption during fasting is not a measure of the amount necessary to support the body; if to a fattening animal, the same amount of protein as excreted during hunger, this is converted to circulating protein and its consumption is measured. In order to maintain an animal in average condition, we must give it approximately 2 to 2.5 times as much protein in its food as is consumed in the body during hunger; when the food has been rich, a much larger quantity is necessary to maintain equilibrium, once established. Fat decrease in protein consumption; hence, lean animals consume it more rapidly than do fat animals. The herbivores can be fattened more rapidly than the carnivores.

3) Effects of Water and Salt on Protein Consumption

Salt causes an increase in the rapidity of the circulation; hence, it also increases protein consumption. It secures advantages in herbivores which may more than counterbalance the increased consumption, provided salt is supplied only in small quantities. Furthermore, salt increases the excretion of urine, since for the excretion of a large quantity of salt more water is necessary; hence, the live weight may rapidly decrease if much salt and little water is given. Water in large quantities increases protein consumption greatly. It is necessary, in order to most advantageously get results, especially in feeding young animals and fattening animals, to avoid everything which involves or leads to the excessive use of water, that is, too watery fodders, too high a temperature in the stable, too much salt, too much movement, *etc*. The normal amount of water in food and drink together may be stated as 4 pounds per pound of dry fodder for cattle and half that quantity for sheep. An exception to this rule can be made for dairy cows, to which a large quantity of water is advantageous, since it increases the secretion of milk. Fat or carbohydrates fed alone do not decrease protein consumption.

4) Feeding with Protein and Fats[25]

All other things being equal, fat decreases protein consumption and, therefore, increases the deposition of flesh in the body. This is most plainly shown if, after the body has been brought into equilibrium, a certain quantity of albuminoids are added to the food. To show this, let me call your attention to the results of the following experiments upon a dog:

Date	Food		Urea	Flesh consumption
	Meat g	Fat g	g	in body g
July 31st	1,000	0	81.7	1,140
August 1st	1,000	100	74.5	1,042
August 2nd	1,000	300	69.3	970
August 3rd	1,000	0	81.2	1,134

As you will see from the preceding table, when the dog was fed with 1,000 grams of meat he was losing daily 140 grams of flesh. The addition of 300 grams of fat prevented this loss and caused a gain of 30 grams per day. The decrease of protein consumption is no greater with a large than with a small ration of albuminoids, provided that the amount of the fat is the same; the addition of fat simply makes it less under the same circumstances. A dog weighing 77 pounds needs 1,500 grams of pure meat daily to keep in good condition; now, if to this dog only 500 grams are given, he will lose 150 grams of flesh and will be wasted away and in wretched condition. If we give 500 grams of meat and 200 grams of fat, the loss of flesh is quickly checked, and if the protein consumption becomes equal to the supply, the animal reserves in good condition. There will be the same amount of flesh formed as may be formed by feeding 1,500 grams of meat, though the animal may be less active and energetic. The same phenomenon, though to a slightly lesser extent, is seen in herbivores. Fat in the food may cause a long continued gain of flesh.

Protein alone causes a gain only for a short time, and soon the loss becomes equal to the supply. Another important fact in connection with this is that the gain of flesh continues longer with a medium ration of protein than with a large one. The total gain up to the time of the beginning of equilibrium is generally no greater and is often less with a large ration than with a medium ration of albuminoids. The amount of fat in fodder for ruminants cannot safely exceed a rather low quantity. A small amount exerts a favorable influence, but a larger amount is often injurious, causing the disturbance of digestion and loss of appetite.

5) Feeding with Protein and Carbohydrates

The action of carbohydrates is analogous to that of fat. Like it, they cannot prevent, and can only decrease protein consumption. Like fat also, they enable animals to subsist or even to gain in flesh with a smaller quantity of albuminoids in the food than would suffice if the ration was composed of pure protein. Many determinations of this have been made, both in carnivores and in herbivores. That carbohydrates decrease protein consumption will be evident from the following experiments:

Fodder Rich in Protein					
	Protein digested lb	Carbohydrate and Fat digested lb	Nutritive ratio	Protein consumption lb	Gain of protein lb
Ox	2.60	10.95	1:4.2	2.14	0.46
Ox	2.51	12.51	1:5	1.83	0.68
Fodder Poor in Protein					
Ox 1	0.82	7.22	1:8.8	0.83	−0.01
Ox 1	0.78	9.99	1:12.8	0.78	0.00
Ox 2	0.89	11.08	1:12.4	0.97	−0.08
Ox 2	0.78	12.12	1:15.6	0.74	0.04

In the majority of cases where the supply of albuminoids is sufficient to cause any production of flesh, the greatest relative gain is produced by rations having a wide nutritive ratio. Many experiments also show that a large proportion of the protein in the feed is applied to the production of flesh when it has a wide nutritive ratio than when it has a narrow one. [p. 1243]

解　題

1. ブルックスの講義と背景

　日本では，その自然条件などから農業の主体は稲作を中心とする作物栽培であり，江戸時代まで畜産が本格的に行なわれることはなかった。もちろん，牛馬は古くから役畜として使われてきたし，鶏も比較的広範囲に飼われていた。牛乳利用についても，奈良・平安時代から江戸時代まで，さまざまな記録に登場する。しかし，畜産が農業の一分野として位置づけられ，本格的に取り組まれるようになるのは，やはり明治以降と言ってよいであろう。明治政府は欧米諸国に学び，畜産奨励策をとり，特に北海道開拓使は乳牛，めん羊，馬などの海外からの導入，普及に積極的に取り組んだ（第1編第1章解題「2. 畜産の導入論」参照）。当然このよう状況の下で，当時の日本に畜産についての体系的な学問，知見があったわけではないし，諸外国の事情もほとんど伝わっていなかったと思われる。札幌農学校が開校され，ブルックスが本講義をした当時の日本は，こうした状況にあったことを前提にしなければならない。

　世界的に見れば，この当時はヨーロッパでは農業革命を通じて農業のなかでの畜産の重要性が高まり，家畜の育種改良が進んだ時代であり，アメリカでは東部での農業確立から西部開拓へと向かい，牛が西部大草原で増加しつつある時代でもあった。また，産業革命の進行，大都市への人口集中に伴い，畜産物の需要が大幅に増加した時代でもあった。

2. 講義録の内容

　本講義録の畜産に関する内容は，大別して3つに分けられる。第1は，農業における畜産の意義，畜産に適した条件，畜産にとっての北海道の評価などについて触れた導入的部分で

ある．第2は，牛およびめん羊の品種，育種・繁殖，一般管理についてであり，いわゆる家畜各論に関する部分である．当時の欧米で，牛とめん羊が最も重要な家畜であったことは事実だが，馬・豚・鶏などにまったく触れていないのはなぜなのか疑問も残る．第3は，いわゆる家畜栄養学に関する部分であり，飼料の消化・吸収・利用などに触れている．

　これ以外に，第6章で牧草や飼料作物の種類，栽培管理とともに，乾草の調製や放牧について取り上げられ，第7章のトウモロコシの項でサイレージ調製についても触れている．牧草・飼料作物などのサイレージとしての調製・利用は，ヨーロッパでは1700年代末から取り上げられており，1800年代後半になるとイギリス，ドイツなどから初期の研究報告が出されている．しかしサイレージ技術がアメリカに導入，普及されたのは遅かった．ブルックスはサイレージ調製を「フランス式貯蔵法」と紹介しているが，これは1877年に，フランスの農民Goffartがサイレージに関する手引書を著し，これがアメリカに紹介されていたためと思われる．ちなみにアメリカで最初のサイロが作られ，サイレージが調製されたのは1873年イリノイ州においてとされている（以上の記述はHenry and Morrison (1928), *Feed and Feeding,* 19th ed. およびWoolford (1984), *The Silage Fermentation*を参照）．ブルックス自身は，サイレージそのものを見たことがなかったかもしれない．いずれにしても，乾草・サイレージなどの自給飼料生産も講義では取り上げられていたが，家畜飼養と一体になった形で紹介されていなかったのは残念であった．また乳・肉・毛などの畜産物の利用・加工などについては講義に含まれていないが，畜産を産業として取り入れようとする場合，その生産物をどう利用するか，そのための知識・技術も不可欠である．当時の日本ではこうした分野は白紙に近い状況だったため，講義に取り入れられればきわめて有用であったであろう．家畜を海外から導入しても，なかなか産業として定着しなかった原因のひとつとして，こうした利用の問題があった．

1) 畜産の意義づけ

　欧米では，古くから農業のなかでの畜産の意義や役割が一般的に認識されていたのに対して，日本では重視されなかった理由について，仏教の影響などにも触れている．同時に，水田稲作が盛んであるが，灌漑に不適当な土地の畜産による利用や有機質肥料供給の意義も指摘している．一般的には，冷涼な気候で，草の生育に適したところが畜産の適地であるとした上で，北海道は畜産に適した地域と評価している．ただし，冬の長いこと，野生動物の危険があることが問題であるとも指摘している．市場からの距離が遠くても，畜産にはあまり不利にならないとしているが，当時の状況から見れば，そもそも日本には畜産物に対する市場がほとんど存在していなかったことについても考慮しなければいけなかったであろう．

2) 畜牛について

　現在では，牛について論ずる場合，乳用牛（乳用種），肉用牛（肉用種），兼用牛（兼用種）と区分して論ずるのが一般的だが，乳生産や肉生産を目的に改良が進んだのは比較的最近であり，従来は役用も含めすべての目的に利用するために牛が飼養されるのが普通であった．ブルックスも，特に乳・肉・兼用などとは区分せず，9品種を取り上げ，品種としての特徴を述べている．

取り上げられている品種は，デボン，ヘレフォード，ウエストハイランド（またはカイロ），エアーシャー，ジャージー，オランダ牛（またはホルスタイン），ギャロウェイ，ロングホーン，ショートホーン（またはダルハム）で，説明はデボン，エアーシャー，ショートホーンなどについてが多く，現在世界的にも，また日本でも最も重要な乳用品種であるホルスタインについての説明はきわめて短い。これは，まだホルスタインが重視されていなかった当時の実情を反映したものであろう。ジャージー，エアーシャーは現在でも世界的に優れた乳用種として各地で飼育されているが，デボン，ショートホーンはごく限られた地域でしか飼育されていない。また，これ以外にも取り上げられてもよいと思われる品種もあるが，なぜか除外されている。いずれも，日本の人々にとっては，初めて聞く名前であったろう。これらの品種のうち，オランダ牛（ホルスタイン）以外はすべてイギリス原産の品種で，18世紀後半から有名な育種家，ロバート・ベークウェルやコーリング兄弟によって改良されたものである。

明治政府は畜産振興のなかでも，特に畜牛に力を入れてきており，各地に種畜場などを設置し，海外から多くの品種を導入していた。明治前半に日本に輸入された牛の品種は，デボン，ショートホーン，ジャージー，エアーシャー，ホルスタイン，ガーンジー，ブラウンスイス，シンメンタールなど多くの品種があり，ブルックスの取り上げた品種と必ずしも一致していない。

札幌農学校では，1877（明治10）年にショートホーン，1878（明治11）年にエアーシャー，1889（明治22）年にホルスタイン，ガーンジーが導入され，教育研究にはもちろんのこと，日本の乳牛改良に大きな貢献をした。特にホルスタインについては，導入後，今日までその系統が絶えることなく維持されてきた。雌牛について言えば，1889年導入の第1号から，2003年10月生まれの第1175号（現北方生物圏フィールド科学センター生物生産研究農場所属）まで続くわが国でも稀にみる牛群である。札幌農学校では，このほか南部牛も導入しており，主として役用に用いられたようである。これらの牛を収容するため，クラークの構想に基づいて北海道における畜産経営の模範になるようなモデルバーン（模範家畜房）が農場に作られた。モデルバーンには牛のほかに，当時，農場で飼育されていた馬や豚が収容されていた。モデルバーンを中心とする「札幌農学校第2農場」の建物群は，明治初期の日本畜産の歴史を示す貴重な遺産として今も保存されており，1969（昭和44）年，国の重要文化財として指定されている。

品種の紹介のなかで，体各部位の形状等について多くの記述がされているが，これは当時から欧米でも強調されていたところで，後述するように日本においても，比較的最近までこうした体型，体格についての知見が重視されていた。また当時の日本の牛は欧米の牛に比べ小型であったが，この体格を改善するため，日本の雌牛にショートホーンの雄牛を交配することを提案している。

また，現在では明確に乳用種とされているホルスタイン，ジャージー，エアーシャーについて，その役用，肉用としての価値を論じたり，逆に肉用種であるヘレフォードについて乳用，役用としての価値を論じたりしているが，これも牛の品種が利用目的によって明確に分

化していなかった当時の時代的背景をうかがわせる。

育種・繁殖や一般管理については，基礎的な学理というよりは，きわめて実践的な内容が主体で，交配のさせ方，分娩時の世話，子牛の育て方，日常的な牛の手入れなど，生産現場ですぐに役立つような内容が多い。しかし，当時でも重要と思われるものとして，飼料の給与法や搾乳についてはなぜか触れられていない。

3）めん羊について

めん羊についても，品種，育種・繁殖，一般管理について説明している。日本とは異なり，世界的に見れば，めん羊はもともと畜産のなかでも重要度は高く，飼養頭数も多かったのに加え，産業革命において紡績工業の原料としての羊毛の需要が増加し，めん羊の重要性が一層増していた。ブルックスが，馬，豚，鶏よりもめん羊を取り上げたのは，こうした背景があったからかもしれない。これに対して，わが国では明治以前にも中国からめん羊が入ってきてはいたが，生産目的では飼育されておらず，生産目的で飼育されたのはやはり明治になってからである。明治初期，政府はめん羊飼育を奨励し，特に北海道開拓使は，海外からのめん羊の導入，貸与に積極的に取り組んだ。しかし，疾病の多発や羊毛・羊肉の需要がほとんどなかったことなどが原因で，明治時代のめん羊奨励事業は失敗に終わっている。その後，大正時代にも「緬羊100万頭計画」が立てられたが，これも挫折した。昭和に入って，軍需羊毛自給のためめん羊飼養頭数は徐々に増加した。第2次大戦後，羊毛不足のためさらに飼養頭数は増加し，1957(昭和32)年に94万頭と最高に達した。しかし，その後の羊毛・羊肉の輸入自由化により急速に減少し，1976(昭和51)年にはわずか1万頭になっている。現在では，肉生産を主目的に2万頭弱が飼養されているにすぎない。

ブルックスが取り上げた品種は，メリノー，レスター，コッツウォルド，リンカーン，サウスダウンの5品種で，最も重要な毛用種であるメリノーについての説明が大半である。なお明治前半の日本に輸入されためん羊は，メリノー，サウスダウン，リンカーン，シュロップシャー，蒙古羊，上海羊などである。札幌でもエドウィン・ダンが真駒内牧場にアメリカなどからめん羊を導入しているが，当時の札幌農学校の農場ではめん羊は飼育されていなかった。

めん羊の育種・繁殖，一般管理についても，きわめて実践的な内容が主で，当時の日本の人々にとっては，やはり初めて聞く内容がほとんどであったろう。

4）家畜の栄養について

家畜栄養については，家畜体の組成，飼料の組成および分析法，消化・吸収，飼料の栄養価評価法などの基礎的な内容で，項目から言えば現在の日本の各大学で行なわれている家畜栄養学の講義内容に近いものと言えよう。

このブルックスの家畜栄養に関する講義内容は，実はアメリカの著名な家畜栄養学者アームスビィ(1853-1921年)の著書 *Manual of Cattle Feeding* (初版1880年，2版1882年)の第1部：家畜栄養の一般原則の内容とほぼ一致する。講義時間の制約のためか省略されている部分もあるが，基本的な構成や引用されている実験データの数値までほとんど一致する。アームスビィは，後に飼料の正味エネルギーという概念を確立した家畜栄養学の歴史に名の

残る学者であるが，当時は家畜栄養学の先進国であったドイツ留学から帰国して間もない新進気鋭として，アメリカの農業試験場で基礎的な，しかし最先端の研究に取り組んでいた時期と思われる。ブルックスの講義が1877-79(明治10-12)年であることから，前述の著書はまだ発行されていなかったことになるが，おそらくその原本に相当するアームスビィの出版物をアメリカで入手し，それをもとに講義をしたものと思われる。いずれにしても，当時としては世界最先端の家畜栄養学の内容と言ってよいであろう。

詳細な内容を見ると，当時はまだ化学が発展過程であったため，タンパク質についての概念が整理されていないなどの制約はあった。しかし，他方では，家畜栄養学特有の粗繊維という概念やその分析法(現在の実験でもまったく同じ方法が使われている)についても紹介されているし，家畜栄養の研究手法についても実験データを示しながら解説するなど，講義レベルの高さがうかがわれる。

3. その後の日本の畜産，畜産学との関連

明治初期，日本政府は欧米からの畜産導入にあたって，牛・めん羊を主体とした大農経営を目指したが，日本の実情にはあわず，成功しなかったことはよく知られている。特に北海道では畜産振興が強調されており，ブルックスの講義もこうした明治政府の方針に合致するものであった。実際の生産現場などでの影響力から言えば，あるいはエドウィン・ダンの影響の方が大きかったかもしれない。ダンは，ブルックスと同時期，北海道開拓使に農業師として雇われ，北海道畜産の基礎作りに多大な貢献をしている。ダンは，札幌農学校第2農場とならんで，北海道酪農・畜産の発展に大きな役割を果たした真駒内牧場を開設し，そこで北海道酪農の育ての親とも言うべき町村金弥を指導している。この町村金弥は，ブルックス講義ノートをとった新渡戸稲造と札幌農学校の同期生であり，このブルックスの講義を受けていたことになる。畜産についての知識がほとんどなかった町村金弥が，ブルックスから初めて畜産について体系的に学んだことは，彼自身はもちろんのこと，後に彼が指導した息子，町村敬貴や宇都宮仙太郎など北海道酪農の先覚者たちに多大な影響を与え，その後の北海道畜産の発展に少なからず貢献したものと思われる。

畜産学の面から見れば，この講義の構成内容は，教育研究で日本の畜産学者に引き継がれていく。例えば，後に農学部畜産学第三講座を担当した井口賢三(1900(明治33)年札幌農学校入学，一度中退するが再入学し，1910(明治43)年東北帝国大学農科大学農学科卒業)は，著書『畜牛之鑑定』(1921年)でブルックスと同様，牛体各部位についての詳細な記述を中心にしているし，また同じく『畜産学』(1931年)の育種・繁殖・管理などについての記述にも共通するところが多い。と同時に，後の畜産学の教育研究で必ず取り上げられている馬，豚，鶏が欠けているのはなぜなのだろうか。特に北海道という地域を考慮した場合，馬についての講義は欠かすことができなかったのではないだろうか。同時期にダンが北海道の馬の改良に努力していたことを併せて考えると，その感が一層強くなる。

家畜栄養学の分野について言えば，やはり後にドイツにおいて飼料評価法としてのデンプン価法を確立した世界的な家畜栄養学者ケルナーが，1881-93(明治14-26)年の間，後に東

京大学農学部へと発展する駒場農学校に在職し，教育研究面から大きな影響を与えたが，ブルックスはケルナー以前にすでに札幌で前述のようなレベルの高い講義をしていたことになる。その後の東京大学と北海道大学における畜産学あるいは家畜栄養学の教育研究の動向を見ると，東京大学では一貫して基礎的な面が強く，北海道大学では今日に至るまで応用面を重視してきている。これもブルックス以来培われてきたものなのであろうか。なお，ブルックスの在任中の卒業生で，畜産学の教育研究への道に進んだ者としては橋本左五郎(1885(明治18)年札幌農学校入学，1889(明治22)年卒業，8期生)がいる。橋本はドイツ留学後，1900(明治33)年に札幌農学校教授になり，畜産学全般を講義し，1907(明治40)年には畜産学科の設置に至っている。橋本は牛乳の加工，特に練乳についての研究を行ない，乳業界に大きな貢献をしている。この橋本を通しても，ブルックスの畜産学についての知見，考え方が，後の世代に引き継がれていったことであろう。

注
1) 表中の ox(複数 oxen)は一般には去勢牛を意味し，役用または肉用に飼育されているものを意味する。また milch-cow は milk-cow の古い表現。
2) 畜産の適地を論ずるにあたって，飼料としての草の生育・利用を中心に述べており，草地を基盤とした草食家畜(牛，馬，めん羊など)による生産を重視していたと思われる。今日の輸入穀物に依存した日本の畜産を見直す上でも重要な視点と言えよう。
3) 畜産についても，生産地から消費地への生産物の輸送がしばしば問題になる。一般的に作物生産物より畜産物の方が，同一重量あるいは容積当たりの価値が高く，輸送に有利な面がある。しかし同時に，乳・肉は腐敗しやすく，交通手段や冷蔵・冷凍設備が未発達の時代には，この点が制約要因となっていた。肉用牛については，かつてアメリカやヨーロッパで，生きたまま生産地から消費地まで大群で歩かせて輸送するという方法もとられていた(例えば，アメリカでは西部大草原からシカゴまで，イギリスではスコットランド北部からロンドンまでなど)。日本でも最近まで北海道の牛乳を首都圏や関西にどのようにして輸送するかは，大きな問題であった。
4) Nii-Kap は，新冠牧馬場(現在の日高支庁静内町にある独立行政法人家畜改良センター新冠牧場の前身)を指し，ここには1872(明治5)年に，7万 ha にも及ぶ御料牧場が設置され，馬の飼育が行なわれていた。当時エドウィン・ダンは，ここを馬の一大育種場に整備するよう提言していた。
5) 熊，鹿など野生動物の危険(特に養羊にとって)を指摘しているが，明治初期，北海道ではオオカミや野犬の被害が大きく，徹底した駆除が行なわれた。その結果，オオカミは絶滅してしまっている(これらについては北海道開拓の歴史に関する著書等を参照されたい)。
6) neat cattle は古い表現であり，現在では牛の総称としては cattle を用いるのが一般的である。
7) 家畜の能力をその外貌から評価しようとする方法は古くからとられてきた(日本では，「相牛法」と言われていた)。各国に家畜の登録協会など家畜改良のための組織ができると，各品種ごとに審査標準が作られた。全体を100点とし，体各部位に点数を割り当てて，採点する方法である。これ以降に記載されている point とは，この点数に相当する。ちなみに，井口賢三『畜牛之鑑定』(1921年)によると米国デボン種倶楽部による体格標準(1886年制定)の牝牛の評点配分は以下の通り：頭8点，頸4点，肩4点，胸郭8点，肋骨8点，背16点，後躯8点，乳房20点，尾2点，脚4点，皮膚8点，生体量2点，相貌8点。
8) 札幌農学校には1878(明治11)年にエアーシャーが導入されており，このデータは導入して間もない時期のものと思われる。
9) 当時はまだ人工授精技術が確立されておらず，交配はすべて自然交配であったため，種雄牛1頭が1年間に交配できる雌牛頭数はきわめて少なかった。現在では人工授精や受精卵移植が普及しており，酪農家などが種雄牛を保有することはほとんどない。
10) 現在では，初乳 colostrum の成分として最も重要なのは免疫グロブリンであり，初乳は母牛から子牛へ免疫物質を移行させる役割を果たしているとされているが，当時はこのことが知られていなかった。

11) 哺乳期子牛の飼養方法に関する研究は，その後きわめて活発に行なわれ，現在では牛乳に代わる液状飼料（代用乳）や固形飼料（人工乳）が広く用いられており，哺乳期間を短縮する早期離乳法も確立されている．この分野では，北海道大学において第2次大戦前の井口賢三の研究以来，現在に至るまで多くの研究がなされている．
12) 基本的な野生羊は4種いるとしているが，1種欠けている．第1番目は *Ovis Urial*，第3番目は *Ovis Canadensis* と思われる．
13) wool は通常羊毛と言われている毛で，綿毛（わたげ）とも言われ，これに対し hair は長く，比較的かたい毛で，粗毛（かたげ）とも言われ，牛，馬などの毛はこれに相当する．また fleece という用語は，刈り取られた1頭分の毛皮状にまとめられた羊毛を意味し，羊毛脂 yolk を含んでいる．
14) この表の数値は誤っていると思われる．
15) Nitrogenous Organic Substances を albuminoids, gelatigenous substances, horny matter に分けているが，現在ではN含有有機物はタンパク質と非タンパク態N化合物に大別され，タンパク質はさらに単純タンパク質，複合タンパク質に分類されるのが一般的である．また後述の飼料成分の項でも，現在とは異なる分類，用語が用いられている．19世紀末ではまだ有機化学が十分発展しておらず，当時の状況を反映した講義内容と言えよう．
16) 1854年に反芻動物が cellulose を消化できることが確認されたとあるが，1854年に何があったか不明である．
17) starch を構成する物質として starch cellulose と granulose をあげているが，現在の amylose と amylopectin に相当する．
18) sugars の分類も現在の単糖類，二糖類などの分類にはなっていない．
19) 1864年にドイツの Weende 農業試験場でヘネンベルグとシュートーマンによって確立された，いわゆる一般分析法と言われる方法で，現在でも使用されている方法と同じである．
20) 単に可消化タンパク質に対する可消化非N物質の比率と述べているが，この Nutritive Ratio 栄養比は飼料のタンパク質とエネルギーの比率を表したものであり，飼料に必要な適正な比率を検討するための指標としての意義がある．
21) Digestible protein: Digestible carbohydrates は Digestible protein: Digestible non-nitrogenous substances の誤り．
22) 図についての記述があるが，図は残っていない．また反芻胃内での飼料の動き，反芻についての記述はあるが，現在では最も重要とされている反芻胃内での微生物による飼料の消化については，当時まったく知られていなかったので，一切触れられていない．発酵が微生物の作用によるとパスツールが発見したのが1863年であり，それから間もない時期であるので当然と言える．
23) 栄養素の体内代謝についての項であるが，当時の科学的知見のレベルをもとにしているため，現在から見るときわめて不正確である．
24) この項で述べられていることは，最新の呼吸試験装置を用いた C，N 出納からタンパク質や，脂肪の体蓄積を検討した研究成果に触れるなど，現在の大学における家畜栄養学の講義としてもきわめて高いレベルのものである．
25) 本項 Feeding with Protein and Fats と次頁 Feeding with Protein and Carbohydrates は，いずれもタンパク質利用についての脂肪および炭水化物併給の影響を指摘しているのだが，これは単に栄養素間の関連ではなくタンパク質利用とエネルギー利用の関連の問題であるが，そうした説明は一切ない．

参考文献
1. 亀高正夫・堀口雅昭・石橋 晃・古谷 修(1987)，基礎家畜飼養学，養賢堂
2. 木村勝太郎(1985)，北海道酪農百年史——足跡と現状及び人物誌，樹村房
3. 日本ホルスタイン登録協会編(1967)，日本の乳牛——ホルスタイン史，同協会
4. 田辺安一編(1999)，お雇い外国人エドウィン・ダン——北海道農業と畜産の夜明け，北海道出版企画センター
5. 蝦名賢造(1963)，牛作り八十年，毎日新聞社

［大久保正彦］

編者・執筆者・校閲者紹介

〈編　者〉

髙井宗宏(たかい　むねひろ)

　1936年生，1960年：北海道大学農学部農業工学科卒業，現在：(社)北海道農業機械工業会専務理事，農学博士，元北海道大学大学院農学研究科教授，専門分野：農業機械学・農業機械化論。主著：札幌文庫61 農学校物語〈共著〉(北海道新聞社，1992)，新版農作業機械学〈共著〉(文永堂出版，1991)，米欧回覧実記の学際的研究〈共著〉(北海道大学図書刊行会，1993)

〈執筆者・校閲者〉[50音順]

梅田安治(うめだ　やすじ)

　1932年生，1955年：北海道大学農学部農業工学科卒業，現在：農村空間研究所長，農学博士，北海道大学名誉教授，専門分野：土地改良学。主著：農地・農村の景観(農業土木新聞社，1990)，土の自然史〈編著〉(北海道大学図書刊行会，1998)

大久保正彦(おおくぼ　まさひこ)

　1938年生，1961年：北海道大学農学部畜産学科卒業，農学博士，北海道大学名誉教授，専門分野：家畜飼育学・畜牧体系学。主著：新編　酪農ハンドブック〈共著〉(養賢堂，1990)，ザ・ルーメン〈共著〉(デーリーマン社，2001)

太田原高昭(おおたはら　たかあき)

　1939年生，1968年：北海道大学大学院農学研究科博士課程単位取得退学，現在：北海学園大学教授，農学博士，北海道大学名誉教授，専門分野：協同組合論・地域経済学。主著：地域農業と農協(日本経済評論社，1979)，北海道農業の思想像(北海道大学図書刊行会，1992)，系統再編と農協改革(農山漁村文化協会，1992)

生越　明(おごし　あきら)

　1937年生，1961年：北海道大学大学院農学研究科修士課程修了，現在：北海道澱粉工業会参与，農学博士，北海道大学名誉教授，専門分野：植物病理学・菌学。主著：Identification of Rhizoctonia Species〈共著〉(APS Press, 1991)，土は求めている〈共著〉(北海道大学図書刊行会，1991)

近藤誠司(こんどう　せいじ)

　1950年生，1977年：北海道大学大学院農学研究科修士課程修了，現在：北海道大学大学院農学研究科教授，農学博士，専門分野：畜牧体系学・家畜管理学・家畜行動学。主著：家畜行動図説〈共著〉(朝倉書店，1995)，ウマの動物学(東京大学出版会，2001)

佐久間敏雄(さくま　としお)

　1932年生，1955年：北海道大学農学部農芸化学科卒業，農学博士，北海道大学名誉教授，専門分野：土壌学。主著：土の自然史〈編著〉(北海道大学図書刊行会，1998)，北海道・自然のなりたち〈共著〉(北海道大学図書刊行会，1994)

但野利秋(ただの　としあき)

　1936年生，1965年：マサチューセッツ州立大学大学院農学研究科修士課程修了，現在：東京農業大学教授，農学博士，北海道大学名誉教授，専門分野：作物栄養学・土壌肥料学。主著：作物栄養・肥料学〈共著〉(文永堂出版，1984)，植物栄養・肥料学〈共著〉(朝倉書店，1993)

中嶋　博(なかしま ひろし)

1941年生，1966年：北海道大学大学院農学研究科修士課程中退，現在：北海道大学北方生物圏フィールド科学センター教授，農学博士，専門分野：植物資源開発学・飼料作物学・植物生殖生理学。主著：未来の生物資源ユーカリ〈共著〉(内田老鶴圃，1987)，生物生産機械ハンドブック〈共著〉(コロナ社，1996)

中世古公男(なかせこ きみお)

1936年生，1963年：北海道大学大学院農学研究科修士課程修了，農学博士，北海道大学名誉教授，専門分野：作物学。主著：植物生産学〈共著〉(文永堂出版，1993)，作物学各論〈共著〉(朝倉書店，1999)

原田　隆(はらだ たかし)

1936年生，1965年：北海道大学大学院農学研究科博士課程中退，農学博士，元北海道大学大学院農学研究科教授，専門分野：園芸学。主著：凍結保存——動物・植物・微生物〈共著〉(朝倉書店，1987)，新版蔬菜園芸〈共著〉(文永堂出版，1996)

ユステス・藤居恒子(ふじい つねこ)

1963年：北海道大学文学部哲学科卒業，1965年：ミシガン大学心理学部修士課程修了，能力開発研究所，産業能率大学を経て，現在：Central Artery/Tunnel Project (BIGDIG) のデータ解析主任，南カリフォルニア大学数理心理学 Ph.D，ボストン在住

由田宏一(よしだ こういち)

1942年生，1966年：北海道大学大学院農学研究科修士課程中退，現在：北海道大学北方生物圏フィールド科学センター教授，農学博士，専門分野：作物学。主著：有用植物和・英・学名便覧〈編〉(北海道大学図書刊行会，2004)

Michael Van Remortel(マイケル・ヴァン・リモーテル)

ミシガン州生まれ。アメリカ・ノートルダム大学 B.A.，北海道大学文学部修士。上智大学，カリフォルニア大学バークレー校でも学ぶ。ユーラシアおよび北アメリカの歴史・歴史地理の研究に従事。主著：*Saint Nikolai Kasatkin and the Orthodox Mission in Japan* (ed.) (Divine Ascent Press, 2003)

ブルックス札幌農学校講義
2004年11月10日　第1刷発行

編　者　　髙井宗宏

発行者　　佐伯　浩

発行所　北海道大学図書刊行会
札幌市北区北9条西8丁目北海道大学構内（〒060-0809）
Tel. 011 (747) 2308・Fax. 011 (736) 8605・http://www.hup.gr.jp

㈱アイワード／石田製本　　　　　　　　　Ⓒ 2004　髙井宗宏
ISBN4-8329-8081-5

W・S・クラーク ―その栄光と挫折―	J. M. マキ 著 高久真一 訳	四六・372頁 定価2400円
William Smith Clark : A Yankee in Hokkaido	J. M. Maki 著	A5・344頁 定価4200円
覆刻 札幌農黌年報	開拓使 発行	菊・全11分冊・平均140頁 定価22000円
朝天虹ヲ吐ク ―志賀重昂『在札幌農學校第貳年期中日記』―	亀井秀雄 松木 博 編著	A5・490頁 定価7500円
北大寮歌「都ぞ弥生」の作詞者 小作官・横山芳介の足跡	田嶋謙三 塩谷 雄 著 大高全洋	四六・220頁・CD付 定価2000円
土は求めている	北海道農業 フロンティア 編 研究会	四六・332頁 定価2400円
土の自然史 ―食料・生命・環境―	佐久間敏雄 梅田安治 編著	A5・256頁 定価3000円
覆刻 農業土木学	鈴木敬策 著	菊・328頁 定価2500円
北海道農業土木史	北海道農業 土木史編集 編 委員会	A5・324頁 定価7000円
有用植物和・英・学名便覧	由田宏一 編	A5・376頁 定価3800円
北海道農業の思想像	太田原高昭 著	四六・274頁 定価2000円
生命(いのち)を支える農業 ―日本の食糧問題への提言―	石塚喜明 著	四六・128頁 定価1600円
日本の農業・アジアの農業	石塚喜明 著	四六・200頁 定価2000円
北大歴史散歩	岩沢健蔵 著	四六変型・224頁 定価1400円
写真集 北大125年	北海道大学 125年史 編 編集室	A4変型・238頁 定価5000円
北大の125年	北海道大学 125年史 編 編集室	A5・152頁 定価900円

〈定価は消費税含まず〉

───── 北海道大学図書刊行会 ─────

Rye	100 bu	97.7	30.5	48.2
Corn	"	88.8	10.3	30.5
Oats	"	64.0	14.0	33.3
Buckwheat	"	68.6	10.0	21.0
Flax Seed	"	59.0	61.0	76.0
Field Beans	"	244.0	72.0	97.0
Potatoes	"	18.8	33.0	10.5
English Turnips	"	10.5	17.6	5.8
Carrots	"	12.3	18.8	6.4
Common Beets	"	10.6	25.3	4.7
Barley	100 lbs	1.5	0.6	0.8

Table Showing the Number of Pounds of Potash, Nitrogen and Phosphoric Acid in Various Cattle Products.

		Nitrogen	Potash	Phosphoric Acid
Milk	100 gal	5.6	1.5	1.6
Cheese	100 lbs	4.5	0.2	1.1
Cattle (Live weight)	1,000 lbs	26.0	1.7	18.6
Sheep	" "	22.4	1.5	12.3
Swine	" "	20.0	1.8	8.8